Practical Aspects of Computational Chemistry III

Jerzy Leszczynski • Manoj K. Shukla
Editors

Practical Aspects of Computational Chemistry III

Springer

Editors
Jerzy Leszczynski
Department of Chemistry
and Biochemistry
Interdisciplinary Center
for Nanotoxicity
Jackson State University
Jackson, MS, USA

Manoj K. Shukla
Environmental Laboratory
US Army Engineer Research
and Development Center
Vicksburg, USA

ISBN 978-1-4899-7444-0 ISBN 978-1-4899-7445-7 (eBook)
DOI 10.1007/978-1-4899-7445-7
Springer New York Heidelberg Dordrecht London

Library of Congress Control Number: 2011940796

Printed on acid-free paper

Springer is part of Springer Science+Business Media (www.springer.com)

Preface

It is a matter of great pleasure to bring the current volume, *Practical Aspects of Computational Chemistry III*. Most of the contributors in this volume have participated in our annual International Conference "Conference on Current Trends in Computational Chemistry (CCTCC)". This volume covers diverse applications of computational methods currently being used in various areas of science and technology. Current volume presents results of state-of-the-science research reviews distributed in the 12 chapters.

The very first chapter explores how computational chemistry methods can be reliably used to explain complicated experimental scanning tunneling microscopy (STM) observations, and this chapter has been contributed by Sonnet and Riedel. In the second chapter, Ivashchenko et al. have reviewed structures and properties of TiN-based heterostructures as revealed by quantum molecular dynamics simulation. Applications of plane-wave density functional theory (DFT) in understanding the complex surface phenomena on metallic surfaces have been detailed by Pushpa in the next chapter. On the other hand, in the fourth chapter, Nhat et al. have presented results of their investigation on geometries, electronic and energetic properties of vanadium doped gold cluster by considering up to 20 gold atoms and adsorption of carbon monoxide on these clusters.

Polyhedral oligomeric silsesquioxane (POSS) cage molecules have wide applications in the material science and catalysis. Contribution comprising POSS cage molecules by Rehman and Gwaltney at the DFT level is presented in the fifth chapter. The magnetic properties of single-walled carbon nanotubes as well as their cross-linking architectures are reviewed by Hagelberg et al. in the next chapter. Golbraikh et al. have discussed how various statistical criteria can be used to predict the modelability of a chemical data set for quantitative structure-activity relationship models in the seventh chapter. The free energy gradient method and its application unrevealing properties of system in solution have been reviewed by Georg et al. in the eighth chapter.

Hydrologic modeling approach to simulate pollution transport has been discussed by Pradhan et al. in the ninth chapter. On the other hand, Majumdar et al. have reviewed current status of research on nerve agents in the next chapter.

Storoniak et al. have discussed and reviewed the molecular factors governing the stability of the valence bound anions of DNA related system in the eleventh chapter. The last chapter reviewed by Wang et al. addresses a fundamental question of origin of life and how computational chemistry methods can play a vital role in addressing such an important question.

With great pleasure, we take this opportunity to thank all contributors for devoting their time and hard work to make this project a success. We are also thankful to all reviewers who devoted their time and efforts in providing valuable comments to each contribution in this volume. We are grateful for the excellent support from the Presidium of the European Academy of Sciences as well as Editors at Springer. Of course, many thanks go to our family and friends; without their support the realization of the book would not have been possible.

Mississippi, USA

Jerzy Leszczynski
Manoj K. Shukla

Contents

Chapter 1
The Scanning Tunneling Microscopy of Adsorbed Molecules on Semiconductors: Some Theoretical Answers to the Experimental Observations

Ph. Sonnet and D. Riedel

Abstract The present chapter depicts various theoretical methods that have been used in the context of experimental studies at the nanoscale. With the scanning tunneling microscopy (STM), we can investigate the manipulation of individual molecules and their electronic properties through electronically induced excitations. To explore in details the surface dynamics of a molecule in a bistable or quadristable motion, the nudged elastic band method is used to describe the energy barrier of the saddle point located along the molecular reaction pathway. When the physisorbed stilbene molecule is studied on the bare Si(100), the tight binding method is associated to the density functional theory (DFT) to simulate the scanning tunneling topographies. The physisorption of a molecule such as the hexaphenyle-benzene at the step edge of the Si(100) show particular lateral movements which diffusion barrier can be described accurately by the dispersive term added to the DFT to describe van der Waals interactions (DFT-D). In order to identify the nature of bonding of a porphyrin molecule adsorbed on a boron doped silicon surface, we have analyzed the Laplacian of the charge density. Finally, we present a full DFT-D study of a bidimensional nanoporous supramolecular network on a silicon surface. We have evaluated the molecule-molecule and molecule-substrate interactions energies that are key parameters to understand the mechanism of formation of these networks. In this context, the simulations of STM images with multi-diffusion with the presence of a tungsten tip show a better agreement with the experimental observations than the Tersoff-Hamann approach.

Ph. Sonnet
Institut de Science des Matériaux de Mulhouse, IS2M, CNRS, UMR 7361,
Université de Haute-Alsace, 3b rue A. Werner, 68093 Mulhouse, Cedex, France

D. Riedel (✉)
Institut des Sciences Moléculaires d'Orsay, ISMO, CNRS, UMR 8214,
Université Paris Sud, 91405 Orsay, Cedex, France
e-mail: damien.riedel@u-psud.fr

J. Leszczynski and M.K. Shukla (eds.), *Practical Aspects of Computational Chemistry III*, DOI 10.1007/978-1-4899-7445-7_1,

1.1 Introduction

The willingness to bring a better understanding to the experimental observations made with scanning tunneling microscopy (STM) has fed continuous interests since the development of such a tremendous tool. From the first demonstration that a surface could be observed at the atomic scale with a metallic tip [1] by using the tunnel effect [2], the group of D. Eigler has demonstrated that the STM can also be used to manipulate matter atom by atom [3]. When one deals with bare metals surfaces, the interpretation of STM images is relatively straightforward as soon as low energies are probed. Yet, semiconductors surfaces are sometimes trickier to analyze, due to their particular electronic properties and their various surface reconstruction phases [4, 5]. Moreover, when molecules are adsorbed on these two types of surfaces, delocalized density of states can induce misleading interpretation of the experimental data and in particular in the identification of molecular conformations and charge states. This stems from the functioning of the STM that does not see directly the atomic position on the surface but rather probe iso-densities of states around them. This same physical effect is therefore used to investigate, for example, the electronic [6], magnetic [7] or optical [8] properties of the studied surface and/or adsorbate as it give access to the local distribution of states of the molecular assemblies. The tunnel current I_t flowing in the STM junction is related, as described by the following expression,

$$I_t \propto \int_{E_f}^{E_f+V} \rho_T(E-V)\rho_s(E)e^{2s\sqrt{2(\phi-E)+V}}dE \tag{1.1}$$

to the local density of state of the tip defined by $\rho_T(E)$ considered at an energy E, the density of states of the surface $\rho_S(E)$, the work function ϕ of the surface and the tip sample separation s [9]. Here, V is the applied voltage in the STM junction and E_f is the Fermi level energy at the surface. The Eq. (1.1) translate very well the power and the versatility of the measurements that can be made with STM in which the local density of states of both the surface (and adsorbate) need to be convoluted to allow the measure of a tunnel current [10]. The third term in Eq. (1.1) leads to the exponential dependency of the current with the tip-surface distance giving rise to a huge sensitivity of I_t in the direction normal to the surface. Hence the measurements performed with the STM allow probing very local effects at metal or semiconductor surfaces.

Like many other tools, the STM presents some limits that can rapidly provide some drawbacks when the electronic properties related to the dynamics of individual molecule are explored. In particular the time scale measured with the STM is often reduced to few microseconds. Additionally, the integration over the probing energy described in Eq. (1.1) underline the weak selectivity of the tunnel electrons energy across the tunnel junction when no resonant tunneling can take place [11]. As we will see in the examples below, STM experiments can be easily combined with theoretical simulations in order to correlate the empirical results to predictable molecular properties. Hence, the molecular relaxation on a surface and the spatial distribution of the subsequent density of state can be computed and compared with STM topographies in order to stress particular molecular adsorption conformations and thus deduce some specific electronic properties. Such comparison has been performed via various simulation packages based on the use of the density functional theory (DFT) [12] in which the Tersoff-Hamman approximation [13] is usually sufficient to describe the spatial distribution of the local density of state at the surface. In a more detailed approach, the reproduction of STM topographies has been explored by using the Bardeen model [14] in which the states of the STM tip are taken into account and a real tunnel current is computed. In this context, the elastic-scattering Quantum chemistry (ESQC) method has raised a new approach for the simulation of STM images. Initially proposed by Joachim and Sautet [15–19] the method offers a mean of studying the transmission of electrons through a defect embedded in an infinite periodic chain. A general algorithm of the ESQC method allowing the simulation of STM image [20] has been developed by Dahlke and Schollwöck [21] that overcome a previous limitation where the defect is located in between two leads of same nature. In this improved version, the effects of several different leads can be computed. Other similar methods have emerged and can be available in various code package such as bSKAN [22] that have been recently improved with an extended Bardeen approach [23].

In this chapter, we will show through various experimental examples that numerical simulations based on the use of the DFT bring complementary and relevant information to the studied physical processes. In the Sect. 1.2, we will talk about a chemisorbed molecule that can be switched between two stable states while the surface dynamics of the molecule cannot be directly probed with the STM. Thus, we show that numerical simulations that use the nudged elastic band (NEB) method explain how the molecule dissociates from the surface before its rotation. In Sect. 1.3, we explore the possibility to reproduce the STM topographies of a functionalized molecule physisorbed on the silicon surface by using the density functional based tight binding (DFTB+) method. The Sect. 1.4 describes a more accurate method based on the dispersion corrected density functional theory (DFT-D) implemented in the Vienna Ab-initio Simulation Package (VASP), to estimate the properties of a hopping physisorbed molecule. Here the charge state of the molecule is described independently with the *Gaussian* code package using hybrid functionals [24]. In the Sect. 1.5, we approach the problem of organometallic molecules adsorbed on the silicon surface for which the role of the molecule-substrate interaction is important. In order to identify the type of interaction, the Laplacian of the charge density can be calculated. Finally, the Sect. 1.6 addresses the problem of the molecule-molecule and molecule-substrate interactions in the case of self-assembled molecular network on silicon surfaces.

1.2 Manipulation of an Individual Molecule Chemisorbed on the Silicon Surface

The control of the manipulation of individual nano-objects (INOs) on surfaces has attracted various studies aimed in particular for the fabrication of nanoscale devices [25–30]. Such control is still of a decisive importance for the development of various domains such as the surface photonics [6, 31, 32], the fabrication of nano-magnetic devices [33] or for the design of molecular circuits for which the interaction between the surface and the adsorbate properties plays a central role [34–37]. In this context, the scanning probe microscopy is of crucial importance to perform molecular manipulation by using either direct or near *contacts* methods in which the STM tip is usually moving with the INO [38, 39], or via electronically induced excitations (EIE) where the STM tip is static and the excited molecule moves across the surface via the relaxation of vibrational energy [40, 41]. On the bare or insulated metal surfaces, various techniques have been developed to achieve such a control on individual atoms or molecules [28, 42–46]. The precise control of the manipulation of INOs involving EIE processes is therefore relevant to reach a transition state related to the chemisorption.

When a biphenyl molecule is adsorbed on a bare Si(100)-2 $\times$ 1 surface at room temperature, the molecule chemisorbs into two symmetrical molecular conformations as shown in Fig. 1.1a, b. The ensuing occupied states STM topographies (Fig. 1.1c, d) of the single molecule exhibit two bright protrusions, one being located at the back-bond row of the Si surface (dark lines in Fig. 1.1c, d), while the second one is centered on the silicon dimer row (light gray lines in Fig. 1.1c, d). These conformations are named S_1 and S_2 as the two stable states of the biphenyl molecule. The STM tip apex can be very precisely located at various positions overlooking the molecule (P_1 or P_2) to proceed to its electronically induced excitation (Fig. 1.1e) [47]. Subsequently to these excitations, the biphenyl molecule undergoes a reversible bistable movement from S_1 to S_2. For each movement, the tunnel current trace is recorded in which a current drop can be detected when the molecule escape from underneath the tip. This experiment can be repeated several times thanks to the reversibility of the movement. The time at which this current drop is detected allows determining an average electronic yield for the switch $S_1 \rightarrow S_2$. We can also detect an intermediate position of the biphenyl molecule in between the S_1 and the S_2 configuration such that it is defined as a transient state named T [45]. Each measured electronic yields of the molecular movement $S_1 \rightarrow S_2$, $T \rightarrow S_1$ or $S_1 \rightarrow T$ are reported in Fig. 1.1f as a function of the surface voltage, (i.e. the energy of the exciting electrons). The electronic yields presented in Fig. 1.1f show an energy threshold that depends of the molecular movement. For example, the movement $S_1 \rightarrow T$ has an electronic threshold at ~2.3 eV whereas the electronic threshold of the $S_1 \rightarrow S_2$ switch is at an energy of ~2.7 eV (Fig. 1.1f). Our ability to manipulate reversibly the biphenyl molecule in these various molecular configurations arises from specific molecular resonances located at each benzene ring of the biphenyl [45]. Hence, we can show that when the excitation is applied at the point P_1, the molecule moves from S_1 to T while an excitation at the point P_2 provokes a bistable movement $S_1 \rightarrow S_2$.

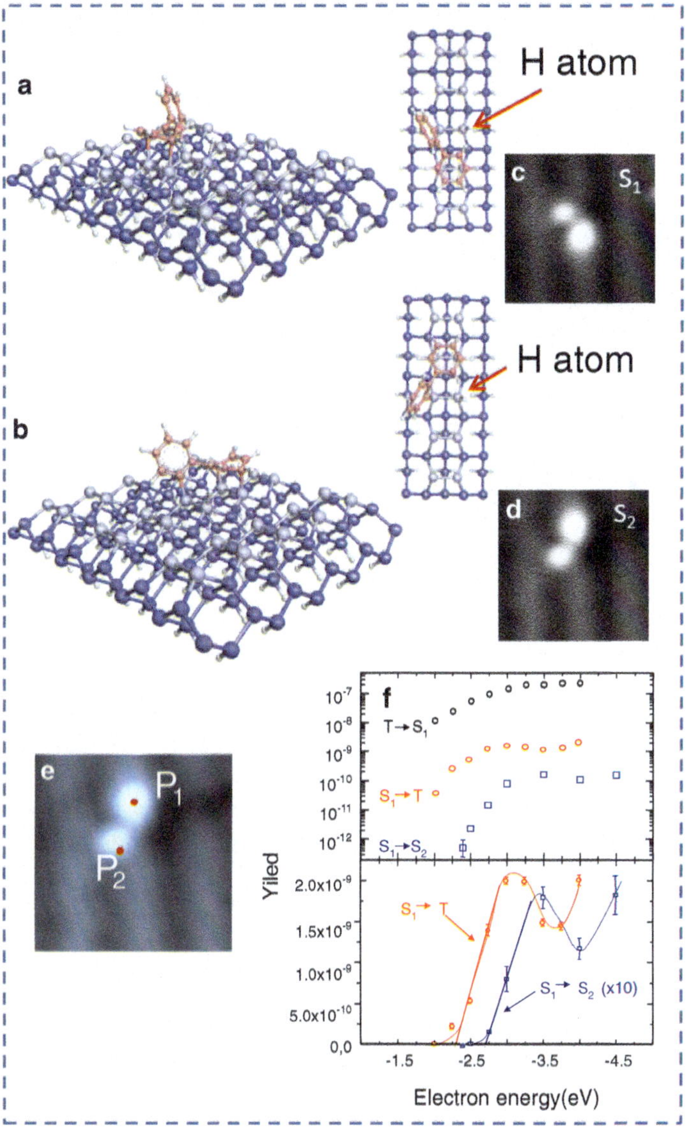

Fig. 1.1 (**a**) and (**b**) Balls-and-sticks views of the two stable adsorption conformations of a biphenyl molecule on the Si(100) surface. The *blue* and *light blue* balls are Si atoms, the *white* balls are hydrogen atoms and the *red* balls are carbon atoms. Beside, a small *top* view of each conformation is sketched. (**c**) and (**d**) (23.1 × 23.1 Å^2) STM topography ($V_s = -2$ V, $I = 200$ pA) of a biphenyl molecule adsorbed on the Si(100) surface for the S_1 and S_2 conformations, respectively. (**e**) (23.1 × 23.1 Å^2) STM topography ($V_s = -2$ V, $I = 200$ pA) where the points excitation location P_1 and P_2 are indicated. (**f**) Yields (per electron) as a function of the surface voltage for molecular switching ($Y_{S1} \rightarrow Y_{S2}$), moving to ($Y_{S1} \rightarrow Y_T$), and moving from ($Y_T \rightarrow Y_{S1}$) to the transient state. The *upper* part is semi-logarithmic and the lower part linear scale (The panel **f** is reproduced with permission from Lastapis et al. (http://www.sciencemag.org/content/308/5724/1000.full.pdf) © 2005 American Association for the Advancement of Science)

A theoretical approach of these experimental results request initially to define precisely the molecular adsorption conformation of the biphenyl molecule on the Si (100) surface. This has been performed using the ab-initio DFT [10, 48] by means of the VASP simulation package [49]. A tight-binding (TB) based STM modeling is used to compute STM images and differential conductance spectra. A four layers silicon slab containing 16 atoms per layers models the surface with the lowest layer passivated by hydrogen atoms. Whereas several local minima have been identified, the structure with the lowest energy is represented in Fig. 1.1a, b as previously reported [50, 51]. The ensuing relaxation show that one of the phenyl rings of the molecule chemisorbs to the Si surface via the breaking of a C-H bond while the residual H atom adsorbs on the neighbor Si of the same silicon dimer (Fig. 1.1a, b). This allows to plot STM images using the TB method that will be described in more details in Sect. 1.3 of this chapter. The use of the TB model is interesting because of its rapidity and increased accuracy compared to Tersoff-Hamman method in which the sole electronic density is plotted. The calculated STM images using this method are presented in Fig. 1.2c–f and reproduce rather well the experimental data (Figs. 1.1c). Because of the presence of a STM tip in the simulations, the tunnel current can be calculated and the differential conductance curves acquired experimentally can be reproduced as shown in Fig. 1.2g. The STS spectra calculated for the bare substrate (Bb and Di are associated to the back-bond and dimer, respectively, in Fig. 1.2g [48]) do not depend on the position of the tip over the surface, for negative biases. The first peak in these two curves has been identified as a surface state related to the π orbital of the Si-Si bonds. The comparison of the spectra obtained over the two phenyl rings of the molecule (noted Mo and Fi in Fig. 1.2b and related to the mobile and fixed phenyl rings, respectively) highlights striking differences. As for the STM images, these differences suggest the contribution of electronic states differently localized inside the molecule as previously reported [45]. The main differences between the Mo and Fi spectra arise for high negative biases. The peak obtained for the Fi position at $V_s = -2.26$ V is not observed when the tip is placed over the Mo phenyl ring. Moreover, at this energy the dI/dV curve over the Mo part exhibits a minimum of intensity; the main peak in this spectrum being obtained for $V_s = -1.82$ V. Thus, the two electronic states probed by STS measurements at different locations over the molecule correspond to strongly localized states on different parts of the molecule that are related to the yield shown in Fig. 1.1f.

To get insight at the manipulation dynamics of the biphenyl molecule on the Si (100) surface, we have used the nudged elastic band (NEB) method as implemented in VASP. Here, we intended to find the possible pathways from the stationary S_1 configuration to the second stationary S_2 configuration that are initially known (Figs. 1.1 and 1.2). Hence, the minimum energy path (MEP) is sought via this method. The potential energy maximum along the MEP is the saddle point energy which gives the activation energy barrier of the transition $S_1 \rightarrow S_2$. Several methods can be used to find reaction paths and the ensuing saddle points [52]. The NEB method is an efficient method for finding the MEP when the initial and final states of a molecular switch are well defined. Although studies of very large

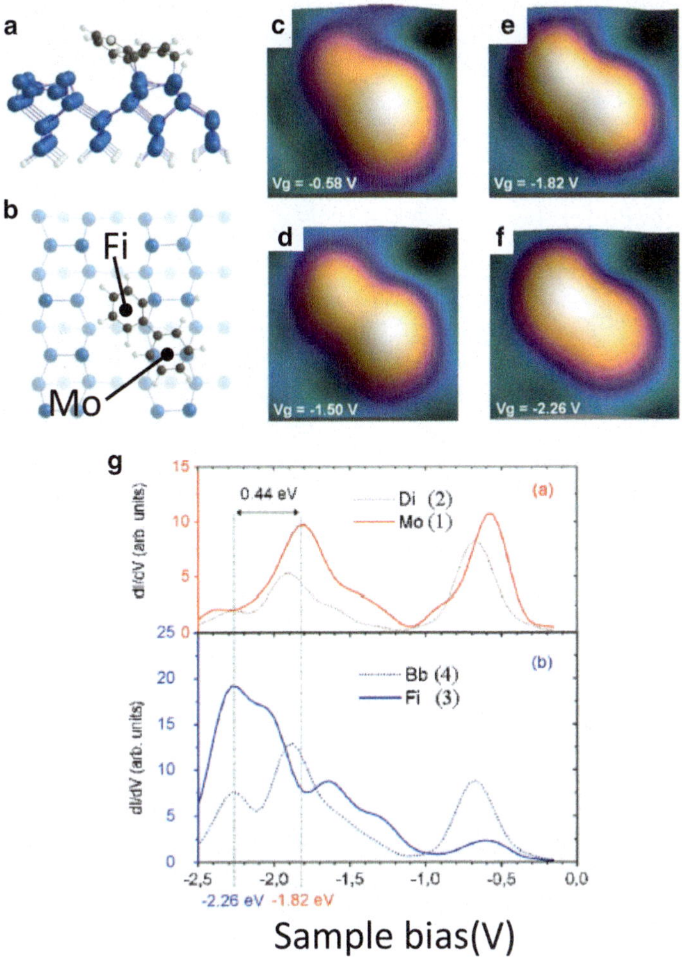

Fig. 1.2 (**a**) side and top views (**b**) of the relaxed biphenyl on Si(100): one phenyl ring is adsorbed in a butterfly configuration whereas there is dissociation between the second ring and a hydrogen atom. This second ring is attached to the surface through a single Si-C chemical bond. The H atom is passivating one of the silicon's dangling bonds. Silicon, carbon, and hydrogen atoms are represented in *blue*, *black*, and *white*, respectively. (**c**)–(**e**) calculated constant current STM images: (**c**) $I_t = 0.12$ nA and $V_g = -\ 0.58$ V (d), $V_g = -1.50$ V (**e**), $V_g = -1.82$ V (**f**), $V_g = -2.26$ V. The color scale used to plot these images is *black* for a low height of the tip and *white* for a high one. (**g**) Calculated STS spectra for the Mo (*plain line*, 1) and Di (*dotted line*, 2) positions of the STM tip over the sample and for the Fi (*plain line*, 3) and Bb (*dotted line*, 4) ones. Numbers (1–4) correspond to different tip positions on the molecule (Reprinted figures with permission from Dubois et al. [50]. Copyright (2013) by the American Physical Society)

systems, including over a million atoms have been previously conducted with classical methods [53], one of the main challenging parts here is to conduct such calculation with a molecule and a surface that counts several hundreds of atoms. The main principle of the NEB method reside in the evaluation of a force projection

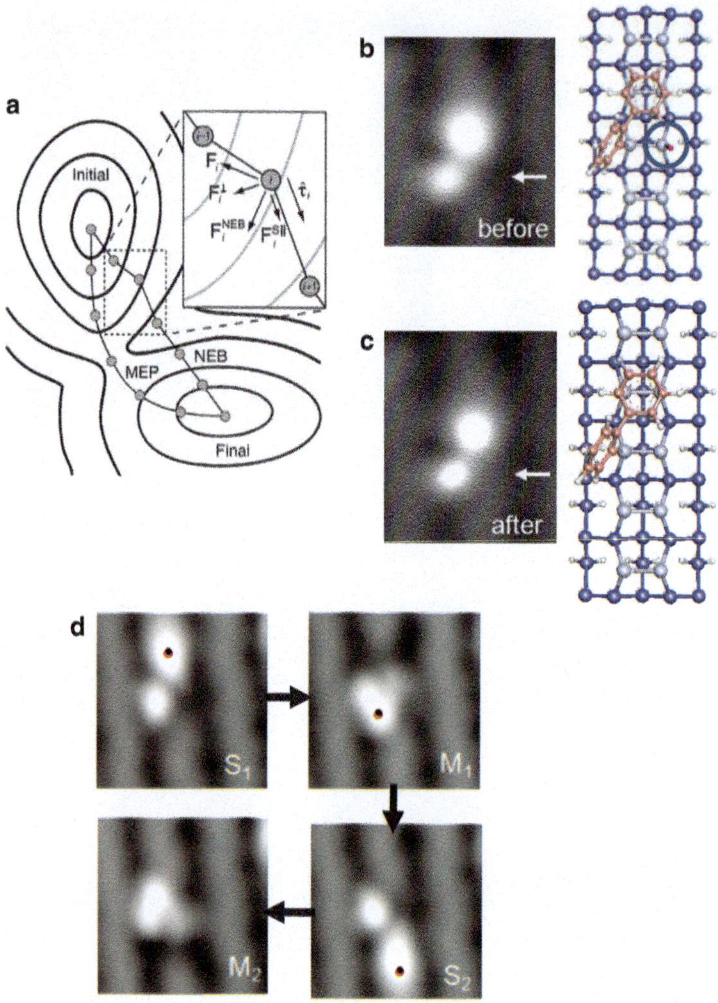

Fig. 1.3 (**a**) Sketch to illustrate the nudged elastic band method where two components make up the nudged elastic band force $\mathbf{F}^{\mathrm{NEB}}$: the spring force $F_i^{//}$, along the tangent τ_i, and the perpendicular force due to the potential $F_i^{\perp}$. The unprojected force due to the potential $\boldsymbol{F}_i$ is also shown for completeness. (**b**) and (**c**) are the (23.1 × 23.1 Å^2) STM topographies of the biphenyl molecule before and after the removal of the side H atom. Beside are the ensuing small balls and stick top views of the Si(100) surface (*blue balls*) and the biphenyl molecule (*red balls*). (**d**) (24 × 24 Å^2) STM topographies series ($V_s = -2$ V, $I = 220$ pA) showing the four states of the quadristable biphenyl molecule (Panel (**a**) is reproduced with permission from Sheppard et al. (http://jcp.aip.org/resource/1/jcpsa6/v128/i13/p134106_s1) © American Institute of Physics)

which ensures that the spring forces do not interfere with the convergence of the elastic band to the MEP, as well as ensuring that the true force does not affect the distribution of images along the MEP (Fig. 1.3a). During the minimization, it is necessary to estimate the tangent to the path at each image and for each iteration in order to decompose the true force and the spring force into components parallel and perpendicular to the path ($F_i^{//}$ and $F_i^{\perp}$ in Fig. 1.3a). Only the perpendicular component of the true force is included, and only the parallel component of the spring force. This force projection is referred to as "nudging." The spring forces then only control the spacing of the images along the band. In the used NEB method [54–56] the total force acting on an image is the sum of the spring force along the local tangent and the true force perpendicular to the local tangent such that

$$F_i^{NEB} = F_{i_{//}}^s - \nabla E(R_i)_{\perp} \tag{1.2}$$

where the true force is

$$\nabla E(R_i)_{\perp} = \nabla E(R_i) - \nabla E(R_i).\tau_i \tag{1.3}$$

and the spring force defined by

$$F_{i_{//}}^s = k(|R_{i+1} - R_i| - |R_i - R_{i-1}|\tau_i) \tag{1.4}$$

This is defined for each images R_0 to R_N along the pathway where E is the energy of the system defined for all the atomic coordinates, τ_i being the normalized local tangent at image i.

This method can be applied on the biphenyl molecule chemisorbed on the Si(100) surface to get insight into the chemical pathways relative to the molecular switch observed experimentally. As mentioned in previous work [57], the strongly chemisorbed bistable (S_1 and S_2 configurations) and quadristable (S_1, S_2, M_1, and M_2 configurations) states, obtained after removing the hydrogen atom adsorbed beside the biphenyl molecule (Fig. 1.3b, c), are characterized by the possibility to activate their reversible movements. As shown in Fig. 1.3d, once the H atom is removed, the biphenyl molecule exhibit four stable conformational states that can be switched randomly from one state to the other via electronically induced processes [58]. Here, we have calculated the energy profiles of these bistable and quadristable movements to compare them. As revealed in Fig. 1.4a, the movement of the biphenyl molecule in its bistable configuration consists of a reversible rotation of the molecule from the S_1 stable state to the equivalent symmetric S_2 state. The rotation, around the fixed phenyl ring bonded to the surface through a Si-C bond, defined by the angle θ (Fig. 1.4a), equals −60° and +60° when the molecule is in the S_1 and S_2 states, respectively. The mobile phenyl ring is bonded to a surface silicon dimer of the same dimer row through two Si-C bonds in the butterfly configuration as described previously. The rotation movement requires breaking the two Si-C bonds of the mobile butterfly phenyl ring in the S_1 state, rotating the molecule from $\theta = -60°$ to +60°, and making two new Si-C bonds of the butterfly ring in the S_2 state. An unstable transient state (T) between the S_1 and S_2 states has been also evidenced in the experiments. This T state is too unstable to be imaged with the STM. However, it could be directly observed between S_1 and S_2 by rapidly scanning the STM tip along a single line [45]. As shown in Fig. 1.4a, the activation

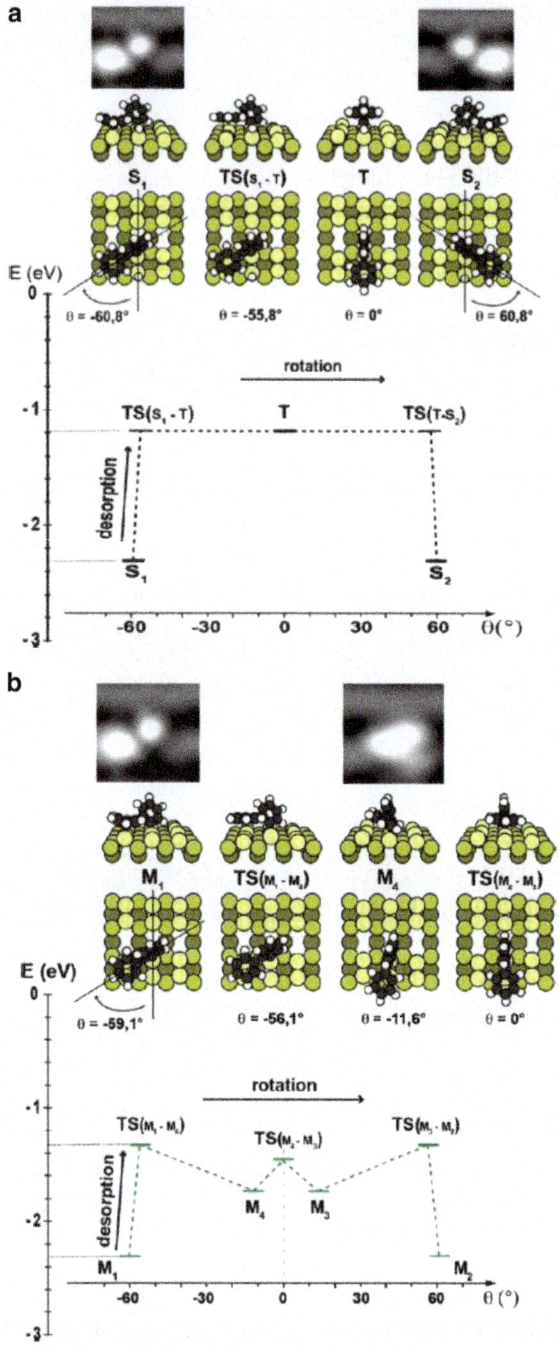

Fig. 1.4 (**a**) Calculated conformational pathways of the biphenyl molecule in its bistable configuration (S_1). Some states along the pathways are shown with top and side views. The rotation angle θ corresponds to the angle between two perpendicular planes to the Si surface: one contains the upright dehydrogenated phenyl ring and one contains the Si dimer linked to H atom and the phenyl fragment. The TS structure corresponds to the desorption of the butterfly ring. STM topographies (17×15 Å^2, $V_S = -2$ V, $I = 240$ pA) of the biphenyl molecule in its S_1 and S_2 states are shown. (**b**) Calculated conformational pathways of the biphenyl molecule in its quadristable configuration. This configuration has four stable states. The (14×13 Å^2) STM topographies, $V_S = -2$ V, $I = 240$ pA) of the biphenyl molecule in each configuration are shown (Reproduced with permission from Riedel et al. (http://pubs.acs.org/doi/pdf/10.1021/ja8101133) © American Chemical Society)

energy of the bistable movement is relatively high (~1.2 eV). Once the butterfly phenyl ring of S_1 has been desorbed, it can freely rotate at constant energy to reach the transient state T and finally readsorb at the S_2 position. From our calculations, several shallow minima at about the same energy as the transition state (TS) and with barrier energies smaller than 0.05 eV are found between S_1 and S_2. For simplicity, these transient states are represented by a single T state in the Fig. 1.4. The shallow minima explain the observed instability of the transient T states experimentally. The absence of any stable transient state is due to the presence of the hydrogen atom detached from the fixed phenyl ring, which passivates the silicon atom dangling bond underneath the transient state. This prevents any attractive interaction between the mobile phenyl ring and the surface from occurring.

In Fig. 1.4b, we show the calculated conformational pathway for the movement of the biphenyl molecule in its strongly chemisorbed quadristable configuration. This configuration is similar to the strongly chemisorbed bistable configuration, except that the hydrogen that passivates the silicon atom dangling bond underneath the fixed phenyl ring is now removed as explained above. In addition to the M_1 and M_2 stable states, which have geometries similar to those of the S_1 and S_2 states (Fig. 1.4b), two new metastable states are found at $\theta = 11.6°$ and the $\theta = 0°$ state becomes a transition state. These new metastable states are stabilized by attractive interaction between the π electrons of the moving phenyl ring and the dehydrogenated silicon dangling bond. These results explain correctly the observation of the four stable states observed in the quadristable configuration. Furthermore, the calculated activation energy for switching the quadristable molecule is found to be lower than for the bistable molecule (1.0 eV instead of 1.2 eV). This explains that the experimental switching efficiency of the quadristable configuration is increased by almost two orders of magnitude as compared to the bistable configuration. We emphasize that all the molecular reactions that have been investigated in this work, i.e., the switching of the bistable and quadristable configurations, have activation energies on the same order of magnitude (in the range 0.8–1.2 eV). This corresponds to the energy required to break the two C-Si bonds of the butterfly phenyl ring.

In conclusion, we have shown that DFT simulations which combine molecular relaxation on semiconductor surfaces, STM images simulation that uses the tight binding method and the search for molecular pathways via the nudged elastic band technique is precise enough and very useful for a detail analysis of STM induced manipulation of chemisorbed molecules at the atomic scale.

1.3 Reversible Switch of an Individual Functionalized Molecule Physisorbed on the Silicon Surface

As we have observed in Sect. 1.2 of this chapter, the manipulation of chemisorbed molecules through EIE processes usually involve transition states which energy barrier have to be overcome in order to initiate the molecular movement over the surface. On the other hand, the same transition state is related to the chemisorption of the molecules with the surface that can offer specific molecular resonances energies useful for the electronic processes induced with the STM. It is especially the case when the molecule has initially no specific function in the gas phase. We have thus explored the possibility to control such excitation on a functionalized molecule that has predefined chemical characteristics in the gas phase such as the isomerization process known to occur in the stilbene-like molecule. In order to preserve these molecular characteristics, we have adsorbed the stilbene molecule on the Si(100) surface [32]. The physisorption of individual molecule on semiconductor surfaces is rarely achieved as usually the low energy surface states offer large opportunities for small molecule to create chemical reactions. The physisorption can be obtained by inducing the adsorption of the molecule at very low surface temperature (12 K). The isomerization and especially the trans-cis photoisomerization reactions of aromatic molecules have been extensively studied in the gas and liquid phases during the past few decades [32]. Stilbene molecules have been, in particular, thoroughly explored because they are considered as a model for trans-cis isomerization studies. Photoexcitation of stilbene in the gas phase shows a competition between the trans-cis isomerization and the photocyclization mechanism forming dihydrophenanthrene [59]. The isomerization process have been studied for molecules adsorbed on surfaces [60–62] by activating the reactions via photon excitation [63], by means of tunnel electrons or with the electrostatic field under the tip of a scanning tunneling microscope (STM) [64]. However, when molecules are adsorbed on surfaces, trans-cis isomerization mechanisms are usually described by isomer conformations that mimic those known in the gas phase. Yet for self-assembled monolayers, molecular islands, or isolated species, steric hindrance involving the surface atoms or the neighboring molecules is expected to have a strong influence on the isomerization reactions rates and pathways as they are known from the gas. New molecular conformations that are unstable in the gas or liquid phases may be stabilized through weak surface interactions. Such surface isomers may lead to unique conformations that can be used to induce further molecular reactions or exhibit new molecular properties.

The Fig. 1.5 shows STM topographies of various stilbene conformations observed on the Si(100)-2 $\times$ 1 surface, following the adsorption of the *trans*-stilbene molecules, on both p-doped and n-doped samples. Gas-phase molecular conformations of *trans*-stilbene (TS) and *cis*-stilbene (CS) are shown for comparison in Fig. 1.5a, b. For each type of dopant (n or p) of the silicon substrate, four different adsorption conformations of stilbene (TS, CS, I_1, and I_2) are observed (Fig. 1.5c, d). From their shape and symmetry, the STM topographies shown in Fig. 1.5e, g, i, k are assigned to the *trans*-stilbene (TS) isomer conformation, whereas the conformations observed in Fig. 1.5f, h, j, and l are assigned to the *cis*-stilbene (CS). Indeed, the empty state topographies ($Vs > 0$) of Fig. 1.5e, i correlate well with the shape of the *trans*-stilbene in the gas phase shown in Fig. 1.5a, whereas the empty state topographies ($Vs > 0$) of the CS conformation (Fig. 1.5f, j) match the shape of the *cis*-stilbene in the gas phase (see Fig. 1.5b). The other observed isomers I_1 and I_2 (see arrows in Fig. 1.5c, d) might be surface-isomer conformations and will be discussed later. The TS conformation in

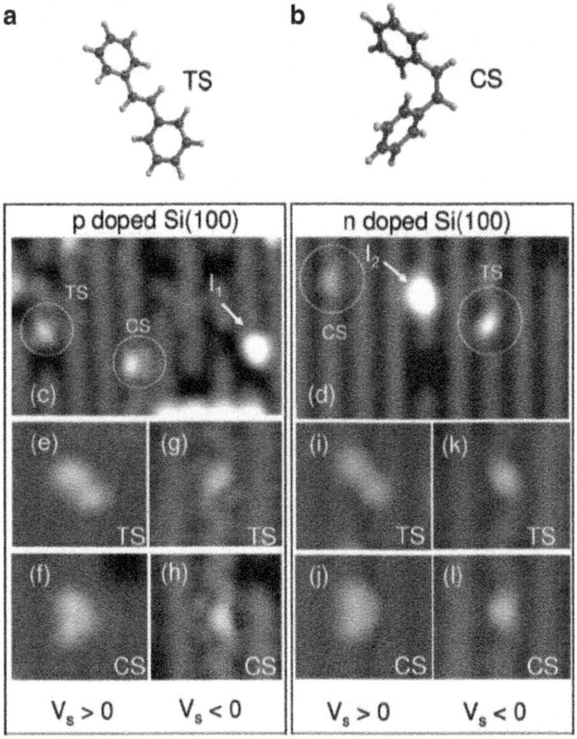

Fig. 1.5 (**a** and **b**) Conformations of trans-stilbene (planar) and cis-stilbene in their gas-phase configuration, respectively. (**c** and **d**) (69 × 44 Å^2) STM topographies (Vs = −1.5 V, I = 110 pA) of stilbene molecules adsorbed on p-doped and n-doped Si(100)-2 × 1, respectively. Circles indicate the trans-stilbene (TS) and cis-stilbene (CS) adsorption conformations. The *arrows* indicate stilbene surface-isomers I_1 and I_2. (**e**, **g**, **i**, **k**) (23 × 23 Å^2) STM topographies of the unoccupied (Vs = +1.5, I = 69 pA) and occupied (Vs = −1.5, I = 69 pA) trans-stilbene conformations on a p-doped (**e** and **g**) and n-doped (**i** and **k**) silicon surface. (**f**, **h**, **j**, **l**) (23 × 23 Å^2) STM topographies of the unoccupied (Vs = +1.5, I = 69 pA) and occupied (Vs = −1.5, I = 69 pA) cis-stilbene on a p-doped (**f** and **h**) and n-doped (**j** and **l**) silicon surface (Reproduced with permission from Riedel et al. (http://pubs.acs.org/doi/pdf/10.1021/ja807498v) © American Chemical Society)

Fig. 1.5e, i appears elongated and oriented at 48° with respect to the dimer rows, while the CS conformation in Fig. 1.5f, j looks folded in a crescent shape. In these unoccupied state (*V*s > 0) STM topographies (Fig. 1.5e, f, i, j), the benzene rings appear as very bright spots in both the TS and the CS conformations. The occupied state STM topographies (*V*s < 0) show only slight differences when the type of dopant in the silicon changes. For p-doped silicon, the benzene rings of both the CS and the TS conformations (Fig. 1.5g, h) appear brighter than for n-doped silicon (Fig. 1.5k, l), while the unoccupied states look almost the same.

The Si(100)-2 × 1 surface has been shown to be very reactive toward molecules with C-C double bonds due to the presence of the silicon dimer π orbitals [65].

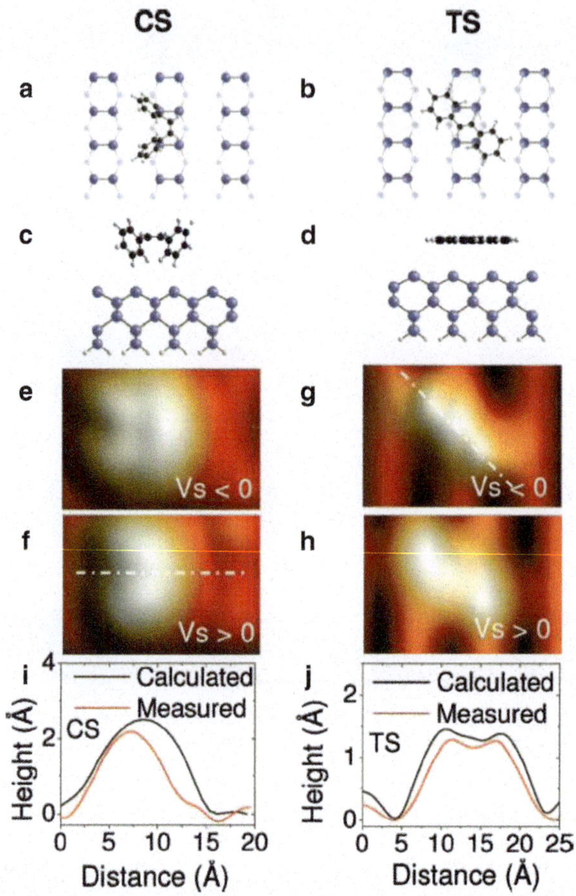

Fig. 1.6 Top views (**a** and **b**) and side views (**c** and **d**) of the conformation of the *cis*-stilbene and *trans*-stilbene molecules relative to the Si(100)-2 × 1 surface used for STM topography simulations. (**e** and **f**) Occupied ($Vs = -1.2$ V) and unoccupied ($Vs = +1.5$ V) calculated STM topographies of the *cis*-stilbene, respectively. (**g** and **h**) Occupied ($Vs = -1.2$ V) and unoccupied ($Vs = +1.5$ V) calculated STM topographies of the *trans*-stilbene molecule, respectively. The STM topographies were calculated at a constant current of 70 pA. (**i** and **j**) Comparison of the measured and calculated profiles (see the *dashed lines* on (**f**) and (**g**)) on a CS and a TS molecule, respectively (Reproduced with permission from Riedel et al. (http://pubs.acs.org/doi/pdf/10.1021/ja807498v) © American Chemical Society)

One might expect stilbene molecules to react similarly. However, from the STM topographies in Fig. 1.5, there are strong indications that physisorption occurs rather than chemisorption of the stilbene molecules on the Si(100)-2 × 1 surface at 12 K. For both cis and trans isomers, the bright protrusion observed in the occupied state STM topographies, which corresponds to the position of the carbon double bond of the stilbene molecule, is always observed to be centered on the dimer row and in a position between two silicon dimers (see Fig. 1.6a–d for the proposed adsorption conformation). If chemisorption occurred during the adsorption at 12 K, this would involve the breaking of the carbon double bond followed by bond formation with the silicon dimers. In this case, the corresponding STM topographies of the molecules, especially the CS conformation, would be expected to be rotated by 90° as compared to what is observed in Fig. 1.6a. Indeed, it is unlikely that chemisorption occurs with a C-C double bond direction of the CS molecules perpendicular to the silicon dimer direction. Furthermore, the physisorbed character is consistent with recent results obtained by synchrotron radiation photoemission in which only a weak π-π bonding between the stilbene and the Si(100) surface was observed at low coverage on Si(100)-2 × 1 at 100 K [66] .

To verify the proposed adsorption conformations in Fig. 1.6a, b, we have simulated the STM topographies of both CS and TS molecules physisorbed on the Si (100)-2 × 1 surface. However, it is known that ab-initio DFT methods, for example in the general gradient approximation (GGA), poorly describe the weak interactions such as van der Waals (VdW) forces. Therefore, adsorption energy calculations of the CS and TS molecules on the Si(100) surface obtained with this method would not be relevant. Although some very recent methods based on VdW-DFT calculations are intended to describe more accurately weakly bonded adsorbates, it is very delicate to build an appropriate VdW functional adapted to physisorbed stilbene molecules on a reconstructed silicon surface [67, 68]. Furthermore, such a method will probably involve a heavy computational cost. Here, we propose an alternative approach that consists of identifying adsorption conformations on the basis of successive comparisons of calculated and experimental STM images. For these calculations, we use a tight binding scheme that correctly reproduces first-principles based density functional electronic structure calculations with a low computational cost. This allows complex systems to be handled with adequate sensitivity as compared to classical DFT methods [48, 69]. To simulate the Si(100) surface, we consider a single slab of 5 planes, each of which contains 24 silicon atoms. The lower plane is passivated by hydrogen atoms. The upper one is reconstructed to form the surface dimers. In the directions parallel to the surface, the system including the Si slab and the molecules is periodic and is built from a supercell corresponding to 12 (3 × 4) dimers. This supercell is large enough to avoid interactions between adsorbed molecules and thus provides a good simulation of STM images. For the same reason, the electronic structure of the interacting system (molecule and surface) is calculated only at $\boldsymbol{k} = 0$, using the self-consistent density functional-based tight binding method developed in the DFTB + code [70, 71]. Within this method, the total energy of the system is defined as:

$$E_0^{TB} = \sum\nolimits_i^{occ.} \langle \psi_i | H_0 | \psi_i \rangle + E_{rep.}, \tag{1.5}$$

where the single-particle wave function Ψ_i can be expanded into a suitable set of localized atomic orbital φ_ν such that

$$\psi_i(\vec{r}) = \sum\nolimits_{\nu} c_{\nu i} \varphi_{\nu}(\vec{r} - \vec{R}_a), \tag{1.6}$$

in which the employed confined atomic orbitals are determined by solving a modified Schrödinger equation [72]. Here, $E_{rep.}$ can be determined as a function of the distance R for the occupied Kohn-Sham orbitals ε_i at the occupied number n_i, by taking the difference between the self consistent field (SCF) local density approximation (LDA) and the ensuing TB band structure energy which can be written for a given reference system,

$$E_{rep.} = \left\{ E_{LDA}^{SCF} - \sum\nolimits_{i}^{occ.} n_i \varepsilon_i(R) \right\} \Big|_{ref.struc.}. \tag{1.7}$$

Within this approach, the parameterization of the Hamiltonian has been validated on various molecular systems, including weakly interacting ones [73]. The tunneling current is then calculated between the STM tip, which is assumed to have a pyramidal apex, and the sample. The elastic scattering approach considers that electrons tunneling process occur through multiple channels, although the electron transport through the STM junction is treated in the coherent regime. It is also possible to include tip effects while the tip geometry has only a weak influence on the results. More complex methods based on the nonequilibrium Green's functions formalism (also known as the Keldysh approach) are not necessary in the present case, because interactions between the tip and the sample remain weak due to the low tunneling current during image acquisition. The initial conformations of the CS and TS molecules are calculated by energy minimization in the gas phase, while the silicon surface is minimized independently. After separate minimizations in the gas phase, the corresponding molecular conformations of the CS and the TS molecules were tested, on the silicon surface (see Fig. 1.6a–d) by small variations of different parameters such as: (i) slight rotations on one or both benzene rings with respect to the benzene-ethene bond axis, (ii) change of the molecule to surface distance, (iii) rotation of the whole molecule with respect to the dimer row direction and translation of the molecule along the dimer rows, and (iv) change of the buckling of the silicon dimers surface. Any of these factors can dramatically modify the result of the calculated image. Numerous adsorption geometries were tested, taking into consideration a variety of rotations and translations of the whole molecule with respect to the dimer row direction. We emphasize that none of the tested conformations matches the I_1 or I_2 STM topographies. This result indicates that the observed stilbene conformations I_1 and I_2 are not simply a change between two different adsorption orientations of the TS or CS conformations. The calculated STM images of the CS and TS conformations that best match the experimental ones are shown in Fig. 1.6 The average molecule-surface distance is estimated to be 4.2 and 3.2 Å in Fig. 1.6c, d for the CS and the TS molecules, respectively. For both the CS and the TS molecules, there is a reasonably good agreement between the calculated (Fig. 1.6) and the experimental (Fig. 1.5) STM topographies for occupied and unoccupied states (see Fig. 1.6e–h) acquired in constant current mode. The observed angle of the CS molecules in the experimental STM topographies can vary (mainly due to silicon dimer buckling) and is estimated to be 71 ± 13°. The influence of the silicon dimer buckling is difficult to reproduce in our STM topography calculations. Thus, the apparent angle between the two benzene rings of the calculated CS STM topography of ~60 ± 3° (Fig. 1.6e) is in reasonable agreement with the observed experimental angles.

Our model reproduces relatively well the variation of the tunnel current with the variations of the tip-surface distance [74]. Therefore, the profiles obtained from the STM topographies of the measured and calculated CS and TS molecules can be compared as shown in Fig. 1.6i, j. Both profiles show very similar tip heights above the stilbene molecules. These calculations support the choice of the orientations of the CS and TS molecules described in Fig. 1.6a–d. Furthermore, the average height of these molecules relative to the silicon surface is too large for chemical bond formation, indicating that they are physisorbed on this surface. However, it should be noted that the calculated STM images of the TS molecules exhibit less contrasted contours than do the experimental ones (see Figs. 1.5g and 1.6h). This is most probably due to the influence of the silicon surface interaction with the stilbene molecule, which changes with the type of dopant (n or p). Unfortunately, this influence cannot be taken into account in our calculations.

Electronic excitation of a stilbene molecule in its TS conformation (the position of the tip is indicated by the dot in Fig. 1.7a) results in the diffusion of the molecule across the surface. After diffusion, the molecule can be found either in the same TS conformation or in a new surface-isomer conformation called I_1 (Figs. 1.7b, c). Similarly, the electronic excitation of a stilbene molecule in its CS conformation (Fig. 1.7d) results in its diffusion across the surface. Most of the time, the diffusion of the CS molecule is accompanied by a change of molecular conformation into another surface-isomer conformation called I_2 (Fig. 1.7e, f). These two new molecular conformations I_1 and I_2 show very similar spherical shapes in the occupied state STM topographies (Fig. 1.7b, e). However, clear differences are visible in the unoccupied state STM topographies (Fig. 1.7c, f). In fact, the bright spot in Fig. 1.7b (conformation I_1) is observed in the middle of the silicon dimer row, while a careful observation of the bright spot in Fig. 1.7e (conformation I_2) shows that it is slightly misaligned with the silicon dimer rows (gray vertical lines). The Fig. 1.7g, h depicts the comparison between the profiles of both surface-isomers I_1 and I_2 after adsorption and after manipulation, respectively. These profiles are very similar, indicating that the surface isomers produced after adsorption and by manipulation are the same. The molecular conformation changes induced by STM manipulation, TS $\rightarrow$ I_1 or CS $\rightarrow$ I_2, are found to be reversible. Furthermore, the I_1 and I_2 surface isomers are unequivocally associated with their initial molecular conformations (i.e., TS and CS, respectively) [32]. Indeed, the TS $\rightarrow$ I_2 or CS $\rightarrow$ I_1 conformation changes are never observed.

From a theoretical point of view, the identification of the I_1 and I_2 conformations is a prohibitively heavy task as the comparison between experimental and calculated STM topographies will require many different configurations to be tested. Furthermore, a precise description of these molecular conformations will probably

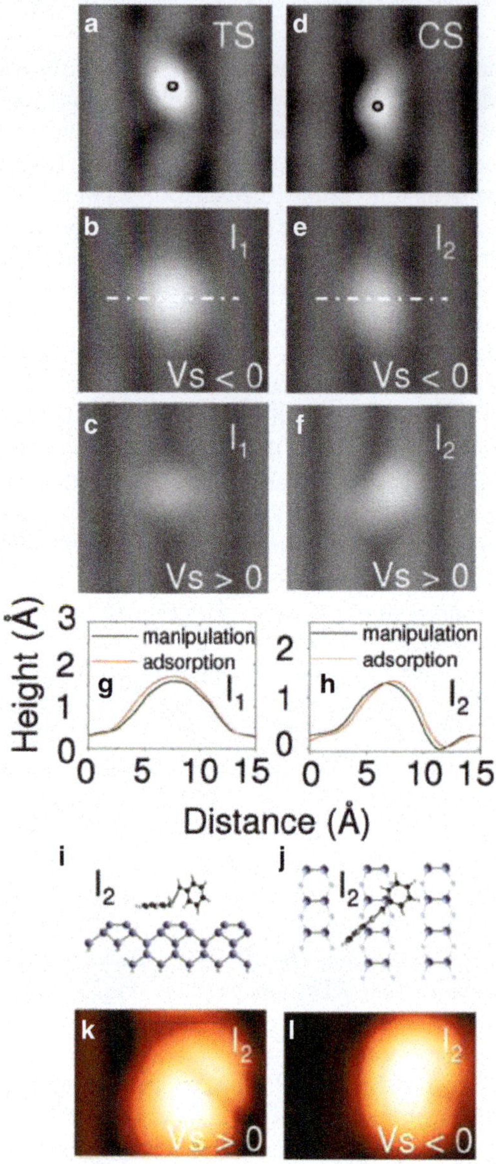

Fig. 1.7 (**a–f**) (19.2×18 Å^2) STM topographies of stilbene molecules adsorbed on a n-doped Si (100)-2 × 1 surface. (**a** and **d**) Occupied state (Vs = −1.5 V, I = 69 pA) of the trans-stilbene and cis-stilbene molecules before a manipulation (the excitation is applied on the *circle* at Vs = −2.5 V, I = 780 pA, for a duration of 3.2 s). (**b** and **e**) Occupied state (Vs = −1.5 V, I = 69 pA) STM topographies of the surface-isomer I_1 of the trans-stilbene molecule after the manipulation in (**a**) and the surface-isomer I_2 of the cis-stilbene molecule after the manipulation in (**d**), respectively. (**c** and **f**) Same image as (**b**) and (**e**) in the unoccupied state (Vs = +1.5 V, I = 69 pA). (**g** and **h**). Profiles measured on I_1 and I_2 surface-isomers STM topographies, respectively (see *dotted lines* in (**b**) and (**e**)), before adsorption and after manipulation. (**i**) Top and (**j**) side views of a suggested conformation of the I_2 surface isomer. (**k** and **l**) Calculated STM topographies of the proposed I_2 surface isomer for, respectively, the unoccupied state (Vs = +1.5 V) and occupied state (Vs = −1.5 V) at I = 70 pA (Reproduced with permission from Riedel et al. (http://pubs.acs.org/doi/pdf/10.1021/ja807498v) © American Chemical Society)

need to take into account the influence of the surface dopant type (n or p), which is difficult to implement in the calculations at this stage. As an example, STM topographies of a possible stilbene conformation analogous to transient state are presented in Fig. 1.7i–l and show clear similarities to the I_2 surface isomer. This stilbene surface isomer is seen to have a different conformation as compared to the CS or TS molecules. Here, the two benzene rings of the stilbene surface isomer are rotated by 90° from each other.

In conclusion, we have seen that the STM topographies of a physisorbed molecule can be described by using the DFTB + method. While STM topographies are shown to describe the main aspects of the molecular conformation that lead to predict that the stilbene molecule is physisorbed on the bare Si(100) surface, the numerical simulation of the STM images appears to be of great help to confirm the experimental observations.

1.4 Lateral Control of the Hopping Diffusion of a Weakly Bonded Molecule

The physisorption of larger molecules has been often explored on metal surfaces, in particular when ultrathin insulating layers are used to electronically decouple the molecule from its substrate [28, 29]. In this context, the control of the manipulation of molecules on the bare silicon surface, known to be chemically very reactive, requests specific functionalized chemical adsorbate that allow their physisorption and thus the control of their lateral manipulation. This has been explored by using hexaphenyl-benzene molecules adsorbed at low temperature (12 K) on the bare Si (100) surface [75]. Here, we show that an individual hexaphenylbenzene (HPB) molecule can be physisorbed at the S_A step edges of the bare Si(100)-2 × 1 surface and exhibit periodic hopping under localized electronically induced excitations processes (Fig. 1.8a). The HPB molecule has been chosen owing to its C_{6v} symmetry with large torsional angles between the central and the peripheral benzene rings. Its non-planar geometry involves the lateral benzene rings to rotate almost 90° relative to the central one and thus results in the orientation of the lateral benzene rings of the HPB nearly perpendicular to the surface. However, due to interactions between the HPB's lateral benzene rings, this angle is reduced to ~65° [76–78]. Such a configuration gives to this molecule a very good stability and a very low probability to chemically bonds with the Si surface [79, 80].

The tip–induced hopping of the HPB molecule can be reversibly controlled along the silicon surface step-edge in two opposite directions without any lateral movement of the STM tip. Hence, the hopping mechanism is induced by the tunnel electrons and can be decomposed in two major steps: The first step of the hopping process induces the formation of an HPB^- anion that triggers the molecular motion into a metastable state. The molecular movement is associated with the electrostatic forces acting in the STM junction on the negatively charged HPB molecule.

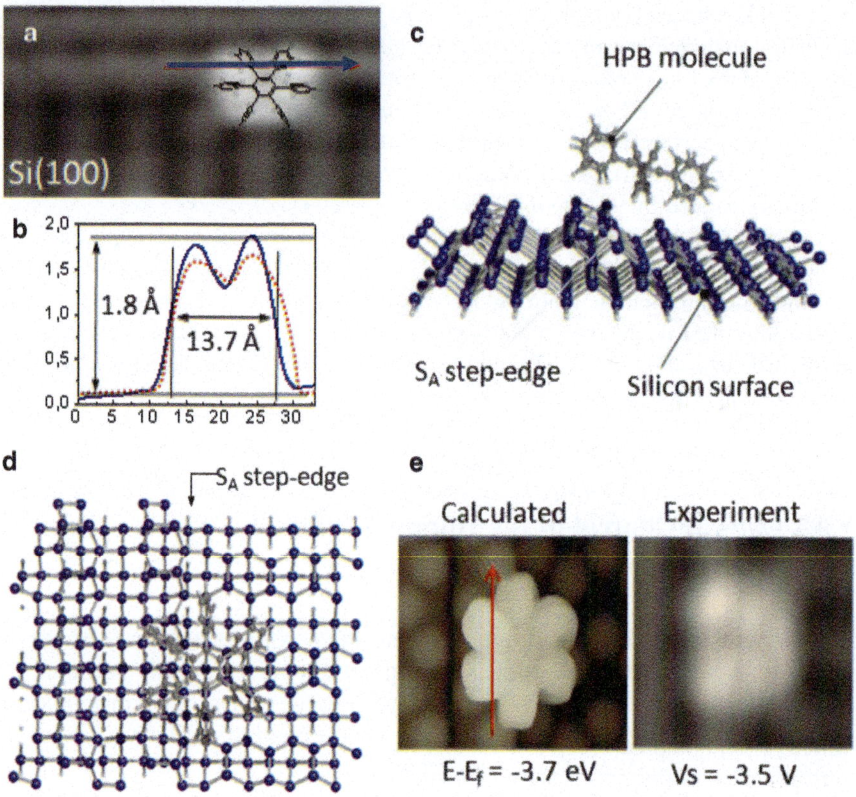

Fig. 1.8 (**a**) Example of a (50 × 26 Å^2) STM topography of an individual HPB molecule ($V_s = -3.5$ *V*, I = 10 pA) adsorbed on the S_A step-edge of the Si(100) surface with the on-scale structure of the HPB molecule overlaid. (**b**) Height profile of the molecule measured along the *blue line* indicated in (**a**), the *doted red curve* is the calculated corresponding profile along the *red line* indicated in (**e**). (**c**) Balls-and-sticks side view of the HPB molecule after its relaxation at the S_A step edge of the Si(100) surface. The *blue*, *gray* and *white balls* are the Si, C and H atoms, respectively. (**d**) Balls-and-sticks top view of the HPB molecule after its relaxation on the Si(100) surface at the S_A step edge. (**e**) Comparison between the calculated (*left*, E-Ef = −3.7 eV) and experimental (*right*, $V_s = -3.5$ V, $I = 52$ pA) STM topographies (2.7 × 2.7 nm^2) of an HPB molecule adsorbed on the Si(100) S_A step-edge (Reproduced with permission from H. Labidi et al. (http://pubs.acs.org/doi/pdf/10.1021/ja807498v) © American Chemical Society)

In a second step, the neutralization of the HPB$^-$ anion allows the molecule to move from its metastable state to its final steady position. We have shown that the initial position of the STM tip is of crucial importance to optimize simultaneously the electronic charging of the HPB molecule and the lateral movement to the desired direction [29]. The density functional theory (DFT-D) calculations of the adsorption site of the HPB molecule on the bare Si(100) surface as well as accurate calculation of the molecular orbital energy of the stationary charged HPB$^-$ anion using TD-DFT methods support our experimental results and reveals the physisorption of the HPB molecule. The calculation of the adsorption conformation and the electronic properties of the HPB molecule configuration adsorbed on the Si(100) surface are performed using the density functional theory method. The Perdew-Burke-Ernzerhof (PBE) exchange-correlation functional [81] with generalized-gradient corrections, projector augmented waves (PAW) [82] and the plane waves as implemented in the VASP code are used [83–85]. The energy cutoff in the simulations is 400 eV and the structure is relaxed until the force on every atom is smaller than 0.02 eV/Å. The integrated local density of states (LDOS) image is obtained following the Tersoff-Hamann approach [86]. The Brillouin zone is sampled using a single k-point at the Γ point which is a reasonable choice due to the large size of the supercell employed. In addition, Van der Waals interactions are computed using the semi-empirical correction of Grimme [87–89]. The used DFT-D method implemented in VASP defines the total energy by

$$E_{DFT-D} = E_{DFT} + E_{Disp.} \tag{1.8}$$

where E_{DFT} is the energy obtained from the usual Kohn-Sham calculation and $E_{Disp.}$ is a dispersion correction containing the C^6R^{-6} dependency. The interaction of each pair of atoms indexed by i and j separated by a distance R is given by

$$E_{Disp.} = -f_{ij}(R)\frac{C_{ij}^6}{R^6} \tag{1.9}$$

where f_{ij} (R) is the damping function which equals one for large values of R and zero for small values of R. The attractive interaction coefficients are fitted into ab-initio data and additional parameters that depend on the exchange-correlation functional used. We determined $E_{Disp.}$ as suggested by Grimme [87]. The pair interaction energies are summed up over all possible combinations of two atoms in the considered system. The resulting dispersive energy is added to the total Kohm-Sham one. This method, based on the use of an additional semi empirical term, raises the problem of transferability. The results obtained with this method are reasonable in the framework in which the parameterization is done. For example, specific parameters have been developed for Au within metal-supported supramolecular domain [90]. Hence DFT-D is very adapted to the present system since the HPB presents only carbon or hydrogen atoms. From the practical point of view where the focus is on robustness and computational speed in DFT simulations, the Grimme approach is known to yield satisfactory results because of the parameterization. In the present case, the silicon surface is described by five layers of silicon atoms with 54 atoms per layer with a step-edge of one Si layer of 24 additional atoms. In order to saturate the dangling bonds with unpaired electrons, the slab is terminated by hydrogen atoms on the underside. The total number of atoms (substrate + molecule) equals 384.

Following a few stabilizing topographies, one of the adsorption configurations of the HPB molecule decorates the S_A silicon step-edges that shows a specific molecular conformation straddled between the upper and lower silicon terraces

(Fig. 1.8a). The measure of the corrugation profile across the HPB molecule along the step edge shown in Fig. 1.8b reveals a apparent height of the molecule of ~1.8 Å [91]. Furthermore, the observed protrusion has a width of ~13.7 Å which is in accordance with the HPB length in gas phase. These preliminary observations already indicate that the HPB molecule is weakly bonded to the S_A step-edge of the Si surface. To further investigate the adsorption configuration of the molecule we have theoretically simulated the molecular conformation and the electronic properties of the HPB molecule at the S_A step-edge.

The calculated conformation of the relaxed HPB molecule at the S_A step edge of the Si(100) surface is presented in Fig. 1.8c, d. In Fig. 1.8c, we can see that the relaxed conformation of the HPB molecule is found to be straddled on the silicon step-edge, similarly to our experimental observation, with an adsorption energy of −0.89 eV. Due to steric hindrance with the silicon surface atoms, the adsorbed HPB molecule encounters slight rotations of the lateral benzene rings that reduce its initial symmetry (C_2). From this conformational simulation, the LDOS distribution of the occupied states of the HPB molecule can be calculated at $E\text{-}E_f = -3.7$ eV and compared with the experimental one (Fig. 1.8e). We can see that there is a rather good reproducibility of the general molecular size and location compared to the surface as well as the internal structure of the molecule. However, the size of the two lateral benzene rings located on each side of the step-edge silicon dimer row is not precisely reproduced. This slight discrepancy can be attributed to STM tip effects that are not accurately taken into account in the simulations using the Tersoff-Haman approximation [22]. A calculated profile can be extracted from the simulated STM image (see the red arrow in Fig. 1.8e) which reproduces the experimental curve with a good accuracy (Fig. 1.8b). A detailed analysis of the simulated conformation of the HPB molecule indicates an average height of ~3.6 Å when the distances between the lowest hydrogen atoms and the tilted silicon dimers are considered. This value further confirms the physisorbed nature of the HPB molecule bonding with the surface and is comparable with the calculated height of physisorbed organic molecules on bare silicon or passivated semiconductor surfaces.

To activate the controlled hopping of the HPB molecules at the S_A step-edge of the Si(100) surface, we place the tip apex at one side above the HPB molecule (see red dot in the upper panel of Fig. 1.9a) and we proceed to its excitation. Each excitation occurs as follow: (i) the STM feedback loop is open and the surface voltage is switched from the scanning voltage (−3.5 V) to − 4.1 V. (ii) The height of the STM tip is then adjusted to control the tunnel current from 50 pA (scanning current) to few nanoamperes. (iii) The surface voltage is kept at −4.1 V during ~1.5 s and then the feedback loop is closed while the surface voltage and the tunnel current are set back to their initial values. (iv) A second STM topography is performed at the same location and the results of the excitation is shown in the lower STM topography of the series in Fig. 1.9. The ensuing excitation of the molecule on the left side induces the lateral shift of the HPB molecule towards the left of the silicon step-edge (Fig. 1.9a). To ease the observation of the molecular movement, a green line crossing

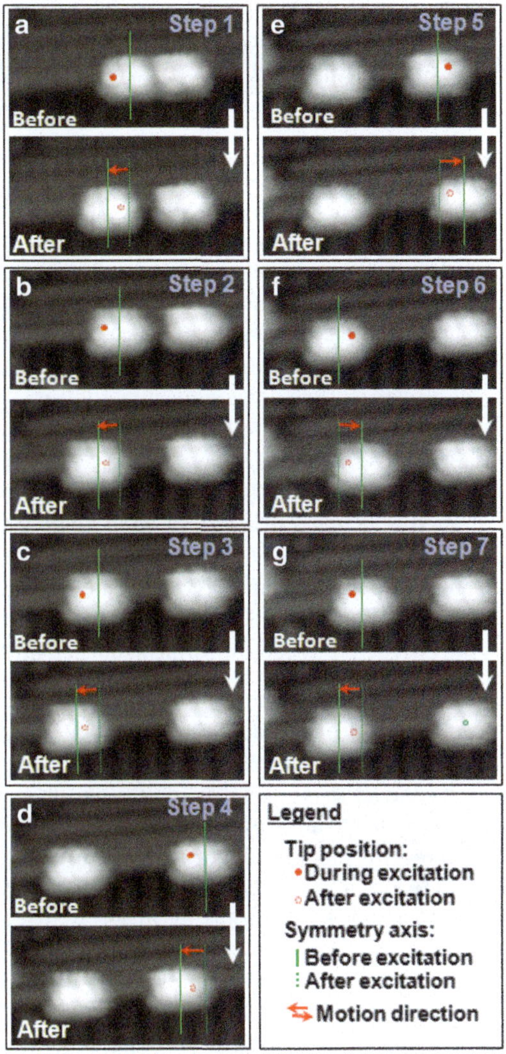

Fig. 1.9 (a–g) ($3.5 \times 8.7\ nm^2$) STM topographies ($V_s = -3.5\ V, I = 16$ pA) series before (*upper image*) and after (*lower image*) various manipulations on two individual HPB molecules at the S_A step edge of the Si(100) surface. The *vertical solid* and *dotted green lines* indicate the location of the centre of the molecule before and after the manipulation, respectively. The *red arrow* indicate the movement direction, the *red filled* and *empty dots* indicate the location of the excitation (STM tip apex) before and after the movement (Reproduced with permission from Labidi et al. (http://pubs.acs.org/doi/pdf/10.1021/ja807498v) © American Chemical Society)

the middle of the HPB in its initial position is drawn (upper panel of Fig. 1.9) with a second dotted green line that crosses the middle of the diffused HPB molecule (lower panel of Fig. 1.9a). The same process can be repeated several times on the same molecule (steps 2 and 3 in Figs. 1.9b, c) and produce exactly the same effect on the molecule, i.e. the HPB shifts towards the left direction when the excitation of the molecule is located at the left side of the HPB molecule. It is important to underline that only the excited molecule is moving as observed on the Figs. 1.9a–c where the neighbor HPB molecule stays motionless during the excitations. This excludes the fact that the electrostatic field in the STM junction is at the origin of the observed hopping of the HPB molecule [92, 93]. Indeed, the second molecule observed in the series of STM topographies of Fig. 1.9 can also be moved to the left of the silicon step-edge with a similar excitation (Fig. 1.9d). Note that the hopping length of the molecule is always constant and periodic with a step of 7.7 Å which is exactly the length separating two dimer rows of the lower silicon terrace. The control of the direction of the molecular hopping can be further tested by reciprocally applying an excitation at the opposite side of the HPB as shown in Fig. 1.9e, f. Here, we wish to stress that the direction of the motion can be controlled by choosing the side where the excitation is applied on the molecule. Finally, the Fig. 1.9g demonstrates that the process is completely reversible. It is also important to point out that we have not observed the hopping of the HPB molecule when the excitation is applied on the silicon surface nearby the molecule or in the center part of the molecule (see the green point in Fig. 1.9g) [29].

If we concentrate now on the directivity of the observed molecular hopping, we can see that the measured electronic yields to activate the HPB hopping are very similar to trigger the left motion ($Y_{left} = 2.4 \pm 0.4 \times 10^{-10}$) or the right motion ($Y_{right} = 1.9 \pm 0.3 \times 10^{-10}$) indicating that the hopping process is completely symmetric as suggested by the configuration of the HPB molecule at the step-edge of the surface. Our experimental results imply that the directivity of the hopping occur once the HPB molecule is negatively charged while attracted by the positively charged STM tip at $V_s < 0$. The lateral movement of the molecule is thus triggered by the following sequence of events:

(i) When the STM tip is located at one side of the upper part of the molecule (here the right side) in a P_1 configuration, the positively charged STM tip attracts the negatively charged HPB^- anion located at the central benzene ring of the HPB^- (Fig. 1.10a). In the present case, the movement is directed to the right due to electrostatic forces while the surface step-edge acts as a guide that prevents the HPB molecule to diffuse on the upper terrace [94].
(ii) The radius of action of the electrostatic field gradient present in the biased STM junction can be rather localized at the silicon surface in particular in the presence of micro-tips [95] and contributes to the attraction of the molecule towards the STM tip [96]. It also acts on the molecule-surface potential energy surface (PES) to lower the energy barrier that separates the two stable positions P_1 and P_2 of the HPB molecule before and after the hopping, respectively. A DFT calculation of the HPB molecule relaxed in the transient position T (Fig. 1.10b) allows deducing an energy barrier height between P_1 and P_2 of 0.39 eV. The estimated value of the TIBB energy (~300 meV at $V_s = -4.1$ V) on the Si(100) surface at the apex position of the STM tip slightly decreases at the transient position T to reach 190 meV. This results in an estimated hopping barrier energy at T of ~0.2 eV.
(iii) The point-charge located at the central benzene ring of the HPB^- anion perturbs very locally the molecule-surface PES and slightly decreases the well at the position P_1 of the molecule compared to the position P_2 [97]. This effect can be represented by an image charge potential and its ensuing attractive force as described in the Antoniewicz processes [29, 98]. This attractive force can be as strong as the one observed on metal surfaces in the case of degenerated semiconductors [99].

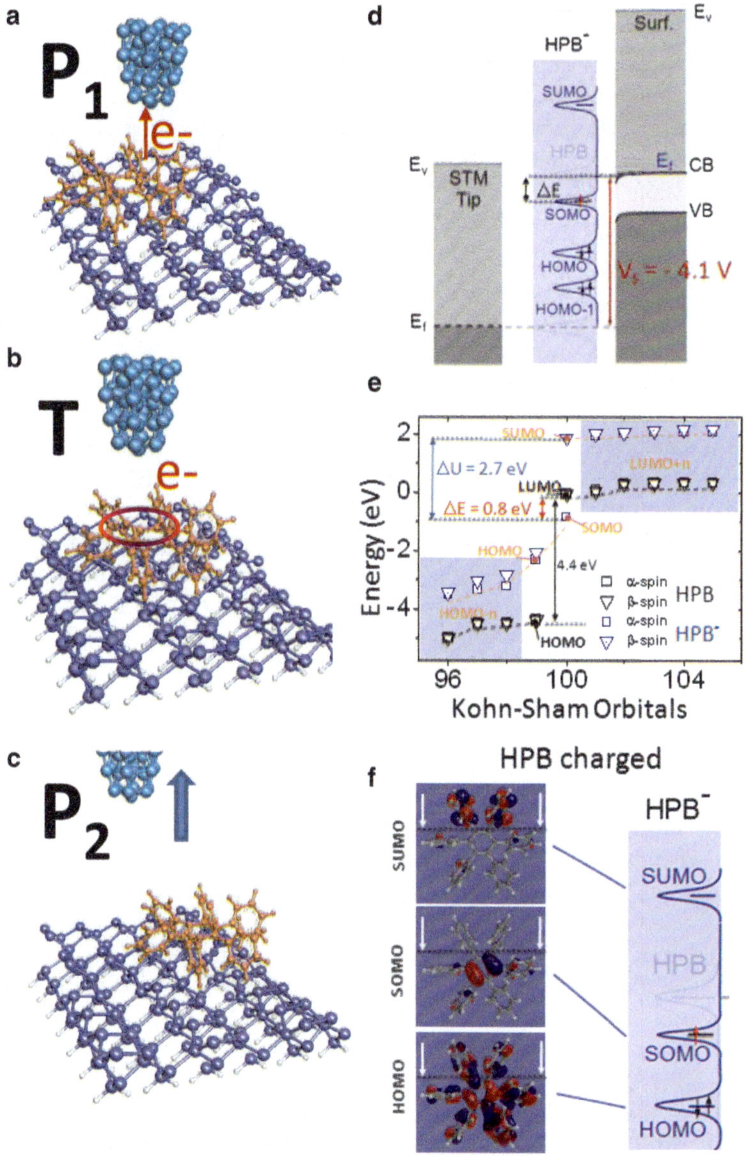

Fig. 1.10 (**a–c**) are series of sketches explaining the lateral manipulation directivity in three steps: in (**a**) the HPB molecule is electronically excited at one side (here on the right side), in (**b**) the molecule is negatively charged and is attracted underneath the STM tip, in (**c**) while the STM tip retracts, the HPB molecule becomes neutral and shifts towards the minimum of the surface potential well. (**d**) Energy band diagram of the STM double junction illustrating the relative energies of the SUMO, SOMO, HOMO and HOMO-1 orbital of the HPB^{-} molecule at $V_s = -4.1$ V as calculated in (**e**). (**e**) Spin polarized calculated Khon-Sham orbitals energies situated within the chemical potential of a neutral (*black*) and negatively charged (*blue*) individual HPB molecule (no surface) which structure was initially relaxed at the S_A step-edge of the Si(100) surface. (**f**) Spin polarized calculation of the spatial distribution of the SUMO, SOMO and HOMO of the HPB- anion related to the corresponding orbitals energies band diagram as presented in (**b**) (Reproduced with permission from Labidi et al. (http://pubs.acs.org/doi/pdf/10.1021/ja807498v) © American Chemical Society)

(iv) At the precise instant when the HPB molecule becomes negatively charged, the combination of these physical effects, i.e. the interaction between the tip and the molecule with the reduction of the hopping barrier height between P_1 and P_2, trigger the lateral hopping of the HPB molecule from the position P_1 towards the P_2 position. However, when the HPB^- moves along the step edge to reach the position T, the distance between the molecule and the tip is slightly shorten and the molecule reaches the maximum of the PES at T with a slightly increased attractive force that stabilizes the HPB^- molecule in front of the STM tip position (Fig. 1.10b).

(v) When the HPB^- molecule arrives at the saddle point T of the lateral motion pathway, it is important to consider that the image charge potential of the HPB^- is moving simultaneously with the molecule resulting in the apparition of a small well at T in the molecule-surface PES [29]. Such configuration stabilizes the molecule and can keep the HPB^- anion in the metastable state T for a relatively long period, as observed experimentally. This is also consistent with the fact that, if the HPB^- anion could reach the position P_2 during the excitation cycle, a symmetric configuration to P_1 would take place and the HPB^- will be attracted back to the metastable point T that clearly appears to be a steady transitory position of the hopping process.

(vi) The fact that the HPB^- molecule spends a relatively long time in its charged state at the transient position T during the excitation period completely explains our experimental findings. Therefore, the first part of the hopping mechanism of the HPB molecule towards the transient T position is triggered by inducing its negatively charge state whereas the complete hopping from T to P_2 still need to be elucidated. Our experimental observation indicates that the second part of the hopping process is related to the neutralization of the molecule. The neutralization of the HPB^- anion can easily occur once the molecule has reached the metastable position T and after the excitation period is finished. Indeed, when the excitation period ends, the surface voltage is set back to the scanning value (−3.5 V) and the STM tip is retracted to the previous scanning current (20 pA). In these conditions, the molecule-surface PES is not anymore perturbed by the STM tip and the energy of the SOMO of the HPB decreases towards the Fermi level of the silicon surface which induces the loss of the extra charge in the HPB^- anion. The neutralization of the HPB^- anion leads to an excess of kinetic energy in the neutral molecule. This supplementary vibrational energy allows the neutral HPB to end the hopping from the metastable state T to the second steady conformation P_2 (Fig. 1.10c).

As described above, during the excitation process, the electron attachment on the HPB involves a complete electronic reordering of the molecule as presented in Figs. 1.10d, e. The additional electron attached to the HPB^- anion partially fill the LUMO of the HPB that splits into two orbitals to create a half-field singly occupied molecular orbital (SOMO) and a singly unoccupied molecular orbital (SUMO) [100]. The energies of the spin polarized Kohn-Sham orbitals ranging within the HOMO and LUMO energies of the neutral HPB are calculated and compared with the energies of the negatively charged HPB^- orbitals as described in Fig. 1.10e. The creation of a SOMO orbital, which energy is lowered by $\Delta E = 0.8$ eV compared to the LUMO of the neutral HPB, is accompanied by the formation of a singly unoccupied orbital (SUMO) which energies is separated from the SOMO by the Coulomb energy ΔU ~2.7 eV (Fig. 1.10e) [98, 101]. Consequently, the fully occupied HOMO of the HPB^- molecule shifts towards lower energies as well as the other fully occupied orbitals (HOMO-n) while the LUMO + n orbitals of the HPB^- are shifted towards higher energies. Each of these processes, i.e. the creation of the SOMO and the energy shift of the HOMOs and LUMOs orbitals of the HPB^- anion strongly reduce the number of conductive channels in the tunnel junction (Fig. 1.10d). It shows that the HPB^- molecular orbitals lying in the energy range between the Fermi levels of the surface and the STM tip does not contribute resonantly in the tunnel electron transport.

The fact that the energy of the SOMO orbital lies within the band gap of the silicon surface is very favorable to keep the HPB negatively charged during all the excitation process. This is related with the physisorption of the HPB on the silicon surface. To further understand the relation between the negatively charged state of the HPB molecule and its lateral movement along the silicon step edge, we have performed the calculation of the charge density of the HOMO and the LUMO of the HPB^- anion. To avoid a random localization of the extra electron in the silicon surface states, the electronic configuration of the negatively charged HPB^- anion is calculated from the geometry of an individual HPB molecule (no surface) which coordinates are initially extracted after its relaxation on the bare Si(100) surface obtained as described above. Then the spin polarized calculations of the HPB^- anion are performed with the Gaussian 03 software package [22] with the TD-DFT method using the B3LYP functional and the 6-31G basis set [102].

Once the HPB molecule catches an additional electron, the ensuing charge density of the two spin-orbitals SOMO and SUMO are shown in Fig. 1.10f. Here we can see that the extra charge is located in the central part of the HPB molecule represented by the SOMO orbital while the SUMO shows a charge density located in the upper part of the molecule. Note that the charge density distribution of the fully occupied HOMO of the HPB^- anion is only slightly modified compared to the one of the neutral molecule. It is important to notice that the localization of the negatively charge in the HPB^- molecule can slightly change the overall molecular conformation and in particular the lateral benzene rings angle. We have however not observed significant changes on the positions of the lateral benzene rings between the neutral physisorbed HPB and the HPB^- anion.

In conclusion, we have shown that the DFT-D method that uses an additional dispersive semi-empirical term is able to correctly predict the adsorption conformation of the physisorbed HPB molecule on the Si(100) surface and to give a acceptable estimation of the adsorption and diffusion barrier energies. Additionally, we have shown that the use of a separated code using localized waves instead of a plane wave basis is suitable to estimate accurately the relative energies of the molecular orbitals of the charged HPB and describe the charge transfer process related to the lateral hooping of the molecule.

1.5 Weakly Bonded Organo-metallic Molecule on the Silicon Surface

Metalloporphyrins continues to draw attention in the scientific community not only because of their biological function, but also due to their current and potential application in many different industries such as catalysis, lubrication and corrosion to the relatively new-born fields of organic and molecular electronics [103–110]. In almost all such cases, devices built with porphyrin derivatives require some methods to provide structure that support the active molecules and hold them in place. A common way to achieve this is to deposit the molecules on a substrate. The flexibility of porphyrin core and puckering modes are key factors that determine and control the functional properties of metalloporphyrins. The unperturbed porphyrin macrocycle is conjugated and therefore expected to be planar. Non-planar geometries arise as a consequence of the electronic and steric effects of peripheral substituents and the nature of metal–ion macrocycle interaction [111–115]. The interactions between the molecule and the surface become very important to characterize because it can have a large effect on the molecule's electronic properties and thus greatly impacts its potential use [116–118]. These features are even more relevant with semiconductors than with metal surfaces because the molecule-semiconductor interactions are usually greater than molecule–metal interactions [119, 120].

Therefore, the use of semiconductor interfaces remains one of the main choices for the development of various devices for molecular electronics. Nevertheless, the Si(111)7 $\times$ 7 or Si(001)2 $\times$ 1 surfaces are known to interact strongly with molecules due to the high reactivity of the silicon dangling bonds. It usually leads to the destruction of the organic molecules and prevents the diffusion of molecules that is absolutely essential to their self-assembly for example. In order to overcome these drawbacks, two main strategies can be proposed. One solution is to passivate the silicon surfaces with hydrogen. A second possibility is to use the Si(111)$\sqrt{3}\times\sqrt{3}$R30°-boron surface as substrate. This surface has the unique particularity to show depopulated dangling bonds due to the presence of boron atoms underneath the top silicon layer. Trivalent elements as boron atoms are located in substitution sites directly underneath the Si adatoms. In this case, one electron is transferred from the Si adatom, which forms three bonds, to the B atom directly below it, which participates in the formation of four bonds. Finally, the passivated Si(111)$\sqrt{3}\times\sqrt{3}$R30°-boron surface exhibits depopulated dangling bonds due to this charge transfer [121–124].

The study of the molecular conformation and the electronic structure of adsorbed porphyrin molecules on semiconductor surface using experimental techniques such as scanning tunneling microscopy (STM) and electronic structure calculations such using DFT have been investigated [125]. In particular, DFT-D method has been used to study the electronic structure of a single porphyrin molecule adsorbed on doped boron silicon surfaces [126].

One of most important questions is to know how the molecule interacts with the surface. When a molecule is adsorbed, the substrate could influence the conformation of the molecule and also its electronic properties. In fact, the formation of covalent bonds between molecule and surface leads the electronic states of molecules and substrate are strongly hybridized and modified. While when the molecule is physisorbed, the molecules keep the intrinsic electronic properties and we have obtained a decoupling between the electronic properties of the molecule and the substrate. In order to achieve more accurate insight into the interaction between the molecule and the surface, the Laplacian of the charge density could be calculated. The Laplacian calculation allows determining if the electronic density between two atoms is shared or separated.

The ensuing DFT simulations use the spatial distribution of the electron density $\rho(\mathbf{r})$ to analyze the bonding and nonbonding interactions present in a chemical system. To understand the type of electronic interactions between the molecules and the substrate, we will investigate the possibility of charge transfer between the molecule and the substrate. This can be obtained by calculating the difference between the charge densities of the fully relaxed structure (molecules + substrate) and the isolated molecules without further relaxation. The calculation will hence show an accumulation or a depletion of electrons density in the binding region between the molecules and the substrate. Thus, a map of the charge densities difference can be visualized. Bader developed a method based on the topology of the total electron density which leads to an atomic classification of properties of the matter [127, 128]. The characteristics of the total electron density topology may be analyzed by a search of the critical points (minima, maxima or saddle points) located at given points $\mathbf{r}$ for which the gradient of the density is zero: $\nabla\rho(\mathbf{r})=0$. Hence, the nature of the bonds is obtained from the Laplacian of the electron density. Whether a function is a minimum or a maximum at an extremum is determined by the sign of the second derivatives at this point. We can calculate the Laplacian $\nabla^2\rho(r)$ of the charge density at this point and also the $\nabla^2\rho(r)$ Laplacian marks the boundaries of regions where $\rho(r)$ is locally upper or lower than its average value in the vicinity of r. Therefore, in the regions where $\nabla^2\rho(r)$ is negative, the electronic density is shared by both nuclei (shared interactions), otherwise, electrons are concentrated in each of the atomic basins separately and the interaction belongs to the closed-shell type (non covalent interaction). The Laplacian of the electronic density $\rho(r)$ will also be represented by 3D contours maps. Thanks to the 3D contour maps, we could determine the type of interaction between surface and molecule (i.e. covalent, ionic, coordination bonding (dative) or electrostatic bonding). Other way can be used to analyze the chemical bonding of the materials: the electron localization function (ELF) representation. ELF is based on the Hartree-Fock pair probability of parallel spin electrons and can be calculated in density functional theory from the excess kinetic energy density due to Pauli-repulsion [129, 130]. This function produces an easily understandable pattern of the chemical bonding and is widely used to describe and visualize chemical bonding in molecules and solids [131]. The ELF is a measure of the probability of finding an electron near another electron with the same spin related to the Pauli Exclusion

Principle [131–134]. The ELF is represented as a contour plot in real space, where different contours correspond to numerical values ranging from 0.0 to 1.0. The upper limit $\eta = 1$ corresponding to perfect localization (i.e. chemical bonding) and the value $\eta = 0.5$ corresponding to electron-gas-like pair probability (i.e. no chemical bonding). High ELF values show that at the examined position the electrons are more localized than in a uniform electron gas of the same density.

In this section, we have applied the calculation of the charge density Laplacian in the case of the adsorption of the Cu-5,10,15,20-tetrakis(3,5-di-tert-butyl-phenyl) porphyrin molecule on Si(111)$\sqrt{3}\times\sqrt{3}$R30°-boron surface (Fig. 1.11a). This calculation allows identifying the nature of the molecule-substrate interaction. The STM study has been realized at room temperature and the STM image of a single Cu-5,10,15,20-tetrakis(3,5-di-tert-butyl-phenyl) porphyrin molecule is presented Fig. 1.11b. The Fig. 1.11c shows the side view of the atomic structure of the slab used for the porphyrin molecule adsorption on a boron doped silicon substrate after DFT-D relaxation. The molecule–surface interaction induces a conformational change of the porphyrin core. In fact, the *C2v*-symmetric saddle configuration is observed due to the steric hindrance coming from the rotation of 3,5-di-tert-butylphenyl rings, a nonplanar deformation of the central porphyrin is induced leading to a tilting of pyrrole rings above and below the central macrocycle. Moreover, the STM study of adsorption of the Cu-porphyrin molecule on Si(111)$\sqrt{3}\times\sqrt{3}$R30°-boron surface has been realized at room temperature.

The STM experiments have shown that the Cu-porphyrin molecule is parallel to the surface without any preferential orientation. However the STM doesn't directly answer to the question: how does the molecule interact with the substrate? In DFT calculations, the charge density gives information concerning the spatial charge distribution in the system. However, in the case of weak charge density, we cannot identify the exact chemical nature of the bonding between molecule-substrate. That is why the Laplacian of the charge density could be determined in the aim to analyze the bonding between the molecule and the substrate. The first step is to determinate the molecular conformation within a DFT-D energetic study presenting a van der Waals correction proposed by Grimme. The simulated STM image, within the Tersoff-Hamann approximation and using a smooth mimic method to simulate a tip, can be performed over the energy window between E_f and (E_f -2 eV). This energy window corresponds to the bias voltage reported in the STM images obtained by El Garah et al. Simulated images are similar to published experimental STM images and this good agreement can validate the atomic model of the Cu-5,10,15,20-tetrakis(3,5-di-tert-butyl-phenyl) porphyrin molecule on Si (111)$\sqrt{3}\times\sqrt{3}$R30°-boron surface.

In order to characterize the interaction between the molecule and the substrate, the Laplacian of the charge density has been calculated. The tridimensional map of the charge density Laplacian is presented in Fig. 1.11d. Herein the analysis of the $\rho(r)$ electron density Laplacian shows that no shared interactions between the surface and the molecule are observed. All theoretical results demonstrate that only a van der Waals or electrostatic interactions delocalized on the entire surface underneath the molecule can be considered i.e. without forming any covalent bond between molecules and substrates.

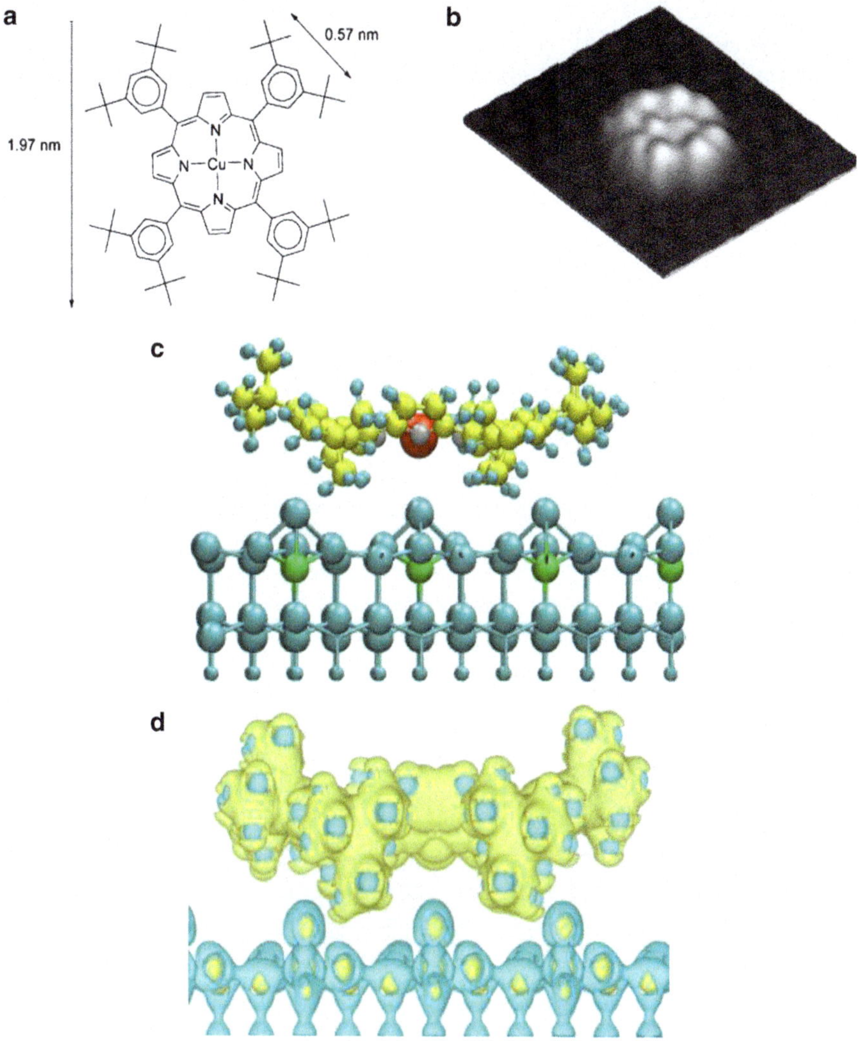

Fig. 1.11 (**a**) Atomic structure of Cu-5,10,15,20-tetrakis(3,5-di-tert-butyl-phenyl) porphyrin molecule (**b**) STM image at room temperature of the Cu-5,10,15,20-tetrakis(3,5-di-tert-butyl-phenyl) porphyrin molecule on Si(111)√3×√3R30°-boron surface (6 × 6 nm^2, Vs = −1.9V, I = 0.6 nA) (Reproduced with permission from Garah et al.(http://onlinelibrary.wiley.com/doi/10.1002/cphc.200900523/abstract) © Willey Interscience) (**c**) and (**d**) Side view of the atomic model, large and small *blue*, *yellow*, *red* and *green* balls correspond to silicon, hydrogen, carbon, copper and boron atoms respectively. (**d**) charge density Laplacian calculation of the Cu-porphyrin molecule on Si(111)√3×√3R30°-boron surface. The isosurface is blue for $\nabla^2\rho(r) < 0$ (i.e. -0.5 e-Å^{-5}) and yellow for $\nabla^2\rho(r) > 0$ (i.e. +0.5 e-Å^{-5}) (Reproduced with permission from Boukari et al. http://onlinelibrary.wiley.com/doi/10.1002/cphc.201200578/abstract) © Willey Interscience)

In conclusion, the Laplacian of electron density identifies the region of space where the electron density is locally concentrated or depleted. It allows for a classification of the nature of chemical bonds and provides a mapping of both the shared electron pairs and the localized electron pairs. The Laplacian of ρ(r) calculations shown that porphyrin molecules do not present any covalent bonds with Si(111)$\sqrt{3}\times\sqrt{3}$R30°-boron surface. DFT-D approximation needs thus to be considered in future works concerning the adsorption of single molecule on passivated Si(111)$\sqrt{3}\times\sqrt{3}$R30°-boron surface.

1.6 Molecule-Surface and Molecule-Molecule Interactions in Molecular Layers

The self-assembling molecules into well-defined dense, ordered and stable layers on different surfaces have received increasing attention over recent years due to potential applications. The formation of supramolecular layers represents a promising strategy for surface functionalization providing well controlled properties such as corrosion resistance [135], surface superhydrophobicity [136] or antifouling coatings [137]. Based on supramolecular engineering, several applications have already been proposed in the molecular electronics or nonlinear optics domains [138, 139].

However, most of the molecular assemblies at surfaces reported to date are stabilized by relatively weak interactions such as hydrogen bonding, van der Waals or electrostatic forces, implying a limited thermal stability. Such systems are governed by the subtle interplay between intermolecular and molecule-surface interactions. The determination of the interaction energies is the key to understand the formation mechanism of these systems [140–142]. Bidimensional molecular self-assemblies on solid surfaces studies concern mostly metallic surfaces, HOPG substrates or, more recently, the epitaxial graphene superstructure. Only few works are devoted to the adsorption on semiconductor substrates, and particularly on silicon surfaces.

We present a numerical simulation study of a large-scale 2D open supramolecular framework on the Si(111)$\sqrt{3}\times\sqrt{3}$R30°- boron surface obtained via the adsorption of the 1,3,5-tri(4′-bromophenyl)benzene as a molecular building block. In order to take the electronic structures of both molecules and substrate into account, DFT calculations would be the most suitable technique. Indeed, DFT simulations have successfully described organic molecular monolayers adsorbed on various substrates. Molecule-molecule and molecules-substrate interactions could be evaluated in the framework state-of-the-art DFT-D method presenting a van der Waals correction proposed by Grimme. The simulated STM image has been performed following different approximations: within the Tersoff-Hamann approximation and within the bSKAN code [20, 143]. A good agreement between STM images obtained by the bSKAN code with experimental ones shows that the presence of the tip is necessary to reproduce the main characteristics of the experimental images.

The Tersoff-Hamann approximation is the simplest approach to modelling STM images and very convenient. It is often remarkable effective for the most metal or semiconductor surfaces problems. However, there are situations in which to take the electronic structure of the tip into account is necessary in order to understand the experiment observations [141]. In these cases, a more sophisticated method could be applied using a treatment of electrons transport through the whole system (the tip and the substrate). So, the scattering method is implemented is the bSKAN code [20].

Theoretical simulations of the electronic properties of the supramolecular structure have been investigated via a simulation of the STM image by using the Tersorff-Hamann approximation. In addition to simulations using VASP, theoretical simulation will be completed by the use of one of the seminal code permitting to calculate the current tunnel by taking into account the presence of the STM tip (i.e. bSKAN code). With this approach, we could envision different STM tips which play a fundamental role on STM images acquisition as demonstrated by Pinto et al. [144]. As already published, the form and chemical nature of the tips could influence strongly the aspect ratio and the resolution of the STM experimental images [145]. The tunneling current is explicitly calculated in the framework of Bardeen approach or scattering method. One of the advantages to use this complete configuration (surface + tip) with an optimized system is that it offers a large set of data that are not accessible with the Tersoff-Hamann approach. For example, it is possible to have good and accurate reproduction of the STS spectroscopy curves measured experimentally which is crucial to give a good description of the electronic properties of adsorbed molecules.

The Tersoff Hamman approximation is nowadays incorporated in every DFT code. In this method the tunnelling current is proportional to the local density of states at the position of the STM tip:

$$I(R) \propto \sum_{E_n > E_f - eV_{bias}}^{E_n < E_f} |\Psi(R,E)|^2. \tag{1.10}$$

In several standard situations, this approximation provides a reliable qualitative picture of the surface topography. In the other complex cases, the STM tip has to be taken into account. In the bSKAN code, the wavefunctions of both surface and tip are determined by precise density functional calculation. The transition process between the surface and the tip is described by perturbation theory. The theoretical models start from the Bardeen approach to the scattering method. In the following, we have described briefly these two approaches. Within the Bardeen model, the tunnelling current is described by the following equation:

$$I = \frac{4\pi e}{\hbar} \sum_{\mu,\vartheta} \left| \int_S (\chi_\vartheta^* \nabla \psi_\mu - \psi_\mu \nabla \chi_\vartheta^*) \right|^2 \delta(E_\vartheta - E_\mu - eV_{bias}), \tag{1.11}$$

where χ_ν are the eigenstates with energy E_ν of the STM tip, ψ_μ the eigenstates of the surface with energy E_μ and V_{bias} is the bias voltage. The integral extends over the

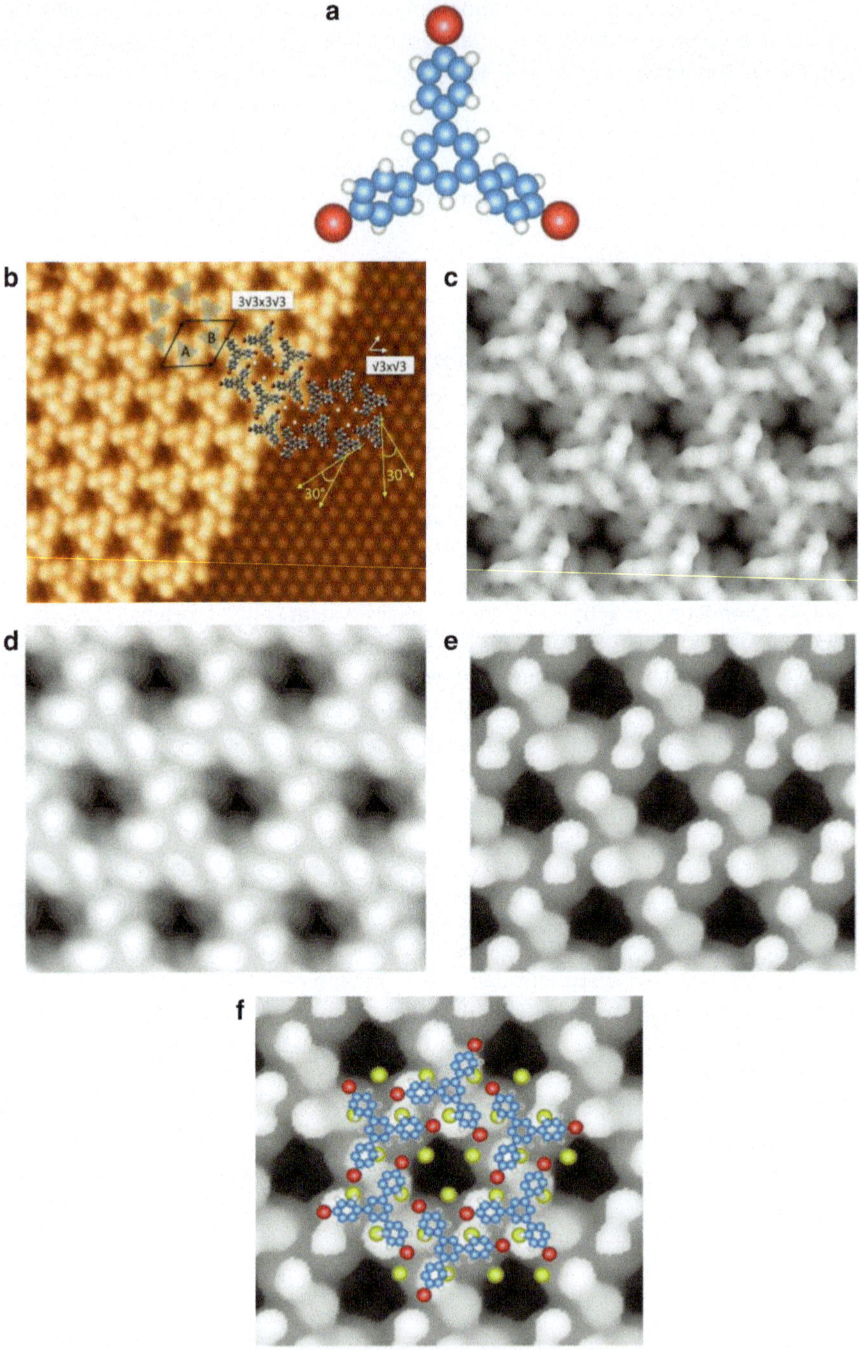

Fig. 1.12 (**b**) (**a**) Atomic structure of the 1,3,5-tri(4′-bromophenyle) benzene noted TBB. Bromine, carbon and hydrogen atoms are represented by *red*, *blue* and *white circles* respectively (Reproduced with permission from Boukari et al. (http://jcp.aip.org/resource/1/jcpsa6/v138/i8/p084704_s1) © American Institute of Physics). (**b**) High-resolution STM image of

separation surface S between the surface and the tip. The summation includes all eigenstates within a given interval from the Fermi level. Within the scattering approach, in the formulation of Landauer-Bütticker formula, the conductance G across the tunnelling junction is given by the equation:

$$G = \frac{e^2}{\pi \hbar} \frac{\mathrm{T}}{R} \approx \frac{2e^2}{h} T \tag{1.12}$$

in a case where R is equal to one. The tunnelling current can be written as follow:

$$I(V_{bias}) = \frac{2e}{h} \int_0^{eV_{bias}} T(E) dE \tag{1.13}$$

The Tersoff-Hamman approximation and the scattering method are both used to model the STM image of a supra-molecular network adsorbed on silicon surface. Experimentally, the adsorption of 1,3,5-tri(4′-bromophenyl)benzene molecules (Fig. 1.12a) on Si(111)√3×√3R30°-boron surface has been realized within the STM [141]. The STM observations show the formation of large islands with a size up to 800 × 800 nm^2.

The STM image in Fig. 1.12b presents a 2D nanoporous network that forms a 3√3 × 3√3 periodicity commensurable with the Si(111)√3×√3R30°-boron substrate. Nanopores contain three small protrusions probably corresponding to the Si adatoms. The unit cell is formed by three paired protrusions. The high resolution STM image shows that the center of the molecule is located between three Si adatoms and the Br-Br axes are rotated by 30° with respect to the rows of the adatoms. Moreover, this molecular network presents a high thermal stability up to 400 K [146, 147]. In order to achieve further insight of the molecule-molecule and molecule-substrate interactions, the energetic study has been realized and STM images have been calculated. The 1,3,5-tri(4′-bromophenyl)benzene molecules on Si(111)√3×√3R30°-boron (denoted SiB) surface are driven by no covalent bond. Indeed the passivation obtained by growing boron atoms on Si(111) surface allows to adsorb aromatic

Fig. 1.12 (continued) 1,3,5-tri(4′-bromophenyl)benzene molecules on Si(111)√3×√3R30°-boron surface (15 × 15 nm^2, $V_s = 2.3$ V, I = 0.037 nA). *Black lines* indicate the lattice parameter (3√3 × 3√3 of the molecular network. Superimposed atomic model and white dots correspond to substrate Si adatoms are represented (Reproduced with permission from Baris et al. (http://onlinelibrary.wiley.com/doi/10.1002/ange.201100332/abstract) © Willey Interscience) (**c**) Simulated STM image at bias voltage of +2.5 V, in the Tersoff Hamann scheme without smoothing (from Eric Duverger, private communication) (**d**) Simulated STM image at bias voltage of +2.5 V, in the Tersoff Hamman scheme with smoothing to mimic a tip (**e**) Simulated STM image at bias voltage of +2.5 V within the bSKAN code with the scattering approach (**f**) Superimposed atomic model of the relaxed TBB network on the simulated STM image. Bromine, carbon, hydrogen, and silicon atoms are represented by *red*, *blue*, *white*, and *yellow* balls, respectively (Reproduced with permission from Boukari et al. (http://jcp.aip.org/resource/1/jcpsa6/v138/i8/p084704_s1) © American Institute of Physics) molecules without reaction; i.e. this passivation prevents the typical cyclo-addition reactions that are induced by silicon dangling bonds.

The adsorption energy per unit cell of the molecular network on the SiB surface is −3.44 eV. The molecule-molecule and molecules-substrate interactions are $E_{mol\text{-}mol} = -0.81$ eV and $E_{mol\text{-}substrat} = -2.66$ eV per unit cell respectively. In this particular case of molecules deposited on a passivated silicon surface, the substrate template plays a major role to achieve the supramolecular network, whereas the molecule-molecule interactions are weaker. Contrarily to the results reported in reference [145] within DFT-GGA approximation, in which the molecule-molecule interactions were found to be +0.16 eV i.e. weakly repulsive, the molecule-molecule interactions are attractive and at long-range within the DFT-D calculations. This specific result shows the importance of taking the dispersive term into account to describe such systems.

DFT calculations within the LDA approximation have equally been performed. In this case, the adsorption energy is −3.21 eV and the molecule-molecule and molecule-substrate interactions are equal to −0.74 eV and −2.52 eV respectively. The energy differences between DFT-D and LDA approximation are 0.23 eV (6.7 %), 0.07 eV (8.7 %) and 0.14 eV (5.2 %) for the adsorption energy, the molecule-molecule and molecule-substrate interactions respectively. The DFT-D approximation adds a correction term that increases the interaction energy in taking the long-range interactions into account which explains that energies are larger than those obtained within LDA approximation as highlighted in different papers [148–150].

Here, we emphasize that the silicon substrate-molecules interaction dominates with respect to the molecule-molecule one. In this way, the approximation used in the gas-phase modeling of supra-molecular assemblies i.e. without considering the substrate, as usually observed in the metal surfaces case [151–155] do not apply to silicon surfaces.

The STM image simulations are performed within the Tersoff-Hamann approximation and in the second step in considering a smoothing of the images. Then, the bSKAN code is used to model more realistic image. On the experimental STM images realized at +2.5 V in reference [145], three triangular joined pair protrusions, associated to the phenyl arms of TBB molecule, were observed. In the calculated LDOS image, in the Tersoff-Hamann approach, taking the van der Waals interactions into account, no pair protrusions were retrieved as however observed in experimental images (Fig. 1.12c).

In a second calculation within the Tersoff-Hamann approximation, a gaussian convolution is used to mimic the finite size of the tip. In the Fig. 1.12d, the protrusions are larger than the case without smoothing. However, the protrusions are not joined by pair as observed in the experimental data. Without taking the STM tip into account in the simulated STM image, the agreement between theoretical and experimental images is not completely convenient. Hence, in order to achieve a clear and direct comparison with the experimental STM data from the paper of Baris et al. [144], we performed a precise STM image calculation of the relaxed structural model corresponding to two molecules adsorbed on the SiB surface, feeding the atoms positions from the VASP code into the bSKAN code. The STM images were simulated using the electronic structure of the converged system, and the scattering approach implemented in bSKAN with a pyramidal tungsten tip constructed with 82 atoms. The calculated image corresponding to the network at bias voltage of +2.5V is shown in Fig. 1.12e. This bias voltage corresponds to the experimental values applied in reference [144].

The calculated STM image that is similar to the experimental one presented in Fig. 1.12b, is clearly showing the presence of triangles, involving three joined pairs of protrusions due to the convolution of the empty electronic states between the tip and the TBB molecules. We also observe the three protrusions located in the nanopores. In order to elucidate the STM observations, we superimpose the atomic models of the relaxed TBB networks on the simulated STM images (Fig. 1.12f). The three pairs of protrusions correspond to the phenyl arms of each TBB molecules and the three protrusions located in the nanopore to the adatoms. As in agreement with the PDOS study, neither the bromine atoms nor the central cycles strongly contribute to the tunneling current.

Our simulated STM image calculated in the framework of the bSKAN code due to the tip presence is in better agreement with experimental STM images than those obtained by the integrated LDOS calculations within the Tersoff-Hamann approximation as reported in Fig. 1.12. In this section, we have shown that non-bonded interactions are essential in self-assembly network of 1,3,5-tri(4′-bromophenyl)benzene molecules on Si(111)$\sqrt{3}\times\sqrt{3}$R30°-boron surface involving no covalent bond between the molecules or between the molecules and the silicon substrate. The van der Waals term suggested by Grimme was added to the total energy, leading to attractive molecule-molecule interactions and emphasizing the major role of the silicon substrate in the network formation. As a consequence, the approximation used in the gas-phase modeling of supra-molecular assemblies without considering the substrate is no more valid in the case of silicon substrates. In future works, DFT-D approximation can be seriously considered when Si substrates are studied.

The STM image simulations within the bSKAN code show a better agreement with the experimental date than the Tersoff-Hamann approach. The presence of the tungsten tip is essential to obtain three paired lobes as experimentally observed. In fact, the spatial dependence of the tip orbital is very important to paired protrusion from parts of two different neighboring molecules.

1.7 Conclusion

In this chapter, we have presented different experimental results that concern the study of the adsorption of various molecules on silicon surfaces using STM. In each case, we have tested various theoretical approaches that investigate atomic-scale simulations of, not only STM images simulations, but also the one that describes the molecular movements or the molecular electronic state. This has allowed a deeper understanding of the experimental observations.

The Sect. 1.2 talks about the chemisorption of individual biphenyl molecules on the Si(100) surface that can be switched between two stable states using electronically induced excitation with the tip of the STM. In this particular case, the calculations of the system energies with the DFT allow to identify the most stable molecular conformation adsorbed on the silicon surface. This work has been completed with simulations of STM images to validate the atomistic model and presenting a good agreement with the experimental STM topographies. In addition, the NEB calculations have allowed evaluating the energetic barrier between the four most stables atomic structures that can be switched randomly.

In the Sect. 1.3, a physisorbed system is studied when the stilbene molecule is adsorbed on the Si(100) surface. In order to avoid a conformational energetic study with a heavy computational cost, we have proposed an alternative approach that consists of to identify adsorption conformations on the basis of successive comparisons of calculated and experimental STM images. For these calculations, we have used a tight binding scheme (DFTB+) that correctly reproduces first-principles based density functional that allow reducing significantly the computational charge. Hence, we have shown that the STM topographies of a physisorbed molecule can be correctly depicted. While STM topographies are shown to describe the main aspects of the molecular conformation that leads to predict that the stilbene molecule is physisorbed on the bare Si(100) surface, the numerical simulation of the STM images appears to be of clear help to confirm the experimental observations.

The third part of the chapter is dealing with the control of the lateral manipulation of single physisorbed molecule at the step edge of the bare silicon surface. The density functional theory with a semi-empirical term proposed by Grimme, that takes van der Waals dispersive interactions into account, has been used to identify the adsorption site of the HPB molecule on the step edge of the bare Si(100) surface. It also permits to give an acceptable estimation of the adsorption and diffusion energies barriers. Moreover, accurate calculations of the molecular orbital energy of the stationary charged HPB^- anion using TD-DFT methods strongly helped describing the stationary charge state of the molecule and hence the charge transfer process related to the molecular hoping process.

A fourth section of this chapter has been devoted to the weakly bonded organometallic molecule adsorbed on silicon surfaces. Here, the calculation of the Laplacian of the electron density can identify the region of space where the electron density is locally concentrated or depleted. It has allowed a classification of the nature of chemical bonds and provides a mapping of both the shared and the localized electron pairs. The calculation of the Laplacian of $\rho(\boldsymbol{r})$ has shown that porphyrin molecules do not present any covalent bond with boron doped silicon surfaces.

Finally, the last part of this chapter depicts a numerical simulation study of a large-scale 2D open supramolecular framework on the boron doped silicon surface. Molecule-molecule and molecules-substrate interactions have been evaluated with state-of-the-art DFT-D methods. The dispersive van der Waals correction proposed by Grimme, has allowed to obtain an accurate molecule-molecule interaction and emphasized the major role of the silicon substrate in the network formation. As a consequence, the approximation used in the gas-phase modeling of supra-molecular assemblies without considering the substrate is no more valid in the case of silicon substrates. The STM image simulations using the bSKAN code show a better agreement with the experimental data than the Tersoff-Hamann approach. In the present case, the presence of the tungsten tip is essential to reproduce the three paired lobes as experimentally observed. In fact, we show that the spatial dependence of the tip orbital is very important to identify joined protrusion arising from parts of two different neighboring molecules.

All these examples have shown that DFT based numerical simulations can answer various questions raised by the STM observations. Therefore, the molecular conformation and the adsorption site are usually the major information that needs a deep understanding to investigate the influence of the molecule-surface interaction. However some other questions can be raised: Is the molecule physisorbed or chemisorbed and how do we clearly define physisorption? In the case of a physisorbed molecule, it is crucial to take the van der Waals interactions into account. Thus, in the case of standard DFT calculation, which does not take into account these interactions, some authors have proposed to add a semi-empirical term to correct the Kohn-Sham energies. This approximation, initiated by Grimme, raises the problem of transferability. Other more precise methods are being developed. In order to identify the nature of interaction between the molecule and the substrate, simple tool could be used based on the charge density Laplacian calculation or ELF function. Concerning the direct comparison with the STM observation, various methods can be used. The Tersoff-Hamman approximation, which is now generally implemented into the most used DFT codes, has allowed obtaining good simulated images without heavy calculations. However more accurate techniques that need additional computational time because they take the STM tip states into account, could be applied (DFTB+, bSKAN,...) successfully. The current calculations codes are powerful enough to allow information on experimental observations, which takes into account a realistic size of the system, without a huge calculation cost. The adequation of these simulations is nowadays of crucial importance for the experimental part where new information can be obtained by comparing both data: simulated STM image, site and conformation of adsorption, energetic barrier, type of interaction. However more accurate methods are being developed that will enable, in the near future, to address new experimental questions such as the charge state of decoupled individual molecule or the simulation of very large systems that counts several thousand atoms.

Used Acronyms

B3LYP	Becke 3-Parameter (Exchange), Lee, Yang and Parr
CS	Cis-Stilbene
DFT	Density Functional Theory
DFTB	Density Functional based Tight Binding
DFT-D	Dispersion-corrected Density Functional Theory

EIE	Electronically Induced Excitations
ELF	Electron Localization Function
ESQC	Elastic-Scattering Quantum Chemistry
GGA	Gradient Generalized Approximation
HOMO	Highly Occupied Molecular Orbital
HPB	Hexaphenyl-benzene
INO	Individual Nano-Objet
LDA	Local Density Approximation
LDOS	Local Density of States
LUMO	Lowest Unoccupied Molecular Orbital
MEP	Minimum Energy Path
NEB	Nudged Elastic Band
PAW	Projector Augmented Wave
PBE	Perdew-Burke-Ernzerhof
PES	Potential Energy Surface
SCF	Self Consistent Field
SOMO	Singly Occupied Molecular Orbital
STM	Scanning Tunnelling Microscopy
SUMO	Singly Unoccupied Molecular Orbital
TB	Tight Binding
TD-DFT	Time Dependant- Density Functional Theory
TIBB	Tip Induced Band Bending
VASP	Vienna *Ab-initio* Simulation Package
VdW	van der Waals

Acknowledgements This work was performed using HPC resources from GENCI-IDRIS and the supercomputer facilities of the Mésocentre de calcul de Franche-Comté. This work has been partly supported by the Agence National de la Recherche (ANR).

References

1. Binning G, Rohrer H, Gerber C, Weible E (1982) Phys Rev Lett 49:57
2. Oppenheimer R (1928) Proc Natl Acad Sci 14:363
3. Eigler DM, Schweizer EK (1990) Nature 344:524
4. Riedel D, Lastapis M, Martin M, Dujardin G (2004) Phys Rev Lett 69:121301
5. Xuea QK, Hashizumec T, Sakuraia T (1999) Appl Surf Sci 141:244
6. Wu XL, Zhou P, Lieber CM (1988) Nature 335:55
7. Loth S, von Bergmann K, Ternes M, Otte AF, Lutz CP, Heinrich AJ (2010) Nat Phys 6:340
8. Riedel D, Delattre R, Borisov AG, Teperik TV (2010) Nano Lett 10:3857
9. Lang N (1986) Phys Rev B 34:5947
10. Rosink JJWM, Blauw MA, Geerligs LJ, van der Drift E, Radelaar S (2000) Phys Rev B 62:10459
11. Riedel D (2010) J Phys Condens Matter 22:264009

12. Hohenberg P, Kohn W (1964) Phys Rev 136:B864
13. Tersoff J, Hamann DR (1983) Phys Rev Lett 50:1998
14. Bardeen J (1961) Phys Rev Lett 6:57
15. Sautet P, Joachim C (1988) Phys Rev B 38:12238
16. Sautet P, Joachim C (1988) Chem Phys Lett 153:511
17. Sautet P, Joachim C (1992) Surf Sci 271:387
18. Repp J, Meyer G, Stojkovic S, Gourdon A, Joachim C (2005) Phys Rev Lett 94:026803
19. Tiwari RK, Otalvaro DM, Joachim C, Saeys M (2009) Surf Sci 603:3286
20. Sautet P, Joachim C (1991) Chem Phys Lett 185:23
21. Dahlke R, Schollwöck U (2004) Phys Rev B 69:085324
22. Hofer WA (2003) Prog Surf Sci 71:147
23. Palotas K, Hofer WA (2005) J Phys Condens Matter 17:2705
24. Frisch MJ, Trucks GW, Schlegel HB, Scuseria GE, Robb MA, Cheeseman JR et al (2004) Gaussian 03, Revision C.02. Gaussian, Inc., Wallingford
25. Kudernac T, Ruangsupapichat N, Parschau M, Macià B, Katsonis N, Harutyunyan SR, Ernst KH, Feringa B (2011) Nature 479:208
26. Sugimoto Y, Abe M, Hirayama S, Oyabu N, Custance Ó, Morita S (2005) Nat Mater 4:156
27. Liljeroth P, Swart I, Paavilainen S, Repp J, Meyer G (2010) Nano Lett 10:2475
28. Swart I, Sonnleitner T, Niedenführ J, Repp J (2012) Nano Lett 12:1070
29. Liljeroth P, Repp J, Meyer G (2007) Science 317:1203
30. Comtet G, Dujardin G, Hellner H, Lastapis M, Martin M, Mayne AJ, Riedel D (2002) Phil Trans R Soc A 362:1217
31. Ruda HE, Polanyi JC, Yang J, Wu Z, Philipose U, Xu T, Yang S, Kavanagh KL, Liu JQ, Yang L, Wang Y, Robbie K, Yang J, Kaminska K, Cooke DG, Hegmann FA, Budz AJ, Haugen HK (2006) Nanoscale Res Lett 1:99
32. Yang J, Deng J, Chandrasekhar N, Joachim C (2007) J Vac Sci Technol B 25:1694
33. Khajetoorians AA, Wiebe J, Chilian B, Lounis S, Blügel S, Wiesendanger R (2012) Nat Phys 8:497
34. Riedel D, Cranney M, Martin M, Guillory R, Dujardin G, Dubois M, Sonnet P (2009) J Am Chem Soc 131:5414
35. Watkins M, Trevethan T, Sushko ML, Shluger A (2008) J Phys Chem C 112:4226
36. Cheng Z, Chu ES, Sun D, Kim D, Zhu Y, Luo M, Pawin G, Wong KL, Kwon KY, Carp R, Marsella M, Bartels L (2010) J Am Chem Soc 132:13578
37. He Y, Ye T, Borguet E (2002) J Am Chem Soc 124:1964
38. Stroscio JA, Eigler DM (1991) Science 254:1320
39. Custance O, Perez R, Morita S (2009) Nat Nanotechnol 4:803
40. Ueba H, Persson BNJ (2004) Surf Sci 566:1
41. Ueba H, Mii T, Lorente N, Persson BNJ (2005) J Chem Phys 123:084707
42. Komeda T, Kim Y, Kawai M, Persson BNJ, Ueba H (2002) Science 295:2055
43. Meyer G, Bartels L, Zöphel S, Henze E, Rieder KH (1997) Phys Rev Lett 78:1512
44. Huang Y, Rettner CT, Auerbach DJ, Wodtke AM (2000) Science 290:111
45. Sainoo Y, Kim Y, Okawa T, Komeda T, Shigekawa H, Kawai M (2005) Phys Rev Lett 95:246102
46. Ohara M, Kim Y, Kawai M (2008) Phys Rev B 78:201405(R)
47. Lastapis M, Martin M, Riedel D, Hellner L, Comtet G, Dujardin G (2005) Science 308:1000
48. Kohn W, Sham LJ (1965) Phys Rev 140:A1133
49. Kresse G, Furthmüller J (1996) Phys Rev B 54:11169. http://cms.mpi.univie.ac.at/vasp/
50. Dubois M, Delerue C, Rubio A (2007) Phys Rev B 75:041302R
51. Mamatkulov M, Stauffer L, Minot C, Sonnet P (2006) Phys Rev B 73:035321
52. McKee ML, Page M (1993) In: Lipkowitz KB, Boyd DB (eds) Reviews in computational chemistry, vol IV. VCH Publishers Inc, New York
53. Rasmussen T, Jacobsen KW, Leffers T, Pedersen OB, Srinivasan SG, Jónsson H (1997) Phys Rev Lett 79:3676

54. Jónsson H, Mills G, Jacobsen KW (1998) In: Berne BJ, Ciccotti G, Coker DF (eds) Classical and quantum dynamics in condensed phase simulations. World Scientific, Singapore
55. Henkelman G, Jónsson H (2000) J Chem Phys 113:9978
56. Henkelman G, Uberuaga BP, Jónsson H (2000) J Chem Phys 113:9901
57. Riedel D, Bocquet M-L, Lesnard H, Lastapis M, Lorente N, Sonnet P, Dujardin G (2009) J Am Chem Soc 131:7344
58. Martin M, Lastapis M, Riedel D, Dujardin G, Mamatkulov M, Stauffer L, Sonnet P (2006) Phys Rev Lett 97:216103
59. Waldeck DH (1991) Chem Rev 91:415
60. Choi BY, Kahng SJ, Kim S, Kim H, Kim HW, Song YJ, Ihm J, Kuk Y (2006) Phys Rev Lett 96:156106
61. Su C, Lee L-X, Chen C-F (2003) Synth Met 137:865
62. Alemani M, Peters MV, Hecht S, Rieder K-H, Moresco F, Grill L (2006) J Am Chem Soc 128:14446
63. Comstock MJ, Levy N, Kirakosian A, Cho J, Lauterwasser F, Harvey JH, Strubbe DA, Fréchet JM, Trauner D, Louie SG, Crommie MF (2007) Phys Rev Lett 99:038301
64. Henzl J, Bredow T, Morgenstern K (2007) Chem Phys Lett 435:278
65. Clemen L, Wallace RM, Taylor PA, Dresser MJ, Choyke WJ, Weinberg WH, Yates JT (1992) Surf Sci 268:205
66. Schmidt PM, Horn K, Dil JH, Kampen TU (2007) Surf Sci 601:1775
67. Rydberg H, Lundqvist BI, Langreth DC, Dion M (2000) Phys Rev B 62:6997
68. Hooper J, Cooper VR, Thonhauser T, Romero NA, Zerilli F, Langreth DC (2008) ChemPhysChem 9:891
69. Dubois M, Delerue C, Allan G (2005) Phys Rev B 71:165435
70. Porezag D, Frauenheim T, Köhler T, Seifert G, Kaschner R (1995) Phys Rev B 51:12947
71. Elstner M, Porezag D, Jungnickel G, Elsner J, Haugk M, Frauenheim T, Suhai S, Seifert G (1998) Phys Rev B 58:7260
72. Eschrig H (1988) The optimized LCAO method and electronic structure of extended systems. Akademieverlag, Berlin
73. Krüger T, Elstner M, Schiffels P, Frauenhein T (2005) J Chem Phys 122:114110
74. Lefebvre I, Lannoo M, Allan G, Martinage L (1988) Phys Rev B 88:8593
75. Labidi H, Sonnet P, Riedel D (2013) J Phys Chem C 117:13663
76. Gagnon E, Maris T, Arseneault PM, Maly KE, West JD (2010) Cryst Growth Des 10:648
77. Wang J, Cooper G, Tulumello D, Hitchcock AP (2005) J Phys Chem A 109:10886
78. Gagnon E, Halperin S, Métivaud V, Maly KE, Wuest JD (2010) J Org Chem 75:399
79. Gross L, Moresco F, Ruffieux P, Gourdon A, Joachim C, Rieder KH (2005) Phys Rev B 71:165428
80. Chiaravalloti F, Gross L, Rieder K-H, Stojkovic SM, Gourdon A, Joachim C, Moresco F (2007) Nat Mater 6:30
81. Perdew P, Burke K, Ernzerhof M (1996) Phys Rev Lett 77:3865
82. Blöchl E (1994) Phys Rev B 50:17953
83. Kresse G, Hafner J (1993) Phys Rev B 47:558
84. Kresse G, Furthmüller J (1996) Phys Rev B 54:11169
85. Kresse G, Furthmüller J (1996) Comput Mater Sci 6:15
86. Tersoff J, Hamann DR (1985) Phys Rev B 31:805
87. Grimme S (2004) J Comput Chem 25:1463
88. Grimme S (2006) J Comput Chem 27:1787
89. Grimme S, Antony J, Schwabe T, Mück-Lichtenfeld C (2007) Org Biomol Chem 5:741
90. Nguyen MT, Pignedoli CA, Treier M, Fasel R, Passerone D (2010) Phys Chem Chem Phys 12:992
91. Bellec A, Ample F, Riedel D, Dujardin G, Joachim C (2009) Nano Lett 9:144
92. Mingo N, Flores F (1998) Surf Sci 395:342
93. Devel M, Girard C, Joachim C, Martin OJF (1995) Appl Surf Sci 87:390

94. Roland C, Gilmer GH (1991) Phys Rev Lett 67:3188
95. Devel M, Girard C, Joachim C (1996) Phys Rev B 53:13159
96. Rogge S, Timmerman RH, Scholte PMLO, Geerligs LJ, Salemink HWM (2001) J Vac Sci Technol B 19:659
97. Teichmann K, Wenderoth M, Loth S, Ulbrich RG, Garleff JK, Wijnheijmer AP, Koenraad PM (2008) Phys Rev Lett 101:076103
98. Guoa H, Saalfrank P, Seidemanc T (1999) Prog Surf Sci 62:239
99. Tran Thoai DB, Zimmermann R, Grundmann M, Bimberg D (1990) Phys Rev B 42:5906
100. Wu SW, Ogawa N, Ho W (2006) Science 312:1362
101. Nguyen TH, Mahieu G, Berthe M, Grandidier B, Delerue C, Stiévenard D, Ebert P (2010) Phys Rev Lett 105:226404
102. Lachaud F, Jeandon C, Beley M, Ruppert R, Gros PC, Monari A, Assfeld X (2012) J Phys Chem A 116:10736
103. Tsuda A, Osuka A (2001) Science 293:79
104. Kadish KM, Smith KM, Guilard R (2000) The porphyrin handbook, vol 1–10. Academic Press, San Diego
105. Heimel G, Romaner L, Zojer E, Bredas J-L (2008) Acc Chem Res 41:721
106. Wan LJ (2006) Acc Chem Res 39:334
107. Weiss PS (2008) Acc Chem Res 41:1772
108. Yang Z-Y, Zhang H-M, Pan G-B, Wan L-J (2008) ACS Nano 2:743
109. Calmettes B, Nagarajan S, Gourdon A, Abel M, Porte L, Coratger R (2008) Angew Chem Int Ed 47:6994
110. Yang Z-Y, Zhang H-M, Yan C-J, Li S-S, Yan H-J, Song W-G, Wan L-J (2008) Proc Natl Acad Sci U S A 104:3707
111. Song XZ, Jaquinod L, Jentzen W, Nurco DJ, Jia SL, Khoury RG, Ma JG, Medforth CJ, Smith KM, Shelnutt JA (1998) Inorg Chem 37:2009
112. Huang Q, Medforth CJ, Schweitzer-Stenner R (2005) J Phys Chem A 109:10493
113. Song X, Jentzen W, Jaquinod L, Khoury RG, Medforth CJ, Jia CJ, Ma JG, Smith KM, Shelnutt JA (1998) Inorg Chem 37:2117
114. Jentzen W, Song XZ, Shelnutt JA (1997) J Phys Chem B 101:1684
115. Jentzen W, Simpson MC, Hobbs JD, Song X, Ema T, Nelson NY, Medforth CJ, Smith KM, Veyrat M, Mazzanti M, Ramasseul R, Marchon JC, Takeuchi T, Goddard WA III, Shelnutt JA (1995) J Am Chem Soc 117:11085
116. Barth JV, Constantini G, Kern K (2005) Nature 437:671
117. Giva PG, DiLabio GA, Pitters JL, Zikovski J, Rezeq M, Dogel S, Hofer WA, Wolkow RA (2005) Nature 435:658
118. Weber-Bargioni A, Auwärter W, Klappenberger F, Reichert J, Lefrançois S, Strunskus T, Wöll C, Schiffrin A, Pennec Y, Barth JV (2008) ChemPhysChem 9:89
119. Hamers RJ, Coulter SK, Ellison MD, Hovis JS, Padowitz DF, Schartz MP, Greenlief CM, Russel JN (2000) Acc Chem Res 33:617
120. Sloan PA, Palmer RE (2005) Nature 434:367
121. Bedrossian P, Meade RD, Mortensen K, Chen DM, Golovchenko JA, Vanderbilt D (1989) Phys Rev Lett 63:1257
122. Avouris P, Lyo I-W, Bozso F, Kaxiras E (1990) J Vac Sci Technol A 8:3405
123. Kaxiras E, Pandey KC, Himpsel FJ, Tromp RM (1990) Phys Rev B 41:1262
124. Shi HQ, Radny MW, Smith PV (2002) Phys Rev B 66:85329
125. El Garah M, Makoudi Y, Palmino F, Duverger E, Sonnet P, Chaput L, Gourdon A, Cherioux F (2009) ChemPhysChem 10:3190
126. Boukari K, Sonnet P, Duverger E (2012) ChemPhysChem 13:3945
127. Bader RFW (1991) Chem Rev 91:893
128. Henkelman G, Arnaldsson A, Jónsson H (2006) Comput Mater Sci 36:354
129. Savin A, Flad HJ, Flad J, Preuss H, von Schnering HG (1992) Angew Chem Int Ed Engl 31:185

130. Kohout M, Pernal K, Wagner FR, Grin Y (2004) Theor Chem Accounts 112:453
131. Becke AD, Edgecombe KE (1990) J Chem Phys 92:5397–5403
132. Savin A, Jepsen O, Flad J, Andersen OK, Preuss H, von Schnering HG (1992) Angew Chem Int Ed Engl 31:187
133. Savin A, Silvi B (1994) Nature 371:683
134. Ooi N, Rairkar A, Lindsley L, Adams JB (2006) J Phys Condens Matter 18:97
135. Felhösi I, Kálmán E, Póczik P (2002) Russ J Electrochem 38:230
136. Bain CD, Whitesides GM (1989) Angew Chem Int Ed Engl 28:506
137. Laibinis PE, Whitesides GM (1992) J Am Chem Soc 114:9022
138. Jortner J, Ratner M (eds) (1997) Molecular electronics: a chemistry of 21st century. Blackwell Science Inc., Oxford
139. Bosshard C, Sutter K, Prêtre P, Hulliger J, Flörsheimer M, Kaats P, Günter P (1995) Organic nonlinear optical materials, vol 1. Gordon and Breach, Amsterdam
140. Yokoyama T, Yokoyama S, Kamikado T, Okuno Y, Mashiko S (2001) Nature 413:619
141. Barth JV, Weckesser J, Cai C, Günter P, Bürgi L, Jeandupeux O, Kern K (2000) Angew Chem Int Ed 39:1230
142. Böhringer M, Morgenstern K, Schneider W-D, Berndt R, Mauri F, De Vita A, Car R (1999) Phys Rev Lett 83:324
143. Briggs GAD, Fisher AJ (1999) Surf Sci Rep 33:1
144. Pinto HP, Enevoldsen GH, Besenbacher F, Lauritsen JV, Foster AS (2009) Nanotechnology 20:264020
145. Wang YL, Cheng ZH, Deng ZT, Guo HM, Du SX, Gao HJ (2012) Chimia 66:31
146. Baris B, Luzet V, Duverger E, Sonnet P, Palmino F, Chérioux F (2011) Angew Chem Int Ed 50:4094
147. Boukari K, Duverger E, Sonnet P (2013) J Chem Phys 138:84704
148. Cuadrado R, Cerdá JI, Wang Y, Xin G, Berndt R, Tang H (2010) J Chem Phys 133:154701
149. Tkatchenko A, Romaner L, Hofmann OT, Zojer E, Ambrosch-Draxl C, Scheffler M (2010) MRS Bull 35:435
150. Mercurio G, McNellis ER, Martin I, Hagen S, Leyssner F, Soubatch S, Meyer J, Wolf M, Tegeder P, Tautz FS, Reuter K (2010) Phys Rev Lett 104:036102
151. Ruben M, Payer D, Landa A, Comisso A, Gattinoni C, Lin N, Collin J-P, Sauvage J-P, De Vita A, Kern K (2006) J Am Chem Soc 128:15644
152. Chen Q, Frankel DJ, Richardson NV (2002) Langmuir 18:3219
153. Miyake Y, Nagata T, Tanaka H, Yamazaki M, Ohta M, Kokawa R, Ogawa T (2012) ACS Nano 6:3876
154. Perdigão LMA, Staniec PA, Champness NR, Kelly REA, Kantorovich LN, Beton PH (2006) Phys Rev B 73:195423
155. Xu W, Kelly REA, Schöck M, Otero R, Lægsgaard E, Stensgaard I, Kantorovich LN, Besenbacher F (2007) Small 3:2011

Chapter 2
Models of the Interfaces in Superhard TiN-Based Heterostructures and Nanocomposites from First-Principles

Volodymyr Ivashchenko, Stan Veprek, Patrice Turchi, and Jerzy Leszczynski

Abstract This chapter reviews structures and properties of various TiN-based heterostructures. The study of such systems is of vital importance in development of new materials with desired optimum properties. After discussion of the current literature, the recently published results of first-principles quantum molecular dynamics (QMD) calculations of heterostructures consisting of one monolayer of interfacial SiNx, SiC, BN and AlN inserted between several monolayers thick slabs of B1(NaCl)-TiN (001) and (111) in the temperature range of 0–1,400 K are reviewed in some detail. The results revealed that SiN(001) exists as pseudomorphic B1-SiN interfacial layer only at 0 K. At finite temperature, this heterostructure transforms into distorted octahedral SiN_6 and tetrahedral SiN_4 units aligned along the {110} directions. At 300 K, the aggregates of the SiN_x units are close to a disordered, essentially amorphous SiN. After heating to 1,400 K and subsequent relaxation at 300 K, the interfacial layer corresponds to a strongly distorted Si_3N_4-like structure. The B1-SiN, Si_3N_4-like SiN and Si_3N_4-like Si_2N_3 interfaces between the TiN(111) slabs are stable in the whole temperature range considered here. The B1-Si_3N_4-like interfaces derived from SiN by the formation of Si-vacancies are unstable at finite temperatures. An estimate of interfacial

V. Ivashchenko (✉)
Institute of Problems of Material Science, National Academy of Science of Ukraine, Krzhyzhanosky str. 3, 03142 Kyiv, Ukraine
e-mail: ivash@ipms.kiev.ua; ivashenko@icnanotox.org

S. Veprek
Department of Chemistry, Technical University Munich, Lichtenbergstrasse 4, D-85747 Garching, Germany
e-mail: stan.veprek@lrz.tum.de

P. Turchi
Lawrence Livermore National Laboratory (L-352), P.O. Box 808, Livermore, CA 94551, USA

J. Leszczynski
Department of Chemistry and Biochemistry, Interdisciplinary Center for Nanotoxicity, Jackson State University, Jackson, MS 39217, USA

J. Leszczynski and M.K. Shukla (eds.), *Practical Aspects of Computational Chemistry III*, DOI 10.1007/978-1-4899-7445-7_2,

formation energies showed that the most favorable configurations of the (111) interfaces are silicon atoms tetrahedrally coordinated to nitrogen. The most stable (001) B1-derived heterostructure with $Si_{0.75}$ N interface consist of both tetrahedrally and octahedrally coordinated silicon atoms. The TiN(001)/B1-SiC/TiN(001) interface exists as pseudo-morphic B1-SiC layer between 0 K and 600 K. After heating to 900–1,400 K and subsequent static relaxations, the interfacial layer corresponds to a strongly distorted 3C-SiC-like structure oriented in the (111) direction in which the Si and C atoms are located in the same interfacial plane. The Si atoms form fourfold coordinated N-Si $\equiv C_3$ configurations, whereas the C atoms are located in the $Ti_2 = C \equiv Si_3$ surrounding. All the (111) interfaces simulated at 0, 300, and 1,400 K have the same atomic configurations. For these interfaces, the Si and C layers correspond to the Si-C network in the (111) direction of 3C-SiC. The Si and C atoms are located in N-Si $\equiv C_3$ and $Ti_3 \equiv C \equiv Si_3$ configurations, respectively. The BN(001) interfacial layer forms a disordered h-BN-like structure consisting of BN_3 units in the whole temperature range considered here. Finally, the B1-AlN(001) interface is found to be stable within the whole temperature range.

Phonon calculations show that the observed modifications of the interfaces are due to dynamical instability of the B1-type (001) and (111) interfacial layers of BN, SiC and SiN driven by soft modes within the given planes. The calculated electronic densities of states (EDOS) of the (001) interfaces suggest that the reconstructed interfaces should be semiconducting.

A comparison with the results obtained by earlier "static" *ab initio* DFT calculations at 0 K shows the great advantage of the QMD calculations that account for the effects of thermal activation of structural reconstructions. The results, which can be understood also without the knowledge of theoretical methods, were used to interpret the available experimental results on TiN-based heterostructures and nanocomposite coatings in order to provide guidance to the experimentalists for the preparation of better coatings.

2.1 Introduction

In this chapter we review our recent results of the first-principles quantum molecular dynamics (QMD) calculations of the TiN/XY heterostructures (XY = one monolayer, 1 ML of SiN_x, SiC, BN, AlN) that model the corresponding nanocomposites. Results of the electronic and phonon spectra of the heterostructures are described. First we present a short review of the experimental and theoretical investigations of these nanocomposites and heterostructures to illustrate the need to perform theoretical studies of the interfaces. In Sect. 2.2 we describe the computational details. In the follow-up three sections, the results of QMD simulations performed on the TiN(001)/XY (XY = SiN_x, SiC, BN, AlN), TiN(111)/SiN_x and TiN(111)/SiC heterostructures are presented and discussed. The emphasis is on the analysis of the SiN_x-based heterostructures since for these nanostructures extensive experimental and theoretical results are available. The conclusions are drawn in the last Section.

2.1.1 nc-TiN/SiN$_x$ Nanocomposites

During the last 20 years, significant progress in the design and understanding of superhard nanocomposites and nanolaminates nc-MN/XY, where MN is a stable and hard transition metal nitride and XY is a covalent nitride such as Si_3N_4 or BN, has been achieved [1–11]. (Note that "nc-" stands for nanocrystalline). Among superhard nanocomposite materials, nc-TiN/SiN$_x$ nanocomposites and heterostructures represent the most studied systems. The nc-TiN/Si_3N_4 nanocomposites exhibit superhardness higher than 65 GPa combined with high-thermal stability and oxidation resistance [1–5, 11]. The large increase of the hardness of the nanocomposites as compared with pure TiN coatings (20–21 GPa) has been explained by the existence of a nanostructure made of 3–4 nm in size TiN nanocrystals, which deform only elastically, connected by about 1 monolayer (1 ML) thick SiN$_x$ interfacial layer around which the plastic flow occurs [12–15]. The interfacial SiN$_x$ layer, that is percolating through the nanocomposite [3], is strengthened by valence charge transfer from the neighbor TiN [14–17], thus reducing grain boundary shear ("sliding"), which is otherwise limiting the achievable hardness enhancement in nano-sized materials to about a factor of ≤ 2 (see Refs. [11, 18] and references therein). Due to the random orientation of the TiN nanocrystals in the fully segregated nanocomposites, the interfacial SiN$_x$ monolayer appears amorphous in XRD [3] as well as in electron diffraction [19]. The thickness of the interfaces strongly influences the strength of the nanocomposite: the highest hardness is achieved with about 1 ML thick SiN$_x$. When the thickness reaches about 2 ML, the hardness enhancement is almost lost [3] due to a weakening of the neighbor Ti-N bonds [20]. A dense, fully segregated 1 ML Si_3N_4-like interfacial layer provides the nanocomposites also with high oxidation and corrosion resistance because it hinders diffusion of oxygen and other corrosive elements along the grain boundaries [5, 18].

Iwamoto and Tanaka investigated crystalline TiN/Si_3N_4 interfaces in nanocomposites deposited by thermal CVD and found that the TiN crystals were aligned to the Si_3N_4 crystals with the [110] direction of B1(NaCl)-TiN (in the following simply referred to as TiN) being parallel to the [0001] direction of the hexagonal h-Si_3N_4 thus forming a semi-coherent $(10\bar{1}0)Si_3N_4 \| (1\bar{1}1)$TiN interface [21]. Several papers recently reported the formation of pseudomorphic B1-related interfaces in TiN/SiN/TiN heterostructures [6, 8–10]. All the papers on heterostructures agree that the 1–2 ML thick interfacial SiN$_x$ layer in the heterostructures is pseudomorphic, albeit in some cases distorted, and that it becomes amorphous when its thickness exceeds several monolayers.

Both the coherent and incoherent TiN/SiN$_x$ interfaces were widely investigated in the framework of different first-principles methods [13–17, 20, 22–25]. The stability of a variety of TiN(111)/Si$_x$N$_y$ heterostructures, including the above mentioned semicoherent $(10\bar{1}0)Si_3N_4 \| (1\bar{1}1)$TiN interface studied by Iwamoto and Tanaka, was theoretically investigated by Hao et al. [16, 17] who also considered the effect of oxygen impurities [22] that strongly degrade the mechanical properties of the nanocomposites [3, 18] and apparently stabilizes the Ti-Si-N solid solution [9, 18]. Density functional theory (DFT) calculations at 0 K of almost 50 different Si_xN_y interfacial configurations sandwiched between two and ten monolayers thick slabs of TiN(111) showed that, under conditions of high nitrogen activity (pressure during the deposition or annealing), the β-like (1 × 3)-Si_2N_3 structure involving tetrahedral N-Si-N bonding was most favorable, while for nitrogen-poor conditions the (1 × 1)-TiSi structure was preferred which involved octahedral Ti-Si-Ti bonding [16, 17]. This is in agreement with the experimental finding that only signals from Si tetrahedrally bonded to nitrogen have been found by XPS in nanocomposites deposited in intense plasmas under a sufficiently high nitrogen pressure, whereas Ti-Si bonds have been found in coatings deposited either in a weak plasma or under low pressure of nitrogen (see Ref. [3] and references therein). The results of Hao et al. are also in agreement with thermodynamic calculations of Zhang et al. [26–28]. Hao et al. have also found that the Si-N bonds in the 1 ML thick β-like Si_2N_3 interface are stronger than in the $[10\bar{1}0]$ oriented bulk β-Si_3N_4, because the decohesion strength of that interfacial layer was larger than that of bulk ideal Si_3N_4 crystal.

Zhang et al. conducted extensive *ab initio* DFT studies on tensile and shear strength, and on the mechanism of the tensile decohesion (that is relevant for crack initiation and growth) and shear deformation (that is relevant for plastic flow) for a variety of TiN/1 ML SiN/TiN heterostructures [13–15]. They have shown that the SiN interfacial layer is strengthened by valence charge transfer from the metallic TiN, and that the weakest links are Ti-N bonds near that layer. Thus decohesion and shear occur within the Ti-N bonds weakened by Friedel-like oscillations of the valence charge density, and not in the interfacial SiN layer [14, 15]. These authors have also shown that the experimentally found loss of the hardness enhancement when the thickness of the interfacial SiN layer increased to $\geq$2 ML [29], resulted from a further weakening of the adjacent Ti-N bonds [20].

Zhang et al. have also studied the stability of different interfaces by displacing the Si atoms within the interfacial layer and following the changes of the total energy calculated at 0 K. They found that the pseudomorphic TiN(111)/B1-SiN and TiN(110)/B1-SiN interfacial SiN structures were, after a structural relaxation to the minimum of total energy under periodic boundary conditions, stable against finite displacements, whereas the TiN(001)/B1-SiN was unstable, showing a local maximum of the total energy [15]. However, this interface could be stabilized by displacement of 12 % in the [110] direction, whereas a displacement along [100] or [010] direction resulted in a saddle point [15]. The semicoherent TiN (111)/β-Si_3N_4–like interface, which has been derived from the SiN-like interface by introducing Si-vacancies, was unstable upon the relaxation to the minimum of the total energy at 0 K, and it underwent structural transformation with the formation of Si-Si bonds [13]. Because no Si-Si bonds were found in well segregated, stable and superhard nanocomposites by means of XPS [3, 5], such a TiN(111)/Si_3N_4-like interface derived from B1-SiN by Si-vacancies is unlikely to occur in the nanocomposites deposited at $\geq$550 °C, which have high thermal stability and oxidation resistance.

Alling et al. have shown that bulk B1-SiN is, within the harmonic approximation at 0 K, inherently dynamically unstable, and also the 1 ML B1-SiN(001) interface is dynamically unstable in its fully symmetric configuration [23]. More recently, Marten et al. [24] showed that this interface is dynamically more stable when distorted in the x and y direction in agreement with the earlier results of Zhang et al. [15]. The small indication of phonon instability around the X point in the Brillouin zone (BZ) indicates that this configuration is slightly unstable in the harmonic approximation at 0 K, but at finite temperature this instability disappears due to anharmonicity of the lattice vibrations [24]. The (111) interface has been found to be dynamically stable [24], also in agreement with the results of the "static" calculations of Zhang et al. [15].

Marten et al. have suggested atomic arrangements of the interfacial phase to explain the apparent epitaxial TiN(001)/SiN_x interfaces [25]. It is shown that the 1 ML B3-derived and 3 ML B3-$D0_{22}$-derived SiN_x interfaces are both dynamically and thermodynamically stable with respect to vacancy formation.

Although these studies provided much in-depth understanding of the heterostructures, their results are limited to DFT calculations at 0 K. Therefore, thermally activated structural reconstruction of these structures, which are likely to occur during deposition and application of nanocomposite coatings on tools at elevated temperatures, were not accessible by these calculations. As mentioned above, Zhang et al. [15] and Marten et al. [24] were forced to artificially shift the silicon atoms at the TiN(001)/1 ML B1-SiN interface to reach a saddle point or a minimum of the total energy. As result of the periodic boundary conditions and symmetry of the system during the "static" DFT calculations, the interfacial SiN layer was distorted in a regular manner keeping periodicity of that distortion. Therefore the question arises about the possible effect of thermal activation on the existence of a more distorted configuration with higher stability. Furthermore one has to ask if it be possible for the B1-SiN(111) interface, that was found to be dynamically stable at 0 K [24], to reconstruct at elevated temperatures, and for the Si_3N_4–like interfaces derived from SiN by the formation of Si vacancies [23, 24] to be stable at elevated temperatures? Finally, it is important to clarify whether the Si_3N_4-, SiN- and Si_2N_3-like interfaces, which are stable at 0 K [17], will be preserved between TiN(111) slabs at finite temperatures.

These questions have been partially resolved by first-principles quantum molecular dynamic calculations (QMD) [30]. The first-principles QMD calculations account for the effect of finite temperature by including the thermal motion of the atoms in the lattice (or a molecule or a liquid). This method automatically accounts for any thermally activated changes of the structure that lead to a lowering of the energy of the system (notice that the entropy is zero in the "static" DFT calculations at 0 K). Ivashchenko et al. [30] carried out QMD calculations of TiN/SiN_x heterostructures at finite temperatures, and relaxed them afterwards to reduce the statistic errors. In such a way, the relaxed atomic configurations were obtained. The study [30] focused on the following selected heterostructures: the dynamically unstable TiN(001)/1 ML B1-SiN that could be stabilized by artificial distortion in the [110] direction as mentioned above, and the TiN(111)/SiN_x heterostructures with one monolayer of interfacial B1-SiN, Si_3N_4-like SiN and Si_3N_4-like Si_2N_3.

2.1.2 TiN/SiC Nanolayered Coatings

In contrast to the TiN/SiN_x nanostructures, the TiN/SiC heterostructures have been studied to a lesser extent. We carried out an extensive literature search on this system, and found only one paper that was devoted to the study of the structural and mechanical properties of TiN/SiC nanolayered coatings [31]. Kong, Dai, Lao *et al.* prepared thin TiN/SiC nanolayered coatings at a relatively low temperature under different conditions of magnetron sputtering [31]. They showed that coherent 3C-SiC interfacial layers have been formed between TiN slabs with the (111) orientation when the thickness of the SiC layers was less than 0.8 nm. When the thickness of the SiC layer was about 0.6 nm, the hardness of the coatings reached a maximum value of about 60 GPa (a relatively small load of 10 mN has been used by the authors).

There are some important differences in the chemistry of the Ti-Si-N and Ti-C-N systems: Whereas pure and stoichiometric TiN and Si_3N_4 are immiscible [32] the Ti-C-N system forms a substitutional solid solution [33] with hardness increasing from that of TiN (21–22 GPa) to that of TiC (40–41 GPa) following the rules of mixtures (see Fig. 2.8 in Ref. [34]). It is therefore of great interest to study the TiN/SiC heterostructure where the carbon is bonded to silicon because the Gibbs free energies of formation at a typical deposition temperature of 800 K are −66.9 kJ/mol for SiC, −174.8 kJ/mol for TiC and −261.2 kJ/mol for TiN (see Ref. [35], pages 634, 640 and 1,542, respectively). It is therefore an open question if the SiC interfacial layer is stable at a final temperature.

To the best of our knowledge, the TiN/SiC interfaces have been theoretically investigated only in Ref. [36]. Given the limited experimental results achieved on the TiN/SiC nanolayered coatings [31], first-principles QMD calculations of the TiN/SiC heterostructures were carried out to understand the role of interfaces on strength enhancement of TiN/SiC nanolayered coatings [36].

2.1.3 TiN/BN Nanocomposites

BN has been used as interfacial material in nc-TiN/BN nanocomposites [37, 38], and in heterostructures [39, 40]. A variety of methods, such as plasma chemical vapor deposition (P CVD e.g. [3, 5, 38, 40]), physical vapor deposition (PVD, e.g. reactive sputtering [41–48], and laser ablation (e.g. [39]) has been used for the deposition of these coatings. It was shown that the B3-BN did not grow hetero-epitaxially on TiN, but that the h-BN tended to "wrap" the TiN nanocrystalline grains [39]. In contrast, the analysis of the P CVD TiN/BN heterostructures revealed that B3-BN had formed on the template TiN layer when the BN layer thickness was sufficiently thin (< 1 nm) [40]. Unlike Si_3N_4, BN cannot form any semicoherent interface with TiN because of the large difference in the atomic radii [3, 38]. Nevertheless, a hardness enhancement by a factor of about 2 has been found in the nc-TiN/BN nanocomposites when the thickness of the interfacial BN was about 1 ML [38]. The thermal stability of these nanocomposites was limited to about 900–1,000 °C [37] and the oxidation resistance was relatively low, similar to that of pure TiN coatings [38]. In contrast, the nc-TiN/Si_3N_4 nanocomposites are thermally stable up to 1,100 °C and oxidation resistant up to ≥800 °C [3].

Atomic configuration, electronic density of states (EDOS) and stress-strain relation for the TiN(001)/BN and TiN(111)/BN interfaces were investigated in Refs. [49]. The results revealed that, by analogy with the TiN/SiN heterostructures [14, 15], Friedel oscillations of bond distances weakened the Ti-N interplanar bonds next to the interface. Large-scale QMD simulations of the TiN(001)/BN heterostructures showed that the BN interface transformed to an amorphous-like h-BN layer consisting of BN_3 units at any temperatures [50].

2.1.4 TiN/AlN Nanolayered Systems

Several publications reported on the epitaxial stabilization of the cubic B1-AlN in TiN/AlN [51, 52] and VN/AlN [53] heterostructures. B1-AlN layers were formed when the layer thickness (d_{AlN}) was below 2 nm, whereas for $d_{AlN} > 2$ nm, the thermodynamically stable hexagonal h-AlN was found [52]. The critical thickness below which the cubic B1-AlN was pseudomorphically stabilized increased up to 6 nm for the VN/AlN heterostructures, where the lattice mismatch of 1.46 % was smaller than in the TiN/AlN one (mismatch of 3.84 %) [53]. The formation of cubic B1-AlN was also found during spinodal decomposition of the metastable $Ti_{1-x}Al_xN$ solid solution at about 900–1,000 °C. This phase however transformed into a thermodynamically stable hexagonal structure accompanied by a decrease of the hardness upon further annealing [54].

QMD calculations of the TiN(001)/AlN heterostructures were carried out in Ref. [50]. It was shown that, the epitaxially stabilized B1-AlN interfacial layer was preserved over all the whole temperature range in agreement with the experiments [51, 52].

2.1.5 Concluding Remarks to Sect. 2.1

It should be emphasized that any calculations of heterostructures with specific interfaces have certain limitations as regards to their implications to superhard nanocomposites with randomly oriented 3–4 nm size TiN nanocrystals, because in the nanocomposites a variety of (hkl) interfaces of small areas of a few nm^2, attached to different neighbor (h′k′l′) planes and terminated by triple junctions have to coexist, which will surely influence the energy of the interfaces in a manner that is difficult to quantify. Therefore, it would be more appropriate to simulate the deposition of the Ti, X and Y atoms forming a solid solution, followed by its segregation with the formation of nanocomposites, before studying the mechanical behavior of such self-organized, stable nanostructures. However in this case, extremely large scale QMD simulations would have to be used, which is impossible within the framework of the present first-principles methods. Such large-scale simulations could only be carried out by using classical MD with empirical potentials. However, reliable potentials for the Ti-X-Y system are not available yet. Therefore, a first-principles QMD study of the stability of the TiN/1 ML XY heterostructures, and their behavior as a function of temperature and composition is presently the only possible extension of the work of Hao et al., Zhang et al. and Marten et al. We present in this review results that provide new and deeper insight into the problem of the stability of the interfaces under consideration at finite temperature.

2.2 Computational Aspects

To perform the calculations of the TiN(001)/1 ML B1-XY (XY = SiN, SiC, BN, AlN) heterostructures, we considered 96-atom supercells constructed of the (2 × 2 × 3) B1(NaCl)-TiN cubic unit cells each containing 8 atoms. The TiN (111)/1 ML B1-SiN and B1-SiC heterostructures consisted of the 54-atom (3 × 3 × 1) supercells, derived from the 6-atom B1-TiN hexagonal unit cells with [1/2 1/2 0], [0 1/2 −1/2] and [1 1 1] as the **a, b** and **c** basis vectors, respectively. The interfacial XY monolayer has been introduced by replacing Ti atoms with X (Si, B, Al) atoms in the central lattice plane of the given supercell. For the TiN (001)/BN heterostructures, we considered two initial configurations: the first heterostructure (BN1) was generated by replacing the Ti atoms with the B atoms. The second heterostructure (BN2) was constructed by substituting N and Ti with B and N, respectively. Thus, there are eight X atoms and eight Y (C,N) atoms in the (001) interface, and nine X atoms and eighteen Y atoms (8 up and 8 down with respect to X) in the (111) interface. (The TiN(111)/1 ML XY heterostructures are "polar" in the c-direction). A similar procedure was considered to construct the initial TiN(111)/Si_3N_4-like SiN (abbreviated as "top-top" in [16, 17]) and TiN(111)/ Si_3N_4-like Si_2N_3 (abbreviated as "Si_2N_3" in [16, 17]) interface, respectively. However in this case, the Si_3N_4-like SiN interface was inserted between two complete TiN(111) units, in which the number of the TiN layers is a multiple of 3 as in perfect TiN. The Si_3N_4-like Si_2N_3 interface was introduced by replacing one TiN layer. The notation and structural characteristics of all the initial heterostructures under consideration are summarized in Table 2.1. All the heterostructures consist of the parallel layers aligned perpendicularly to the c-direction, and one of them is the XY interfacial layer. To clarify a possibility of the formation of stable Si_3N_4 configurations by introducing Si vacancies we removed two Si atoms from both the B1-SiN interfacial planes.

The present calculations were done using the first-principles pseudo-potential DFT MD method as implemented in the quantum ESPRESSO code [55] with

Table 2.1 Structure, composition, number of atoms in supercells (N_a), number of the layers in the TiN unit (N_L), supercell configurations (with respect to the simple cubic TiN unit cell), symmetry of an interface and configuration of Si atoms at an interface (Si-N_i) of all the initial heterostructures under consideration

Denotation	Structure	N_a	N_L	Supercell	Symmetry of interface	X-Y_i
SiN(001)	$Ti_{40}N_{40}$(001)/Si_8N_8	96	5	2 × 2 × 3	B1 (NaCl-type)	Si-N_6
$Si_{0.75}$N(001)	$Ti_{40}N_{40}$(001)/Si_6N_8	94	5	2 × 2 × 3	B1 (NaCl-type)	Si-N_6
SiN(111)	$Ti_{18}N_{18}$(111)/Si_9N_9	54	4	3 × 3 × 1	B1 (NaCl-type)	Si-N_6
$Si_{0.78}$N(111)	$Ti_{18}N_{18}$(111)/Si_7N_9	52	4	3 × 3 × 1	B1 (NaCl-type)	Si-N_6
TOP(111)	$Ti_{27}N_{27}$(111)/Si_9N_9	72	6	3 × 3 × 1	Si_3N_4-like SiN	Si-N_4
Si_2N_3(111)	$Ti_{45}N_{45}$(111)/Si_6N_9	105	10	3 × 3 × 2	Si_3N_4-like Si_2N_3	Si-N_4
SiC(001)	$Ti_{40}N_{40}$(001)/Si_8C_8	96	5	2 × 2 × 3	B1 (NaCl-type)	Si-C_6
SiC(111)	$Ti_{18}N_{18}$(111)/Si_9C_9	54	4	3 × 3 × 1	B1 (NaCl-type)	Si-C_6
BN(001)	$Ti_{40}N_{40}$(001)/B_8N_8	96	5	2 × 2 × 3	B1 (NaCl-type)	B-N_6
AlN(001)	$Ti_{40}N_{40}$(001)/B_8N_8	96	5	2 × 2 × 3	B1 (NaCl-type)	Al-N_6

periodic boundary conditions. The generalized gradient approximation (GGA) of Perdew, Burke and Ernzerhof [56] has been used for the exchange-correlation energy and potential, and the Vanderbilt ultra-soft pseudo-potentials were used to describe the electron-ion interaction [57]. In this approach, the orbitals are allowed to be as soft as possible in the core regions so that their plane-wave expansion converges rapidly [57]. The non-linear core corrections were taken into account as described in [55]. The criterion of convergence for the total energy was 10^{-6} Ry/formula unit ($1.36 \cdot 10^{-5}$ eV). To speed up the convergence, each eigenvalue was convoluted with a Gaussian with a width $\sigma = 0.02$ Ry (0.272 eV).

All the initial structures were optimized by simultaneously relaxing the supercell basis vectors and the atomic positions inside the unit cells using the Broyden-Fletcher-Goldfarb-Shanno (BFGS) algorithm [58]. The QMD simulations of the initial relaxed heterostructures were carried out in the temperature range of 10–1,400 K with fixed unit cell parameters and volume (NVT ensemble, i.e., constant number of particles-volume-temperature) for 2.7–4 ps. Below, we will analyze the heterostructures generated at 0 K (denoted L0), at 300 K (called further in the text "low temperature" and denoted LT) and at 1,400 K (called "high temperature" and denoted HT). In all the QMD calculations, the time step was 20 atomic units (a.u., about 10^{-15} s). The system temperature was kept constant by rescaling the velocity, and the variation of the total energy was controlled during each QMD time step. During the initial 1–1.5 ps, all structures reached closely their equilibrium state and, afterwards, the total energy of the equilibrated structures varied only slightly around the equilibrium value with a small amplitude of 0.005 eV/atom. We verified the convergence of the final QMD equilibrated structure with respect to the simulation time and the total energy of an equilibrated structure by choosing two B1-SiN(001) heterostructures at later stages of simulation at 1,400 K with simulation time and total energy differences of 0.2 ps and 0.01 eV/atom, respectively. Their pair correlation functions and atomic configurations practically coincided, and the total energy difference was only about 0.27 mRy/unit cell (0.038 meV/atom; not shown here). Thus, the resultant relaxed structures are only very weakly sensitive to structural changes during the last stage of the QMD simulation.

After QMD equilibration, the geometry of all LT and HT heterostructures was optimized by simultaneously relaxing the supercell basis vectors and the atomic positions inside the unit cells using the BFGS algorithm [58]. The cut-off energy for the plane-wave basis $E_{cut} = 30$ Ry (408 eV) and the Monkhorst-Pack [59] (2 2 2) mesh (4 *k*-points) were used. The relaxation of the atomic coordinates and of the unit cell were considered to be complete when the atomic forces were less than 1.0 mRy/Bohr (25.7 meV/Å), the stresses were smaller than 0.05 GPa, and the total energy during the structural optimization iterative process was changing by less than 0.1 mRy (1.36 meV).

The calculations of the amorphous silicon nitrides a-SiN and a-Si_3N_4 were carried out by analogy with those of the HT nanolayered systems, with only small differences. The initial 96-atom and 112-atom samples of SiN and Si_3N_4, respectively, were constructed on the basis of the $(2 \times 2 \times 2)$ β-Si_3N_4 supercell. These initial structures were heated up to 5,000 K, where the systems showed a diffusive behavior characteristic of liquids. QMD simulations of the amorphous samples were carried out in the NVT ensemble by taking into account the sample density ρ_s of β-Si_3N_4 (3.12 g/cm^3). The liquids were cooled down to 300 K within about 3.5 ps. The final amorphous samples were equilibrated at 300 K for 2.0 ps and then relaxed. In molecular dynamics simulations of the large-scale systems, only the Γ point was taken into account in the BZ integration, and the cut-off energy for the plane-wave basis was set to 24 Ry (~326 eV). The application of a reduced E_{cut} in large-scale QMD calculations was validated by Wang, Gudipati, Rao et al. [60] who successfully used a plane wave energy cutoff of 348.1 eV for the pseudopotential calculations of the elastic constants C_{ij} of the TiN/Si_xN_y superlattices.

The above-described pseudopotential procedure was used to study the phonon spectra of the B1-type XY in the framework of the density-functional perturbation theory (DFPT) described in Ref. [61]. The first-principles DFPT calculations were carried out for a (8 8 8) q-mesh, and afterwards the phonon densities of states (PHDOS) were computed using a (12 12 12) q-mesh by interpolating the computed phonon dispersion curves. Both the EDOS and PHDOS were calculated with the tetrahedron method implemented in the "Quantum-ESPRESSO" code [55].

For the Ti-Si-N system, the heat of formation of the bulk materials, H^f, has been calculated as $H^f = E_{tot} - \Sigma n_i E_i$, where E_{tot} is the total energy of the bulk compound with n_i atoms of all involved elements *i* (Si, Ti and N) and E_i is the total energy of the bulk Si and Ti atom, and half of the energy of N_2 molecule, respectively. The conditions of the calculations for bulk Ti and Si are summarized in Table 2.2. The total energy and equilibrium bond length of N_2 molecule were calculated using the extended two-atom cubic cell. The bond length of N_2 molecule was in agreement with the experimental value (1.098 Å) within 1 %.

Table 2.2 Structural and energetic characteristics of the computed phases in comparison to those of other experimental (in parentheses) and theoretical (in braces) investigations taken from Ref. [17, 42] that were used in this work for the interpretation of the heterostructures

Phase	Symmetry	N	*k*-mesh	Lattice parameters a, b, c (Å)	H^f (eV/f.u.)
TiN	Fm-3 m	2	(8 8 8)	4.249 (4.24)[a] {4.239–4.275}[a]	−3.46 (−3.46)[b] {−3.43, −3.56}[b]
β-Si_3N_4	$P6_3/m$	14	(4 4 12)	7.670, 7.670, 2.930 (7.607, 7.607, 2.911)[b] {7.652, 7.652, 2.927}[b]	−6.85 (−6.44 ± 0.83 to −8.73 ± 0.03)[b], {−7.81}[b] {−9.89}[e]
$TiSi_2$-C49	Cmcm	6	(4 4 4)	3.538, 13.418, 3.590 (3.62, 13.76, 3.60)[b] {3.571, 13.573, 3.556}[b]	−1.69 (−1.77 ± 0.09)[f]
$TiSi_2$-C54	Fddd	6	(4 4 4)	8.259, 4.795, 8.521 (8.269, 4.798, 8.553)[b] {8.270, 4.801, 8.553}[b]	−1.71
Ti	$P6_3/mmc$	2	(8 8 6)	2.913, 2.913, 4.614 (2.951, 2.951, 4.683)[c]	–
Si	Fd-3 m	2	(8 8 8)	5.465 (5.43)[d]	–

[a]Ref. [62]
[b]Ref. [17]
[c]X-ray powder diffraction file [044–1294]
[d]X-ray powder diffraction file [027–1402]
[e]Ref. [27]
[f]Experimental data for $TiSi_2$ [63]

The interfacial formation energy E^f, that yields information about the relative stabilities of various interfacial structures in the TiN/Si_xN_y heterostructures, is calculated as

$$E^f = S^{-1} \times \{E_H(N_{TiN}, N_I) - E_L(N_{TiN}) - n_{Si}E_{Si} - n_N E_N\}, N_I = n_{Si} + n_N, \quad (2.1)$$

where E_H is the total energy of the heterostructure under consideration, N_{TiN} and N_I are the numbers of atoms in the TiN slab and Si_xN_y interface, respectively, and $E_L(N_{TiN})$ is the total energy of the TiN unit. To calculate E_L we used two approaches: in approach A), $E_L(N_{TiN}) = N_{TiN} \times E_{TiN}$, where E_{TiN} is the total energy of bulk TiN per atom. For the approach B), the TiN unit is constructed by means of: i) a removal of the interface with interfacial area S from the TiN/Si_xN_y heterostructure, and ii) optimization of the resultant TiN structure as described above; n_S *and* n_N are the numbers of interfacial silicon and nitrogen atoms, respectively; E_{Si} and E_N are the energies (per atom) of the Si and N interfacial atoms.

Equation (2.1) is similar to Eq. (3) in Ref. [17], where slab structures were considered. In our case, the expression (2.1) is applied to periodic bulk structures. We used the approximations for values of E_{Si} and E_N as suggested by Hao et al. [17], who considered E_{Si} and E_N for so-called nitrogen-rich and nitrogen-poor (Ti-rich) conditions. For N-rich conditions, $E_{Si} = 1/3\{E_{tot}(\beta - Si_3N_4) - 2E_{tot}(N_2)\}$, $E_N = 1/2E_{tot}(N_2)$, where $E_{tot}(\beta - Si_3N_4)$ and $E_{tot}(N_2)$ are the total energy of bulk β-Si_3N_4 and half the energy of an N_2 molecule, respectively. For the N-poor conditions, $E_{Si} = 1/2\{E_{tot}(TiSi_2) - E_{tot}(Ti)\}$, $E_N = E_{tot}(TiN) - E_{tot}(Ti)$, where $E_{tot}(TiSi_2)$, $E_{tot}(Ti)$, $E_{tot}(TiN)$ are the total energies of bulk $TiSi_2$-C54, hexagonal Ti and TiN, respectively [17].

In order to interpret the structures of the interfaces, the structural parameters of the Ti-Si-N phases and the most stable XY structures were calculated: B1-XY (space group Fm-3 m), B3-XY (space group F-43 m), h-XY (space group $P6_3/mmc$) and B4-XY (space group $P6_3/mc$). The resulting structural parameters and heat of formation of the above mentioned crystalline solids are summarized in Tables 2.2 and 2.3. One can see that our results compare well with the published experimental and theoretical values, whenever available, for all compounds. Also our results for a-Si_3N_4 (not shown in Table 2.2) compare well with those of Giacomazzi and Umari [67], except of a small difference in the pair correlation function (PCF): our average Si-N bond length of 1.77 Å is somewhat larger than the value of 1.73 Å reported in Ref. [67], and the experimental values of 1.73–1.75 Å reported in Refs. [68]. This is due to the different approximations for the exchange-correlation potential used: GGA in our investigation, and LDA in Ref. [67]. It is well known that the GGA usually overestimates bond lengths, whereas the LDA underestimates them [55].

In large-supercell calculations, the reduced energy cut-off and the mesh of k-points were selected in order to minimize computing time without compromising accuracy, after it was verified that this choice yielded correct values for the lattice parameter a and bulk modulus B of TiN: For the 96-atom sample with $E_{cut} = 24$ Ry (~326 eV) and the one k- point, we obtained a = 4.228 Å and B = 276 GPa, whereas with $E_{cut} = 30$ Ry (~408 eV) and the (2 2 2) mesh, the lattice constant and bulk modulus were 4.248 Å and 278 GPa, respectively, which is a satisfactory agreement. For the 2-atom TiN cell with $E_{cut} = 30$ Ry (~408 eV) and the (8 8 8) mesh, our calculated lattice constant was a = 4.249 Å and bulk modulus B = 273 GPa. The calculated lattice constants are close to the experimental value of 4.24 Å and comparable to other theoretical results of 4.239–4.275 Å [62] (cf. Table 2.2), and the values of bulk moduli agree with the experimental and theoretical values of 264–326 GPa [62]. Also the calculated heat of formation for the 2-atom TiN cell is in excellent agreement with the experimental value of H^f (cf. Table 2.2). We compared also the total energy of pure TiN obtained for the 96-atom unit cell with E_{cut} =30 Ry (~408 eV) and (2 2 2) mesh with 2-atom cell, $E_{cut} = 30$ Ry and the (8 8 8) mesh. The total energy difference was only about 0.3 mRy/formula unit (~4 meV/atom), i.e. very small. The calculated surface energy σ (for details see Ref. [17]) for the 5-layer (2 × 2) TiN(001) and 10-layer (3 × 3) TiN(111) slabs

Table 2.3 Lattice parameters (a, b, c), X-Y bond length (R), and total energy (E_T) of the computed XY phases in comparison to those of other experimental and theoretical investigations

XY phase	Lattice parameters a, b, c (Å)	R (Å)	E_T (eV/atom)
h-BN	2.510, 2.510, 6.713[a]	1.449[a]	−178.583[a]
	2.504, 2.504, 6.661[b]	1.446[b]	
	2.491, 2.491, 6.378[c]	1.438[c]	
B3-BN	3.621[a]	1.568[a]	−178.517[a]
	3.616[d]	1.566[d]	
	3.592[c]	1.555[c]	
B4-BN	2.552, 2552, 4.222[a]	1.327[a]	−178.499[a]
	2.55, 2.55, 4.215[e]		
	2.531, 2.531, 4.189[c]		
	2.544, 2.544, 4.208[f]		
B1-BN	3.499[a]	1.750[a]	−176.760[a]
B4-AlN	3.116, 3.116, 4.996[a]	1.571[a]	−221.478[a]
	3.111, 3.111, 4.980[g]		
	3.114, 3.114, 4.992[h]		
B3-AlN	4.384[a]	1.898[a]	−221.457[a]
	4.365[i]	1.890[i]	
	4.380[h]	1.897[h]	
h-AlN	3.302, 3.302, 4.147[a]	1.906[a]	−221.364[a]
B1-AlN	4.052[a]	2.026[a]	−221.310[a]
	4.045[j]	2.023[j]	
	4.052[h]	2.026[h]	
B3-SiC	4.378[a]	1.896[a]	−142.948[a]
	4.359[k]	1.888[k]	
B4-SiC	3.092, 3.092, 5.063[a]	1.671[a]	−142.945[a]
h-SiC	3.173, 3.173, 4.798[a]	1.832[a]	−142.498[a]
B1-SiC	4.060[a]	2.030[a]	−142.211[a]
	4.080[l]	2.040[l]	
B3-SiN	4.301[a]	1.862[a]	−200.476[a]
B4-SiN	3.312, 3.312, 4.636[a]	1.530[a]	−200.328[a]
h-SiN	3.467, 3.467, 4.111[a]	2.002[a]	−200.215[a]
B1-SiN	4.251[a]	2.126[a]	−200.112[a]

[a]The present Quantum ESPRESSO-GGA calculations
[b]X-ray powder diffraction file [073–2095]
[c]CASTEP-LDA calculations [25, 64]
[d]X-ray powder diffraction file [087–1983]
[e]X-ray powder diffraction file [049–1327]
[f]CASTEP-GGA calculations [29, 65]
[g]X-ray powder diffraction file [065–0832]
[h]DFT-GGA calculations [6, 66]
[i]X-ray powder diffraction file [080–0010]
[j]X-ray powder diffraction file [046–1200]
[k]X-ray powder diffraction file [029–1129]
[l]X-ray powder diffraction file [049–1623]

are 0.074 eV/Å^2 and 0.203 eV/Å^2, respectively. These values are in good agreement with those computed in Ref. [17] for the 9-layer (1 × 1) TiN(001) slab (0.084 eV/Å) and 8-layer (1 × 1) TiN(111) slab (0.214 eV/Å). We used larger size of the x-y slab surface as compared to those considered by Hao et al. [17], which explains why our surface energies are slightly lower than those obtained in Ref. [17].

2.3 Characterization of the Structure of the Calculated Interfaces

2.3.1 Structural Properties of SiN and Si_3N_4-Like Si_2N_3 Interfaces

The geometry optimization of the initial B1-SiN heterostructures at 0 K preserved the hetero-epitaxial arrangement for all interfaces in a similar way as described by Zhang et al. [14, 15] (not shown here; one should keep in mind that the symmetric (001) interface corresponds to a local maximum of the total energy [15] and is dynamically unstable [24]). The geometry of the (001) interface was preserved, but the nitrogen atoms above and below the interface were slightly shifted by 0.26 Å toward this interface. However, at finite temperatures, the (001) interfacial structure significantly changed, whereas there were no significant changes in the (111) interface. In Fig. 2.1 we show the atomic configurations of the high-temperature TiN/1 ML SiN/TiN heterostructures after QMD equilibration at 1,400 K and subsequent relaxation. Figure 2.2a shows the corresponding change of the total energy (E_T) of the TiN(001)/SiN heterostructure during the QMD calculations at 10 K, which is sufficient to trigger the structural transformation. The temperature-dependent evolution of E_T for the TiN(001)/SiN heterostructure is shown in Fig. 2.2b. One can see that the structural transformation and concomitant decrease of the total energy occurs at 10 K. Figure 2.3 shows the arrangement of the atoms located in the (001) interfacial plane or close to it for the low– and high-temperature B1-SiN(001) heterostructures.

Because the TiN(001)/B1-SiN structure is strongly influenced by temperature, we need to consider this temperature effect in detail. We note significant atomic re-arrangement caused by thermal motion of the interfacial atoms that occurs predominantly within the interfacial SiN layer. There is an almost symmetrical, "down" and "up" shift of the N atoms in the layer just above and below the SiN (001) interfacial layer, an almost random shift of the Si atoms within the interfacial plane, and a breaking of about half of the Si-N bonds. This leads to the formation of a β-Si_3N_4–like units that are represented by the SiN_4, and of SiN_6 units with Si atoms distorted along the (110) directions (cf. Fig. 2.3). One can see from Fig. 2.3 that, along with the new Si_3N_4–like units, some of the original sixfold coordinated

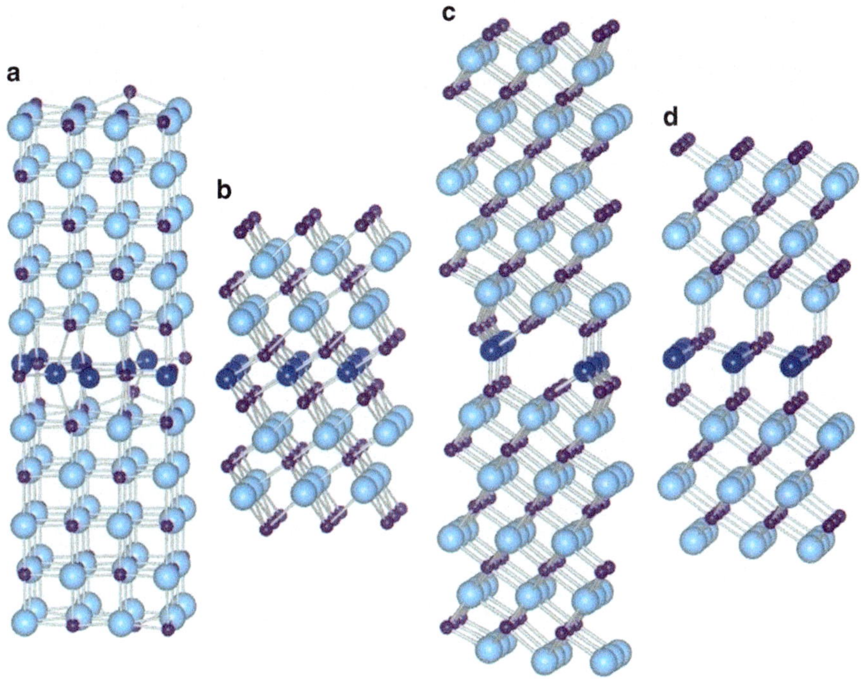

Fig. 2.1 Atomic configuration of the high-temperature TiN/1 ML SiN/TiN heterostructures: HT-SiN(001) (**a**), HT-SiN(111) (**b**), HT- Si_3N_4-Si_2N_3(111) (**c**) and HT-Si_3N_4-SiN(111) (**d**). The atomic configurations were computed using the Si-N and Ti-N bond length cut-offs of 2.5 Å. Ti-big *blue-grey*, Si-*medium dark blue* and N-*small purple*

B1-SiN units are still present albeit distorted. We note that only two SiN_4 clusters are present in the interface at 300 K (see LT-SiN(001)) for the Si-N bond length cut-off up to 2.5 Å (which is too long for a covalent bond in Si_3N_4), whereas at high temperature the number of such units increases from two to four (see - HT-SiN(001)). The thermal energy supplied to the TiN(001)/B1-SiN heterostructure during annealing at 1,400 K enhances the formation of the stable SiN_4 units. At slow cooling rate the influence of the Ti-N surrounding around the (001) interface is enhanced and the distortion of the SiN_4 units as well as the total energy slightly increase (not shown here for brevity). This result is similar to that of Zhang et al. who found that the (001) interface lowers its energy by distortion of the Si network so that these atoms become essentially fourfold coordinated [15]. The important difference is the fact that whereas the intentional distortion was regular in the "static" (at 0 K) DFT calculations of Zhang et al. due to the symmetry constrains imposed on their calculations, our present first-principles QMD calculations show that the interfacial layer is distorted randomly, which decreases the strain energy as discussed below.

We also note that the thermal energy supplied to the TiN(001)/B1-SiN heterostructures during evolution at 300 K is not sufficient for a complete relaxation of the

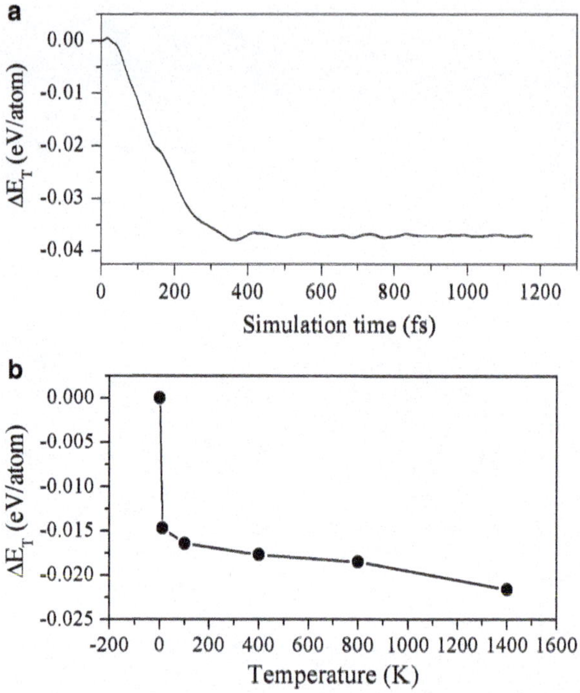

Fig. 2.2 (**a**) Total energy $\Delta E_T = E_T(t)\text{-}E_T(0)$ as a function of the calculation time for the TiN (001)/SiN heterostructure. The calculations were carried out at 10 K. (**b**) Total energy $\Delta E_T = E_T(t)\text{-}E_T(0)$ as a function of temperature for TiN(001)/SiN heterostructure

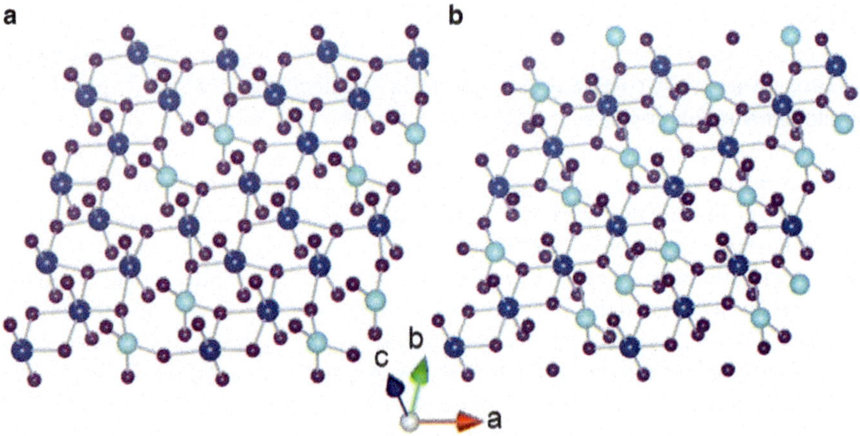

Fig. 2.3 Bond configurations in and around the interfacial planes for LT-SiN(001) (**a**) and HT-SiN(001) (**b**). The atomic configurations were computed using the Si-N bond length cut-off of 2.5 Å. Si-medium dark blue, N-small purple, the light-colored large circles are the fourfold coordinated Si atoms (Reprinted with permission from Ivashchenko et al. [30]. Copyright 2012 by the American Physical Society)

system. This interface is characterized by an atomic rearrangement within the interface and a shift of N atoms in the neighborhood as well as in distant TiN planes (cf. LT-SiN(001) in Fig. 2.1). At 1,400 K, the distortions within the interfacial SiN planes are more pronounced, and the shifts of the atoms with increasing distance from it are somewhat relaxed. We note that significantly slower cooling of the (001) heterostructure (HLT-SiN(001)) equilibrated at 1,400 K results in very similar interfacial structures (not shown here).

Since the relaxed heterostructures are obtained by quenching the different heterostructures to zero temperature, it is interesting to compare their total energy. It was found that the total energy decreased in the following sequence: relaxed zero-temperature (001) heterostructure (as reference) L0-SiN(001) 0.0000 eV/atom → LT-SiN(001) −0.0077 eV/atom → HLT-SiN(001) −0.0103 eV/atom → HT-SiN (001) −0.0193 eV/atom, i.e. the high temperature heterostructure HT-B1-SiN(001), which has the largest random distortion, has the lowest total energy.

This result may appear surprising because in macroscopic systems the ordered, crystalline phase is more stable than the disordered, amorphous one (we shall see later that the disordered (001) interface has similar pair correlation function as for amorphous SiN_x). However, the thermodynamical behavior of nano-sized materials shows differences as compared with the macroscopic ones. For example, the melting point T_m decreases with decreasing size d approaching, in the case of Au, CdSe and other materials, only few 100 °C for crystallite size of about 3 nm [69, 70]. In the case of compact thin films of nanocrystalline silicon, Veprek et al. reported transition to amorphous phase when the crystallite size reached about 3 nm [71]. These authors found that this crystalline-to-amorphous transition is due to increasing elastic strain energy of the nc-Si with decreasing crystallite size, which is seen as a dilatation of the lattice constant, and is driven by the tensile stress within the grain boundaries. The tensile stress within the grain boundaries originates from the density deficit present there. As it has been shown by a simple calculation in Ref. [71], the system reduces this energy by forming amorphous network that eliminates the grain boundaries, and as such it reduces the elastic strain energy originating from the grain boundaries in the nc-Si. Thus, the amorphous phase becomes more stable than the crystalline one for nano-sized compact films, in contradiction to what is happening in a macroscopic system. By analogy, the random distortion of the SiN interfacial layer obtained in the present QMD calculations reduces the strain as compared to the regularly distorted one in the calculations by Zhang et al.

In contrast to the (001) interface, the (111) interface is stable upon heating to 1,400 K and subsequent static relaxation (see HT-SiN(111), HT-Si_3N_4-SiN(111) and HT-Si_3N_4-Si_2N_3 (111) in Fig. 2.1). The structures of the LT- and HT-(111) interfaces are similar to that of the initial (111) interface at 0 K. The total energy of the HT-heterostructures relatively to that of the initial one at 0 K shows only very small changes: it increases by 0.0049 eV/atom for HT-Si_3N_4-SiN(111), by 0.0014 eV/atom for HT-B1-SiN(111), and decreases by 0.0034 eV/atom for HT-Si_3N_4-Si_2N_3(111). Because of the surprisingly high stability of the B1-SiN (111) interface we carried out QMD simulations for the (111) interface by raising the temperature up to 3,000 K (i.e., close to the melting point of TiN of 3,223 K [72]), and found no breaking of old bonds or formation of new bonds. Except for some small distortions, there were no significant changes of this structure. To make sure that the high thermal stability of these (111) interfaces is not any artifact of our QMD calculations, we performed QMD calculations for the TiN(111)/B1-SiC heterostructures under the same conditions and found that the B1-SiC interfacial layer transforms to 3C-SiC already above 600 K (see below). Thus, the stability of the B1-SiN(111) interface is intrinsic to this configuration in the SiN-TiN (111) interface. It remains an open question if it is due to hetero-epitaxial stabilization, or due to a different thermodynamics at the nanoscale, or to kinetic limitations because it is well documented that transformations, that require a final nucleation volume, are hindered in nano-sized materials. For these reasons the 1 ML B1-SiN(111) interface should not be excluded when considering the nanocomposites. With this regards, it would be of great interest to conduct experimental studies on the TiN(111)/B1-SiN heterostructures.

To analyze the structural evolution of the interfaces in more details, we computed the pair correlation functions (PCF) and the bond-angle distribution, g(Θ), for all heterostructures under consideration. The results are shown in Figs. 2.4 and 2.5. We see that the first PCF peak corresponding to the Ti-N nearest neighbor (NN) distance in the (001) interface is located around 2.11 Å that is slightly lower than the Ti-N bond length in pure TiN (2.12 Å). The broadening of this peak can be attributed to small variations in the bond distance as results of the Friedel-like oscillations around the SiN interfaces (SiN(001) and SiN(111)), or to distortions around the Si_3N_4-like interfaces (Si_3N_4-SiN(111) and Si_3N_4-Si_2N_3(111)) (cf. Fig. 2.2).

The above mentioned shift of the N and Si atoms in the TiN(001)/B1-SiN heterostructures leads to the formation of the many-peak structure of the Ti-Si nearest distances indicating a strengthening of the Ti-Si interaction. This type of interactions shows up in the PCF as a peak associated with the Ti-Si "nearest neighbor" correlations at 2.75–2.82 Å. These values compare well with the calculated and experimental Ti-Si bond lengths of 2.54–2.76 Å found in $TiSi_2$ (cf. Table 2.2). Also this finding is in qualitative agreement with the results of Zhang et al. because in the case of the (001) interface, there are two different (011) planes running perpendicularly to that interface: "planes I" which pass through Si atoms and "planes II" that pass through N-atoms within that interfacial layer, thus causing a "phase shift" of the Friedel-like oscillations between these planes (see Fig. 2.2 in [15]). One can see from Fig. 2.4 that the splitting of the Ti-Si peak is absent in the PCF of the SiN(111) and Si_3N_4-SiN(111) heterostructures. Owing to the symmetry of the Si_2N_3 interface, the strong Ti-Si "nearest neighbor" correlations are shown as a peak at 2.4 Å in the Ti-Si PCF of the HT-Si_3N_4-Si_2N_3(111) structure.

Because the Si-N bond length of 1.75–1.77 Å in β-Si_3N_4 is much shorter than the Ti-N bond length of 2.12 Å in pure TiN, there is a tensile misfit stress within the interfaces. This stress and the tendency of silicon to attain its fourfold coordination to nitrogen as in Si_3N_4, which is the most stable configuration in the Si-N system, are the main factors that cause the reconstruction of the hetero-epitaxial layers. Figure 2.4 shows that the Si-N bonds are shorter than the Ti-N bonds in all

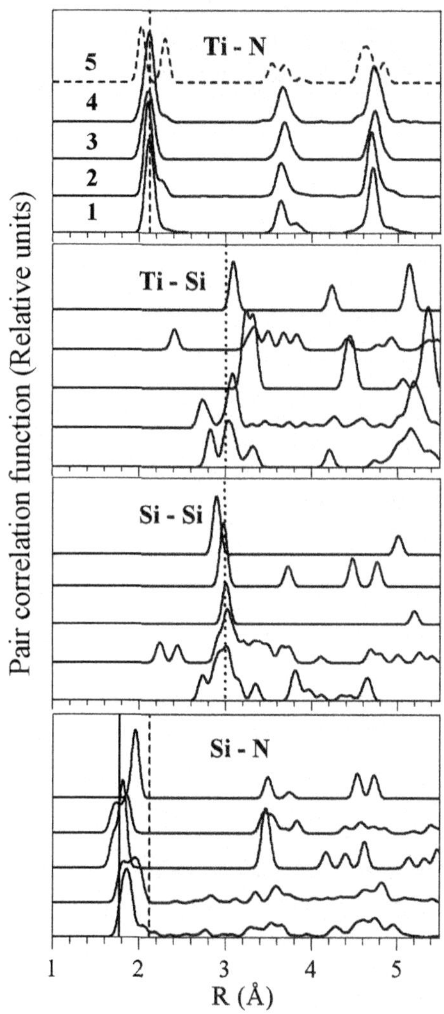

Fig. 2.4 Pair correlation functions for the low- and high-temperature 1 ML SiN interfaces: 1-LT-SiN(001), 2-HT-B1-SiN(001), 3-HT-Si_3N_4-SiN(111), 4-HT-Si_3N_4-Si_2N_3(111) and 5-HT- SiN(111). The *broken* and *dotted vertical lines* show the Ti-N and Ti-Ti distances in TiN, respectively. The *solid vertical line* corresponds to the Si-N bond length in β-Si_3N_4 (Reprinted with permission from Ivashchenko et al. [30]. Copyright 2012 by the American Physical Society)

heterostructures, and this tendency to shortening is more pronounced for the HT-SiN(001), HT-Si_3N_4-SiN(111) and HT- Si_3N_4-Si_2N_3(111) interfaces which have almost the equilibrium Si-N bond length of 1.77 Å, seen as the shoulders in the Si-N PCF peak. The one-peak Si-N nearest neighbor correlation with the elongated Si-N bonds of 1.85–1.95 Å is seen in all interfaces. An inspection of the atomic arrangement and the PCF for the HT-SiN(001) clearly shows that the bond lengths of 1.76–1.83 Å correspond to the Si-N bonds in the distorted β-Si_3N_4–like units, whereas the peak at 1.95 Å corresponds to the Si-N bond lengths in the SiN_6 units. The shortening of the Si-N bonds causes an elongation of the Ti-N ones, which is more pronounced for the bonds between the interfacial N atoms bonded to Si and the Ti atoms next to it than between Ti and N atoms far away from the interface (cf. Figs. 2.1 and 2.3).

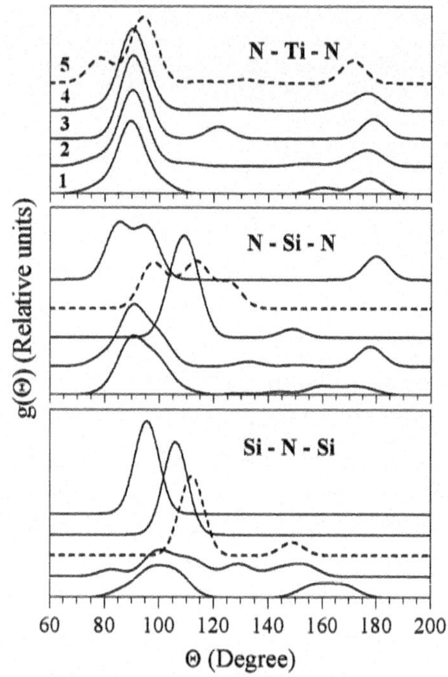

Fig. 2.5 Bond angle distribution g(Θ) for the low- and high-temperature 1 ML SiN interfaces: 1-LT-SiN(001), 2-HT-SiN(001), 3-HT-Si_3N_4-SiN(111), 4-HT-Si_3N_4-Si_2N_3(111) and 5-HT- SiN(111). The bond angle distributions were computed using the Si-N bond length cut-off of 2.2 Å. The Ti-N bonds were cut at 2.5 Å (Reprinted with permission from Ivashchenko et al. [30]. Copyright 2012 by the American Physical Society)

For the B1-SiN interfaces, the average lengths of the Ti-N bonds adjacent to the interfacial SiN layer are 2.275 and 2.298 Å for the (001) and (111) interfaces, respectively. The average length of the Ti-N bonds between the next neighbor planes parallel to the interfaces is about 2.076 and 2.025 Å for the (001) and (111) heterostructures, respectively. Thus, the broadening of the PCF peaks of the Ti-N next-near neighbor peaks seen in Fig. 2.4 is due to this bond length distribution. The Ti-N bond length increases when moving away from the Si_3N_4-like SiN interface. For the Si_3N_4-like Si_2N_3 interface, we did not identify any regularity in the change of Ti-N bonds in the c-direction (not shown).

It is interesting to compare our results for the B1-SiN based interfaces with those of Zhang et al. who, using a static *ab initio* DFT at 0 K, found fairly regular Friedel-like oscillations of the bond lengths damped with increasing distance from the (111) interface [14]. In the case of the (001) interface, these oscillations were "phase shifted" in planes perpendicular to that interface and passing either through Si or N atoms in that interface [15]. These authors identified the longest Ti-N bonds close to the interface as the weakest links in the heterostructures (and by analogy in the nanocomposites). They could show that for all three interfaces, (111), (001) and (110), studied in their work, the ideal decohesion and shear occurred within the longest and therefore weakest Ti-N bonds and not in the SiN interfacial layer. Our results obtained with more general first-principles QMD at 300 K and at elevated temperatures fully confirm the weakening of the elongated Ti-N bonds adjacent to the SiN interfaces. The disorder due to thermal activation introduces some broadening of the near neighbor peaks in the PCF. However, the structural instability of the (001) interface discussed above calls for re-considering the stress-strain calculations at finite temperature using the first-principles QMD. Comparing, as an example, the (001) interface one sees that the longest Ti-N bonds adjacent to the SiN layer of 2.206–2.275 Å are shorter than those of 2.35–2.37 Å reported in Ref. [15] for the DFT calculations at 0 K. Also the longest Ti-N bond lengths of 2.298 Å for the (111) interface are shorter than those of 2.32 Å reported in Ref. [14]. This is likely related to a more efficient relaxation of the structure in the present QMD calculations as seen also in the decrease of the total energy discussed above. Because the longest bonds are the weak links where decohesion and shear occurs [14, 15], it would be interesting to see if the somewhat shorter bond lengths found in the present work will reflect larger ideal strengths of the interfaces. This is the subject of our present work.

Let us now examine the PCF for the Si-Si bonds and distances. As one can see in Fig. 2.4, there is one peak at a distance of about 3 Å in all the (111) heterostructures that corresponds to the second nearest-neighbor (Si-N-Si) correlations in the interfaces. In contrast, this peak splits into several shoulders in the case of the LT- and HT-SiN(001) interface due to the random distortion as discussed above. This splitting is much more pronounced for the HT-SiN(001) interface with the appearance of a new peak around 2.35 Å, that are close to the Si-Si bond length in crystalline silicon. This can be understood on the basis of thermodynamic instability of SiN which, in the case of closed system simulated by the QMD, should decompose according to $4SiN \rightarrow Si_3N_4 + Si$ with a high Gibbs free energy of the reaction of -6.44 ± 0.83 to -8.73 ± 0.03 eV/formula unit (cf. Table 2.2). This value is in a reasonable agreement with the result of Zhang and Veprek who, using a combined DFT and thermodynamic calculations, reported Gibbs free energy for that reaction to be -136.3 kJ/mol atoms (*i.e*., 9.89 eV/formula unit) [27]. Obviously, there is a kinetic activation barrier for this reaction that must occur at room temperature. Therefore this peak is absent in the PCF of LT-SiN(001) interface (see Fig. 2.4).

In Fig. 2.5, we show the distributions of the N-Ti-N, N-Si-N and Si-N-Si bond angles. The N-Ti-N angles are between 90° and 94°, which are very close to the angle 90° for pure TiN. For the (001) interfaces, the N-Si-N angles are around 90° and 101°, *i.e.*, close to the angles for the octahedral (90° in B1-SiN) and tetrahedral (109.7° in h-Si_3N_4) configurations, respectively. In the case of the B1-SiN(111) interface, there are two peaks in the N-Si-N bond angle distribution located at 85° and 95°. Because the sixfold octahedral coordination is preserved in these interfaces, such a bond angle splitting indicates its asymmetry, which can be seen by careful observation of Figs. 2.1: the N-Si-N angles oriented up and down (i.e. where the N-atoms are further bonded to Ti only) are clearly larger than those oriented to the left or right (i.e. where the N-atoms are bonded to the second neighbor Si and Ti). The tetrahedral coordination of the Si atoms in the HT-Si_3N_4-SiN(111) interface is confirmed by the single peak in the $g_{N\text{-}Si\text{-}N}(\Theta)$ distribution around 109.7°. The distorted tetrahedral configuration of the Si atoms in the Si_3N_4-Si_2N_3(111) interface is seen as two peaks localized at 98° and 113°. The Si-N-Si bond angle distribution in the HT-SiN(111) interface shows one relatively symmetric peak at about 97°. The Si-N-Si bond angle in the LT-SiN(001) interface shows a

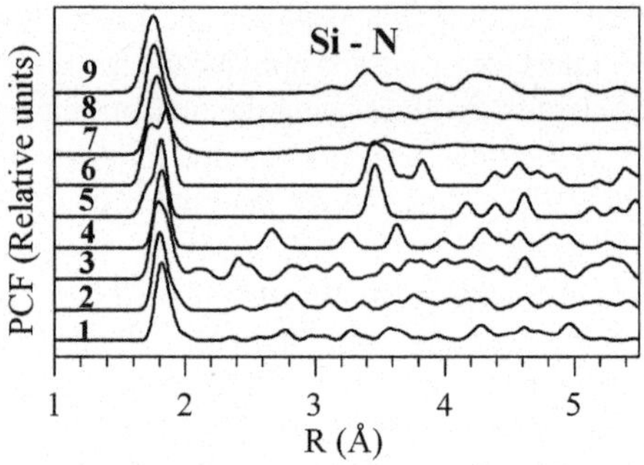

Fig. 2.6 Si-N pair correlation functions for the interfaces with the fourfold coordination of silicon atoms. Denotation: 1-LT-SiN(001), 2-HT-SiN(001), 3-the structure was obtained by slow cooling B1-SiN(001) from 1,400 to 300 K, 4-LT-$Si_{0.75}N(001)$, 5-HT-Si_3N_4-SiN(111), 6-HT- Si_3N_4-$Si_2N_3(111)$, 7-a-SiN, 8-a-Si_3N_4 and 9-β-Si_3N_4. The PCFs of the LT-SiN(001), HT-SiN(001) and HT-$Si_{0.75}N(001)$ were computed for the heterostructures in which the chain of the sixfold coordinated Si atoms was removed (Reprinted with permission from Ivashchenko et al. [30]. Copyright 2012 by the American Physical Society)

broad distribution with a main peak around 101°, whereas in the HT-SiN(001) interface, the distribution is even broader and the main peaks are located around 101° and 128°. Such a distribution shows that nitrogen atoms form mostly the NSi_4 and NSi_3 coordinated units, respectively. The Si-N-Si angles around 106° and 112° in the HT-Si_3N_4-$Si_2N_3(111)$ and HT-Si_3N_4-SiN(111), respectively, indicates the distorted fourfold configurations of the N atoms in the Si-N network of these heterostructures.

To obtain a deeper insight into the structure of the (001) interfaces we calculated the PCF and g(Θ) functions for β-Si_3N_4, a-SiN and a-Si_3N_4. The results are shown in Figs. 2.6 and 2.7 in comparison with those for the (001) interfaces. A comparison of the PCFs and $g_{Si\text{-}N\text{-}Si}(\Theta)$ functions enables us to infer that the Si-N network of the LT(001) interface is more similar to that of a-SiN than to the others. However, the positions of the main peaks in the PCF and $g_{N\text{-}Si\text{-}N}(\Theta)$ for both structures differ by 0.08 Å and 7°, respectively, which can be attributed mostly to the presence of a larger number of sixfold coordinated Si atoms in the interface than in the amorphous sample. Therefore, in a first approximation, the LT-B1-SiN(001) interface can be roughly considered as an over-coordinated a-SiN-like structure.

As shown above, in the HT-SiN(001), there are two groups of Si atoms: one forms the SiN_4 coordinated units like in β-Si_3N_4 with the Si-N bond lengths in the range of 1.77–1.85 Å and another one that forms the SiN_6 units with interatomic distances around 1.95 Å. To confirm the formation of the β-Si_3N_4-like units we computed the $g_{N\text{-}Si\text{-}N}(\Theta)$ and $g_{Si\text{-}N\text{-}Si}(\Theta)$ of the HT-SiN(001) heterostructure using the bond length cut-off of 1.9 Å thus eliminating the contributions of Si-N nearest neighbor correlations with the larger bond lengths. These functions are shown as the curves with broken lines in Fig. 2.7. Obviously, the broad peaks at about 99° and 127° in the N-Si-N and Si-N-Si bond angle distributions, respectively, can be attributed to 106° and to 121° in the corresponding bond angle distributions of a-Si_3N_4.

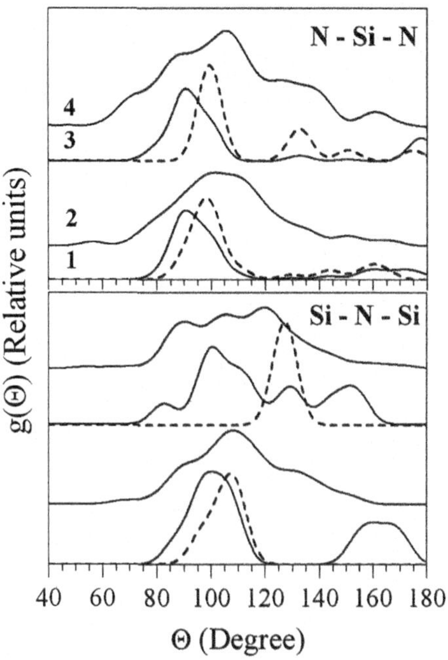

Fig. 2.7 Bond angle distribution g(Θ) for: 1-LT-SiN(001), 2-a-SiN, 3-HT-SiN(001) and 4-a-Si_3N_4. The bond angle distributions were computed using the Si-N bond length cut-off of 2.2 Å. The *curves* with *broken lines* correspond to the g(Θ) functions for HT(001) obtained using a Si-N bond length cut-off of 1.9 Å (Reprinted with permission from Ivashchenko et al. [30]. Copyright 2012 by the American Physical Society)

To summarize the results of this section, we showed that the pseudomorphic B1-TiN(001)/1 ML SiN/TiN heterostructure is, at finite temperature, unstable in its highly symmetric arrangement in agreement with earlier results [15, 23] which were obtained by means of DFT at 0 K. At finite temperature, it transforms into a complex interfacial structure with disordered B1-SiN- and β-Si_3N_4–like units. The B1-TiN(111)/1 ML B1-SiN/TiN, B1-TiN(111)/1 ML Si_3N_4-like SiN/TiN, and B1-TiN(111)/1 ML Si_3N_4-like Si_2N_3/TiN heterostructures are stable in the temperature range of 0–1,400 K.

Obviously, the interfacial structures of TiN/Si_xN_y nanocomposites and nanolaminates are much more complex than as it has been considered so far. One has to keep in mind that in nanocomposites consisting of 3–4 nm, randomly oriented TiN nanocrystals of a fairly regular shape [19] connected by about 1 ML thick Si_3N_4–like interfacial layer many different (hkl) interfaces have to coexist [3, 4, 18, 29]. Moreover, the early DFT and the present QMD calculations can consider only closed systems whereas the system is open during the deposition, i.e. the number of nitrogen atoms in the system is not constant. Therefore "nitriding" will occur whenever energetically and structurally preferred. For example, the structural reorganization of the fcc-B1-TiN/1 ML SiN/TiN heterostructure with the SiN_4 and SiN_6 units will be – due to the "nitriding" – transformed so that only the β-Si_3N_4-like SiN_4 units will remain. Therefore studies on heterostructures provide only a limited, although valuable insight into the nature of the interfaces in the nanocomposites. In the following section we discuss only briefly the possibility of the formation of Si_3N_4–like interfaces derived from the SiN ones by means of the formation of Si vacancies.

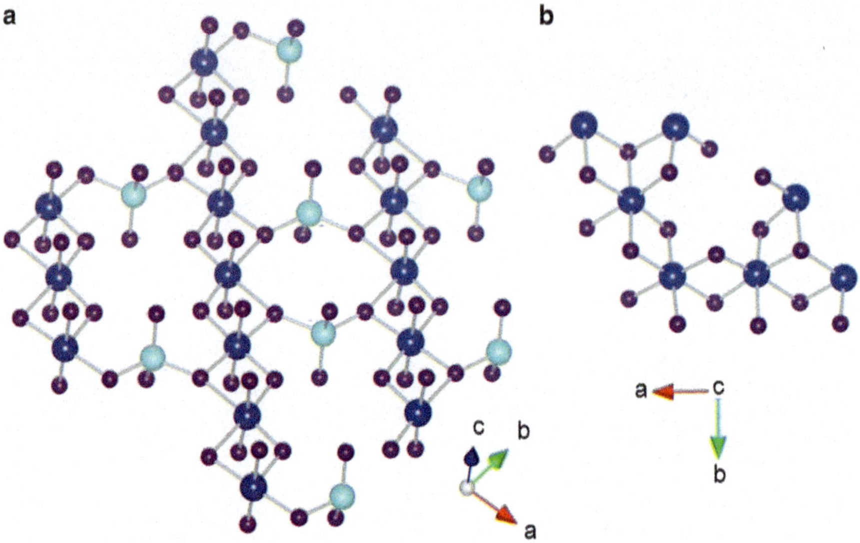

Fig. 2.8 Bond configurations at and around the interfacial planes of the LT-$Si_{0.75}N(001)$ (**a**), and LT-$Si_{0.78}N(111)$ (**b**) heterostructures. In Fig. 2.8a, the interface lies in the **a–b** plane. The atomic configurations were computed using a Si-N bond length cut-off of 2.5 Å. Si-*medium dark blue*, N-*small purple*, the *light large circles* are the fourfold coordinated Si atoms

2.3.2 Si_3N_4-Like Interfaces Derived from SiN by the Formation of Si Vacancies

Figure 2.8 shows the bond configurations at and around the interfacial planes of the TiN(001)/B1-$Si_{0.75}$ N (a) and TiN(111)/B1-$Si_{0.78}$ N (b) heterostructures at 300 K. (For an easier understanding, the TiN slabs above and below the SiN_x interfaces have been removed.). Figure 2.8a shows the Si_3N_4-like interfacial configurations derived from the TiN/B1-SiN/TiN heterostructure by the formation of Si vacancies at finite temperature of 300 K. The pair correlation functions and bond angle distributions of the substoichiometric heterostructures are shown in Figs. 2.9 and 2.10, respectively. The comparison of the atomic configurations in Figs. 2.3 and 2.8 shows that, for the TiN(001)/SiN heterostructure, the presence of silicon vacancies does not lead to any increase in the number of the SiN_4 units at the interface. The chains of the SiN_6 units are still present.

Let us analyze the temperature-induced changes in the energetics and the structures of the TiN(001)/1 ML B1-$Si_{0.75}$ N(001) heterostructure. The total energy decreases in the following sequence: relaxed zero-temperature (reference) L0-$Si_{0.75}$ N(001) heterostructure 0.0000 eV/atom → LT-$Si_{0.75}$ N(001) −0.0054 eV/atom → HT-$Si_{0.75}$ N(001) −0.0096 eV/atom. The lowering in the total energy with increasing temperature is accompanied by structural changes of the interface (cf. Fig. 2.9b).

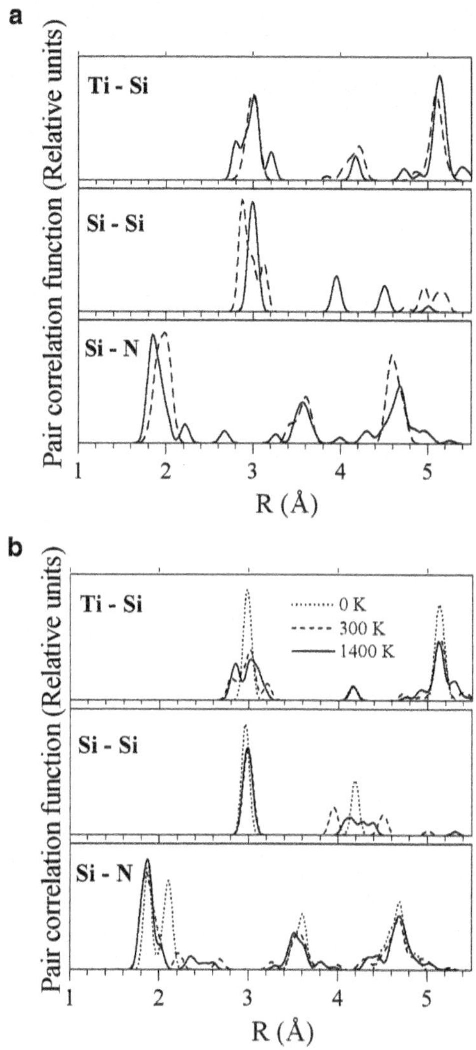

Fig. 2.9 (**a**) Pair correlation functions (PCF) for the LT-$Si_{0.75}N(001)$ (*solid line*) and LT-$Si_{0.78}$ N (111) (*broken line*) heterostructures. (**b**) Pair correlation functions for the $Si_{0.75}N(001)$ heterostructure as functions of temperature

One can see from Fig. 2.9a that the thermally-assisted motion of the interfacial atoms causes a shortening of the Si-N bond lengths, and this leads to a strengthening of the Ti-Si interactions.

In the case of the $Si_{0.78}$ N(111) heterostructure, the nitrogen atoms shift towards the neighbor silicon and away from the titanium atoms. As a result, a shoulder at 1.95 Å at the peak of Si-N correlations appears, and the peak of the Ti-N correlations

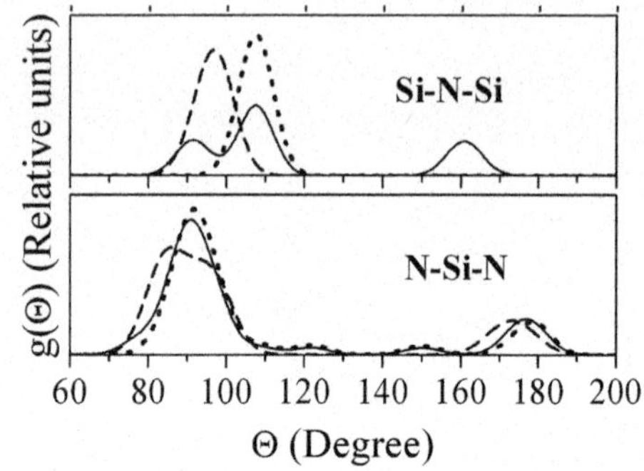

Fig. 2.10 Bond angle distribution g(Θ) for the LT-$Si_{0.75}$ N(001) (*solid* and *dotted lines*) and LT-$Si_{0.78}$ N(111) (*broken line*) heterostructures. The bond angle distributions were computed using the Si-N bond length cut-off of 2.1 Å (*dotted line*), 2.2 Å (*broken line*) and 2.4 Å (*solid line*)

shifts towards higher distances by 0.025 Å. The distorted octahedral coordination of the Si atoms is preserved (cf. Figs. 2.8b and 2.10). Thus, the deficit in the silicon atoms in the (111) interface also does not promote any formation of stable Si_3N_4–like configurations at finite temperature.

The calculation of the formation energy of silicon vacancies in the LT-SiN heterostructures was performed using the equation $E_v^f = \frac{1}{2}(E_{vac} - E_{ideal} + 2E_{Si})$, where E_{vac} and E_{ideal} are the total energies of the heterostructures with two Si vacancies and without them, respectively, and E_{Si} is the total energy of bulk Si. The values of E^f_v for the LT-SiN(001) and LT-SiN(111) heterostructures were found to be −2.090 and −0.461 eV, respectively, which is close to the values of −1.59 and −0.65 eV obtained by Marten et al. in their calculations at 0 K [24, 25]. It follows that both interfaces are unstable with respect to the formation of Si vacancies. This finding and the results of previous calculations [24, 25] clearly indicate that, for N-rich conditions, the 1:1 Si-to-N ratio for the TiN(001)/ 1 ML SiN_x interfaces is unfavorable in the heterostructures. A careful examination of the atomic configuration and pair correlation functions shows that the observed strengthening of the interfaces caused by Si vacancies is due to a reduction in the coordination of the N atoms at the interfaces, a shortening of the Si-N bond length in the SiN_6 units from 1.95 to 1.85 Å (LT-$Si_{0.75}$ N(001)) and a smaller distortion of the SiN_6 units.

Thus, despite the fact that a formation of Si vacancies at the B1-SiN interfaces does not lead to any increase in the number of tetrahedrally coordinated Si atoms compared to the stoichiometric interfaces, a formation of Si vacancies in the LT-SiN(001) and LT-SiN(111) is energetically favorable. However, in an open system under the conditions of deposition with excess of nitrogen, the stable stoichiometric β-Si_3N_4-like interfacial layers will form (cf. Table 2.4).

2.3.3 TiN/SiC Interfaces

The TiN(001)/SiC heterostructures were investigated at 0, 300, 600, 750, 900, and 1,400 K, whereas the TiN(111)/SiC heterostructures have been studied at 0, 300,

Table 2.4 Interfacial formation energies H^f (in eV/Å^2) estimated for the N-rich and N-poor conditions using Approaches A and B

	Approach A		Approach B	
Structure	N-rich	N-poor	N-rich	N-poor
LT-B1-SiN(001)	0.279	0.509	0.127	0.356
HT-B1-SiN(001)	0.264	0.493	0.111	0.341
LT-B1-$Si_{0.75}N$(001)	0.156	0.426	0.003	0.274
HT-B1-$Si_{0.75}N$(001)	0.150	0.421	−0.002	0.268
LT-B1-SiN(111)	0.292	0.573	0.142	0.423
LT-B1-$Si_{0.77}N$(111)	0.205	0.525	0.057	0.378
LT-Si_3N_4-SiN(111)	0.227	0.489	0.227	0.489
LT- Si_3N_4-Si_2N_3(111)	0.165	0.490	0.026	0.351

Reprinted with permission from Ivashchenko et al. [30]. Copyright 2012 by the American Physical Society

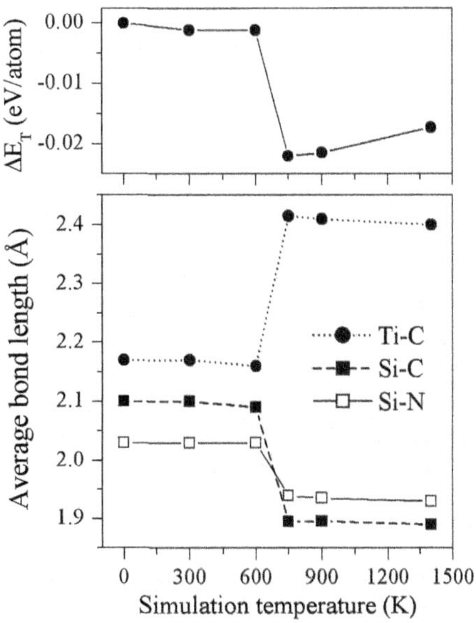

Fig. 2.11 Total energy ($\Delta E_T = E_T(T) - E_T(0)$) and average bond lengths for the TiN(001)/SiC heterostructure, determined from the positions of the corresponding first peaks of the pair correlation functions, as functions of simulation temperature

and 1,400 K [36]. Figure 2.11 shows the dependence of the total energy (E_T) and the structural parameters of the TiN(001)/SiC heterostructure on temperature. We note the pronounced changes in the energy and in the structural properties occur between 600 and 750 K. The initial TiN(111)/SiC heterostructure formed after static relaxation at 0 K remains practically unchanged with increasing temperature up to 1,400 K. The total energies of 0.000, −0.079 and −0.098 meV/atom of the L0-SiC(111), LT-SiC(111) and HT-SiC(111) heterostructures, respectively, are, within accuracy of the calculations, essentially the same, and significantly smaller than those seen in Fig. 2.11.

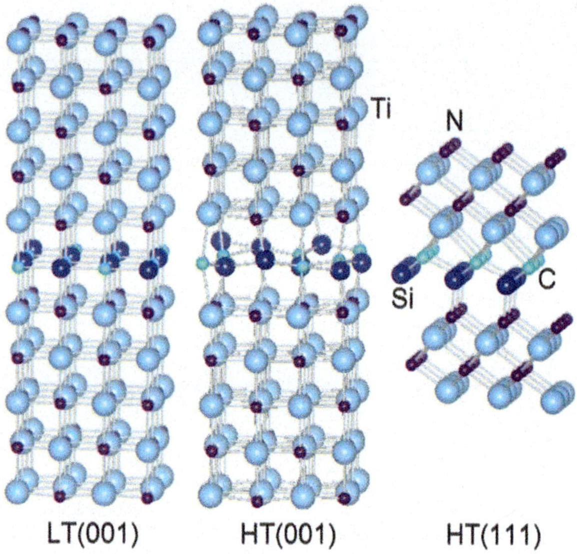

Fig. 2.12 Atomic configuration of the low- (LT) and high-temperature (HT) TiN/1 ML SiC/TiN heterostructures. Ti-*big blue-grey*, Si-*medium dark blue*, C-*small green-blue* and N-*small purple*. The atomic configurations were computed using Si-C, Ti-C and Ti-N bond length cut-offs of 2.5 Å (Reprinted with permission from Ivashchenko et al. [36]. Copyright 2012 by the American Physical Society)

Figure 2.12 shows the structure of the TiN(001)/SiC heterostructures equilibrated at 300 K (LT(001)) and 1,400 K (HT(001)) after subsequent relaxation, as well as the TiN(111)/SiC heterostructure equilibrated and relaxed at 1,400 K (HT (111)). Because the (001) heterostructures are strongly influenced by temperature, we will focus on them. One can see that the LT(001) heterostructure has an epitaxial B1-SiC interfacial layer preserved up to 600 K (see also Fig. 2.11). A further increase in the temperature leads to a significant change in the atomic arrangement caused by thermal motion of the atoms within and near the interface. We note a shift of the N atoms in the layer just above and below the (001) interface. The redistribution of the atoms results in the formation of 3C-SiC–like units, consisting of Si in the tetrahedral N-Si $\equiv C_3$ coordination. The C atoms form fivefold coordinated $Ti_2 = C \equiv Si_3$ configurations. The atomic arrangement around the (001) interface in the LT(001) and HT(001) heterostructures is shown in Fig. 2.13. We can see the Si-C_4 and C-Si_4 B1-SiC–like units in the LT(001), and the Si-C_3 and C-Si_3 3C-SiC–like units in HT(001).

Figure 2.14 shows PCF for the LT(001), HT(001), and HT(111) heterostructures. The bond-angle distribution (g(θ) for these high-temperature heterostructures is shown in Fig. 2.15. The PCF of the LT(001) displays the peaks of the nearest neighbor (NN) Ti-C, Ti-N, Si-C, Si-N correlations at 2.1–2.2 Å, and of the NN Ti-Si and Si-Si correlations at 3.0 Å, which are very close to the corresponding lengths of the metal-nonmetal and metal-metal correlations in TiN, respectively. The average Ti-C bond length of 2.4 Å in the HT-SiC(001) heterostructure is larger than that in TiC (2.09 Å), whereas the Ti-N bond lengths are close to those in TiN. The NN Si-Si correlations occur at a distance of ~3 Å, that is close to that of the second nearest neighbor correlations in the SiC epitaxial layer of the LT-SiC(001) heterostructure. In contrast, additional Ti-Si bonds with $l_{Ti\text{-}Si}$ ~ 2.8–2.9 Å are formed

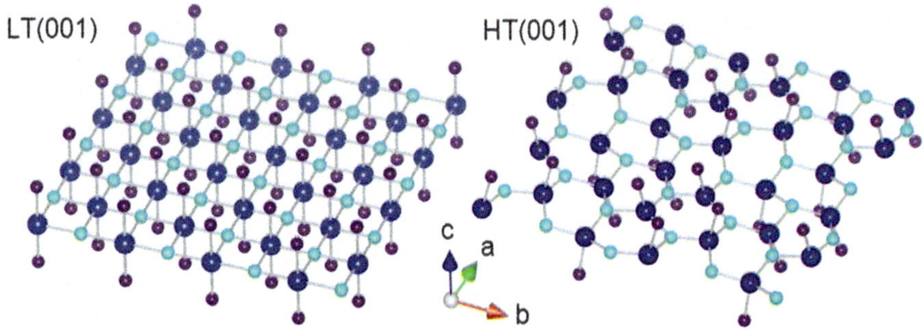

Fig. 2.13 Si-C-N bond configurations in and around the interfacial planes for LT(001) and HT (001). The atomic configurations were computed using Si-N and Si-C bond length cut-offs of 2.5 Å. The two Ti atoms that are neighbor of the C atoms and site above and below the interfaces are not shown. Ti-*big blue-grey*, Si-*medium dark blue*, C-*small green-blue* and N-*small purple*. (Reprinted with permission from Ivashchenko et al. [36]. Copyright 2012 by the American Physical Society)

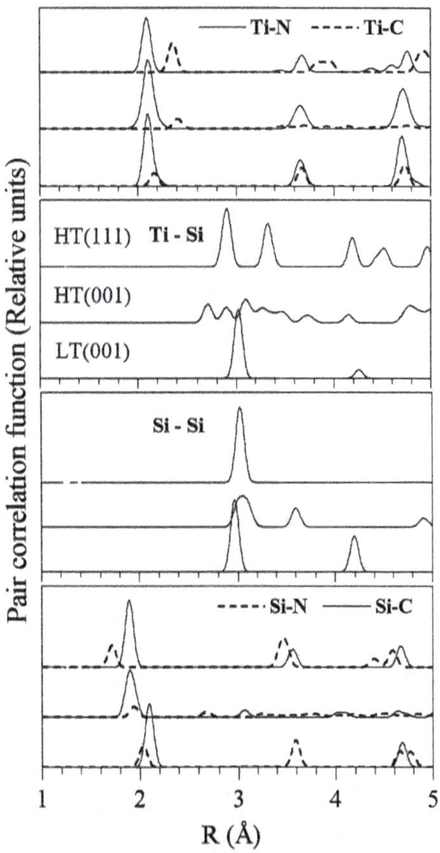

Fig. 2.14 Pair correlation functions for the low- and high-temperature 1 ML SiC interfaces: LT-SiC(001) (*lower curves*), HT-SiC (001) (*middle curves*) and HT-SiC(111) (*upper curves*) (Reprinted with permission from Ivashchenko et al. [36]. Copyright 2012 by the American Physical Society)

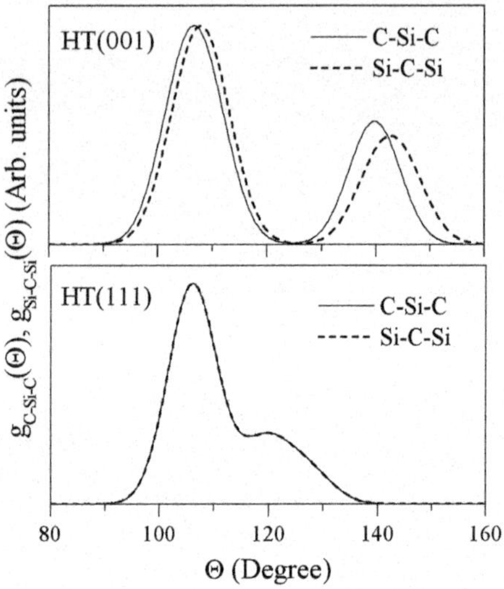

Fig. 2.15 Bond angle distribution g(Θ) for the high-temperature 1 ML SiC interfaces. The bond angle distributions were computed using a Si-C bond length cut-off of 2.5 Å (Reprinted with permission from Ivashchenko et al. [36]. Copyright 2012 by the American Physical Society)

around the interface of the HT-SiC(001) heterostructure. The peak of the NN Si-C correlations is located at a distance that is equal to l_{SiC} = 1.89 Å in 3C-SiC. The Si-N bonds ($l_{Si\text{-}N}$ = 1.9 Å) are elongated compared to those in β-Si_3N_4 ($l_{Si\text{-}N}$ = 1.76–1.8 Å). The bond angle distribution of the HT-SiC(001), g(θ), in Fig. 2.15, shows a main peak at ~109°, which means that the interfacial Si and C atoms form a Si-C network similar to that in 3C-SiC. The occurrence of a minor peak at 140° indicates that this network is strongly distorted. These findings show that the B1-SiC–like interfacial layer in the LT-TiN(001)/SiC heterostructures transforms into a 3C-SiC-like interface above 600 K. This interface can be viewed as two Si(111) and C(111) polar layers in 3C-SiC uniaxially deformed in the (111) direction.

Let us now discuss the (111) heterostructure. Figure 2.13 shows that the Si atoms are tetrahedrally coordinated and form Si-C_3N_1 configurations like the Si-C_4 clusters in 3C-SiC. The Si-C and Si-N bond lengths are 1.89 Å and 1.77 Å, respectively (cf. Fig. 2.14). The C atoms have six neighbor forming C-Si_3Ti_3 configurations (cf. Fig. 2.12). The Ti-C bonds are strongly elongated (l_{TiC} = 2.4 Å) as compared to those in B1-TiC (l_{TiC} = 2.09 Å). The Ti-N network is weakly distorted by the incorporation of the SiC interface into the TiN(111). The Si-Si and Ti-Si correlations at 2.35 Å and 2.7–2.9 Å, which are present in silicon and $TiSi_2$, respectively are absent in the (111) heterostructures (cf. Fig. 2.14). The main peak of g(θ) in Fig. 2.15 is located at ~109°, and a small maximum is observed at 120°. Accordingly, the Si-C interfacial layer in the (111) heterostructures is close to that of 3C-SiC and the distortion of the interfacial layer gives rise to a small peak around 120°. In Fig. 2.16 we compare the perfect 3C-SiC structure with a fragment of the Si-C-N network in the (111) interface. We see that three neighbor N(111), Si(111), and C(111) layers are similar to the corresponding layers in 3C-SiC when one C-layer is substituted by one N-layer. This conclusion is consistent with the structural analysis of the (111) interface done above.

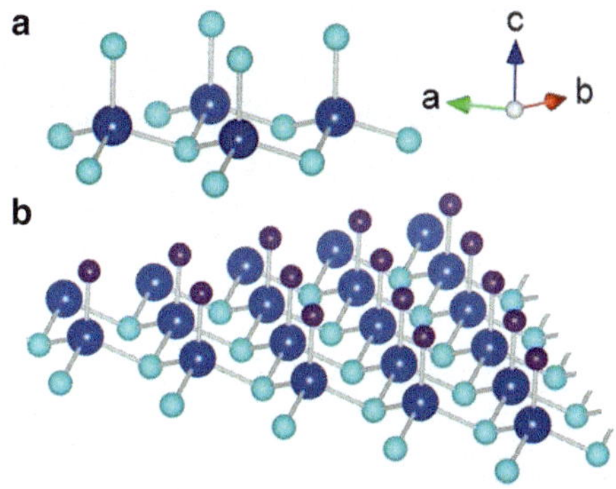

Fig. 2.16 Atomic configuration for 3C-SiC (**a**) and Si-C-N bond configurations in and around the interfacial planes for HT-SiC(111) (**b**). The atomic configurations were computed using Si-N and Si-C bond length cut-offs of 2.5 Å. The three Ti atoms that are neighboring to the C atoms and site below the C plane are not shown. Ti-*big blue-grey*, Si-*medium dark blue*, C-*small green-blue* and N-*small purple*. (Reprinted with permission from Ivashchenko et al. [36]. Copyright 2012 by the American Physical Society)

We also calculated the interfacial formation energy E^f (2.1) for the heterostructure under consideration [36]. The values E^f for the HT-TiN(001)/SiC and HT-TiN(111)/SiC heterostructures are 0.148 eV/(atom × Å^2) and 0.029 eV/(atom × Å^2), respectively, which implies that, in the TiN/SiC nanocomposite and nanolayered coatings, the formation of the TiN(111)/3C-SiC-like interfaces will be preferable already at relatively low temperature. Unfortunately, the TiN/SiC nanolayered coatings in which the TiN(111)/3C-SiC-like interfaces have been predominantly observed [31] were deposited close to room temperature and never annealed.

2.3.4 TiN(001)/BN Interfaces

As already mentioned above, two initial configurations, BN1(001) and BN2(001), were considered for this interface. Both the relaxed heterostructures L0-BN1(001) and L0-BN2(001) were very similar, although the total energy of the second heterostructure was slightly higher by 0.017 eV/atom. Both the L0-BN1(001) and HT-BN1(001) heterostructures are shown in Fig. 2.17, and the atomic configuration around the HT-BN1(001) interface is presented in Fig. 2.18. The pair correlation function (PCF) and bond angle distribution (g(Θ)) for the HT-BN1(001) and HT-BN2(001) heterostructures are presented in Fig. 2.19. An analysis of the results presented in Figs. 2.17, 2.18, and 2.19 shows that the BN interface can be considered as a set of h-BN like sheets connected by BN_3 units. All the boron atoms are threefold coordinated. The HT-BN2(001) network is more disordered than that of HT-BN1 (001) (cf. Fig. 2.19). The B-N and Ti-B correlations with the average bond lengths of

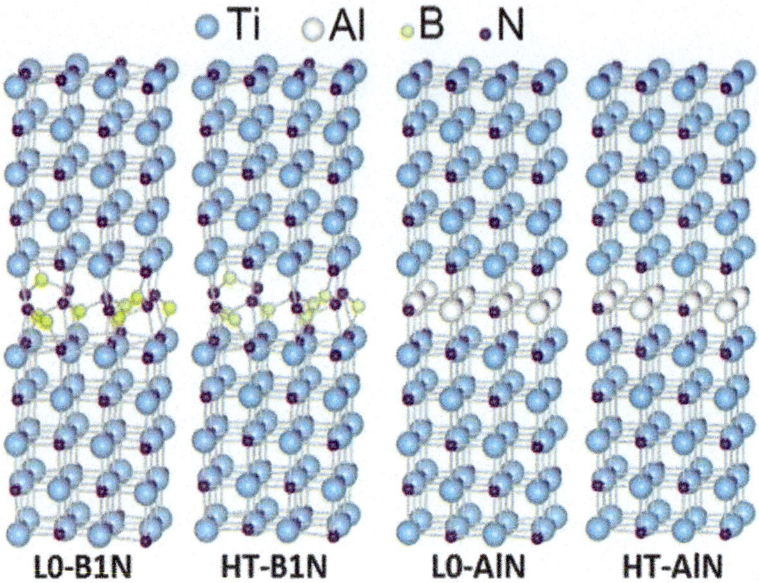

Fig. 2.17 Atomic configurations of the TiN(001)/(BN1, AlN) heterostructures at different temperatures

Fig. 2.18 B-N network at and around the HT-BN1 (001) interface). The dashes denote the distorted h-BN –like sheets

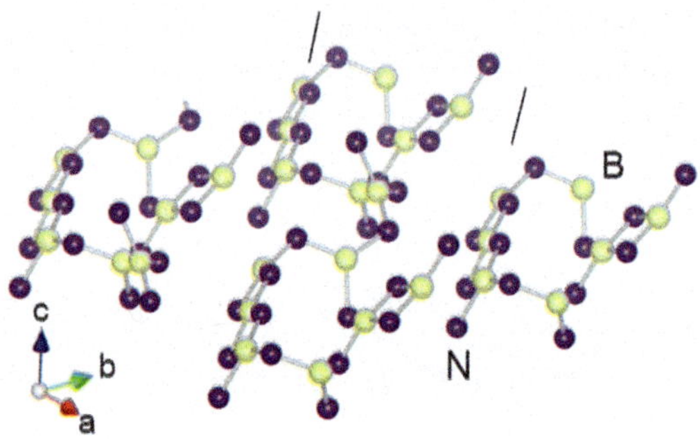

1.5 Å and 2.4 Å, respectively, are seen in Fig. 2.19. These bond lengths are very close to those in h-BN (1.45 Å), TiB_2 (2.37 Å) and TiB (2.34–2.53 Å) that were obtained in the framework of our present DFT calculations (not shown here). The peak at about 3 Å in PCF of Ti-B corresponds to a correlation between B-atoms at the bottom of the interfacial layer and Ti-atoms in the TiN slabs above the interface. These results show that the TiN(001)/BN interface consist of disordered h-BN-like units connected by NBN groups, and it is close to an amorphous state at any temperatures considered here. This conclusion is consistent with the experimental findings for TiN/BN and TiN/BNC heterostructures [39, 74] and nanocomposites [38]. In the nanocomposites, there were always some Ti-B bonds seen in X-ray photoelectron spectroscopy (XPS) even for the stoichiometric system. It has been suggested that the large mismatch

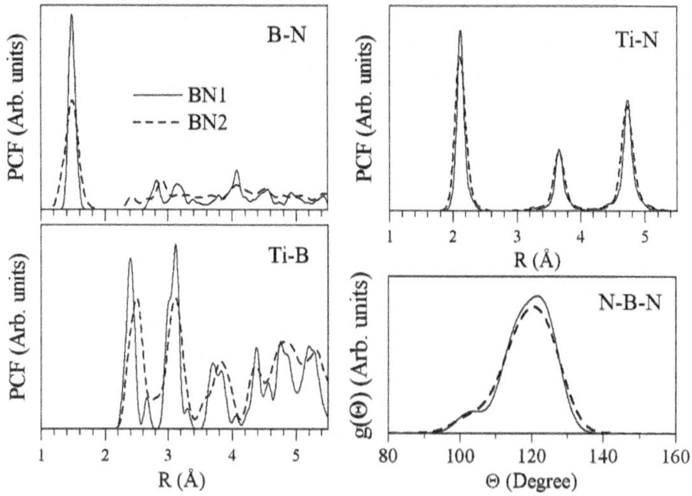

Fig. 2.19 Pair correlation function (PCF) and bond angle distribution (g(Θ)) for the HT-BN1 (001) (*solid line*) and HT-BN2(001) (*dashed line*) heterostructures

between the TiN and BN leads to incoherent interface with vacancies within the BN interfacial layer where the Ti-B bonds form (see Fig. 2.12a in [38]). This explanation is supported by the present QMD calculations.

Our results are in disagreement with the static DFT modeling of the TiN(001)/BN interface at 0 K presented in Ref. [49] that predicted that this interface should be epitaxial. According to our QMD results, this interface is not hetero-epitaxial, but it exhibits a distorted h-BN-like structure within the whole temperature range of 0–1,400 K [50]. This discrepancy is obviously due to the use of a too small supercell in the static DFT calculations reported in Ref. [49], and, therefore, it does not allow full relaxation of the lattice. Besides, the large difference in the atom size between titanium (1.47 Å) and boron (0.98 Å) clearly excludes any hetero-epitaxy in this system as discussed already in Ref. [38].

The incoherent nature of the TiN/BN interface rises the question of the mechanism that is responsible for the hardness enhancement found in the nc-TiN/BN nanocomposites, because it must be different from that found in the nc-TiN/Si_3N_4 ones, where it has been attributed to the absence of dislocation activity in 3–4 nm small TiN nanocrystals and a reduction of grain boundary shear due to the formation of pseudomorphic SiN_x interfacial layer that is strengthened by valence charge transfer from TiN [13–15, 20] . A close look at Fig. 2.7 in Ref. [38] reveals that the maximum hardness is achieved when the size of the TiN nanocrystals is around 10 nm, i.e. at the "strongest size" [73], where the strengthening due to a decrease in dislocation activity with decreasing crystallite size (Hall-Petch effect) becomes dominated by a softening caused by an increasing of grain boundary shear, as discussed recently in Ref. [11]. Thus, the fact that only a limited hardness enhancement by a factor of ≤ 2 has been found in this system [38], can be understood. The relatively high hardness of 40 GPa found in this system is a consequence of high intrinsic hardness of TiN, but not of the formation of a strong semicoherent interfacial layer.

The incoherent TiN/BN interface also explains the insensitivity of the hardness of nc-TiN/BN nanocomposites to oxygen impurities (see Fig. 2.12 in Ref. [18]), whereas the strengthened Si_3N_4-like interfacial layer is strongly sensitive to oxygen already at a level of $<$2,000 ppm ($<$ 0.2 at.%) [11, 75]. The presence of Ti-B bonds, which is another consequence of the incoherent interface in this system, together with the low melting point of boron oxide of about 460 °C also explains the low oxidation resistance of these nanocomposites reported in Ref. [37].

2.3.5 *TiN(001)/AlN Heterostructures*

In Fig. 2.17 we show the atomic configurations of the low and high-temperature TiN(001)/AlN heterostructures. As already mentioned above, the interfacial AlN layer in the TiN(001)/1 ML AlN/TiN heterostructures corresponds to a slightly distorted B1-AlN at any temperatures between 0 K and 1,400 K [50]. This is understandable in terms of small differences in atomic radii between Ti (1.47 Å) and Al (1.43 Å), and in agreement with the experimental results [53, 76]. This is also reflected by the relatively small fluctuation of bond lengths in the direction perpendicular to the interface as shown by the following sequence (the bond lengths are given in Å):

$$\mathbf{Al_I} - 2.026 - \mathbf{N} - 2.194 - \mathbf{Ti} - 2.144 - \mathbf{N} - 2.144 - \mathbf{Ti} - 2.194 - \mathbf{N} - 2.026 - \mathbf{Al_I}$$
$$\mathbf{N_I} - 2.158 - \mathbf{Ti} - 2.095 - \mathbf{N} - 2.112 - \mathbf{Ti} - 2.112 - \mathbf{N} - 2.095 - \mathbf{Ti} - 2.158 - \mathbf{N_I},$$

where Al_I and N_I are the aluminium and nitrogen atoms located at the interface between TiN and AlN interfacial layer. One notices that the Al-N bond distance within the AlN interfacial layer (i.e. parallel to the interface) is 2.098 Å.

2.4 Dynamical Stability of the Interfaces

In order to understand the origin of the thermal instability of the B1-XY layers in TiN(001)/XY heterostructures we calculated the phonon spectra of the B1-XY phases with lattice parameters corresponding to (a) B1-TiN (a_{TiN}) and (b) to the XY phase in equilibrium (a_0) [50]. Then the question is whether the B1-XY pseudomorphic phases in TiN(001)/B1-XY heterostructures are dynamically stable or unstable when they have the lattice parameter of the prototype TiN. Let us recall that "dynamical instability" means the occurrence of imaginary ("negative") frequencies in the calculated phonon spectra [77].

The calculated phonon densities of states for the BN, AlN, SiC and SiN phases with the B1-structure are shown in Fig. 2.20. The dependences of the frequencies of

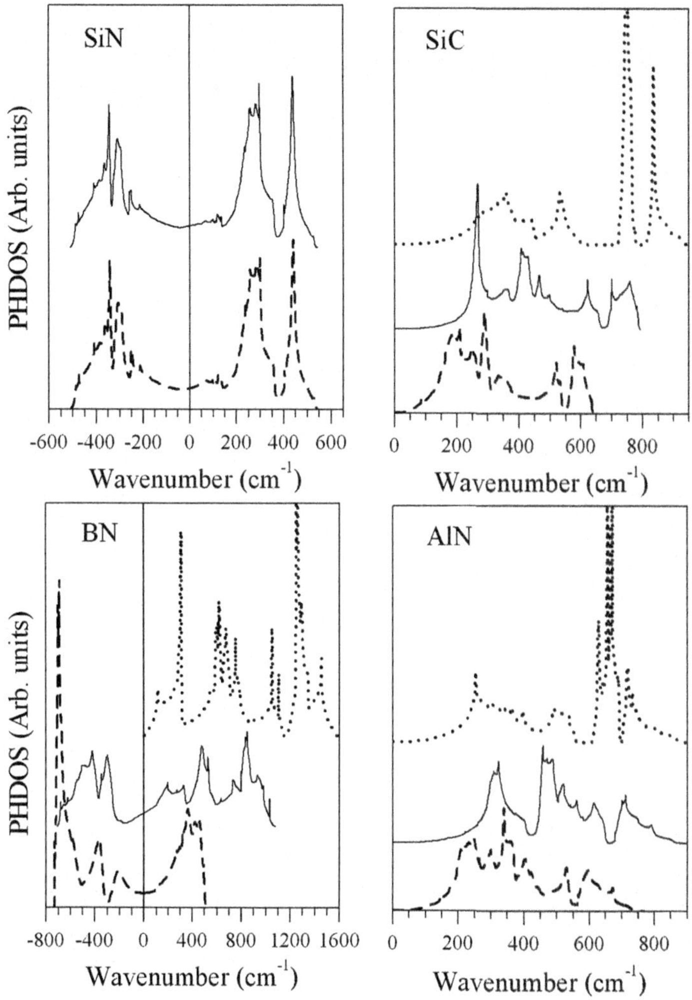

Fig. 2.20 Phonon density of states (PHDOS) for B1-BN, B1-AlN, B1-SiC and B1-SiN structures with the lattice parameters in equilibrium, $a = a_0$ (*solid line*) and with $a = a_{TiN} = 4.232$ Å (*dashed line*). For the sake of comparison, the PHDOSs for the most stable XY phases (h-BN, B4-AlN, B3-SiC) in equilibrium are presented in the corresponding figures (*dotted line*)

soft modes on the lattice parameter are shown in Fig. 2.21. We see that the PHDOS of B1-BN and B1-SiN have a wide range of negative (imaginary) frequencies which means that these phases with the B1-structures are dynamically unstable. The phonon spectra of B1-AlN and B1-SiC contain only positive frequencies, although some softening of acoustic phonon modes is observed for both phases with the lattice parameter of TiN. This softening is seen to be larger for B1-SiC than for B1-AlN. To summarize, the B1-BN and B1-SiN layers should be unstable as interfacial layers between the TiN(001), whereas the B1-AlN and B1-SiC

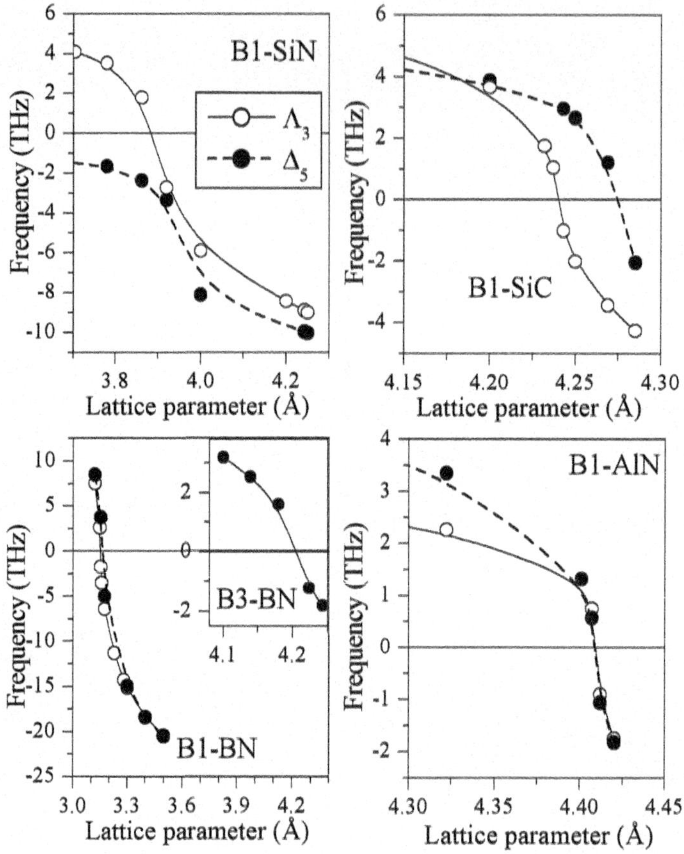

Fig. 2.21 Frequencies of soft phonon modes Λ_3 [2π/a(1/8,1/8,1/8)] and Δ_5 [2π/a(0,0,1/4)] as functions of lattice parameters for B1-XY. The results for B3-BN are presented in the insert

interfacial layers are stable and can form. We should recall that the dynamic instability of the B1-SiN(001) has been reported also by Alling et al. who used the static DFT [23].

It has been shown above that the TiN(111)/B1-SiC interface was unstable at 0 K [36], whereas the TiN(111)/B1-SiN interfacial layer was stable up to 3,000 K [30]. For the (001) interfacial layers, we found the opposite picture: B1-SiC(001) is stable at low temperature whereas B1-SiN(001) is unstable. From Fig. 2.21 we see that, for B1-SiC, the condensation of the soft Λ_3 mode, which corresponds to atomic vibrations within the (111) plane, occurs at smaller lattice parameters than the condensation of the Δ_5 mode which corresponds to atomic vibrations within the (001) plane. This means that the reconstruction of the (111) interface will occur at lower temperature than the reconstruction of the (001) interfacial layer, in agreement with earlier results [36]. For B1-SiN, the Δ_5 mode has negative frequencies for a wide range of lattice parameters, whereas the condensation of the soft Λ_3 mode occurs at a lattice parameter of about 3.88 Å and the frequencies are negative at higher values of a. The QMD calculations of the TiN(111)/B1-SiN heterostructure at various temperatures showed that the Si-N bond lengths were in the range of 1.9–1.96 Å that corresponded to a_{SiN} of about 3.8–3.92 Å in B1-SiN, and did not depend on temperature [30]. However, in this range of lattice parameters, the frequency of the Λ_3 mode is changing from negative to positive (cf. Fig. 2.21), i.e. the (111) interfacial layer should be dynamically stable, which is in agreement with the earlier finding that the TiN(111)/B1-SiN–like interface is stable even at high temperatures. According to the conventional thermodynamics at macroscale, it should undergo the reaction $4SiN \rightarrow Si_3N_4 + Si$. The possible explanation of this contradiction can be found in the need for nucleation of the new phase that requires a simultaneous motion and rearrangement of several atoms that is difficult to take place within the 1 ML thin interfacial layer. The TiN(001)/B1-SiN interfacial layer reconstructs already at low temperature because the frequencies are negative within the whole range of lattice parameter [30].

The phonon spectrum of B3-BN for $a < 4.206$ Å ($a < a_{TiN}$) does not contain any soft modes (cf. Fig. 2.21). This means that the B3-BN interface can be quite stable in TiN(001)/BN nanolayers, provided the TiN layers are sufficiently thin, e.g. 5 nm as in Ref. [40], or contain impurity atoms (V, O) which also cause a reduction in the lattice parameter of the TiN slabs. In the nc-TiN/BN nanocomposites the maximum hardness has been found for 1 ML of BN. Although the thermal stability of these nanocomposites was slightly lower (up to 900–1,000 °C) than that of the nc-TiN/Si_3N_4 ones (up to 1,100–1,200 °C), it was fairly high when considering the large lattice mismatch. The present QMD calculations provide a natural explanation.

Because the phonon frequencies are positive for lattice parameter $a \leq a_{TiN}$, the TiN(001)/1 ML AlN/TiN heterostructures are stable in agreement with the QMD results shown in Fig. 2.17. Considering the dependence of the frequencies of the soft modes on the lattice parameter shown in Fig. 2.21 we note that the B1-AlN interfacial layer should be stable also in B1-VN(001)/AlN/VN heterostructures ($a_{VN} = 4.14$ Å) since the phonon frequencies are, for this value of the lattice parameter, positive. This is in agreement with the reported experimental results in Refs. [44, 53]. However, it should be metastable in B1-NbN(001)/AlN heterostructures ($a_{NbN} = 4.4$ Å), because, for the lattice parameter of NbN the phonon frequency is approaching zero. Thus, the VN/AlN/VN heterostructures and nanocomposites are promising candidates for superhard coatings whereas NbN/AlN ones are unlikely to reach high hardness.

We note that the maximum frequency of the soft modes is reached for the X-Y bond length $R_{X\text{-}Y}$ ($R_{X\text{-}Y} = a_{XY}/2$ in the case of B1-XY) approaches that of the most stable phase of the XY system. The bond lengths of these phases are about 1.5 Å for h-BN, 1.9 Å for B4-AlN, 1.89 Å for 3C-SiC and 1.75 for β-Si_3N_4. For the interfacial layers these lengths increase approximately by 21–22 % for B1-BN and B1-SiN, and by 7 % for B1-AlN and B1-SiC. An increase in the bond length leads to a decrease in frequencies and then to a condensation of the soft modes for systems with eight valence electrons, and to an increase of the negative (imaginary) frequencies for SiN with nine valence electrons (cf. Fig. 2.21). Therefore the interfacial B1-XY layers with eight electrons can be stabilized at high pressure whereas the B1-SiN with nine electrons is dynamical unstable at any pressure.

To complete this section we should note that the suggested explanation of the thermal instability for the interfaces based on the dynamical stability of the bulk materials does not allow for the availability of Ti-N-Y chains, aligned perpendicularly to the interfacial layer in the heterostructures. Such chains cause an asymmetry in the vibration of the interfacial atoms. As a consequence, soft modes can appear. Nevertheless, despite of the fact that our model is qualitative, it accounts for the thermal stability of the interfaces under consideration in agreement with experiments evidences.

2.5 Electronic Structure and Chemical Bonding of the Calculated Heterostructures

2.5.1 Electronic Density of States of the B1-XY (XY = BN, AlN, SiC, SiN) Phases

The next issue to be addressed is the origin of condensed modes in the phonon spectra of the B1-XY phases. To answer this question we calculated the total EDOS of the B1-XY structures shown in Fig. 2.22. The boding s- and p-states can accommodate eight electrons. For β-Si_3N_4, BN, AlN and SiC which have eight valence electrons the Fermi level ($E_F = 0$ eV) lies between the bonding and anti-bonding states. However, in the case of SiN that has nine valence electrons, E_F lies in a local maximum of the density of anti-bonding states (see Fig. 2.22). This causes a strong screening of vibrations and, consequently leads to a softening of phonon modes.

In summary, the thermal stability of the B1-XY interfaces correlates well with the dynamic stability of the B1-XY phases. Judging from the EDOSs presented in Fig. 2.22, one can see that dynamical instability of the interfaces cannot be assigned to the peculiarities of the electronic density of states, but is caused by the large differences between the X-Y bond lengths in the most stable configurations and in the heterostructures. Besides, the origin of the dynamical instability in B1-SiN can be attributed to the high electronic density of state within the anti-bonding states [23].

2.5.2 Electronic Density of States and Chemical Bonding of the Heterostructures

Given the peculiarities of the electronic density of states of B1-XY phases, let us analyze the densities of states related to the (001) interfaces. We calculated the total and local density of states associated with TiN slabs and (001) interfaces.

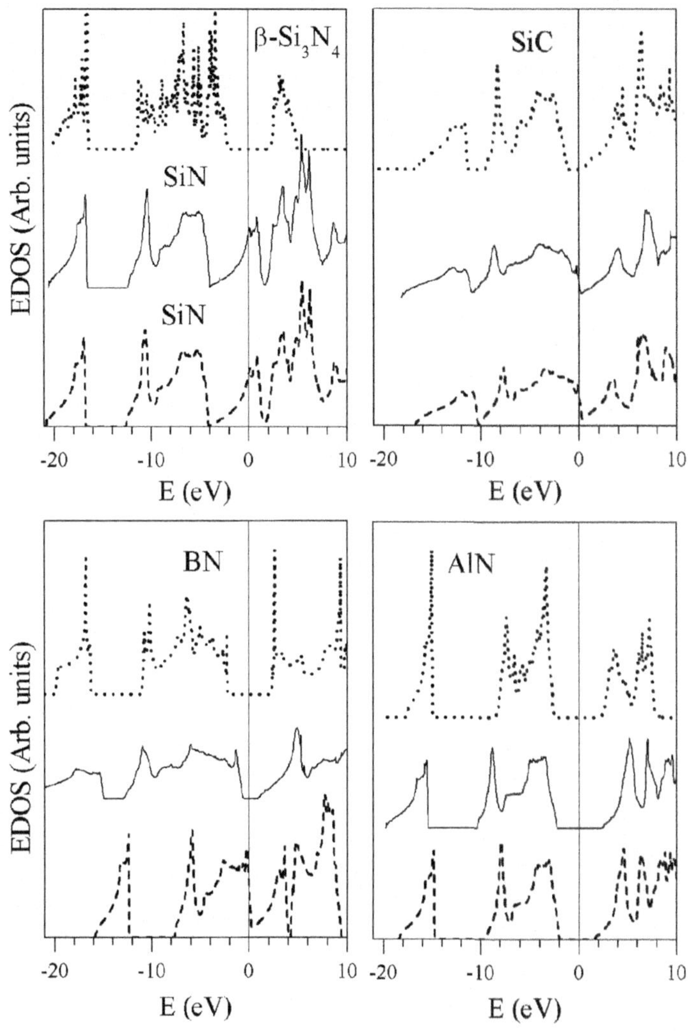

Fig. 2.22 The total electronic density of states (EDOS) of the B1-XY phases (XY = SiN, SiC, BN, AlN) with the lattice parameters at equilibrium, a = a_0 (*solid line*) and with a = a_{TiN} = 4.232 Å (*dashed line*). For the sake of comparison, the EDOS of β-Si_3N_4, B3-SiC, h-BN and B4-AlN are presented in the corresponding figures (*dotted line*). The *vertical lines* denote the Fermi level

In Fig. 2.23a–d we show only the local EDOS of the interfaces, since the TiN-based EDOS changes insignificantly in comparison with the EDOS of the bulk TiN. We note only that the TiN-related EDOSs of the heterostructures under investigations differ from the EDOS of the bulk TiN by additional small peaks at the low- and high-energy slopes of the main peak located around −5 eV (cf. Fig. 2.23d). These peaks originate mostly from the N-atoms of the TiN layers adjacent to the XY interfaces.

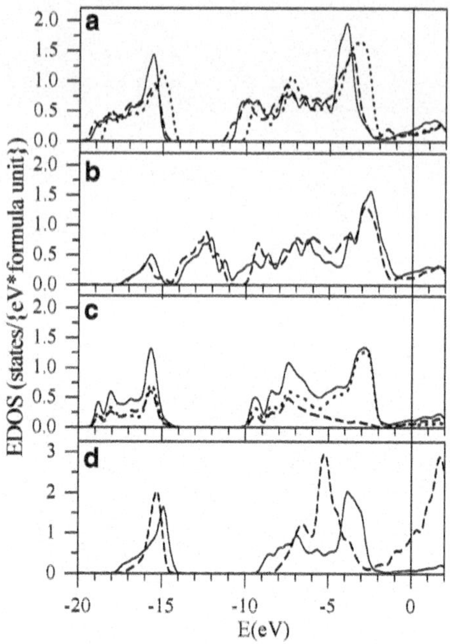

Fig. 2.23 Local density of states (EDOS) related to the (001) interfaces in: LT-SiN (*solid line*), HT-SiN (*dashed line*) and HT-$Si_{0.75}N$ (*dotted line*) (**a**); LT-SiC (*solid line*), HT-SiC (*dashed line*) (**b**); HT-BN (*solid line*), B-partial EDOS in HT-BN (*dashed line*) and N-partial DOS in HT-BN (*dotted line*) (**c**); HT-AlN (*solid line*) (**d**). For a comparison, the EDOS of the bulk TiN is also shown (*dashed line*) (**d**). The *vertical lines* denote the Fermi level

Figure 2.23 shows that, for all the interfaces, the energy gap forms around the Fermi energy. However, gap states are present owing to the defect structure of the interfaces. Since these gap states are localized, interfaces should exhibit conductivity typical of semiconductors. The energy gaps widen and gap states decrease after the thermally-induced structural transformations due to the formation of stable atomic configurations at the interfaces (tetrahedral XY_4 for SiC, Si_3N_4 and AlN, and h-BN-like BN_3 for BN). In the previous section we assumed that the dynamical instability of B1-SiN could be caused by the high EDOS at the Fermi level. The latter results show that the EDOS related to the L0-SiN(001) interface does not show any peaks around the Fermi energy. Therefore we expect that the main cause for the instability of the B1-SiN(001) interface should be found in the large difference between the Si-N bond lengths in the Si_3N_4 structure and TiN(001)/B1-SiN heterostructure. Nevertheless, we do not exclude that, for the multi-layered B1-SiN(001) interfaces, when the interface structure approaches the structure of the bulk B1-SiN, the high EDOS at the Fermi level can be the main cause of dynamical instability for such interfaces.

One can see from Fig. 2.23 that the formation of Si vacancies on the B1-SiN (001) interface leads to a decrease in the number of gap states and to a shift in the bands toward high energies. An analysis of the local EDOSs related to interfaces (we present here only the B- and N-related local EDOS for the H-BN(001)) shows that the nitrogen atoms form mostly the peak just below the Fermi level (cf. Fig. 2.23c).

To better illustrate the charge distribution in the HT heterostructures, we calculated the Löwdin charges by means of the Löwdin population analysis implemented in the "Quantum ESPRESSO" code [55]. (The Löwdin charges reflect the changes in the valence charge density on the atoms upon formation of chemical bond. For example, upon the formation of an Si-N bond where the valence charge density moves from silicon to the more electronegative nitrogen, the charge on Si decreases and that on N increases: the larger the charge density changes the stronger the bond. For more details about the Löwdin population analysis and charges see Ref. [78]). Because the plane-wave basis is not orthogonal, the Löwdin symmetric orthogonalization scheme was applied to convert it back to the orthogonal one [79]. The Löwdin charges (Q_i) (as well as other conventional atomic charges) do not satisfy the sum rule [55]. Therefore, the analysis of a charge state on the same ion in different structures should be done in a comparative way for similar bond configurations.

We calculated the charges Q_i for the ions localized on the (001) interfaces, the TiN layers adjacent to the interfaces and in the centre of the TiN slabs. For the TiN adjacent layers and slabs, we found $Q_{Ti} \sim 2.8$ and 2.9, respectively, $Q_N \sim 6.0$. For the (001) interfaces the following charges were obtained: $Q_{Si} \sim 2.76$, $Q_N \sim 5.94$ (HT-SiN); $Q_{Si} \sim 2.97$, $Q_C \sim 5.06$ (HT-SiC); $Q_B \sim 2.74$, $Q_N \sim 5.61$ (HT-BN); $Q_{Al} \sim 2.13$, $Q_N \sim 5.97$ (HT-AlN); For the bulk TiN, $Q_{Ti} = 2.85$, $Q_N = 6.02$. We recall that the B and Al atoms have a similar configuration of valence orbitals (ns and np) occupied by three valence electrons. However, on the (001) interface, the ionicity of the B-N bonds is lower compared with that of the Al-N bonds. Also, the positive charge of the Si ions is higher in the HT-SiN(001) than in the HT-SiC(001) interfacial layer. Since the Löwdin charges of the Ti ions in the TiN layers located above and below the interfaces are less than those within the TiN slabs and bulk TiN, one understands the weakening of the Ti-N bonds close to the and XY layers.

Let us now consider the charge states in the Ti-Si-N systems in more details. The calculated Löwdin charges localized on the Si and N ions are: $Q_{Si} = 2.50$, $Q_N = 6.02$ in β-Si_3N_4. For the HT-SiN(001) interface, the calculated average charges are: Q_{Si} ~2.73 in the Si-N_4 configurations with short bond lengths, $Q_{Si} \sim 2.77$ in the octahedral configurations and $Q_{Si} \sim 3.14$ in the SiN_4 units with long bond lengths, whereas Q_{Ti} ~2.8, Q_N ~5.94 for all configurations. For the HT-SiN(111) interface, Q_{Si} ~2.75, Q_{Ti} ~2.97 and Q_N ~5.99. A comparison of the charge distributions in both heterostructures shows that two factors determine the charges localized on Si atoms: the higher the coordination by nitrogen and shorter the Si-N bond, the lower the Q_{Si}. For a given coordination of silicon to nitrogen the following simple rule applies: the shorter the average bond length is, the lower Q_{Si} and stronger the Si-N bonds are. It is also consistent with an increase in the average charge Q_{Si} up to about 3.0 in a-SiN in comparison with the corresponding charge of Q_{Si} ~2.5 in β-Si_3N_4 caused by a reduction in the average coordination number from 4 to 3 and the Si-N bond length. Q_N decreases insignificantly from 6.0 in β-Si_3N_4 to 5.94 in a-SiN. A low charge on Si means a high stability of that SiN_x configuration. The most stable configuration is the

SiN_4-like one as in β-Si_3N_4. One can see that, whereas the Löwdin charges on Ti and N are similar in TiN and the HT heterostructures, the charges on silicon in the heterostructures are higher than in β-Si_3N_4, thus indicating that the Si-N bonds in these heterostructures are weaker than in β-Si_3N_4, in agreement with the results of Hao et al. [16, 17]. However, because the Löwdin charge on Si in a-SiN of about $Q_{Si} \sim 3$ is higher than those on most of the atoms Si in the interfaces, the interfacial SiN_x layers are strengthened as compared to a-SiN in agreement with the results of the static DFT calculations at 0 K of Zhang et al. [13–15].

2.5.3 Valence Charge Density Distribution Around the TiN/SiN Interfaces

In this section we investigate the distribution of the valence charge in the TiN, β-Si_3N_4 as well as in the HT-B1-SiN(001) and HT- B1-SiN(111) heterostructures to clarify a possible origin of the lattice distortions and stability of these heterostructures. These heterostructures were selected because the silicon atoms combine six- and four-fold coordination. Figure 2.24a, b show the atomic structure and the valence charge density (VCD) distribution in the b-c planes of the HT-SiN(001) and HT-SiN(111) heterostructures. Obviously, not all atoms are within the same plane (cf. also Fig. 2.1), as seen *e.g.*, in Fig. 2.24a by the different apparent diameters of the atoms within that plane. For the HT-SiN(111) heterostructure, we have chosen the b-c plane consisting of the Si, N and Ti atoms that are located in this plane. In the unreconstructed epitaxial heterostructures, the Si-N bonds are relatively weak because of their large lengths. However, at finite temperatures the atoms are allowed to move and the distances between Si and N atoms decrease so that strong Si-N covalent bonds form. This is reflected in the appearance of a charge polarization towards N atoms (cf. Fig. 2.24a, b). The shifted N atoms interact strongly with Si and to a lesser extent with Ti atoms. Such a redistribution of the valence charge around the interfaces leads to the elongation and weakening of the Ti-N bonds near the interfacial SiN layer as discussed above.

The high stability of the (111) interface in comparison with the (001) one is related to the fact that only Si-atoms are present within the (111) interfacial plane, whereas both Si and N atoms are within the (001) one. In the latter case, the two different atoms with significantly different electronegativities (1.8 for Si and 3 for N, both larger than that of Ti of 1.5 [72]) cause a "phase shift" of the Friedel-like oscillations in neighbor (110) planes perpendicular to that interface and running either through Si or through N atoms within that interface, as discussed by Zhang et al. From the results of their DFT calculations at 0 K [14, 15]. These fairly regular "phase shifts" obtained by DFT at 0 K appear in the present first-principles QMD calculations as an almost random distortion of the Si and N atom positions, re-distribution of valence charge and resulting bond breaking within and near the interfacial layer, as seen in Figs. 2.1 and 2.24a. In the case of the HT-SiN(111) heterostructures, N atoms should

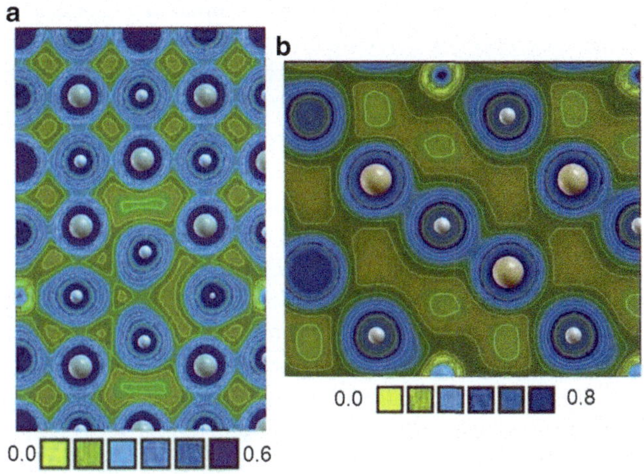

Fig. 2.24 Valence charge density distributions for HT-SiN(001) (**a**) and HT-SiN(111) (**b**) heterostructures in the **b–c** plane. Numerals denote the valence charge density scales in electrons/Bohr^3. The large balls are Ti, middle balls Si, and small balls N atoms. The cut-off distance of the Si-N and Ti-N bonds is 2.5 Å

move towards the Si interface to form stable SiN_4 configurations. However, the strong Ti-N bonds nearby prevent this motion so that the distorted SiN_6 coordination is preserved and the N atoms are only slightly shifted towards the interfacial Si layer (see Fig. 2.24b).

2.6 Conclusion

We performed first-principles quantum molecular dynamics calculations of the TiN (001)/1 ML B1-derived XY (XY = SiN, SiC, BN, AlN), TiN(001)/1 ML B1-derived $Si_{0.75}$ N, TiN(111)/1 ML B1-derived SiN, TiN(111)/1 ML B1-derived $Si_{0.77}$ N, TiN(111)/1 ML Si_3N_4-derived SiN, TiN(111)/1 ML Si_3N_4-derived Si_2N_3 and TiN(111)/1 ML B1-derived SiC heterostructures at various temperatures. It has been shown that the B1-SiN(001) containing hetero-epitaxial layers were preserved during structural optimization at 0 K in agreement with the earlier "static" DFT calculations of Zhang *et al.* However, the B1-SiN(001) interfacial layer underwent significant reconstruction already at 10 K, and when equilibrated at 300 K, it reconstructed to an over-coordinated a-SiN-like structure. After heating it up to 1,400 K and subsequent static relaxation, the number of the Si_3N_4–like units increased. The interface consists of the octahedral SiN_6 and distorted tetrahedral SiN_4–like units aligned along the (110) direction. In contrast, the B1-SiN(111) interface was stable at 300 K and after heating to 1,400 K and 3,000 K and subsequent relaxation, it only showed small distortions of the atomic positions from their fully symmetric ones and preserved local coordination.

A formation of Si vacancies at zero and finite temperatures in the (001) and (111) interfacial B1-SiN layers is energetically favored. Both interfaces are unstable with respect to formation of Si vacancies due to a reduction of the coordination of the N atoms and Si atoms (in the case of the (001) interface) at the interfaces, a shortening of the Si-N bond length in the SiN_6 units, and a smaller distortion of the SiN_6 units.

An estimate of the interface formation energy shows that, among all the interfaces under consideration, the TiN(001)/B1-derived $Si_{0.75}$ N interface with the six- and four-fold coordinated Si atoms, as well the TiN(111)/Si_3N_4-derived interfaces composed of tetrahedrally N-coordinated Si atoms are most stable.

According to our results obtained at finite temperatures, the 1 ML B1-SiN(001) interface can be stable at high temperatures under nitrogen-poor conditions. We also showed that for both N-poor and N-rich conditions, the B1-derived $Si_{0.75}$ N interface between the TiN(001) slabs is most favorable, whereas the Si_3N_4-derived interfaces are most stable between the TiN(111) slabs. The TiN (001)/B1-derived Si_xN_y interfaces cannot exist as hetero-epitaxial layers because they consist of distorted octahedral SiN_6 and tetrahedral SiN_4 units aligned along the (110) direction. Furthermore, despite the fact that according to our QMD calculations, the TiN(111)/B1-derived Si_xN_y interfaces were not favoured under any conditions, they are stable even at high temperatures. This stability may be of kinetic origin because decomposition of SiN into Si_3N_4 and Si requires a nucleation of new phases that is difficult to occur within one monolayer of SiN inserted between TiN slabs.

An analysis of the generated TiN(001)/SiC and TiN(111)/SiC heterostructures at various temperatures shows that: The B1-SiN(001) interface hetero-epitaxial layers are preserved up to 600 K; Above 600 K, the B1-SiC –like interfacial layer in the TiN(001)/SiC heterostructures transforms into the 3C-SiC-like one. This interfacial layer consists of two distorted Si(111) and C(111) polar layers in 3C-SiC uniaxially deformed in the (111) direction; The interface in the TiN(111)/SiC heterostructure is stable in the temperature range of 0–1,400 K. Three neighboring N(111), Si(111) and C(111) layers in the TiN(111)/SiC heterostructure are identified as the corresponding layers in 3C-SiC, provided one C-layer is substituted by one N-layer. The formation of the TiN(111)/3C-SiC-like interfaces is more energetically favorable than the TiN(001)/3C-SiC-like interfaces in the temperature range of 0–1,400 K.

Our studies reveal that the TiN(001)/BN interface forms an amorphous-like h-BN layer consisting of BN_3 units at any temperatures. For TiN(001)/AlN heterostructures, the epitaxially stabilized B1-AlN interfacial layer is preserved over all the whole temperature range.

The thermal stability of the interfaces under consideration has been studied in terms of the dynamics stability of the B1-XY phases that form the interfaces with different lattice parameters a. The phonon calculations showed that B1-BN and B1-SiN interfacial layers are dynamically unstable in a wide range of lattice parameters, particularly close to that of TiN (a_{TiN}). For this reason these interface are drastically modified already during the static relaxation or at 0 K. The hetero-epitaxially stabilized B1-AlN interfacial layer is dynamically stable up to high temperatures because the phonon spectra of the bulk B1-AlN do not exhibit any soft modes for lattice parameters up to a = 4.4 Å. In the case of the B1-SiC interfacial layer, the condensation of the phonon modes occurs for lattice parameters that are slightly above a_{TiN}. Therefore this structure becomes dynamically unstable at high temperatures, and it transforms into the 3C-SiC one above 600 K. It was further shown that B1-BN and B1-SiC interfacial layers could be stabilized at low value of a ("high pressure"), whereas the B1-SiN interfacial layer was dynamically unstable at any pressure due to the high density of states at the Fermi level.

In the electronic densities of states (EDOS) related to the (001) interfaces, the energy gap forms around the Fermi energy. However, gap states are present owing to the defect structure of the interfaces. Since these gap states are localized, interfaces should exhibit a conductivity typical of semiconductors. The gap states decrease after the thermally-induced structural transformations due to the formation of stable atomic configurations at the interfaces (tetrahedral XY_4 for SiC, SiN and AlN, and graphite-like XY_3 for BN). These results point to the fact that the instability of the B1-SiN(001) and B1-BN(001) interfaces is not related to the specific EDOS, but is caused by the large difference between the X-Y bond lengths in the most stable structures and the corresponding heterostructures.

The calculated Löwdin charges (Q_i) for the TiN(001)/XY heterostructures indicate that the N ions on the SiN-, BN- and AlN-based interfaces have lower charges than those in the TiN slabs. Also, the Löwdin charges of the Ti ions in the TiN layers located above and below the interfacial layers are lower compared to Ti-charges inside the TiN slabs. Such charge distribution is supposed to lead to a decrease in the cohesion strength between the TiN and XY layers.

The results of the present theoretical study are consistent with the available experimental data on the corresponding heterostructures and nanocomposite coatings. They provide deeper insight into the choice of new systems that promise to form superhard heterostructures and nanocomposites, and help to identify other systems that are less suitable. We hope that this study will be useful to those who are conducting experimental work in the design of advanced nano-layered heterostructures and of nanocomposite coatings.

Acknowledgments This work was supported by the STCU contract, No. 5539. The work of P. T. was performed under the auspices of the U. S. Department of Energy by the Lawrence Livermore National Laboratory under contract No. DE-AC52-07NA27344. The authors are grateful to the directorate of the Summery Institute at Jackson State University for financial support and the possibility to perform large-scale calculations. We also thank to Dr. M.G.J. Veprek-Heijman for critical reading of the manuscript and many useful comments.

References

1. Veprek S (1999) J Vac Sci Technol A 89:2401
2. Veprek S, Niederhofer A, Moto K, Bolom T, Männling H-D, Nesladek P, Dollinger G, Bergmaier A (2000) Surf Coat Technol 133-134:152
3. Veprek S, Veprek-Heijman MGJ, Karvankova P, Prochazka J (2005) Thin Solid Films 476:1

4. Veprek S, Zhang RF, Veprek-Heijman MGJ, Sheng SH, Argon AS (2010) Surf Coat Technol 204:1898
5. Veprek S, Reiprich S (1995) Thin Solid Films 268:64
6. Söderberg H, Odén M, Molina-Aldareguia JM, Hultman L (2005) J Appl Phys 97:114327
7. Chen YH, Lee KW, Chiou W-A, Chung YW, Keer LM (2001) Surf Coat Technol 146-147:209
8. Hultman L, Bareño J, Flink A, Söderberg H, Larsson K, Petrova V, Odén M, Greene JE, Petrov I (2007) Phys Rev B 75:155437
9. Flink A, Beckers M, Sjölén J, Larsson T, Braun S, Karlsson L, Hultman L (2009) J Mater Res 24:2483
10. Söderberg H, Odén M, Flink A, Brich J, Persson POA, Beckers M, Hultman L (2007) J Mater Res 22:3255
11. Veprek S, Veprek-Heijman MGJ (2012) Thin Solid Films 522:274
12. Prilliman SG, Clark SM, Alivisatos AP, Karvankova P, Veprek S (2006) Mater Sci Eng A 437:379
13. Veprek S, Argon AS, Zhang RF (2007) Philos Mag Lett 87:955
14. Zhang RF, Argon AS, Veprek S (2009) Phys Rev Lett 102:015503
15. Zhang RF, Argon AS, Veprek S (2009) Phys Rev B 79:245426
16. Hao S, Delley B, Veprek S, Stampfl C (2006) Phys Rev Lett 97:086102
17. Hao S, Delley B, Stampfl C (2006) Phys Rev B 74:035402
18. Veprek S (2011) J Nanosci Nanotechnol 11:1
19. Christiansen S, Albrecht M, Strunk P, Veprek S (1998) J Vac Sci Technol B 16:19
20. Zhang RF, Argon AS, Veprek S (2010) Phys Rev B 81:245418
21. Iwamoto C, Tanaka S-I (1998) J Am Ceram Soc 81:363
22. Hao S, Delley B, Stampfl C (2006) Phys Rev B 74:035424
23. Alling B, Isaev EI, Flink A, Hultman L, Abrikosov IA (2008) Phys Rev B 78:132103
24. Marten T, Isaev EI, Alling B, Hultman L, Abrikosov IA (2010) Phys Rev B 81:212102
25. Marten T, Alling B, Isaev EI, Lind H, Tasnádi F, Hultman L, Abrikosov IA (2012) Phys Rev B 85:104106
26. Zhang RF, Veprek S (2006) Mater Sci Eng A 424:128
27. Zhang RF, Veprek S (2008) Thin Solid Films 516:2264
28. Zhang RF, Veprek S (2007) Phys Rev B 76:174105
29. Veprek S, Veprek-Heijman MGJ (2007) Surf Coat Technol 201:6064
30. Ivashchenko VI, Veprek S, Turchi PEA, Shevchenko VI (2012) Phys Rev B 85:195403
31. Kong M, Dai J, Lao J, Li G (2007) Appl Surf Sci 253:4734
32. Rogl P, Schuster JC (1992) Phase diagrams of ternary boron nitride and silicon nitride systems. ASM International, The Materials Information Society, Materials Park
33. Toth LE (1971) Transition metal carbides and nitrides. Academic, New York
34. Veprek S, Haussmann M, Reiprich S, Shizhi L, Dian J (1996) Surf Coat Technol 86-87:394
35. Chase MW, Davies CA, Downey JR, Frurip DJ, McDonald RA, Syverud AN (1985) JANAF thermochemical tables, 3rd ed., J Phys Chem Ref Data 14 (1856 pages)
36. Ivashchenko VI, Veprek S, Turchi PEA, Shevchenko VI (2012) Phys Rev B 86:014110
37. Karvankova P, Veprek-Heijman MGJ, Zawrah MF, Veprek S (2004) Thin Solid Films 467:133
38. Karvankova P, Veprek-Heijman MGJ, Azinovic D, Veprek S (2006) Surf Coat Technol 200:2978
39. Mendez JM, Qu BD, Evstigneev M, Prince RH (2000) J Appl Phys 87:1235
40. Lee SH, Nam KH, Lim J-W, Lee J-J (2003) Surf Coat Technol 174–175:758
41. Holleck H, Schier V (1995) Surf Coat Technol 76–77:328
42. Helmersson U, Todorova S, Barnett SA, Sundgren J-E, Markert LC, Greene JE (1987) J Appl Phys 62:481
43. Shinn M, Hultman L, Barnett SA (1992) J Mater Res 7:901
44. Mirkarimi PB, Hultman L, Barnett SA (1990) Appl Phys Lett 57:2654
45. Sproul WD (1996) Science 273:889

46. Barnett SA (1993) Physics of thin films. In: Francombe MH, Vossen JL (eds) Mechanic and dielectric properties, vol 17. Academic Press, Boston
47. Barnett SA, Madan A (1998) Phys World 11:45
48. Barnett SA, Madan A, Kim I, Martin K (2003) MRS Bull 28:169
49. Wang BJ, Xie F, Jia K (2011) Appl Mech Mater 110–116:1020
50. Ivashchenko VI, Veprek S (2013) Thin Solid Films 545:391
51. Ziebert C, Ulrich S (2006) J Vac Sci Technol A 24:554
52. Madan A, Kim IW, Cheng SC, Yashar P, Dravid VP, Barnett SA (1997) Appl Phys Lett 78:1743
53. Kim JW, Li Q, Marks LD, Barnett SA (2001) Appl Phys Lett 78:892
54. Mayrhofer PH, Hörling A, Karlsson L, Sjölen J, Larsson T, Mitterer C, Hultman L (2003) Appl Phys Lett 83:2049
55. Giannozzi P, Baroni S, Bonini N, Calandra M, Car R, Cavazzoni C, Ceresoli D, Chiarotti GL, Cococcioni M, Dabo I, Dal Corso A, Fabris S, Fratesi G, de Gironcoli S, Gebauer R, Gerstmann U, Gougoussis C, Kokalj A, Lazzeri M, Martin-Samos L, Marzari N, Mauri F, Mazzarello R, Paolini S, Pasquarello A, Paulatto L, Sbraccia C, Scandolo S, Sclauzero G, Seitsonen AP, Smogunov A, Umari P, Wentzcovitch RM (2009) J Phys: Condens Matter 21:395502
56. Perdew JP, Burke K, Ernzerhof M (1996) Phys Rev Lett 77:3865
57. Vanderbilt D (1990) Phys Rev B 41:7892
58. Billeter SR, Curioni A, Andreoni W (2003) Comput Mater Sci 27:437
59. Monkhorst HJ, Pack JD (1976) Phys Rev B 13:5188
60. Wang S, Gudipati R, Rao AS, Bostelmann TJ (2007) Appl Phys Lett 91:081916
61. Baroni S, De Gironcoli S, Dal Corso A, Gianozzii P (2001) Rev Mod Phys 73:515
62. Isaev EI, Simak SI, Abrikosov IA, Ahuja R, Vekilov KY, Katsnelson MI, Lichtenstein AI, Johansson B (2007) J Appl Phys 101:123519
63. Topor L, Kleppa OJ (1986) Metallurg Mater Trans A 17:1217
64. Cai Y, Zhang L, Zeng Q, Cheng L, Xu Y (2007) Solid State Commun 141:262
65. Yu B-R, Zeng Z-Y, Guo HZ, Chen X-R (2007) Commun Theor Phys 48:925
66. Saib S, Bouarissa N (2005) Eur Phys J B 47:379
67. Giacomazzi L, Umari P (2009) Phys Rev B 80:144201
68. Aiyama T, Fukunaga T, Niihara N, Hirai T, Suzuki K (1079) J Non-Cryst Solids 33:131
69. Guisbiers G, Ganguli D (eds) (2010) Size effects in metals, semiconductors and inorganic compounds. Trans Tech Publications, Stafa-Zurich
70. Roduner E (2006) Nanoscopic materials. RSC Publishing, Cambridge
71. Veprek S, Iqbal Z, Sarott F-A (1982) Philos Mag 45:137
72. Greenwood NN, Earnshaw A (1984) Chemistry of elements, Pergamon, Oxford [Quoted after the German translation Chemie der Elemente, 2nd. ed. (VCH Verlagsgesellschaft, Weinheim, 1990)]
73. Argon AS, Yip S (2006) Philos Mag Lett 86:713
74. Chen X, Wang Z, Ma S, Ji V (2010) Diam Relat Mater 19:1336
75. Veprek S, Karvankova P, Veprek-Heijman MGJ (2005) J Vac Sci Technol B 23:L17
76. Stueber M, Holleck H, Leiste H, Seemann K, Ulrich S, Ziebert C (2009) J Alloy Comp 483:321
77. Dove MT (1993) Introduction to lattice dynamics. Cambridge University Press, Cambridge
78. Szabo A, Ostlund NS (1989) Modern quantum chemistry. McGraw-Hill, New York
79. Löwdin PO (1950) J Chem Phys 18:365

Chapter 3
Investigating Complex Surface Phenomena Using Density Functional Theory

Raghani Pushpa

Abstract Surfaces of some materials exhibit vastly different structure than the bulk-truncated atomic structure due to the reduced surface atomic coordination that leads to surface dangling bonds and/or surface stress. In order to relieve the stress, partial dislocation networks are formed on some surfaces, changing the surface periodicity to tens or hundreds of nanometers. By using Pt(111) and Ag/Ag/Pt(111) surface reconstructions as examples, we demonstrate how such large length scale pattern formation can be investigated using "classical" models parameterized within density functional theory. We then present an example of determining the magnetic state of ultrathin films on semimetallic substrates. It is possible that such magnetic films also get reconstructed due to the mismatch in lattice spacing between the film and the substrate. Although the pattern formation on magnetic films can be studied using similar type of models, determining the magnetic state of such superstructures requires further modeling efforts.

3.1 Introduction

For many applications such as catalysis and microelectronic devices, surface structure plays a crucial role. A surface can be thought of having been created by a cleavage process or by a growth process on a bulk material. Due to the reduced atomic coordination, the surface can have dangling bonds [1], and/or it can be under tensile or compressive stress [2]. To relieve this strain, the surface structure can change in two ways: (i) surface atoms can move vertically while retaining the surface periodicity, known as surface relaxation; (ii) surface atoms can move horizontally in the surface layer, changing the surface periodicity, known as the surface reconstruction.

R. Pushpa (✉)
Department of Physics, Boise State University, 1910 University Dr., Boise, ID, USA
e-mail: pushparaghani@boisestate.edu

J. Leszczynski and M.K. Shukla (eds.), *Practical Aspects of Computational Chemistry III*, DOI 10.1007/978-1-4899-7445-7_3,

In the surface reconstruction process, some surface atoms end up at positions away from the minima of the bulk potential, and surface dislocations are formed. Many such dislocations combine together and form dislocation networks. Such dislocation networks are observed on bulk truncated homoepitaxial surfaces [3, 4] as well as on the epitaxial layers of one material deposited onto the surface of another, known as the heteroepitaxial surfaces [5, 6]. Pattern formation on surfaces has attracted considerable interest due to the possibility of creating self-assembled, highly ordered nanostructure superlattices [5], as the periodicity of such dislocation patterns can be tens or hundreds of nanometers. A well-known example of the homoepitaxial surface reconstruction is the (7×7) reconstruction of the Si(111) surface [7]. The reconstructed surface has a very different atomic geometry compared to the bulk truncated one, leading to a change in the surface conductivity compared to that of the bulk [8]. Similarly, many other properties of surfaces such as electronic structure [8], surface magnetism [9], and conductivity [10] can be significantly different from those of the bulk.

Density functional theory (DFT)-based *ab initio* methods have proven to be the most reliable theoretical tools to investigate structural properties [11]. However, large length scales involved in the surface reconstruction make it virtually impossible to study such systems by exclusively using DFT-based methods due to the high computational cost involved. Although there have been a few attempts along these lines that use huge computational resources [12, 13], it is not a widespread practice. For these reasons, pattern formation in strained epitaxial layers is often investigated using various “classical” models; one such example is the Frenkel-Kontorova (FK) model that has been widely used [14, 15]. In this chapter, we will present two examples of how the FK model combined with DFT can be used to predict dislocation patterns on a homoepitaxial Pt(111) surface and in heteroepitaxial Ag films grown on the Pt(111) surface.

However, if the films deposited on a substrate are composed of magnetic atoms, then accurately determining the magnetic state of the film using classical models could be challenging and one needs to rely only on the DFT calculations for the accurate description of the magnetic state. As a consequence, this imposes a limit on the cell sizes that can be studied and restricts us from studying magnetic properties of reconstructed surfaces at large length scales.

In this chapter, we first provide a brief introduction to the density functional theory and the FK model that has been used to investigate the reconstruction of homoepitaxial and heteroepitaxial surfaces – Pt(111) and Ag/Ag/Pt(111), respectively. Later, we present our results on dislocation patterns formed on these surfaces and compare them with experimental observations. In the following section, we provide an example of determining the magnetic state of Ti monolayers deposited on the Cu_2N/Cu(100) substrate without taking into account the reconstruction of the monolayer that might be present in experiments due to the lattice-mismatch between the film and the substrate. We conclude with a brief discussion about the models and future research challenges in the field.

3.2 Methodology

The primary driving force for the reconstruction of metallic surfaces, both homoepitaxial and heteroepitaxial, is the surface stress caused by the difference in the equilibrium lattice parameter of the surface from that of the bulk substrate. To relieve this stress, surface atoms move away from their bulk-truncated positions. This movement is, however, opposed by the fact that it costs energy when surface atoms lose registry with the substrate. Competition between the two processes of relieving stress and losing registry with the substrate can give rise to long-range superstructures on the surface. The presence or absence of such superstructures involves a rather delicate balance between the various contributions to the total energy. Hence, we calculate these factors from DFT and incorporate them into the FK model to predict the geometry of minimum energy superstructures on surfaces.

3.2.1 Density Functional Theory

The quantum behavior of electrons is described by the Schrödinger equation (SE). The number of electrons in a material is generally so high that the SE cannot be solved exactly. There are two primary approaches used to solve the SE: electron wav-function based approach and electron-density based approach. DFT is an electron density based approach where the many-electron problem is mapped to a single-electron problem and electron density (instead of the many electron wave function) is used to solve the SE. This approach decreases the complexity of the problem to a great extent. As a result, DFT has been successfully used to calculate a variety of structural, electronic, thermal, magnetic, and other properties of metals, semiconductors, and insulators from first principles. Although the complexity of the many-body problem has been greatly reduced in DFT, even with the present-day super computers, it is still difficult to study large systems (thousands of atoms in a supercell) or complex phenomena by carrying out only *ab initio* calculations. Hence, we adopt the strategy of formulating a model whose parameters are obtained from DFT for studying phenomena at large length scales. Besides computational feasibility, modeling also provides an insight into the operative mechanisms of a certain phenomenon.

3.2.2 Frenkel-Kontorova Model

Frenkel-Kontorova model in its original form [16], consists of a one-dimensional (1D) chain of atoms connected by harmonic springs of equilibrium length b, sitting in a sinusoidal substrate potential of periodicity a. To study a realistic 2D surface such as the Pt(111) surface, this model is generalized to two dimensions, with a two-dimensional (2D) layer of atoms interacting via a more realistic anharmonic potential V_{film} and sitting in 2D potential due to bulk atoms V_S. The Hamiltonian would then be given by

$$H = \sum_i V_S(r_i) + \sum_{i,j} V_{film}(|r_i - r_j|) + \mu N, \tag{3.1}$$

where, $r_i = (x_i,y_i)$ is a two-dimensional position vector of the ith atom in the surface layer. The third term in this equation corresponds to the energy cost of adding extra atoms to the surface. N is the total number of atoms in the surface supercell, and μ is the chemical potential of extra atoms getting incorporated into the surface; its value depends upon the nature of the reservoir, which could be bulk, step edge, adatom, etc.

There are certain limitations of the FK model in the present form. For example, it does not account for: (i) the exchange of atoms that may be present at an interface of heteroepitaxial surfaces [17], (ii) the specific surface states that could exist on metallic surfaces and can modify the interactions between adatoms [18], (iii) the charge transfer that could take place at the interface between a film and a substrate [19], and (iv) the change in the film potential with the change in the coordination number of the film. Nonetheless, this model has proven to be instructive for studying the domain pattern formation. A similar approach had been used to study the reconstruction of Au(111) [20], Cu/Ru(0001) [15], and Ag/Pt(111) [21], and the results were found to match the experimental findings.

The potentials V_{film}, V_S and μ are obtained from *ab initio* DFT calculations by studying the energetics of compressing the surface layer, making surface stacking faults, and extracting atoms from various sites. These calculations are done on a much smaller surface cell of size ~ 10 × 10Å by using Quantum-Espresso [22] code. We used a plane-wave basis with a cut-off of 20 Ry, ultrasoft pseudopotentials, and the local-density approximation to the exchange-correlation potential. Surface calculations have been carried out using supercells with nine layers of atoms, separated by a vacuum of seven-layer thickness. A Monkhorst-Pack grid corresponding to 27 **k** points in the irreducible part of the surface Brillouin zone is used to sample reciprocal space for calculations using a (1 × 1) surface cell; the grid is varied commensurately when using larger surface cells. As a preliminary check, lattice constants, bulk moduli, and cohesive energies of Pt are found to be in agreement with the experimental values. Details can be found in Ref. [23].

3.3 Surface Reconstruction

Once the extended FK Hamiltonian has been determined from DFT calculations, one can obtain the positions of the surface Pt or Ag atoms that minimize the total energy, i.e., surface equilibrium configuration. Global optimization of a potential-energy surface (PES) with a large number of local minima is a highly nontrivial task. We use a conjugate-gradient technique for the energy minimization for the Pt(111) and a quantum simulated annealing for the Ag/Ag/Pt(111).

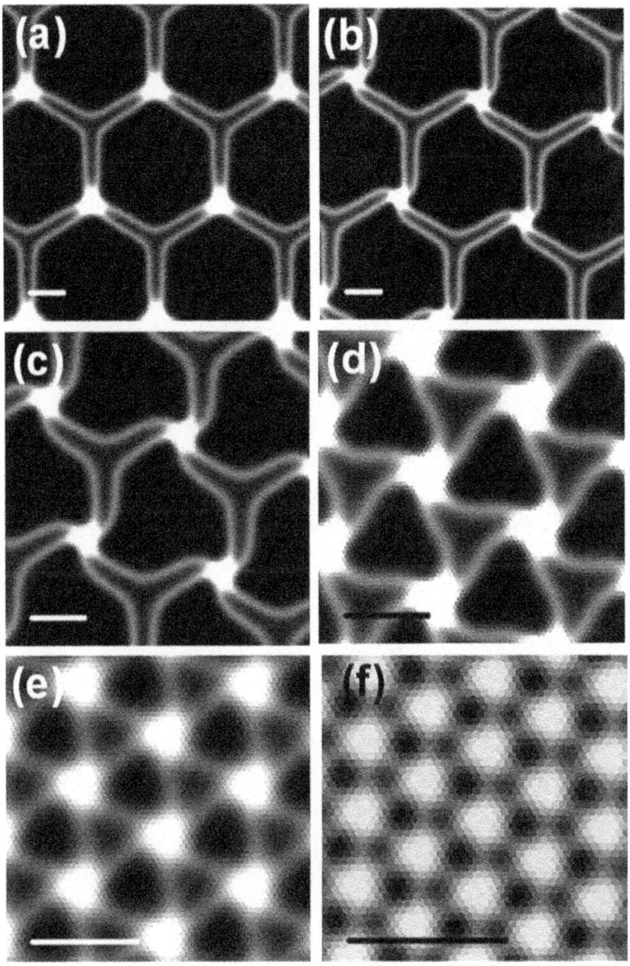

Fig. 3.1 Variation of dislocation patterns on the Pt(111) surface as the surface density increases from (**a**) 2.9 % to (**f**) 21 %. Atoms are shaded according their height; *black*, *dark gray*, *light gray*, and *white* areas correspond to regions where the surface atoms sit at fcc, hcp, bridge, and top sites, respectively. The *black/white lines* in each image correspond to a length of 50 Å [23]

3.3.1 Reconstruction of Pt(111)

We find that the Pt(111) surface is under tensile stress and isotropically compressed surface structures are lower in energy than bulk-truncated (unreconstructed) surface structures. The reconstructed surface structures as a function of the surface layer density are shown in Fig. 3.1. These are simulated STM (Scanning Tunneling Microscope) images obtained by using surface corrugation and 2D Fourier expansion. For the Pt(111), we also include an additional degree of freedom in the FK model, which is the rotation of the surface layer with respect to the substrate.

In these images, black, dark gray, light gray, and white areas correspond to regions where the surface atoms sit at face-centered cubic (fcc), hexagonal close pack (hcp), bridge (points between fcc and hcp), and top sites, respectively. Figure 3.1a shows the honeycomb reconstruction of the surface when surface density increases by 2.9 %. Large hexagonal fcc domains are separated from narrow hcp domains by domain walls that form a honeycomb network. Alternate vertices of the hexagons are extremely bright; atoms here sit at top sites. When we rotate the surface with respect to the substrate, keeping the surface density fixed, the surface energy is found to further decrease at a very small rotation angle of 0.14°, and the bright vertices in Fig. 3.1a are converted to bright rotors, which are indeed observed in experiments [24, 25]. The similarity between Fig. 3.1b and the experimental STM images [24, 25] suggests that the surface layer indeed gets locally rotated with respect to the substrate. It is not known if the electronic structure of the surface changes significantly with the rotation of the surface. To understand this, one would need to use DFT alone to calculate the electronic structure of the reconstructed surface, which can be a formidable task to compute. At the same time, it would be very interesting to look at the effect of such rotors on the catalytic activity of the surface, which again requires DFT calculations. The catalytic activity of such reconstructed surfaces can also be analyzed experimentally. As the surface layer becomes denser, the honeycomb transforms into a pattern of interlocking wavy triangles, shown in Fig. 3.1c, d, for surface density increments of 4.0 % and 6.8 %, respectively. In experiments, one can change the surface density by controlling the chemical potential of the Pt atoms. Such structures have been observed on dense terraces of Pt(111) [25]. Na/Au(111) [26] and 3 monolayers (ML) of Cu on Ru(0001) [6] were also found to have such surface reconstructions. As the chemical potential is lowered and the surface density is increased still further, the wavy domain walls straighten out, resulting in the "bright star" pattern shown in Fig. 3.1e, for a 10.2 % increment in surface density. This pattern has not been observed on the Pt(111) surface as it requires a very low chemical potential and high adatom coverage; however, it is been observed on Ni/Ru(0001) [27], presumably due to a large mismatch in the lattice spacing between Ni and Ru.

The triangular domains of the bright star transform into hexagons, resulting in the Moiré pattern shown in Fig. 3.1f, when the density in the surface layer is increased by 21 %. The density distribution on the surface is now almost uniform. This pattern is also observed experimentally, e.g., for 4 ML of Cu/Ru(0001) [6]. Thus, by using FK type model, we find that the domain patterns on these surfaces depend on both the surface density and rotation angle. A similar sequence of dislocation patterns was found on a heteroepitaxial Cu/Ru(0001) surface as a function of the number of overlayers using FK model [15]. In our work, dislocation patterns on the homoepitaxial Pt(111) surface change as a function of the chemical potential, surface density, and rotation angle of the surface with respect to the substrate.

3.3.2 Reconstruction of Ag on Ag/Pt

Reconstruction of the Ag monolayer on the Ag/Pt(111) surface was studied using a vertically extended FK model, yet another extension of the FK model with respect to the one used for the Pt(111) surface. The total energy of the system was defined by the Eq. (3.1) without the chemical potential term in it. The position vector r_i becomes a 3D vector of the ith Ag atom instead of the 2D vector as in the case of the Pt(111). The sum over pairs $\langle i,j \rangle$ is extended over neighbors within a certain cutoff distance, which is determined so as to include the ring of first nearest-neighbor (NN) atoms. The vertical extension of the substrate potential [28] allows for stress relaxation via small vertical displacements along the z-direction; it was obtained by calculating the hcp, bridge, and top-site stacking-fault energies as a function of the distance between the surface layer and the substrate. In contrast, these stacking fault energies for the Pt(111) surface were calculated at a fixed distance from the substrate.

Unlike in the case of the Pt(111) surface reconstruction where the external stimulus for inducing the reconstruction was the chemical potential of the Pt atoms that can get incorporated into the surface and drive the reconstruction [24], here the external stimulus is temperature that is found to induce the reconstruction of the surface in experiments [29]. Therefore, in this case, we use a variant of the simulated thermal annealing, which is one of the most popular nonbiased general-purpose optimization methods, in which thermal fluctuations allow the system to escape metastable states in order to achieve the real global minimum. A recent innovation is that of *simulated quantum annealing*, in which quantum fluctuations interact with thermal fluctuations in this task [30]. Replica-pinned quantum annealing (RPQA) is a technical variant that has provided the most unbiased results to date for the determination of the global minimum-energy structures of Lennard-Jones clusters [31]. Our work was the first time that RPQA was applied to study the minimum energy configurations of a heteroepitaxial surface in Ref. [32]. RPQA thoroughly explores the area around the best replica so far, thus removing defects faster than other methods. As in any other minimization technique, it is impossible to be completely sure to have achieved the true global minimum, but after an exhaustive search, a large degree of confidence can be assumed.

After calculating the energy of the unreconstructed and reconstructed states of various configurations, we found that the striped and triangular configurations (shown in Figs. 3.2 and 3.3, respectively) are lower in energy than the unreconstructed state. In the striped configuration, stress is relieved only along one direction. Conversely, in the triangular configuration, stress is relieved in an isotropic manner, creating lines of defects following the symmetry axes of the substrate, as it is shown in Fig. 3.3. In order to allow the relaxation of the surface bonds, some atoms end up close to bridge and top sites. The density of this configuration decreases by 7.8 % with respect to the unreconstructed case. This is an example of an expansive reconstruction as opposed to the compressive reconstruction of the Pt(111) surface where the surface density increases upon reconstruction.

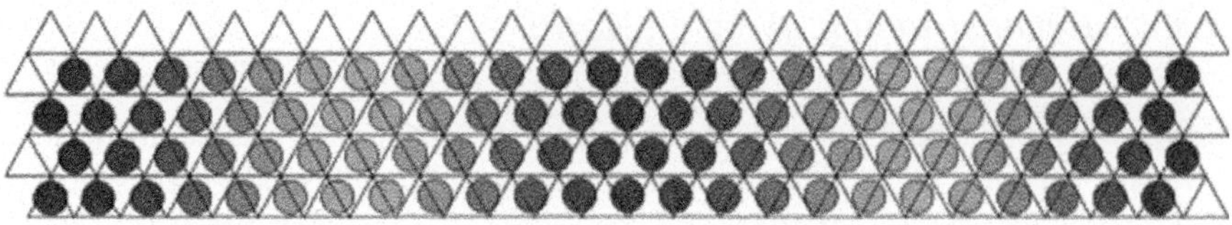

Fig. 3.2 Top view of the uniaxial strain-relief pattern of Ag on the Ag/Pt(111) surface. Up- and down-facing triangles represent fcc and hcp sites, respectively. Atoms are shaded according to their height. Periodic length of the pattern is ~71 Å [32]

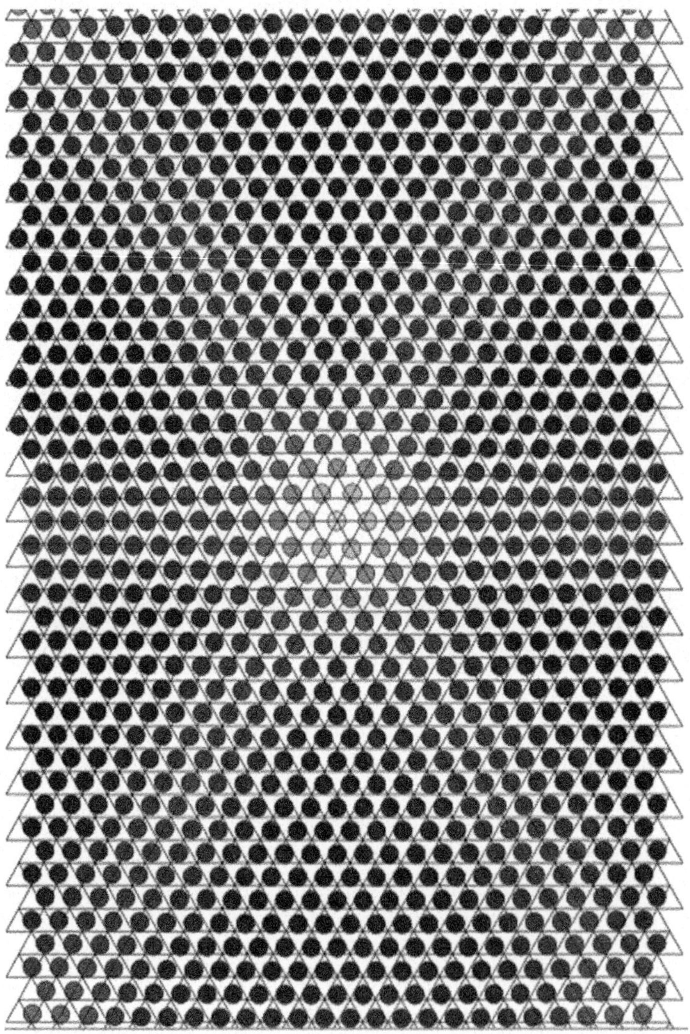

Fig. 3.3 Triangular dislocation network of Ag/Ag/Pt(111) surface corresponding to a 7.8 % decrease in the surface density. Atoms are shaded according to their height [32]

The striped configuration of unidirectional strain-relief was indeed observed in STM experiments at 340 K by Brune et al. [29], whereas a trigonal network configuration was observed at 800 K. It is known that, under annealing, the system can find the global energy minimum of the PES. Thus, trigonal configuration could be the global minimum whereas the striped configuration, a long-lived metastable state. The reasons for its long lifetime could be a large energy barrier and a small energy difference between the two configurations. In the trigonal network configuration however, atoms do not sit at bridge or top sites as in the case of the triangular configuration found in our simulations. Upon using the TN structure as the starting configuration in the energy minimization algorithm, the system relaxes back to the triangular structure. This could mean that we may need to improve the FK model with respect to its limitations described earlier. The likely assumption to fail here is the independence of the film potential from the coordination number. The lower coordination of a site in a metallic film will increase its binding strength with the neighbors. This effect has been accounted for in more complex models, such as the glue model [33].

3.4 Magnetism of Ti Films on Cu_2N/Cu(100) Surface

A contrasting work that we are going to discuss is determining the magnetic state of thin films deposited on substrates. In this project, we looked at the monolayer of Ti deposited on a Cu_2N/Cu(100) surface; a single Cu_2N layer was used to separate the Ti monolayer from the conduction electrons of the Cu substrate, so that the spin magnetic moments of Ti are not Kondo screened by the conduction electrons of the metal substrate [34–36]. In Fig. 3.4, we show the top view of the Cu_2N/Cu(100) surface. As shown in the figure, along the x-axis, a hollow site separates two surface Cu atoms, and along the y-axis they are separated by a N atom. These directions will be referred to as hollow-axis and N-axis, respectively. Ti atoms are found to have the highest adsorption energy on top of Cu atoms. At one-monolayer coverage, Ti forms a square "lattice" on the surface; to determine its magnetic state, we use a five-layer symmetric slab of Cu_2N/Cu(100) with a Ti monolayer on the top and bottom of the slab.

The exchange coupling (J) is calculated using the Ising model ($H = J\, S_1 \cdot S_2$); where J can be equated to the energy difference between the ferromagnetic and antiferromagnetic configurations as,

$$2S^2 J = E_{\uparrow\uparrow} - E_{\uparrow\downarrow}.$$

S is the magnitude of spins on Ti atoms and J is the exchange coupling. $E_{\uparrow\uparrow}$ and $E_{\uparrow\downarrow}$ are the total energies calculated from DFT when spins are aligned and antialigned on the nearest neighbor Ti atoms. The energy of the ferromagnetic configuration ($E_{\uparrow\uparrow}$) in the above equation will be the total energy of the (1×1) unit cell shown by black lines in Fig. 3.4. To obtain $E_{\uparrow\downarrow}$, we design three

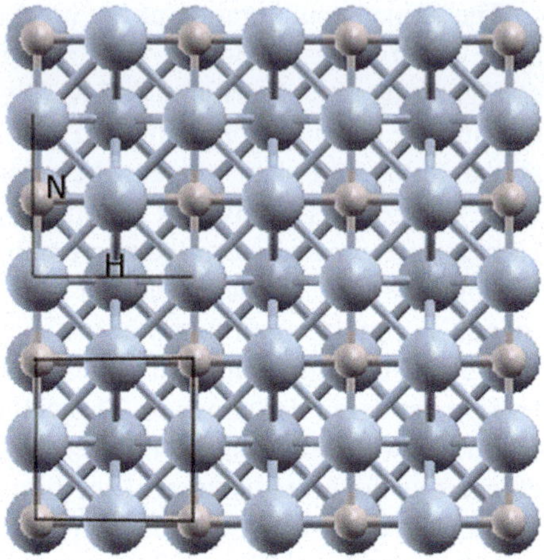

Fig. 3.4 Top view of the $Cu_2N/Cu(100)$ surface, *small pink* and *large gray* spheres represent N and Cu atoms, respectively. The square drawn by *black lines* is the surface unit cell of size 3.7 Å [36]

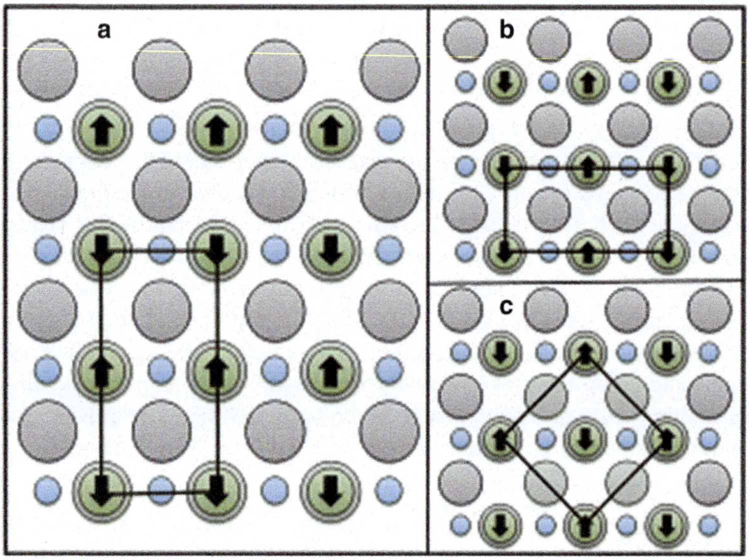

Fig. 3.5 Schematic diagram showing spin configurations in Ti monolayer deposited on the $Cu_2N/Cu(100)$ surface. *Circles* with *arrow* signs are Ti atoms on top of Cu atoms; *gray* and *blue circles* are the Cu and N atoms, respectively [36]

different configurations with (1 × 2), (2 × 1), and $\left(\sqrt{2} \times \sqrt{2}\right)$ unit cells shown in Fig. 3.5a–c, respectively. Arrow signs in the figure indicate the relative direction of spins on Ti atoms. In Fig. 3.5a, spins are aligned along the N-axis and antialigned along the hollow-axis, and in Fig. 3.5b spins are antialigned along the N-axis and aligned along the hollow-axis. Configuration Fig. 3.5c is a checkerboard configuration with spins antialigned along both the N- and hollow-axes. Subtracting the energy of the ferromagnetic configuration from the energy of the above configurations gives the exchange coupling of Ti atoms along the hollow-axis, the N-axis, and in the checkerboard configurations, respectively, assuming that there are only nearest neighbor interactions.

Our results indicate that the exchange coupling between Ti atoms across the N-atom is ferromagnetic, i.e., $E_{\uparrow\uparrow}$ is lower than $E_{\uparrow\downarrow}$ (along N-axis) by 16.1 meV. However, the exchange coupling across a hollow site is antiferromagnetic, i.e., the energy of $E_{\uparrow\downarrow}$ (along the hollow-axis) is lower than $E_{\uparrow\uparrow}$ by 106.8 meV. Thus, the antiferromagnetic coupling along the hollow-axis is much stronger than the ferromagnetic coupling along the N-axis. The checkerboard pattern (Fig. 3.5c) is more favored over the pure ferromagnetic state with an energy difference of 77.8 meV; however, it is less favorable than the hollow-axis antiferromagnetism (Fig. 3.5a), presumably due to the energy disadvantage of having antialigned spins along the N-axis. Thus, the configuration in Fig. 3.5a is found to be the ground state, and we term it as "spin-striped" state. These ferromagnetic stripes should be observable in experiments in large enough lattices.

Notice that while determining the magnetic state of the film, we did not include any reconstruction of the Ti film that might occur due to the lattice mismatch between the film and the substrate. To date, it's not known whether such films will reconstruct.

3.5 Discussion and Future Work

In summary, we have shown two diverse applications of DFT: one to study phenomena at large length scales with DFT-parameterized approximate models and the other to study phenomena at short length scales using DFT exclusively. In the former case, the accuracy is traded for the efficiency, and in the later case, the efficiency is traded for the accuracy, inline with the nature of the problem. Surface dislocation pattern formation is an example of large length scale surface phenomena. Both Pt (111) and Ag/Ag/Pt(111) surfaces display large length scale pattern formation under different conditions. The type of patterns formed may impact the catalytic activity of these surfaces, which requires further investigation. Such surface structures have also been used to grow regular arrays of magnetic nanoparticles or of semiconducting materials, which can be utilized for various device applications [5, 37, 38].

The diverse set of dislocation patterns on Pt(111) [23, 39], Ag/Ag/Pt(111) [32], and Cu/Ru(0001) [15] obtained using FK model schematically agree well with those obtained in STM experiments. It may seem that similar mechanisms are responsible for various domain pattern formation; however, recently [17] it was observed that an exchange of atoms takes place at the interface of the Ag film and the Pt substrate, causing a gradual transition in lattice constant between the two elements. It was also suggested that the dislocations of the trigonal network are buried in the Pt interface layer instead of being exclusively in the top layer. In another paper [18], it was suggested that long-range interactions mediated by surface states become anisotropic on strained surfaces, and these anisotropic interactions facilitate the formation of dislocation networks in heteroepitaxial overlayers. To include such an interface mixing and the surface states, one would require quantum-mechanical calculations, as the bonds are being broken and reformed in the mixing process and the origin of surface states is quantum mechanical in nature. This requires either going beyond the FK model or effectively including these effects in the present form of the model, which may be challenging given the quantum mechanical nature of the phenomena.

We also presented an example of determining the magnetic state of thin-films on metallic or semimetallic substrates. It is possible that these magnetic films get reconstructed due to the mismatch in lattice spacing between the film and the substrate as in the case of Ag on Pt(111). Although superstructure formation in the magnetic films can be studied using similar type of models, determining the magnetic state of such superstructures requires further modeling efforts. Developing models to study the magnetic state of reconstructed homoepitaxial or heteroepitaxial films can be very instructive. The magnetic ground state of such systems is expected to be non-trivial due to the coexistence of strained, partially strained and strain-relieved regions on the same surface. These regions of varying strain could have drastically different magnetic properties; it is known that the non-magnetic surfaces can become when strained [40]. Strain is also expected to influence the exchange coupling among magnetic atoms.

Acknowledgements I gratefully acknowledge the financial support from the Research Corporation Cottrell college science award and National Science Foundation CAREER award (DMR-1255584). I wish to acknowledge Shobhana Narasimhan, Barbara Jones, Javier Rodriguez-Laguna, Silvia N. Santalla, and Jesus Cruz for collaborating on these projects.

References

1. Srivastava GP (1997) Theory of semiconductor surface reconstruction. Rep Prog Phys 60:561–613
2. Needs RJ, Mansfield M (1989) Calculations of the surface stress tensor and surface-energy of the (111) surfaces of iridium, platinum and gold. J Phys Condens Matter 1:7555–7563
3. Barth JV, Brune H, Ertl G, Behm RJ (1990) Scanning tunneling microscopy observations on the reconstructed Au(111) surface: atomic structure, long-range superstructure, rotational domains, and surface defects. Phys Rev B 42:9307–9318
4. Sandy AR, Mochrie SGJ, Zehner DM, Grübel G, Huang KG, Gibbs D (1992) Reconstruction of the Pt(111) surface. Phys Rev Lett 68:2192–2195
5. Brune H, Giovannini M, Bromann K, Kern K (1998) Self-organized growth of nanostructure arrays on strain-relief patterns. Nature 394:451–453
6. Günther C, Vrijmoeth J, Hwang RQ, Behm RJ (1995) Strain relaxation in hexagonally close-packed metal-metal interfaces. Phys Rev Lett 74:754–757
7. Binnig G, Rohrer H, Gerber C, Weibel E (1983) 7 × 7 Reconstruction on Si(111) resolved in real space. Phys Rev Lett 50:120–123
8. Uhrberg RIG, Kaurila T, Chao YC (1998) Low-temperature photoemission study of the surface electronic structure of Si(111)7 × 7. Phys Rev B 58:R1730–R1733

9. Chen LJ, Wu RQ, Kioussis N, Blanco JR (1997) First principles investigations of 4d magnetism on C(0001). J Appl Phys 81:4161–4163
10. Heike S, Watanabe S, Wada Y, Hashizume T (1998) Electron conduction through surface states of the Si(111)-(7 × 7) surface. Phys Rev Lett 81:890–893
11. Pushpa R, Narasimhan S, Waghmare U (2004) Symmetries, vibrational instabilities, and routes to stable structures of clusters of Al, Sn, and As. J Chem Phys 121:5211–5220
12. Brommer KD, Needels M, Larson BE, Joannopoulos JD (1992) Abinitio theory of the Si (111)-(7 × 7) surface reconstruction – a challenge for massively parallel computation. Phys Rev Lett 68:1355–1358
13. Wang Y, Hush NS, Reimers JR (2007) Simulation of the Au(111)-(22 x root 3) surface reconstruction. Phys Rev B 75:233416
14. Narasimhan S, Vanderbilt D (1992) Elastic stress domains and the herringbone reconstruction on Au(111). Phys Rev Lett 69:1564–1567
15. Hamilton JC, Foiles SM (1995) Misfit dislocation structure for close-packed metal-metal interfaces. Phys Rev Lett 75:882–885
16. Frenkel J, Kontorowa T (1938) Physikalische Zeitschrift der Sowjetunion 13:1–10
17. Aït-Mansour K, Brune H, Passerone D, Schmid M, Xiao W, Ruffieux P et al (2012) Interface-confined mixing and buried partial dislocations for Ag bilayer on Pt(111). Phys Rev B 86:085404
18. Howe JD, Bhopale P, Tiwary Y, Fichthorn KA (2010) Patterns in strained-layer heteroepitaxy: beyond the Frenkel-Kontorova model. Phys Rev B 81:121410
19. Altieri S, Tjeng LH, Sawatzky GA (2001) Ultrathin oxide films on metals: new physics and new chemistry? Thin Solid Films 400:9–15
20. Takeuchi N, Chan CT, Ho KM (1991) Au(111): a theoretical study of the surface reconstruction and the surface electronic structure. Phys Rev B 43:13899–13906
21. Hamilton JC, Stumpf R, Bromann K, Giovannini M, Kern K, Brune H (1999) Dislocation structures of submonolayer films near the commensurate-incommensurate phase transition: Ag on Pt(111). Phys Rev Lett 82:4488–4491
22. Giannozzi P, Baroni S, Bonini N, Calandra M, Car R, Cavazzoni C et al (2009) QUANTUM ESPRESSO: a modular and open-source software project for quantum simulations of materials. J Phys Condens Matter 21:395502
23. Pushpa R, Narasimhan S (2003) Reconstruction of Pt(111) and domain patterns on close-packed metal surfaces. Phys Rev B 67:205418
24. Bott M, Hohage M, Michely T, Comsa G (1993) Pt(111) reconstruction induced by enhanced Pt gas-phase chemical potential. Phys Rev Lett 70:1489–1492
25. Hohage M, Michely T, Comsa G (1995) Pt(111) network reconstruction: structure, growth and decay. Surf Sci 337:249–267
26. Barth JV, Behm RJ, Ertl G (1994) Mesoscopic structural transformations of the Au(111) surface-induced by alkali-metal adsorption. Surf Sci 302:L319–L324
27. Meyer JA, Schmid P, Behm RJ (1995) Effect of layer-dependent adatom mobilities in heteroepitaxial metal film growth: Ni/Ru(0001). Phys Rev Lett 74:3864–3867
28. Rodríguez-Laguna J, Santalla SN (2005) Vertically extended Frenkel-Kontorova model: a real space renormalization group study. Phys Rev B 72:125412
29. Brune H, Röder H, Boragno C, Kern K (1994) Strain relief at hexagonal-close-packed interfaces. Phys Rev B 49:2997–3000
30. Lee YH, Berne BJ (2000) Global optimization: quantum thermal annealing with path integral Monte Carlo. J Phys Chem A 104:86–95
31. Gregor T, Car R (2005) Minimization of the potential energy surface of Lennard-Jones clusters by quantum optimization. Chem Phys Lett 412:125–130
32. Pushpa R, Rodríguez-Laguna J, Santalla SN (2009) Reconstruction of the second layer of Ag on Pt(111): extended Frenkel-Kontorova model. Phys Rev B 79:085409
33. Ercolessi F, Tosatti E, Parrinello M (1986) Au (100) surface reconstruction. Phys Rev Lett 57:719–722

34. Hirjibehedin CF, Lin CY, Otte AF, Ternes M, Lutz CP, Jones BA et al (2007) Large magnetic anisotropy of a single atomic spin embedded in a surface molecular network. Science 317:1199–1203
35. Hirjibehedin CF, Lutz CP, Heinrich AJ (2006) Spin coupling in engineered atomic structures. Science 312:1021–1024
36. Pushpa R, Cruz J, Jones B (2011) Spin and exchange coupling for Ti embedded in a surface dipolar network. Phys Rev B 84:075422
37. Voigtländer B, Meyer G, Amer NM (1991) Epitaxial growth of thin magnetic cobalt films on Au(111) studied by scanning tunneling microscopy. Phys Rev B 44:10354–10357
38. Chambliss DD, Wilson RJ, Chiang S (1991) Nucleation of ordered Ni Island arrays on Au(111) by surface-lattice dislocations. Phys Rev Lett 66:1721–1724
39. Pushpa R, Narasimhan S (2003) Double stripe reconstruction of the Pt(111) surface. Bull Mater Sci 26:91–96
40. Pushpa R, Ghosh P, Narasimhan S, de Gironcoli S (2009) Effective coordination as a predictor of adsorption energies: a model study of NO on Rh(100) and Rh/MgO(100) surfaces. Phys Rev B 79:233406

Chapter 4
Growth Mechanism, Energetics and CO Affinities of Vanadium Doped Gold Clusters, Au_nV with n = 1−20

Pham Vu Nhat, Jerzy Leszczynski, and Minh Tho Nguyen

Abstract We report a comprehensive review on the geometric, electronic and energetic properties and the CO adsorption on bimetallic clusters Au_nV in the range of $n = 1-20$, obtained using density functional theory computations (BB95 and B3LYP functionals in conjunction with the pseudo-potential cc-pVDZ-PP basis set for metals and the full-electron cc-pVTZ basis set for non-metals). The effects of the vanadium dopant on the properties of gold clusters are analyzed in detail.

4.1 Introduction

Contrary to the peculiar inertness of its bulk, gold exhibits either in the nanoscale form or when finely dispersed on metal oxide surfaces an enhanced catalytic activity for many gas-phase reactions such as CO oxidation, propylene epoxidation, NO reduction, water-gas shift and methanol synthesis [1–4] etc.. . . Gold clusters in addition present with potential applications in devices with dimension of the order of nanometers [5]. Owing to good biocompatibility, facile synthesis, and conjugation to a variety of bimolecular systems, gold nano-particles are found to exhibit a prospective use in biochemical sensing and detection [6]. It could be argued that

P.V. Nhat
Department of Chemistry, Can Tho University, Can Tho, Vietnam

J. Leszczynski (✉)
Department of Chemistry and Biochemistry, Interdisciplinary Center for Nanotoxicity, Jackson State University, Jackson, MS 39217, USA
e-mail: jerzy@icnanotox.org

M.T. Nguyen (✉)
Department of Chemistry, University of Leuven, B-3001 Leuven, Belgium

Institute for Computational Science and Technology (ICST), Ho Chi Minh City, Vietnam
e-mail: minh.nguyen@chem.kuleuven.be

J. Leszczynski and M.K. Shukla (eds.), *Practical Aspects of Computational Chemistry III*, DOI 10.1007/978-1-4899-7445-7_4,

gold is amongst the most characterized transition metal cluster systems, in terms of both experimental and theoretical aspects [7–13].

Previous studies revealed one of the key findings that pure gold clusters exhibit planar structures in small sizes, and a crossover from planarity to non-planarity is likely to occur at the size of Au_{13} for neutrals [13].A large number of studies were also carried out for doped gold clusters, with dopant atoms ranging from light elements to transition metals. Introduction of impurity atoms not only induces geometric and electronic changes, but also strongly influence the physical and chemical properties of host gold systems [14–18].

In fact, even though neither the 13-atom gold neutral nor its charged species favor a high symmetry icosahedral (I_h) structure [13, 19], a number of icosahedral Au_{12}Mclusters, in which the impurity atom is usually a transition metal, have been found. The existence of platonic solid I_h clusters $Au_{12}M^x$ with $M^x = V^-$, Nb^-, $Ta^{0/-}$, W, Re^+... was first theoretically predicted and then experimentally confirmed by several groups [16, 17, 20, 21]. Especially, a series of isoelectronic gold-caged neutrals $Au_{14}M$ (M = Zr, Hf) and anions $Au_{14}M^-$ (M = Sc, Y) were computed to have HOMO–LUMO gaps even larger than those of $Au_{12}W$ and Au_{20} clusters, which hold a large measured gap of >1.6 eV [22]. Several groups also investigated the anionic Au_{16}^- cage encapsulating Cu, Fe, Co, Ni, Ag, Zn, In,... species, and confirmed that the impurity greatly modifies the geometric and electronic properties of the host [23–25]. Recently, the presence of Y was predicted to enhance considerably the thermodynamic stability of yttrium-doped gold Au_nY clusters [26].

The reactivity of gold clusters toward CO also has attracted considerable attention in recent times since such materials exhibit surprisingly high catalytic activity for the CO oxidation reaction [27]. In the CO combustion catalyzed by gold (CO + O_2 + Au), the CO adsorption on Au_n clusters is a fundamental step determining the catalytic mechanism. There have therefore been an increasing number of studies on Au_nCO complexes in both neutral and charged states, and in different environments [28–35]. Prior investigations revealed that irrespective of the charged state, CO is likely to attach to low-coordinated gold atoms, and due to the dominance of a charge transfer from CO to gold over a π-back donation, the adsorption strength obeys the ordering cations > neutrals > anions. The chemical bonding in these carbonyl complexes can normally be understood by the Blyholder model [36], which describes the metal–CO bond as arising from donor–acceptor interactions.

As mentioned above, introduction of impurity leads to significant changes in geometries with respect to pure gold clusters. Consequently, the reactivity of Au-based catalysts can be tuned, and it is necessary to extend investigations of CO adsorption on binary clusters. A large amount of work has been performed on the catalytic mechanism of CO combustion by doped gold clusters including $Au_nCu_m^+$ [37], $Au_nAg_m^{-/+}$ [38–42], Au_nX_m (X = Cu, Ag, Pd, Pt) [37, 43–45], $Au_{12}X$ (X = V^-, W) [46–48], Au_nX (X = H, Li, Na; $n = 1{-}7$) [49], Au_nY ($n = 1{-}9$) [50] and MAu_n^+(M = Ti, Fe, Au) [51]....

For Au_nPt_m, it was found that CO adsorption on the Pt site is much stronger than that on the Au site, and if only Au sites are available for CO adsorption, the strongest adsorption occurs at ~25 % Pt composition [45]. Similarly, a replacement of Au in pure gold clusters by Pd atoms significantly improves their ability to attach

CO [43]. In contrast, the presence of Cu, Ag tends to reduce such activity of gold clusters, and a systematic decrease of CO adsorption affinity with increasing silver or copper content was observed [40, 43].

Recently, Lin et al. [50] also found a drop of reactivity of doped gold clusters with CO after one gold atom is replaced by one yttrium for most of the sizes considered. Furthermore, the net charges of CO in Au_nYCO complexes are consistently negative, in such a way that CO basically plays as a global electron acceptor. For $Au_nCu_m^+CO$ complexes, Zhao et al. [37] however found an electron density flow from CO to metals, and the CO binding energies to clusters decrease with respect to the increasing the number of copper atoms as the amount of electron transfer decreases. It is thus expected that the greater the charge transfer involved the larger the binding energy is.

Concerning vanadium as a dopant, the $Au_{12}V^-$ anion cluster was suggested to be an improved catalyst for CO oxidation [46]. Recently, a number of kinetic parameters of CO adsorption on Au_nV clusters in both neutral and cationic states were experimentally determined by Le et al. [52], but our knowledge about adsorption and combustion on these bimetallic clusters is still limited. In this context we carried out a detailed examination on the adsorption of a single CO molecule on a series of vanadium-doped gold clusters Au_nV with $n = 1$–20. The main purpose is to probe the CO adsorption behavior on these binary metallic clusters, including the binding sites, energies and interaction mechanism. In addition, a comparison of the calculated results obtained for doped and pure gold systems allows us to evaluate the effects of the V dopant. By doing this, we can understand in some detail how the gold systems are perturbed by the presence of a dopant having a large number of unpaired electrons on *d* shells. The chemical bonding is also examined to probe further the mechanism of interactions between CO and clusters. For this purpose, first we set out to analyze the theoretical results of a series of vanadium-doped gold clusters Au_nV up to $n = 20$ in order to systematically emphasize the evolution of their structural, electronic and energetic properties. Basic energetics including dissociation energies (D_e) and binding energy per atom (BE) as a function of cluster sizes are determined and calibrated from available experimental data.

4.2 Computational Methods

All calculated results reported in this chapter are performed using density functional theory (DFT) with the aid of the Gaussian 09 suite of program [53]. The meta-GGA BB95 [54, 55] functional, in conjunction with the correlation consistent cc-pVDZ-PP and cc-pVTZ basis sets, is employed for both geometry optimizations and energetic calculations. The basis set with an effective core potential (ECP) cc-pVDZ-PP [56] is applied for both Au and V metals, while the all electrons cc-pVTZ basis set is used for C and O. The former (PP) set already includes relativistic effects that are very crucial in treatment of heavy elements such as gold. Harmonic vibrational frequencies are also calculated to confirm the character of optimized geometries as local minima on the potential energy surface.

Table 4.1 Theoretical and experimental results of bond length R (Å), dissociation energy D_e (eV), ionization energy IE (eV), electron affinity EA (eV) and vibrational frequency ω_e (cm^{-1}) for Au_2 and AuV

	Au_2					AuV				
Method	R_e	D_e	IE	EA	ω_e	R_e	D_e	IE	EA	ω_e
B2PLYP	2.51	1.48	8.44	1.41	180	2.48	1.98	7.55	0.82	236
B3LYP	2.50	1.96	9.28	1.89	166	2.51	2.48	8.26	1.10	219
B3P86	2.51	2.11	9.83	2.39	177	2.49	2.45	7.58	1.53	227
PW91	2.52	2.31	9.44	1.97	173	2.45	2.66	7.08	1.13	232
BB95	2.51	2.29	9.33	1.80	172	2.47	2.57	6.97	1.02	228
B3PW91	2.52	2.00	9.21	1.82	174	2.50	2.28	7.03	1.01	223
BP86	2.52	2.27	9.53	2.08	173	2.47	2.71	7.18	1.17	231
BPW91	2.52	2.18	9.34	1.91	172	2.48	2.50	7.06	1.07	227
M06	2.58	2.17	9.10	1.77	160	2.50	2.39	7.05	1.23	211
M062X	2.55	1.54	8.62	1.61	164	2.55	2.29	6.94	0.98	210
PBE	2.52	2.29	9.37	1.91	172	2.47	2.65	7.06	1.06	228
PBE0	2.52	2.03	9.10	1.73	176	2.50	2.29	7.32	0.94	223
TPSS	2.51	2.28	9.23	1.79	178	2.47	2.70	6.89	0.90	232
TPSSh	2.51	2.18	9.14	1.72	179	2.48	2.55	6.85	0.85	228
B98	2.53	2.12	9.18	1.71	165	2.50	2.64	6.58	1.31	219
CCSD(T)	2.50	2.15	9.18	1.78	187	2.48	2.76	8.04	1.16	143
Exptl	2.47[a]	2.29[a]	9.20[b]	1.92[c]	191[a]		2.51[a]			
		(±0.02)	(±0.21)				(±0.09)			

[a]Taken from Ref. [76]
[b]Taken from Ref. [77]
[c]Taken from Ref. [78]

The rationale for the selection of the BB95 functional is given below. The main reason for choosing the pseudo-potential (PP)-based correlation consistent basis sets cc-pV*a*Z-PP (a = D, T, Q, 5) is that they are obviously economic in terms of computing times but suitable for describing the chemical bonding of the coinage metals when yielding some of the most accurate carried out to date [57].

Functionals with low or zero amount of Hartree–Fock exchange are found to be generally more consistent than the hybrid counterparts for description of bonding between transition metals [58, 59]. However, application of traditional GGAs in describing Au–Au interactions is still a matter of debate as these functionals are seemingly heavily biased towards 2D clusters [60]. Hence, in order to calibrate the performance of computational methods employed, we first make benchmark computations for the Au_2 and AuV dimers, along with the adsorption energy of CO on Au^+ cation, using various types of functionals, namely four pure GGAs BP86 [54, 61], BPW91, PW91 [62], PBE [63]; six hybrid-GGAs B3P86, B3PW91, B3LYP [64–66], B2PLYP [67], PBE0 [68, 69], B98 [70]; two meta-GGAs TPSS [71], BB95; and three meta hybrid-GGAs TPSSh [72], M06 [73], M06–2X [73], and the coupled-cluster theory CCSD(T) method [74, 75]. The basis sets employed are cc-pVTZ-PP for metals and cc-pVTZ for non-metals. The computed results are summarized in Tables 4.1 and 4.2, which also include the available experimental data for the purpose of comparison.

Table 4.2 Theoretical and experimental results for the CO adsorption energy on Au^+

Method	CO adsorption energy (eV)
BP86	2.60
PW91	2.66
BPW91	2.50
PBE	2.64
B98	2.10
B3PW91	2.19
B3LYP	2.08
B3P86	2.28
B2PLYP	1.47
PBE0	2.21
TPSS	2.45
BB95	2.51
TPSSh	2.29
M06	1.92
M062X	1.63
CCSD(T)	2.04
Experiment	2.08 ± 0.15[a]

[a]Taken from ref.[79]

For Au_2, all functionals tested tend to overestimate the bond length. The deviations from the experiment vary from 0.04 (TPSSh) to 0.10 Å (M06). Consistent with the tendency for overestimation of bond length, the $D_e(Au_2)$ values obtained from hybrid functionals are too low as compared to the experimental value of 2.29 ± 0.02 eV [76]. Generally, hybrid and hybrid-meta GGAs are not quantitatively accurate for computing bond length and bond energy between gold atoms as they are likely to underbind them. On the contrary, pure and meta-GGAs perform much better for such parameters. The computed $D_e(Au_2)$ values using BB95, PBE and TPSS are around 2.28–2.29 eV, and thus in excellent agreement with experiment (Table 4.1). Similarly, functionals containing a certain amount of Hartree-Fock exchange consistently yields the longer equilibrium (r_e) distance for AuV than those with zero or lower Hartree–Fock exchange. Unfortunately, experimental information on the bond length of the mixed diatomic species is not available for a benchmark.

Concerning the dissociation energy of AuV, BPW91 calculation provides more reliable result than any other functional considered. The $D_e(AuV) = 2.50$ eV using BPW91 compares well with the experimental value of 2.51 ± 0.09 eV [76]. Several functionals that also yield $D_e(AuV)$ values lying within the error margin (±0.09 eV) include the B3LYP, B3P86, BB95 and TPSSh.

The experimental values $IE(Au_2)$ [77] and $EA(Au_2)$ [78] are also well reproduced by B3LYP, BB95 and TPSSh, but significantly overestimated by B3P86. Based on the agreement between the measured data and our computed results for both Au_2 and AuV, a non-hybrid functional, namely the meta-GGA BB95, appears to be an advantageous choice for the systems containing these elements. However, this functional tends to overestimate the adsorption energy of CO on Au^+ [79], where the hybrid B3LYP appears to describe it better (see Table 4.2). Therefore, we select the two BB95 and B3LYP functionals for most calculations carried out in this study.

As for a further calibration of the basis set employed, some molecular properties of the dimers Au_2 and AuV are also computed using the basis sets with diffuse functions. The results are found to be marginally modified by the basis sets. For example, at the BB95/aug-cc-pVTZ-PP level (aug stands for diffuse functions) the IE and EA values of Au_2 correspond to 9.34 and 1.82 eV, respectively, as compared to the values of 9.33 and 1.80 eV at the BB95/cc-pVTZ-PP level. Similarly, the adsorption energies of CO on Au^+ cation computed using the aug-cc-pVTZ-PP basis set in conjunction with either the BB95 or B3LYP functional are practically equal to those obtained using the cc-pVTZ-PP basis set.

In the present chapter, we report on CO adsorption energy, which is defined as a measure of the strength of CO adsorption to a specific Au_nV cluster, as follows:

$$E_{ads} = E(CO) + E(Au_nV) - E(Au_nVCO)$$

where E(CO), E(Au_nV), E(Au_nVCO) are the total energies of CO, Au_nV and Au_nVCO, respectively. Hence, as for a convention, a positive value of E_{ads} corresponds to an exothermic adsorption. In addition, this parameter can be used to evaluate the relative stability of a specific carbonyl complex.

The natural bond orbital (NBO) charges of atoms are computed using the NBO 5.0 code [80]. We exploit NBO charges for electron population analysis instead of Mulliken charges because the former are expected to be more reliable [43]. The densities of states (DOS), which are used to assign the contributions of atomic orbitals to a particular molecular orbital [81], are plotted using the PyMolyze-2.0 program [82].

4.3 Results and Discussion

As for a convention, the optimized structures of Au_nV species are denoted as **X-Y**, where **X** is the cluster size ranging from 1 to 20, and **Y** = **I, II, III**, … being the number of isomers located with decreasing stability. For Au_nVCO complexes, the structures considered are denoted as **Xn-Y**. The letter GS stands for the ground electronic state.

4.3.1 *Growth Pattern of Au_nV*

Figures 4.1 and 4.2 display the shape of the MOs. The lowest-energy structures of bimetallic clusters Au_nV are presented in Figs. 4.3, 4.4, 4.5, and 4.6. They adopt planar forms up to $n = 5$, and a 2D–3D transition is likely to start at $n = 6$.

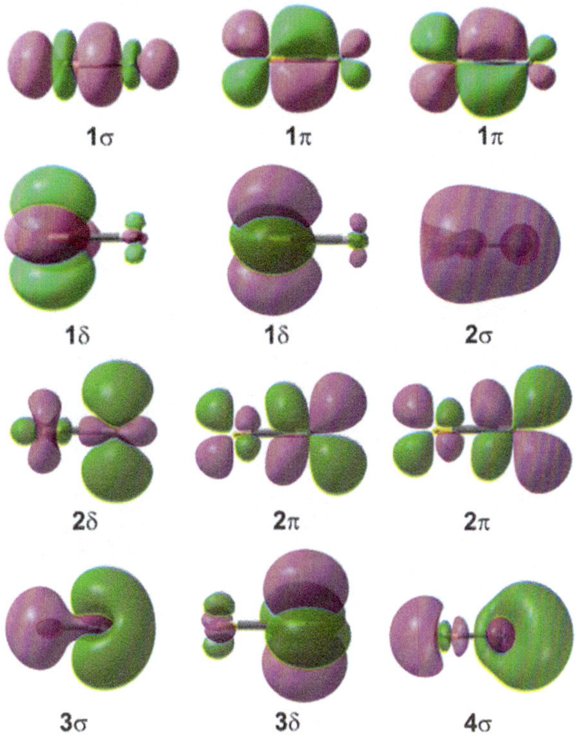

Fig. 4.1 Shapes and symmetries of frontier MOs in Au-V (BB95/cc-pVDZ-PP). The gold nucleus is on the *left*

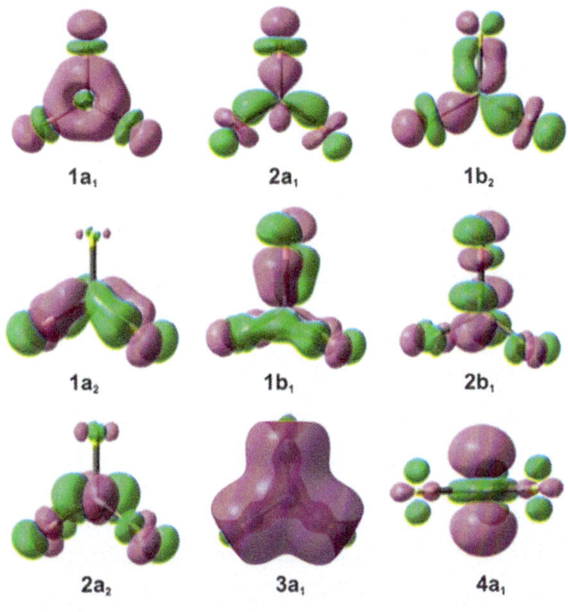

Fig. 4.2 Shapes and symmetries of frontier orbitals and MOs generated from the interactions of valence orbitals of V with Au_3 in Au_3V (BB95/cc-pVDZ-PP)

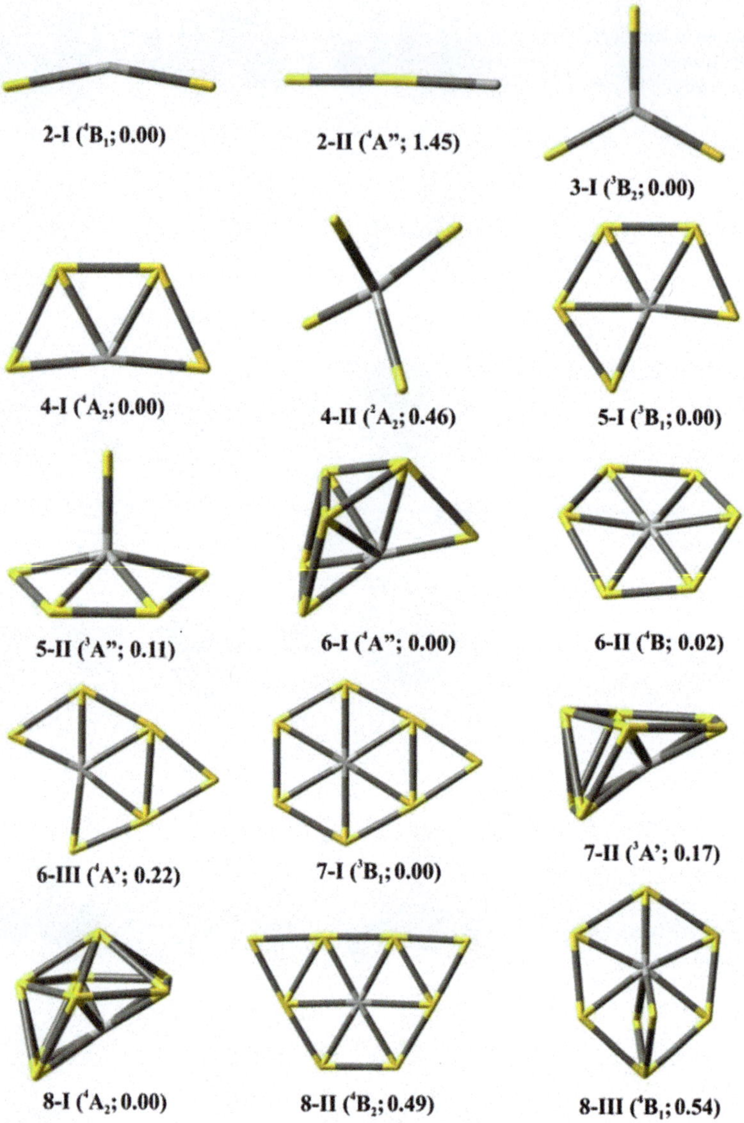

Fig. 4.3 Lower-lying structures considered for Au_nV clusters, $n = 2$–8, and their electronic states and relative energies (eV) at the BB95/cc-pVDZ-PP + ZPE level

Clusters larger than Au_8V apparently prefer 3D structures, in which V atom is endohedrally doped to maximize its coordination number. From a growth mechanism point of view, we find that replacement of one Au in Au_{n+1} species by one V atom to form Au_nV complexes generally results in a significant atomic rearrangement. Except for Au_2V and Au_4V, the doped Au_nV and the pure Au_{n+1}

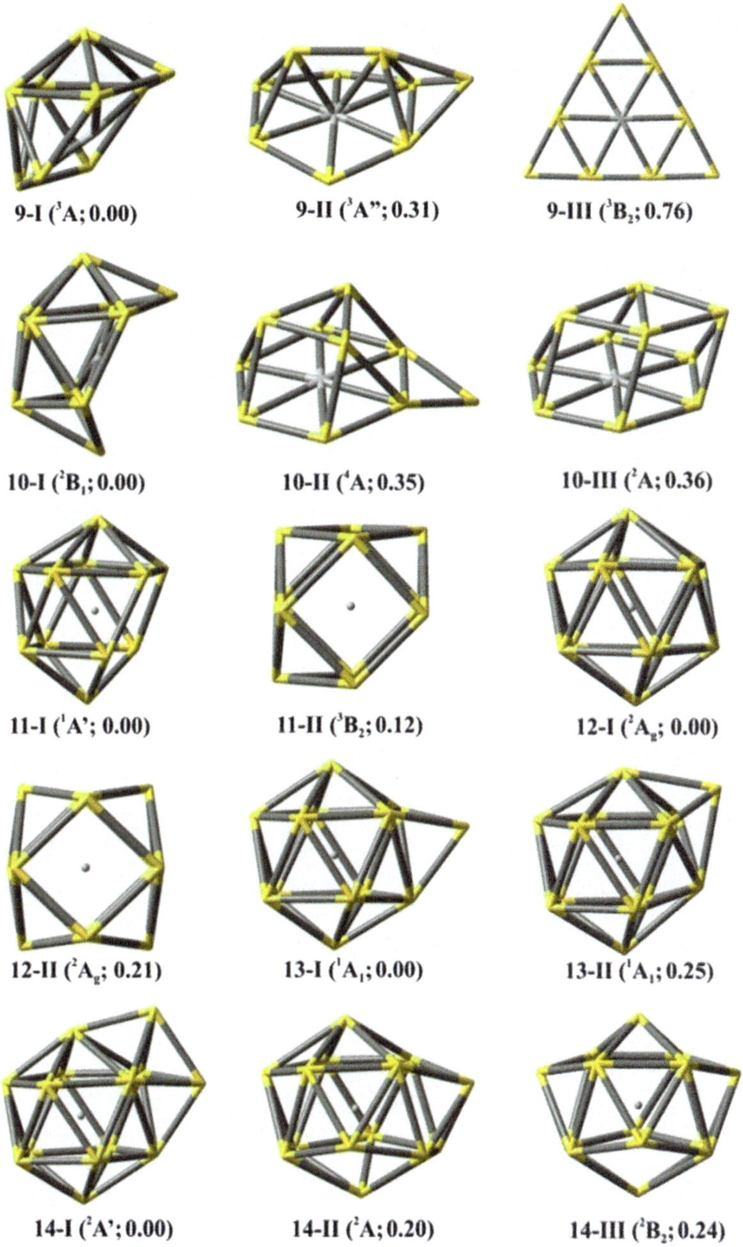

Fig. 4.4 Lower-lying structures considered for Au_nV clusters ($n = 9-14$) and their electronic states and relative energies (eV) at the BB95/cc-pVDZ-PP + ZPE level

Fig. 4.5 Lower-lying structures considered for Au_nV clusters, $n = 15-17$, and their electronic states and relative energies (eV) determined at the BB95/cc-pVDZ-PP + ZPE level

clusters bear a completely different shape in their ground state. The equilibrium structures of the doped clusters are detailed in the following sections.

AuV. The GS of the smallest binary species AuV is a quintet state with the orbital configuration $^5\Delta$:...$(2\sigma)^2(2\delta)^1(2\pi)^1(2\pi)^1(3\sigma)^1$. The highest singly occupied orbital (SOMO) 3σ (given in Fig. 4.1) is an anti-bonding orbital and composed mainly of 4 *s*, $3d_{z2}$ (V). Other singly occupied orbitals include one δ and two π MOs, and they all have non-bonding or anti-bonding character and are largely contributed from V(*d*: d_{xy}, d_{xz}, d_{yz}) orbitals as well. Hence, four unpaired electrons are mostly located on the *d*(V) orbitals. The interaction that forms chemical bonding in AuV occurs between the $3d_{z2}$, $3d_{xz}$, $3d_{yz}$, 4 *s* AOs of V and the same symmetry counterparts of Au, giving rise to 1σ, 1π and 2σ bonding orbitals.

Fig. 4.6 Lower-lying structures considered for Au_nV clusters ($n = 18-20$) and their electronic states and relative energies (eV) at the BB95/cc-pVDZ-PP + ZPE level

The lower spin singlet and triplet states are 2.24 and 0.83 eV above GS, respectively. The optimal bond length Au–V varies from 2.47 (BB95), 2.50 (M06) to 2.51 Å (B3LYP), while that of Au_2 covers the range from 2.52 (BB95), 2.55 (B3LYP) to 2.58 Å (M06). Replacement of Au by V thus results in a mild decrease of bond length. In addition, due to the larger EA of the Au element

(2.31 eV) than that of V (0.53 eV) [77, 83], the Au–V bond is relatively ionic with a positive charge centered at V (+0.38 electron).

Au_2V. The lowest-energy structure of the trimer Au_2V is a C_{2v} form **2-I** with a $^4B_2$ GS, and an equilibrium Au–V distance of *ca.* 2.44 Å (Fig. 4.3). The latter is slightly shorter than those in AuV (~2.47 Å) and in Au_3 (~2.55 Å). The low-spin $^2B_2$ state of this form lies at 0.83 eV above GS. The second isomer**2-II** is slightly bent (C_s symmetry) and its high spin state $^4A''$ is *ca.* 1.45 eV above the $^4B_2$ GS. As the GS of the vanadium atom possesses a [Ar] $3d^34s^2$ configuration, it consequently uses nine orbitals of 3*d*, 4 *s* and 4*p* shells that are available for making chemical bonds. In addition, empty *d*-AOs of V can act as electron acceptor by combining with orbitals of the gold backbone. Therefore, Au_nV is expected to be more stabilized if the coordination number of V is getting higher. Indeed, isomer **2-I** in which the V atom has a coordination number of two, is much more stable than **2-II** (by ~1.45 eV), in which V has a coordination number of one (Fig. 4.3).

Au_3V. In terms of crystal field theory, when placed in a D_{3h} ligand field, the five *d* orbitals of V will be split into three different levels as follow:

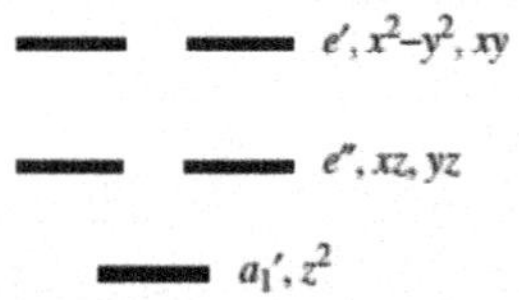

Thus, starting from the higher symmetry D_{3h} structure, Au_3V with four *d* electrons and a triplet state will have two electrons in a_1', one in e" and the other in e'. In this case, the energy of the a_1' orbital is substantially lower than those of e" and e' orbitals which are very close in energy. Hence, a triplet state is favored over the singlet or quintet states. However such degenerate state is not stable upon Jahn-Teller distortions and the molecule undergoes a geometry relaxation to remove that degeneracy. The resulting stable structure **3-I** of Au_3V (Fig. 4.3) is thus slightly distorted from D_{3h} to C_{2v} with V atom occupying a central position. Again, this system also prefers a high-spin $^3A_2$ GS, and the quasi-degenerate $^3B_1$ state is only 0.02 eV higher. We are not able to locate the triplet states with two equivalent electrons in either e' or e" orbitals, as all geometry optimizations invariably converge to the $^3A_2$ or $^3B_1$state.

The shapes and symmetries of frontier molecular orbitals in Au_3V, resulting from the interactions between valence AOs of V and Au atoms are shown in Fig. 4.2. It is clear that, of the five *d*(V) AOs, only $3dz^2$ is not involved in bonding orbitals. In Au_3V complex, the $3dz^2$(V) becomes the lowest unoccupied orbital $3a_1$ (Fig. 4.2). The interactions between other 3*d*(V) AOs with the symmetry-matching ligand orbitals yield b_2, a_1, a_2 and b_1 bonding and anti-bonding MOs. Two unpaired electrons are located in the$3a_1$ and $2a_2$ orbitals. Previously Lin et al. [26] found that the tetramer Au_3Y exhibits an ideal D_{3h} form as it possesses a $^1A_1$' GS.

Au_4V. The most stable form of Au_4V **4-I**, with C_{2v} point group and $^4A_2$ GS, is built up either from replacing one Au in GS of Au_5 [84–86] by V, or from attaching one Au atom to **3-I**. The low-spin $^2A_1$ state of this form is less favored than GS by 0.57 eV. The first 3D structure **4-II**, which is distorted from a tetrahedral to D_{2d} symmetry, has a doublet $^2A_2$ state and is 0.46 eV higher in energy (Fig. 4.3).

Au$_5$V. Two isomers **5-I** and **5-II** are predicted to be the most stable forms of the hexameric cluster. The lowest energy **5-I** can be considered as arising from addition of one Au atom to the lowest-energy isomer of the pentamer **4-I**, while the other **5-II** is greatly distorted from a C_{5v} shape. According to the applied level of theory, the $^3B_1$ state of **5-I** is only ~0.11 eV more stable than the $^3A''$ state of **5-II** (Fig. 4.3).

Au$_6$V. Previous investigations [87, 88] revealed that Au$_6$V has a planar structure with V atom located at the center of an Au$_6$ ring. Our current calculations however find the 3D structure **6-I** to be the lowest-energy form of Au$_6$V, while the isomer **6-II** built upon the Au$_6$ ring is computed to be nearly degenerate, being separated by only 0.02 eV. In addition the latter is not perfectly planar and exhibits a C_2 configuration rather than a D_{2h} arrangement as previously predicted by the GGA BP86 functional [89]. The lower spin states $^2A''$ of **6-I**and ^2B of **6-II**, respectively, are located at 0.45 and 0.15 eV higher in energy than the GS. The next lower-lying isomer is **6-III**, which can be viewed as a result of substituting one Au in GS of Au$_7$ [13] and is around 0.22 eV above GS (Fig. 4.3).

Au$_7$V. As in Au$_6$V, there exist several configurations, i.e. **7-I** and **7-II**, competing for the GS of Au$_7$V. The planar C_{2v} shape **7-I**, which is built upon **6-II** by capping one Au on an edge of the Au$_6$ ring, is predicted as the global minimum. The second most stable isomer **7-II**, being only 0.15 eV higher energy than **7-I**, has a C_s 3D conformation and comes from **6-I**. In addition, such a form can be regarded as a result of cutting one gold atom from the lowest-energy structure of the larger size (Fig. 4.3).

Au$_8$V. The most stable isomer found for the nonamer Au$_8$V can be viewed as a fraction of an icosahedron (structure **8-I** in Fig. 4.3). The high spin $^4A_2$ state of this form is predicted to be GS, while the corresponding low-spin $^2B_1$ state is less stable by ~0.30 eV. Other local minima **8-II** and **8-III** reached by adding two gold atoms to **6-II** are relatively less stable than the GS. In fact, the electronic states $^4B_2$ of **8-II** and $^4B_1$ of **8-III** are around 0.50 and 0.54 eV above the GS, respectively.

An interesting point found here is that from this size the 3D shapes are apparently more favored than the 2D conformations. In addition, it should be stressed that all results mentioned above are obtained using the BB95 functional. Other approaches can give a different energy landscape. We find, for example, BP86 and B3LYP functionals still heavily biased towards planar structures up to $n = 9$.

Au$_9$V. As in preceding systems, the global minimum of Au$_9$V **9-I** is reached by attaching one extra gold atom to the most stable isomer of the smaller size **8-I**, and prefers to stay in a high spin ^3A state as well. The 2D structure **9-III** is now significantly less stable than **9-I**, being separated by 0.76 eV above. The second most stable form is also a non-planar structure **9-II** (Fig. 4.4). Its triplet $^3A''$ state is computed to be 0.31 eV higher in energy than GS.

Au$_{10}$V. Previous investigations on Au$_n$Ti recorded a 2D–3D transition at $n = 8$ [90]. For Au$_n$V species, the crossover from planarity to non-planarity is likely to be initiated at Au$_5$V. However, planar geometries are still located as local minima up to $n = 9$, and $Au_{10}V$ represents the first size whose lower-lying structures do not have two-dimensional shape. The three most stable isomers of $Au_{10}V$, i.e. **10-I**, **10-II** and **10-III**, are all non-planar and exhibit low spin GS (Fig. 4.4). The $^2B_1$ state of **10-I** is predicted to be the GS, but only 0.11 eV more stable than the corresponding high-spin $^4B_1$ state. Isomers **10-II** and **10-III** are evaluated to be around 0.35–0.40 eV above the GS.

$Au_{11}V$. The most stable form of $Au_{11}V$is predicted to be an incomplete icosahedron **11-I** with C_s point group and low spin $^1A'$ state (Fig. 4.4). Such a conformation can also be acquired by capping an extra gold atom on **10-I**. The corresponding triplet state $^3A'$ is computed to be only 0.04 eV higher than the singlet GS. Within the error margin of computational methods, these states can be regarded as degenerate. It has been pointed out that cuboctahedra and icosahedra are very closely related and a minor distortion of an icosahedron eventually leads to a cuboctahedron [91]. Thus the dodecamer $Au_{11}V$ can also exist as an incomplete cuboctahedron**11-II**. However, at the BB95 level, the latter is predicted to be a transition state with a tiny imaginary frequency of $15i$ cm^{-1} and ~0.12 eV above **11-I**.

$Au_{12}V$. Starting from **11-I** and adding an extra Au, we obtain the GS**12-I** (Fig. 4.4). Previously, the I_h icosahedral structure of the anion $Au_{12}V^-$ was reported to be significantly preferred over the O_h cuboctahedron [20, 46]. In addition, the anion has a closed-shell $^1A_{1g}$ state with ten electrons on the fivefold degeneracy (h_g) HOMO [92]. Removing one electron from $Au_{12}V^-$ is thus accompanied with a geometric distortion due to a Jahn-Teller effect. Consequently, **12-I** is characterized by a slightly distorted icosahedron with C_i point group, instead of a perfect I_h form, as it possesses an open-shell but low spin electronic configuration ^2A_g. As in $Au_{11}V$, the cuboctahedral shape **12-II** is not a genuine local minimum with an imaginary frequency of around $15i$ cm^{-1} and lies ~0.20 eV above the icosahedrons **12-I** (BB95).

$Au_{13}V$. From attachment of one gold atom to the icosahedron, two different isomers **13-I** (C_{3v}) and **13-II**(C_{2v}) are built upon (Fig. 4.4). Nevertheless, to a certain extent they possess a remarkable energy difference, in which the low-spin $^1A_1$ state of **13-I** is ~0.25 eV below the corresponding low-spin state of **13-II**. Interestingly, both **13-I** and **13-II** can be viewed as super-atomic molecules, containing $Au_{12}V^-$ and Au^+ units. Indeed, NBO analysis shows that the $Au_{12}V$ moiety of **13-I** bears a partially negative charge of –0.22 electron, and the capping gold atom has a positive charge of +0.22 electron. Likewise, another super-atomic molecule of transition metal clusters was recently recorded for $Cu_{17}Sc$, whose constituents are $Cu_{16}Sc^+$ and Cu^- [93].

$Au_{14}V$. Again, a growth path going from **13-I** leads to the most stable cluster with 14 gold atoms **14-I**. The next low-lying structure **14-II** is computed to stay at around 0.20 eV higher in energy than the GS $^2A'$ of **14-I**. Previous investigations [22, 90] on several gold-caged clusters $Au_{14}M$ (M = Ti, Zr, Hf) showed that these systems exhibit similar geometries, that is, 14 outer gold atoms constitute a hollow cage with D_{2d}symmetry, while the dopant atom is encapsulated in the gold cage. For $Au_{14}V$, however, this form (cf. **14-III** in Fig. 4.4) is predicted to be a transition state with an imaginary frequency of 15*i* cm^{-1} and a slight distortion leads it to **14-II**.

$Au_{15}V$. Previously, the sixteen-atom species $Cu_{15}Sc$ is predicted to favor a Frank-Kasper (FK) polyhedron [93]. For $Au_{15}V$, however, this form, namely **15-IV** in Fig. 4.5, is found as a transition state with a corresponding 70*i* cm^{-1} imaginary frequency and lies around 0.17 eV above GS. Upon relaxation, two structures **15-I** and **15-II** are reached, in which the former is located as the global minimum. The former can also be considered as the derivative of **14-I** and it is only 0.04 eV lower in energy than the latter.

Other stable conformations including **15-III** and **15-V** are obtained from the fivefold symmetry core $Au_{12}V$ by adding three extra Au atoms (Fig. 4.5). These are also quite close in energy with the GS, being separated by only 0.13–0.18 eV. Noticeably, from this size, the clusters invariably prefer a high spin state as the GS, i.e. triplet and quartet states for even- and odd-electron systems, respectively.

$Au_{16}V$. As in smaller systems, several structures with comparable energy content are detected for this species. A growth algorithm from **15-II** yields the most stable form **16-I** (Fig. 4.5). This is a chiral cluster whose non-superposable mirror image is **16-II**. Two forms **16-I** and **16-II** are utterly isoenergetic. Previously, the lowest-energy structure found for anion Au_{34}^- is also chiral [94]. Other meta-stable isomers**16-III** and **16-IV** are located to be around 0.08–0.13 eV above the GS and both can be considered as the continuation of the icosahedral evolution from **15-I**. In addition, the heptadecamer $Au_{16}V$ can appear as a distorted FK **16-V**, whose energy is computed to be ~0.18 eV above the GS. A recent study [24] on sixteen-gold atoms doped with a transition-metal atom revealed that the anionic states$Au_{16}M^-$ (M = Fe, Co, Ni) prefer to appear as the distorted tetrahedral Au_{16}^- cage encapsulating the dopant element.

$Au_{17}V$. We identify two nearly degenerate structures **17-I** and **17-II**competing for the GS of $Au_{17}V$ (Fig. 4.5). Among them, the global minimum **17-I** is a capped derivative of the distorted FK **16-V**. It has a threefold rotational axis and exhibits a triplet ^{3}A GS. More particularly, similar to $Cu_{17}Sc$ [93], the cluster can be regarded as a superatomic molecule composed of $Au_{16}V^+$ and Au^- entities. The latter is coming from the low-lying isomer **16-IV**of the smaller size and lies only 0.02 eV above **17-I**.

On the other hand, a growth path from **16-III** leads to the incomplete double icosahedron **17-IV**, whose energy is predicted to be ~0.20 eV higher than the GS. In addition, other possible structures of the octadecamer $Au_{17}V$ include **17-III** and **17-V**, which stay around 0.15 and 0.30 eV, respectively, above **17-I** (Fig. 4.5).

$Au_{18}V$. At the level employed, we indentify two nearly isoenergetic structures, i.e. **18-I** and **18-II** shown in Fig. 4.6, competing for the GS. While **18-I** is derived from the most stable form of $Au_{17}V$, the other **18-II** is reached by adding three Au atoms to the Frank-Kasper **15-IV**. The quartet ^{4}B state **18-I** is predicted to be most stable, but it is only 0.10 eV lower than the ^{4}A" state **18-II**. They are thus energetically degenerate and each of them can emerge as the GS. We also consider several structures based on the icosahedral evolution and obtain two low-lying isomers **18-III** and **18-IV**, in which the latter can be viewed as a double icosahedron with the dimer AuV encapsulated. They are predicted to be around 0.26–0.35 eV less stable than the GS.

$Au_{19}V$. The most stable form **19-I** is reached from **18-II** by adding an extra Au atom, while the isomer **19-VI** based on the capping of **18-I** is significantly less stable, being around 0.25 eV above **19-I** (Fig. 4.6). On the other hand, a continuing trend of icosahedral growth from **18-IV** yields another meta-stable form, i.e. **19-II**, which is only 0.03 eV less table than **19-I**. Other low-lying conformations, namely **19-III**, **19-IV** and **19-V** shown in Fig. 4.6, are predicted to be 0.10–0.24 eV higher in energy than the GS and all can be regarded as derivatives of the preceding clusters. These results, together with those observed for previous systems, lend a support for the view that in terms of structural evolution mechanism *V-doped gold clusters favor a successive growth algorithm* [95].

$Au_{20}V$. As shown in Fig. 4.6, two lowest-energy structures **20-I** and **20-II** are generated by capping one extra gold atom on **19-III** and **19-II**, respectively. Although their geometries are quite different from each other, they are extremely close in energy and can appear as competing GSs. The quartet $^4B_2$ state of the C_{2v} shape **20-I** is computed to be only 0.01 eV more stable than the corresponding ^4A state of **20-II**. The growth path from **19-I** gives rise to another local minimum, i.e. **20-IV**, but it is predicted to lie around 0.17 eV above **20-I**. Recent studies [11, 96] confirmed the distinct and remarkable stability of the tetrahedral structure Au_{20} that supports for the high possibility of synthesizing its derivatives $Au_{20}M$ (M = metal elements). However, for $Au_{20}V$, such a capped tetrahedron (**20-VII**) is located around 0.45 eV above **20-I**.

4.3.2 Thermodynamic Stabilities of Au_nV Clusters

To obtain more quantitative insights into Au_nV clusters, we evaluate the change of binding energy per atom (BE), the second-order difference of energy (Δ^2E), and the one-step fragmentation energy (E_f) with respect to the clusters sizes. For these clusters, such parameters of a specific system Au_nV are defined by these equations:

$$BE = [nE(Au) + E(V) - E(Au_nV)]/(n+1)$$

$$\Delta^2E = E(Au_{n+1}V) + E(Au_{n-1}V) - 2E(Au_nV)$$

$$E_f = E(Au_{n-1}V) + E(Au) - E(Au_nV)$$

where $E(Au_nV)$ is the total energy of the lowest-lying Au_nV cluster.

For the sake of consistency, optimal geometries (see Fig. 4.7), and then the BE values, of pure gold clusters are recomputed using the BB95/cc-pVDZ-PP level. The graph of BEs as a function of cluster size is plotted in Fig. 4.8. This parameter can be regarded as the energy gained in assembling a definite cluster from isolated Au and V constituents.

Fig. 4.7 Optimal geometries, electronic states and binding energies per atom (eV) of the pure Au_n clusters ($n = 3-21$) at the BB95/cc-pVDZ-PP + ZPE level

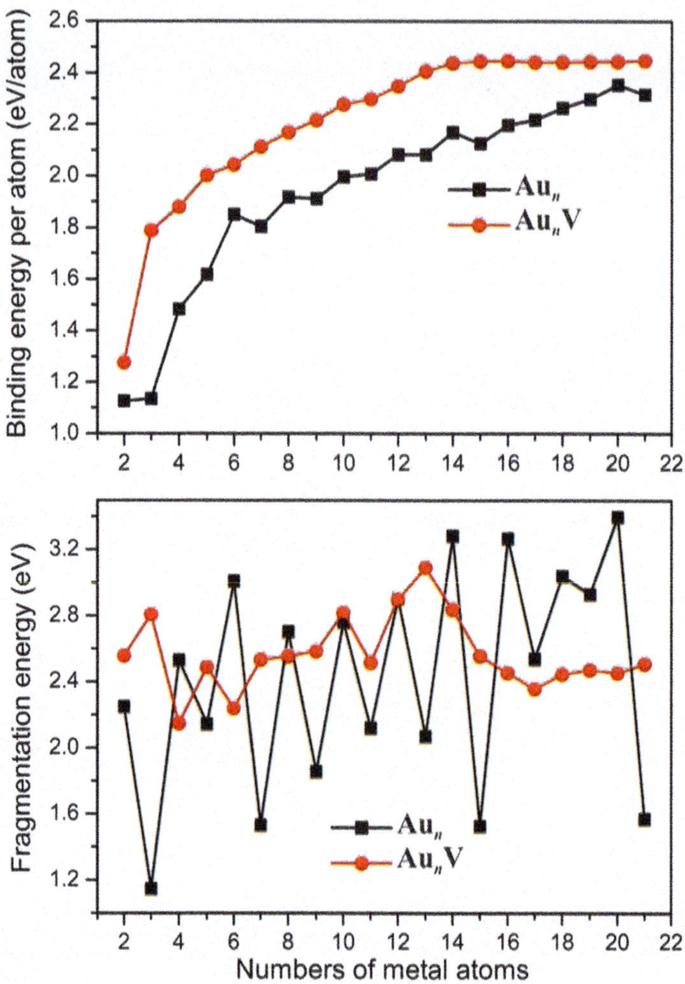

Fig. 4.8 Comparison of binding energies per atom (*upper*) and one-step fragmentation energies (*lower*) with respect to cluster sizes. The results are obtained at the BB95/cc-pVDZ-PP + ZPE level

For the purpose of comparison, the binding energies per atom of the most stable Au$_n$V and Au$_n$ clusters are also computed and simultaneously illustrated in Fig. 4.8. There have been a great number of studies on structures of small pure gold clusters [97, 98]. Nevertheless, in the current work, we take the results from the recent reports [13, 99], in which the predicted structure of Au$_7$ is consistent with a combined experimental and theoretical vibrational investigation [11]. Previously, Wang et al. [100] used a local density approximation to calculate the BEs up to Au$_{20}$. Experimental data have not been systematically reported yet.

As shown in Fig. 4.8, the BEs of the pure clusters roughly exhibit a gradual growth and reach to the maximal value of 2.35 eV/atom for Au_{20} and Au_{21}, whereas those of doped-clusters steadily increase up to 2.45 eV/atom at $Au_{14}V$ and then insignificantly modify. In general, the average binding energy of Au_{n+1} is slightly smaller than that of the isoatomic Au_nV.

Let us take the dimers AuV and Au_2 as for an example. The BE values are predicted to be 1.28 and 1.14 eV/atom, respectively, as compared to the experimental values of 1.26 ± 0.09 eV/atom for AuV and 1.15 ± 0.02 eV/atom for Au_2 [76]. This indicates a stronger interaction between atoms in AuV cluster as a result of the relativistic effects.

It has previously been shown that a relativistic stabilization is increased for the AuX bond if X atom is more electropositive than gold [101, 102]. Due to the fact that the electro-negativity of V is smaller than that of Au, the ionic contribution to the bond Au–V is augmented by relativistic effects and, as expected, the dissociation energy is subsequently increased.

As discussed above, the optimal structure of Au_nV at a certain size is normally generated from that of the smaller one by adding an extra gold atom. Accordingly, the energy gain in incorporating an Au to the smaller size can be considered as the embedding energy (EE). Such parameter can also be characterized by the one-step fragmentation energy (E_f), i.e. the energy needed to detach one gold atom from Au_nV, giving rise to Au_{n-1} V. The E_f values for Au_nV clusters calculated at the BB95/cc-pVDZ-PP + ZPE level are plotted in Fig. 4.8.

Among the Au_nV species considered, Au_2V and $Au_{12}V$are characterized by the highest E_f values, implying a particular thermodynamic stability even though they have open-shell electronic structure. The calculated E_f value of Au_2V is ca 2.8 eV as compared to 2.6 and 2.2 eV of AuV and Au_3V, respectively. The trimer is found especially stable since both Au and V are expected to fulfill their outer shell, in which V donates two electrons for making chemical bonds. A similar trend was also observed for Au_2Mn, Au_2Fe, and Au_2Zn [103–105]. However, Au_2Cr and Au_2Ag were found to be extremely unstable [83, 106], as chromium and silver are willing to provide only one electron on the utmost orbital.

Also stable are the clusters at $n = 9$, 11 and 13. It is worth noting that $Au_{12}V$ possesses an odd number of electrons. Accordingly, its particular stability could be due to a dominant *geometric effect*, in which the *stability is determined more by the arrangement of atoms rather than by the number of electrons*. In other words, the cluster is stabilized as it can form a compact symmetric structure yielding an enhanced stability [107]. In fact, as discussed above, the most stable form found for $Au_{12}V$ is slightly distorted from a highly symmetric icosahedron.

For the sake of comparison, the E_f values of pure gold clusters up to Au_{21}, calculated at the BB95/cc-pVDZ-PP + ZPE level, are also displayed in Fig. 4.8, along with those of Au_nV clusters. As reflected in Fig. 4.8, the E_f's of the smallest pure clusters, i.e. Au_2 and Au_3, are much smaller than those of AuV and Au_2V. This indicates that a chemical bond between Au and V is stronger than that between two Au atoms.

Regarding the systems containing from four to sixteen metal atoms, the replacement of Au by V is found to increase greatly the stability of Au_n with odd number n,

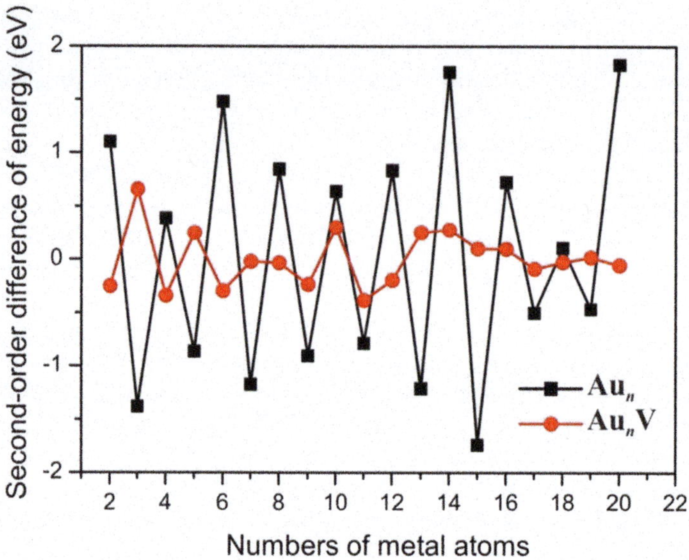

Fig. 4.9 Second-order difference of energy for Au_{n+1} and Au_nV as a function of cluster size. The results are obtained at the BB95/cc-pVDZ-PP + ZPE level

while a reverse result holds true for even-numbered size systems. For instance, the E_f's of Au_{14} and Au_{15} are 3.3 and 1.5 eV, respectively, as compared to 2.9 and 2.6 eV of $Au_{13}V$ and $Au_{14}V$. From Au_{17}, introduction of V is observed to accompany with a decrease of the dissociation energy, except for Au_{21} whose D_e value (1.57 eV) is much smaller than that of $Au_{20}V$ (2.52 eV).

The second-order difference of energy (Δ^2E) is additionally an important indicator that measures the relative stability of clusters. In particular, peaks in the graph of Δ^2E as a function of cluster sizes were found to be correlated well with peaks in the experimental mass spectra [108]. As illustrated in Fig. 4.9, our calculations show local maxima at $n = 2, 4, 9, 12, 13$ and 18, implying that these clusters are more stable than their immediate neighboring ones. However, the energy difference between them is not very large. This is also in agreement with the analysis based on embedding energies given above. For pure gold system (given also in Fig. 4.9), there exist extreme odd-even fluctuations, in that a cluster with an even number of atoms is more stable than the odd-numbered ones.

In agreement with the above analysis based on the fragmentation energies, replacing an Au atom in odd-numbered gold clusters by vanadium element invariably enhances the system stability. A reversed fashion also holds true for the even-numbered gold clusters. Thus, doping a gold cluster with impurity causes significant modifications not only on molecular structures, but also on the relative stabilities.

An NBO charge analysis reveals that the dopant V is willing to donate as well to accept electrons from the gold host located outside. For systems up to $n = 6$, there exists an electron density flow from V to Au atoms as the net charges of the former is consistently positive (see Table 4.3). Amongst these, the highest stability of Au_2V

Table 4.3 Natural charges (electron) distributed on V of Au_nV clusters ($n = 1-20$) computed at the BB95/cc-pVDZ-PP level

Isomer	NBO charge of V	Isomer	NBO charge of V
1-I	0.38	**9-I**	−0.83
2-I	0.78	**10-I**	−1.40
3-I	0.71	**11-I**	−2.11
4-I	0.62	**12-I**	−2.59
4-II	0.44	**13-I**	−2.62
5-I	0.40	**14-I**	−2.32
5-II	0.33	**15-I**	−2.08
6-I	0.16	**16-I**	−1.79
6-II	0.14	**17-I**	−1.73
7-I	0.21	**18-I**	−1.66
7-II	−0.20	**19-I**	−1.83
8-I	−0.49	**20-I**	−1.70

correlates well with the largest charge transfer from V. On the other hand, from Au_7V, the electron back-donation from Au atoms to V is observed, as the latter is always negatively charged in Au_nV. Thus, owing to its empty 3*d* orbitals, the vanadium atom can also play as an electron acceptor. In this case, species with greater charge transfer, i.e. $Au_{12}V$ and $Au_{13}V$, are found to be much more stable than others.

Electron donation comes from a combination of the *d* orbitals of vanadium with those of the gold framework. Hence, it is more effective if V is endohedrally confined and the host is large enough to accommodate the dopant element. However, moving to larger sizes, the gold cage expands, so interactions of encapsulated V valence orbitals with those of the skeletons outside become less effective. Indeed, as shown in Table 4.3, the electron transfer in these species is less significant as compared to the similar procedure in smaller systems. Consequently, the systems with $n = 12–14$ are predicted to be more stable than ones with $n > 14$.

An interesting point here is that the charge transfer in V-doped gold clusters is found to correlate well with the magnetic property. As discussed above, systems $Au_{12}V–Au_{14}V$ tend to exhibit the lowest possible spin ground states, i.e. singlet or doublet. The smaller and larger clusters, on the contrary, are predicted to favor the high spin ground states, except for $Au_{10}V$ and $Au_{11}V$, whose low and high spin states are nearly isoenergetic. Thus there exists a close relation between the magnetic property and the charger transfer. The preference to the low spin states of $Au_{12}V–Au_{14}V$ species is manifested in their significant electron transfer.

4.3.3 CO Adsorption on Au_nV Clusters

4.3.3.1 Equilibrium Structures of Au_nVCO Adducts

The most stable forms of Au_nVCO adducts are present in Figs. 4.10, 4.11, 4.12, and 4.13, along with their CO adsorption energies computed at both BB95 and B3LYP levels. In general, for clusters smaller than $Au_{11}V$ when both Au and V sites are

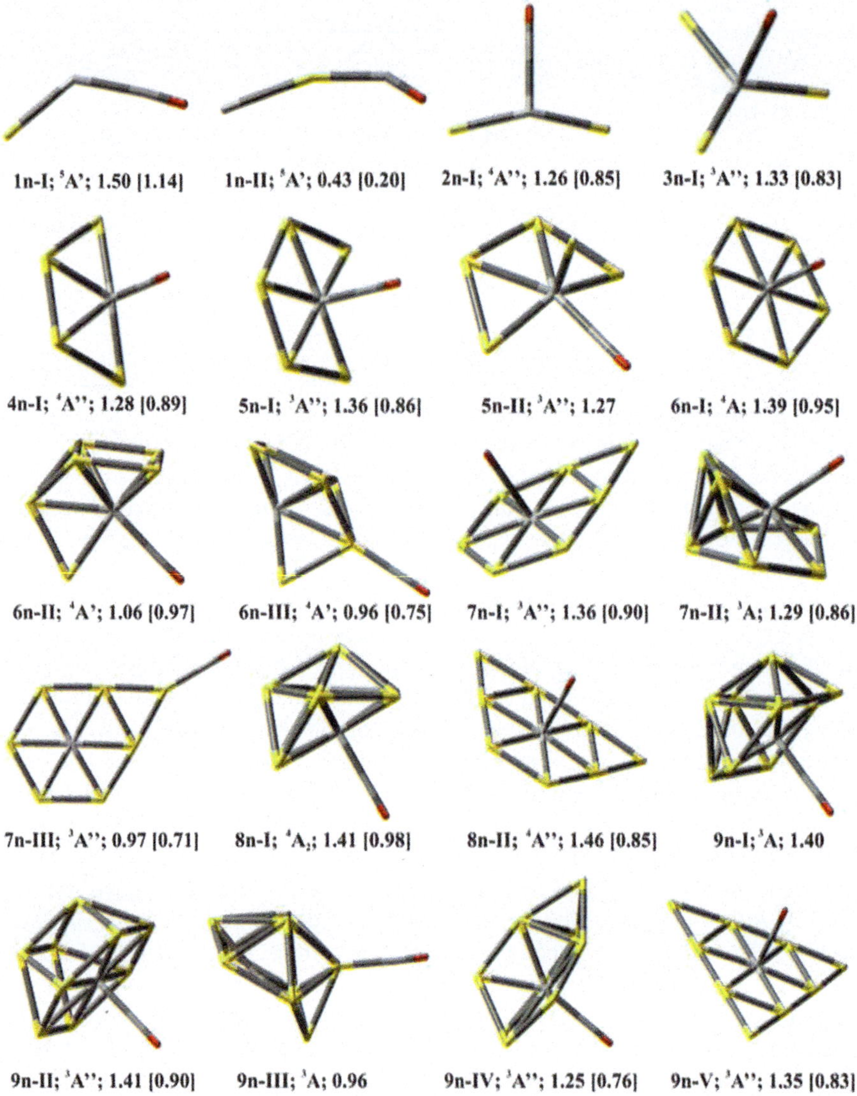

Fig. 4.10 Lower-lying structures considered for Au_nVCO complexes ($n = 1{-}9$) their electronic states and CO adsorption energies. The values in *square brackets* [eV] are computed using the B3LYP functional

exposed, the latter is consistently more favorable to adsorb CO than the former (see Fig. 4.10). In the next section, we discuss the shapes of these species in some detail.

n = 1. For the dimer AuV, CO prefers to anchor on the V site, forming the complex **1n-I** (Fig. 4.10) with C_s symmetry and a high spin $^5A'$ ground state. In AuVCO complex, the V–C and C–O bond lengths are around 2.00 and 1.16 Å,

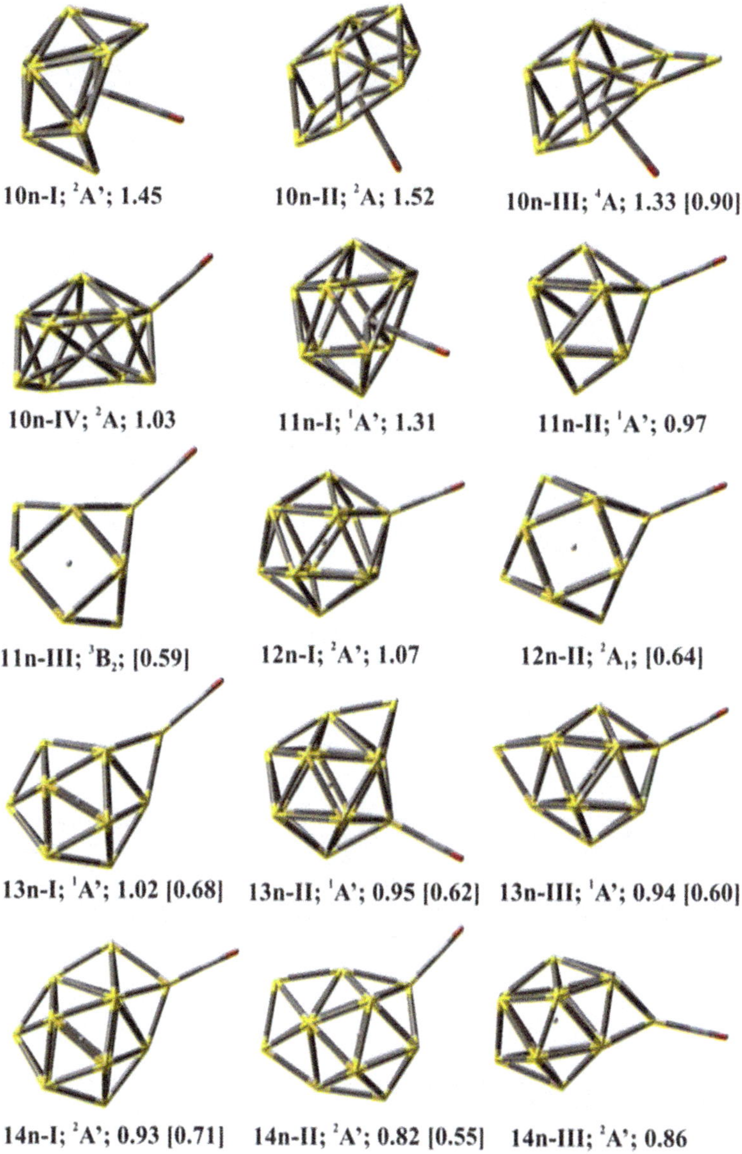

10n-I; $^{2}A'$; 1.45 — 10n-II; ^{2}A; 1.52 — 10n-III; ^{4}A; 1.33 [0.90]

10n-IV; ^{2}A; 1.03 — 11n-I; $^{1}A'$; 1.31 — 11n-II; $^{1}A'$; 0.97

11n-III; $^{3}B_{2}$; [0.59] — 12n-I; $^{2}A'$; 1.07 — 12n-II; $^{2}A_{1}$; [0.64]

13n-I; $^{1}A'$; 1.02 [0.68] — 13n-II; $^{1}A'$; 0.95 [0.62] — 13n-III; $^{1}A'$; 0.94 [0.60]

14n-I; $^{2}A'$; 0.93 [0.71] — 14n-II; $^{2}A'$; 0.82 [0.55] — 14n-III; $^{2}A'$; 0.86

Fig. 4.11 Lower-lying structures considered for Au_nVCO complexes ($n = 10-14$), their electronic states and CO adsorption energies. The values in *square brackets* [eV] are computed using the B3LYP functional

respectively. The V–CO linkage is thus relatively longer than a typical M–CO distance in metal carbonyls (often <1.80 Å), but it is about equal to a metal-alkyl bond. CO can also bind to the Au atom of AuV, though the corresponding complex is much less stable than **1n-I** (>1.0 eV). The attachment of CO on V is thus more energetically favored than on Au.

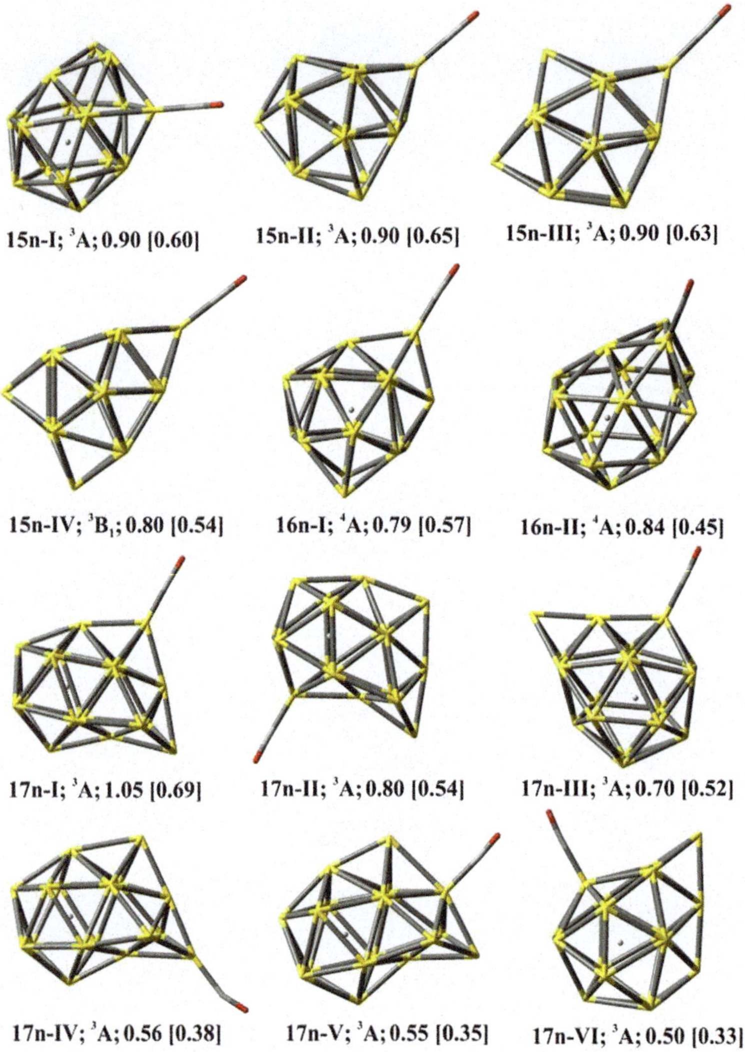

Fig. 4.12 Lower-lying structures considered for Au_nVCO complexes ($n = 15-17$) and their electronic states and CO adsorption energies. The values in *square brackets* [eV] are computed using the B3LYP functional

n = 2. The lowest-energy structure of Au_2VCO **2n-I** has C_s symmetry in which CO directly attaches to V atom of AuVAu (Fig. 4.10). The V–CO distance in Au_2VCO is ~1.99 Å, compared to 2.00 Å of AuV–CO. However, its CO binding energy of 1.26 eV is smaller than that of AuVCO (1.50 eV, BB95).

n = 3. The global minimum found for Au_3VCO, i.e. **3n-I** in Fig. 4.10, is a triangular pyramid (C_s) with a V–C bond length of 1.97 Å. Such geometry is

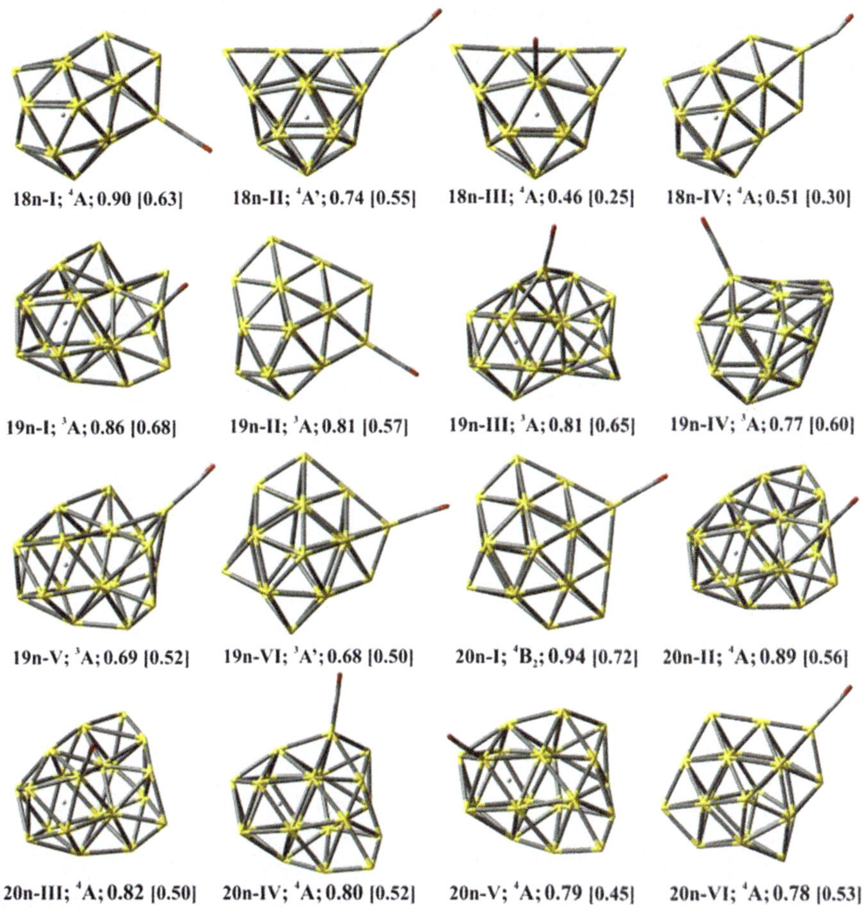

Fig. 4.13 Lower-lying structures considered for Au_nVCO complexes ($n = 18–20$) and their electronic states and CO adsorption energies. The values in *square brackets* [eV] are computed using the B3LYP functional

slightly distorted from C_{3v} symmetry and it is the first 3D structure of mono-carbonyl complexes located in this work.

n = 4 and 5. Optimal structures of both Au_4VCO **4n-I** and Au_5VCO **5n-I** adducts (Fig. 4.10) have also 3D shapes and formed by CO attacking the dopant V. The former has a V–C distance of around 1.99 Å and a CO binding energy of 1.28 eV, while these values of the latter amount to ~1.98 Å and 1.36 eV, respectively.

n = 6 and 7. By adding one CO molecule to the second lowest-energy Au_6V **6-II** through the dopant site, we obtain the most stable form Au_6VCO **6n-I**. The complexes built upon the most stable Au_6V **6-I**, i.e. **6n-II** and **6n-III** in Fig. 4.10, are calculated to be around 0.32–0.42 eV above **6n-I**. The optimal structure **7n-I** of

Table 4.4 Ground and lower-lying states of Au_nVCO complexes, with $n = 1-14$, and relative energies (RE) computed using the BB95 and B3LYP functionals

			RE (eV)					RE (eV)	
n	Isomers	State	BB95	B3LYP	*n*	Isomers	State	BB95	B3LYP
1	**1n-I** (C_s)	$^5A'$	0.00	0.00	10	**10n-I** (C_s)	$^2A'$	0.00	
	1n-II (C_s)	$^5A'$	1.07	0.95			$^4A''$	0.75	
2	**2n-I** (C_s)	$^4A''$	0.00	0.00		**10n-II** (C_1)	^2A	0.28	
3	**3n-I** (C_s)	$^3A''$	0.00	0.00			^4A	0.59	0.75
4	**4n-I** (C_s)	$^4A''$	0.00	0.00		**10n-III** (C_1)	^4A	0.47	0.00
5	**5n-I** (C_s)	$^3A''$	0.00	0.00		**10n-IV** (C_1)	^2A	0.53	
6	**6n-I** (C_2)	^4A	0.00	0.00	11	**11n-I** (C_s)	$^1A'$	0.00	
	6n-II (C_s)	$^4A'$	0.32	0.10			$^3A'$	0.78	0.75
	6n-III (C_s)	$^4A'$	0.42	0.29		**11n-II** (C_s)	$^1A'$	0.34	
7	**7n-I** (C_s)	$^3A''$	0.00	0.00			$^3A''$	0.36	
	7n-II (C_1)	^3A	0.24	0.66		**11n-III** (C_{2v})	$^3B_2$	0.54	0.00
	7n-III (C_s)	$^3A''$	0.40	0.19	12	**12n-I** (C_1)	^2A	0.00	0.56
8	**8n-I** (C_{2v})	$^4A_2$	0.00	0.75		**12n-II** (C_{2v})	$^2A_1$	0.37	0.00
	8n-II (C_s)	$^4A''$	0.44	0.00	13	**13n-I** (C_s)	$^1A'$	0.00	0.00
	8n-III (C_s)	$^4A''$	0.44	0.75		**13n-II** (C_s)	$^1A'$	0.07	0.05
9	**9n-I** (C_3)	^3A	0.00	0.70		**13n-III** (C_s)	$^1A'$	0.08	0.07
	9n-II (C_s)	$^3A''$	0.01	0.21	14	**14n-I** (C_s)	$^2A'$	0.00	0.00
	9n-IV(C_s)	$^3A''$	0.46	0.21		**14n-II** (C_s)	$^2A'$	0.11	0.17
	9n-V (C_s)	$^3A''$	0.81	0.00		**14n-III** (C_s)	$^2A'$	0.26	0.30

the larger system Au_7VCO can be viewed as arising from attachment of one more gold atom to **6n-I**, while the next low-lying conformation **7n-II** with the 3D backbone Au_7V is located at around 0.24 eV (BB95) above.

n = 8 and 9. These systems appear to be extremely sensitive with the functionals employed. Indeed, at the BB95 level, the form **8n-I** which is built upon**8-I**, is computed to be 0.44 eV lower in energy than **8n-II** which comes from a planar configuration of Au_8V **8-III** (Fig. 4.10). At the same level, the most stable form **9n-I** of Au_9VCO is obtained by adding one more gold atom to **8n-I** and possesses a threefold axis. A slight distortion of **9n-I** leads to **9n-II**, whose energies are separated from each other by only 0.01 eV. The isomer **9n-V** coming from the 2D Au_9V **9-III** is 0.81 eV less stable than **9n-I** (Fig. 4.10).

However, B3LYP computations depict a much different energy landscape. Accordingly, two forms **8n-II** and **9n-V** become the global minima. The **8n-I** is now a local minimum, being 0.75 eV above **8n-II**, while **9n-I** is characterized at this level to be a transition state with an imaginary frequency of $25i$ cm^{-1}. The latter functional is thus significantly favor the 2D conformations of the Au_nV moiety. Further information on relative energies at both levels of calculation is given in Table 4.4.

n = 10. At the BB95 level, the global minimum **10n-I** of $Au_{10}VCO$ is predicted as a derivative of **9n-I** (Fig. 4.11). This adduct is found to exhibit a doublet $^2A'$ GS. Attaching the CO molecule to the dopant atom of **9n-II** gives rise to the lower-lying isomers **10n-II** and **10n-III**. These forms are computed to be around 0.28–0.47 eV (BB95) above the GS. Although **10n-II** is less stable than **10n-I**, its CO binding energy (1.52 eV) is somewhat larger than that of **10n-I** (1.45 eV). In addition, it is interesting to note that B3LYP calculations confirm **10n-III** as the global minimum. The **10n-II** is now a first-order saddle point with an imaginary frequency of 20*i* cm^{-1} and 0.75 eV above **10n-III**. We could not locate both **10n-I** and **10n-IV** by using the B3LYP.

n = 11. Again, a growth pattern going from **10n-I** yields the most stable $Au_{11}VCO$ **11n-I**, while anchoring of the CO on the top-side gold atom of an incomplete icosahedron $Au_{11}V$ leads to **11n-II** (Fig. 4.11). Both prefer to exist in a low-spin $^1A'$ state and are separated from each other by 0.34 eV. This is also the last cluster of the series in which the impurity is the most favorable site for binding carbonyl group.

The next isomer **11n-III**, which is built upon an incomplete cuboctahedron, is confirmed to be a transition state as it has a tiny imaginary frequency of 10*i* cm^{-1} and is ~0.54 eV less stable than **11n-I**. However, B3LYP predicts **11n-III** as the lowest-energy form of $Au_{11}VCO$, while **11n-I** is 0.75 eV above and has also an imaginary frequency of 27*i* cm^{-1}.At this level, CO preferentially adsorbs on a vertex gold atom rather than on the V atom. As a result, the CO absorption energy abruptly drops in going from $Au_{10}V$ (0.90 eV) to $Au_{11}V$ (0.59 eV), computed using the B3LYP functional.

n = 12. From $Au_{12}V$, CO clearly favors an attachment on Au atoms as the dopant V is now located completely inside the gold cage. We find two forms competing for the ground state of $Au_{12}VCO$, i.e. **12n-I** and **12n-II**, with CO molecule capping on one apex gold atom of an icosahedron and of a cuboctahedron, respectively (Fig. 4.11). The latter is located at 0.37 eV above the former and has a small imaginary frequency of 15*i* cm^{-1} at BB95 level. However, B3LYP calculations reverse the energy ordering, in that the **12n-I** is now not even a local minimum and lies 0.6 eV above **12n-II**. These observations can be understood as cuboctahedra and icosahedra are very closely related and they can easily interconvert to each other [91].

Previously, Graciani et al. [46] examined a multiple CO adsorption on the anion $Au_{12}V^-$ using the plane-wave pseudo-potential DFT calculations. These authors reported an average binding energy of about 1.0 eV. Nevertheless, this value cannot directly be compared to the present values for the neutral $Au_{12}VCO$, due in addition to the previous use of a different method that we cannot calibrate.

n = 13 and 14. Contrary to $Au_{12}VCO$, several genuine minima with small energy separations are located for $Au_{13}VCO$ and $Au_{14}VCO$ at both levels of theory, BB95 and B3LYP. For instance, using BB95 functional, two forms **13n-II** and **13n-III** are predicted to be 0.07 and 0.08 eV, respectively, above the most stable form **13n-I**. In this case, all these conformations can be considered as candidates for the GS of the complex $Au_{13}VCO$ (Fig. 4.11).

Similarly, some nearly iso-energetic structures of $Au_{14}VCO$ are also detected. Three isomers **14n-II**, **14n-III** and **14n-III** are calculated to be at 0.11–0.26 eV above the most stable form **14n-I** (Fig. 4.11). One noticeable and common point here is that the most stable form of these complexes is based on the core of an icosahedral cluster $Au_{12}V$ and all exhibit the lowest possible spin state as their GS.

n = 15 and 16. Adding the CO molecule to the iso-energetic forms **15-I** and **15-II** of the bimetallic species $Au_{15}V$ through the lowest-coordinated gold atoms gives rise to the most stable isomers **15n-I** and **15n-II**, respectively, as shown in Fig. 4.12. These structures are also nearly degenerate, being separated by only 0.04 eV (BB95 value). Other conformations located for $Au_{15}VCO$ include **15n-III** and **15n-IV**, which are predicted to stay around 0.18–0.19 eV above **15n-I**.

The most stable forms of $Au_{16}VCO$, i.e. **16n-I** and **16n-II**, are also built up from attaching CO to the lowest-lying isomers **16-I** and **16-III** of $Au_{16}V$. At the BB95/cc-pVDZ-PP level, the **16n-I** is computed to be 0.04 eV more stable than **16n-II**, but the former somewhat has a lower CO binding energy than the latter. In addition, as in smaller systems, the favorite sites for CO attacking are the low-coordinated gold atoms.

n = 17. The energetically optimized configuration obtained for $Au_{17}VCO$ **17n-I** can be viewed as the derivative of **17-II**, instead of **17-I** (Fig. 4.12). The **17n-I** also exhibits a particularly high stability in comparison with other structures detected, as the second most Stable **17n-II** is computed at 0.25 eV above. The local minimum coming from the most stable cluster **17-I** of $Au_{17}V$, i.e. **17n-III** in Fig. 4.12, is around 0.33 eV higher in energy than **17n-I**. Other possible structures including **17n-IV, 17n-V** and **17n-VI** are located from 0.50 to 0.53 eV above GS.

n = 18. Two isomers **18n-I** and **18n-II**, which can be considered to be degenerate with an energy gap of 0.07 eV, are candidates for the global minimum of $Au_{18}VCO$. While the former is the product of CO adsorption on **18-II**, the latter is built upon **18-I**. More interestingly, though these forms have comparable energy content, the CO binding energy of **18n-I** (0.90 eV) is quite larger than that of **18n-II** (0.74 eV), computed using BB95 functional. The next isomers, namely **18n-III** and **18n-IV** in Fig. 4.13, are computed to be around 0.35–0.39 eV less stable than **18n-I**.

n = 19. Relevant structures located for $Au_{19}VCO$ are also plotted in Fig. 4.13. Amongst these, four isomers **19n-I, 19n-II, 19n-III**, and **19n-IV** with comparable energies are competing for the GS. Both structures **19n-I** and **19n-III** are derived from the attachment of CO to **19-II**, whereas the CO addition to the lowest-energy cluster **19-I** yields either **19n-II** or **19n-IV**. The relative energies of **19n-II, 19n-III**, and **19n-IV** with respect to the lowest-lying structure **19n-I** are predicted to slightly vary in the range 0.02–0.06 eV (see Table 4.5). In addition, it is worth noting for systems from $n = 17$ that CO molecule apparently prefers to anchor on the apex gold atoms of metal clusters, forming the corresponding adducts with compact structures.

In order to probe the effects of impurity on the ability to attract CO of the host clusters, we now explore again the optimal structures of the Au_nCO complexes resulting from addition of CO to the lowest-energy pure gold clusters. Optimized geometries of the Au_nCO species, and then CO adsorption energies, are computed using the BB95 and B3LYP functionals in conjunction with the same cc-pVDZ-PP basis set for Au and the cc-pVTZ basis set for C and O.

Table 4.5 Ground and lower-lying states of Au_nVCO complexes, with n = 15–20, and relative energies (RE) computed using the BB95 and B3LYP functionals.

			RE (eV)					RE (eV)	
n	Isomers	State	BB95	B3LYP	n	Isomers	State	BB95	B3LYP
15	**15n-I** (C_1)	^3A	0.00	0.10	18	**18n-III** (C_1)	^4A	0.35	0.40
	15n-II (C_1)	^3A	0.04	0.00		**18n-IV** (C_s)	$^4A'$	0.39	0.25
	15n-III (C_1)	^3A	0.18	0.10	19	**19n-I** (C_1)	^3A	0.00	0.15
	15n-IV (C_{2v})	$^3B_1$	0.19	0.10		**19n-II** (C_1)	^3A	0.02	0.00
16	**16n-I** (C_1)	^4A	0.00	0.00		**19n-III** (C_1)	^3A	0.05	0.21
	16n-II (C_1)	^4A	0.04	0.24		**19n-IV** (C_1)	^3A	0.06	0.07
17	**17n-I** (C_s)	$^3A'$	0.00	0.46		**19n-V** (C_1)	^3A	0.18	0.30
	17n-II (C_s)	$^3A'$	0.25	0.72		**19n-VI** (C_s)	$^3A'$	0.32	0.22
	17n-III (C_s)	$^3A'$	0.33	0.00	20	**20n-I** (C_{2v})	$^4B_2$	0.00	0.03
	17n-IV (C_s)	$^3A'$	0.49	0.80		**20n-II** (C_1)	^4A	0.06	0.00
	17n-V (C_s)	$^3A'$	0.50	0.82		**20n-III** (C_1)	^4A	0.13	0.08
	17n-VI (C_1)	^3A	0.53	0.86		**20n-IV** (C_1)	^4A	0.15	0.10
18	**18n-I** (C_1)	^4A	0.00	0.11		**20n-V** (C_1)	^4A	0.16	0.12
	18n-II (C_s)	$^4A'$	0.07	0.00		**20n-VI** (C_s)	$^4A'$	0.16	0.22

Previously, several groups [29, 33, 35, 50] also employed DFT calculations to predict the CO adsorption on both neutral and charged Au_n clusters containing up to ten gold atoms. CO adsorption on the larger clusters, i.e. Au_{11}, Au_{12} and Au_{13}, was also evaluated [31], but their structures are much different from the lowest-energy isomers recently reported [13]. In this work, for the purpose of comparison, we extend the investigations up to the size of Au_{21} and the results are shown in Figs. 4.14 and 4.15.

In general agreement with previous studies, our results point out that CO prefers to be attached at the on-top sites of pure gold clusters. The CO binding energies are positive for all clusters, implying that the adsorption reactions are exothermic. The BB95 values agree well with previous predictions [29, 32, 33, 46, 50], while those obtained from the B3LYP are consistently lower. Although the measured CO adsorption energies were reported in the literature for some small anions Au_n^- (n = 1–3) [39], and a series of cations Au_n^+ (n =1–65) [79], no experiment is available for the neutral systems.

Initial structures for Au_nCO geometry optimization are constructed by attaching CO to the optimal structures of pure gold clusters through low-coordinated gold atoms. We find that attachment of CO causes rather negligible structural changes for Au_n species. In fact, the Au–Au bond length of Au_2CO is ~2.51 Å as compared to ~2.52 Å in free Au_2.

For larger systems, free gold clusters and gold frameworks in Au_nCO almost bear the similar shape and geometrical parameters. The free CO distance is also nearly conserved after adsorption. These support for the prediction that, owing to its lone pair of electrons, *carbon monoxide tends to react with metal clusters via facile association reactions,* in which either molecular or dissociative chemisorptions are involved, rather than breaking molecular bonds and then making stronger new

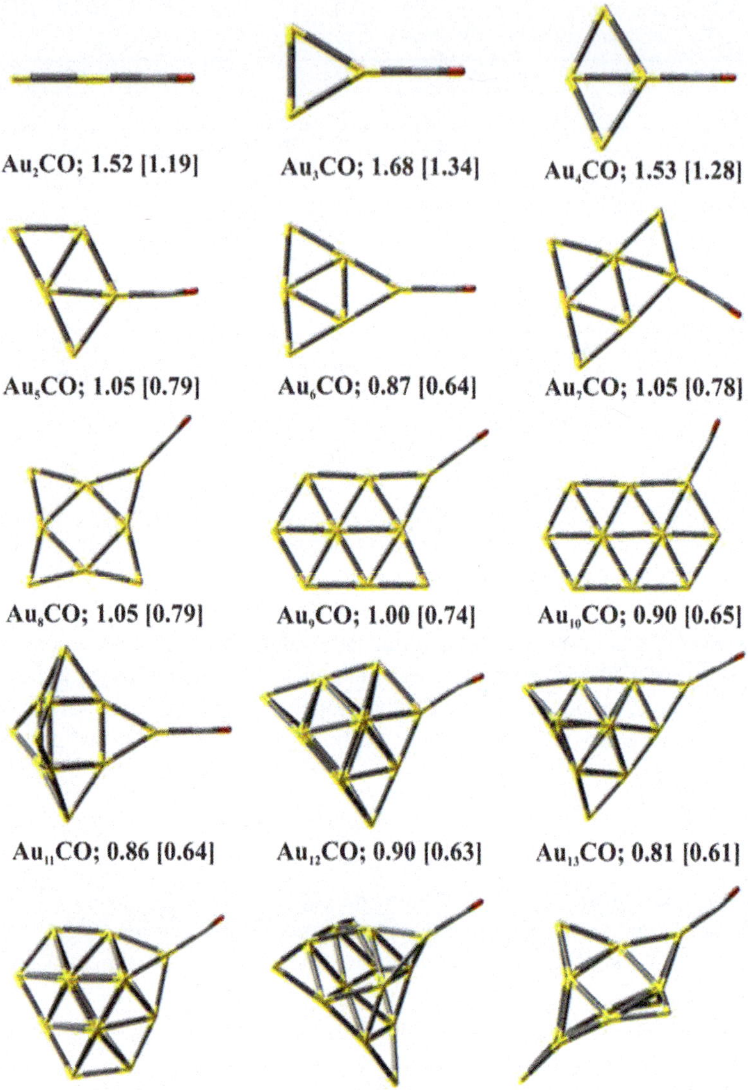

Fig. 4.14 Lower-lying structures considered for the Au$_n$CO complexes (n = 2–15) and their electronic states and CO adsorption energies. The values in *square brackets* [eV] are computed using the B3LYP functional

bonds between the resulting fragments and the metal surface [109]. Similar observations also hold true for the adsorption of several amino acids and DNA bases on small gold clusters [110]. The property to retain the original form of adsorbed molecules makes such clusters of great interest for the use in biochemical sensing and detection [111].

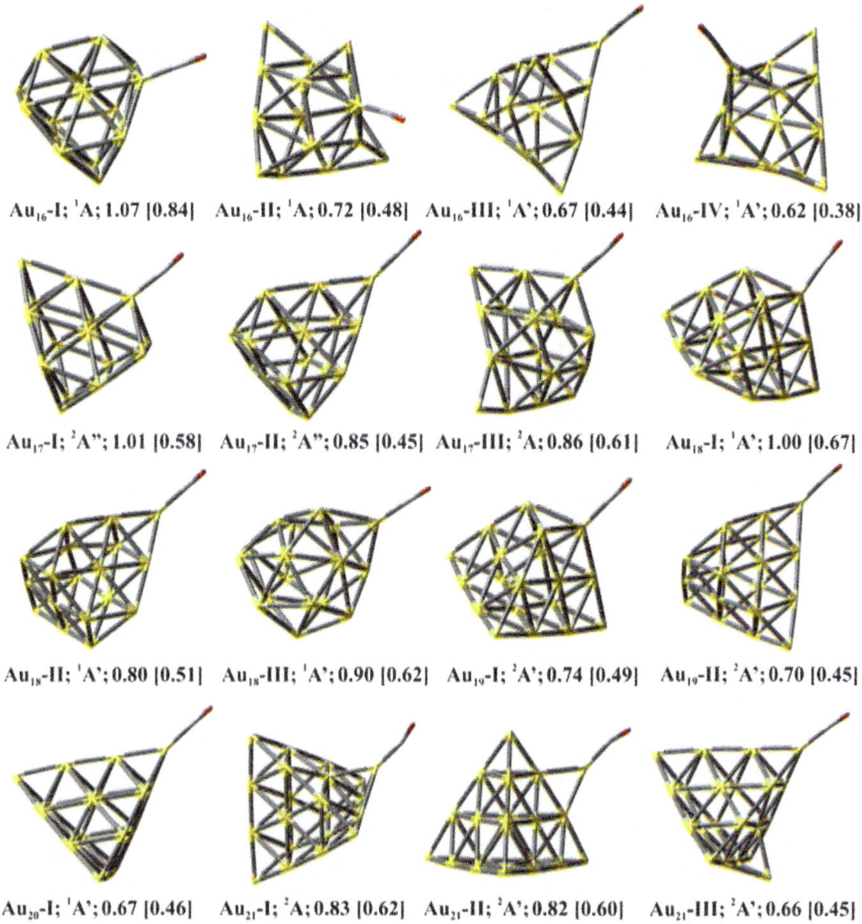

Fig. 4.15 Lower-lying structures considered for the Au_nCO complexes ($n = 16-21$) and their electronic states and CO adsorption energies. The values in *square brackets* [eV] are computed using the B3LYP functional

4.3.3.2 Effects of Vanadium on the CO Adsorption

Binding Energies and Charge Populations. In recent papers, Lin and co-workers [50] confirmed that the CO affinity of gold clusters in general decreases upon replacement of one gold atom by one yttrium (Y). As previously pointed out [26], the presence of Y draws an increase of electron density on gold atoms due to the charge transfer from Y to Au. Such increase leads to a shift of the lowest unoccupied molecular orbitals towards higher energies and thereby weakens the σ-donor bond of CO and gold atoms [40].

Similarly, the adsorption energies of CO on the pure gold clusters in cationic states are also greater than those on the neutral and anionic states because the

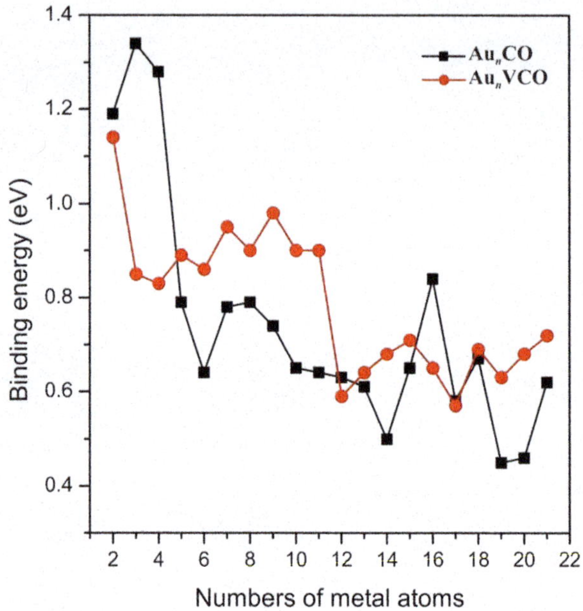

Fig. 4.16 CO adsorption energies on Au$_n$V and Au$_n$ clusters as a function of cluster size. The results are obtained at the B3LYP/cc-pVDZ–PP level

CO → metal electron donation is typically more important than the metal → CO back donation [29]. In other words, the CO molecule tends in these cases to play its classical role as a nucleophile, and thus its attack to positively charged sites is invariably favored.

Figure 4.16 illustrates the variations of CO adsorption energies on both V-doped Au$_n$V and pure Au$_n$ systems. As stated above, though the BB95 functional appears to be more reliable than the B3LYP in predicting geometries and energetics of pure and V-doped gold clusters, it tends to overestimate the CO absorption energy, where the latter appears to describe somewhat better. Hence, we here report the B3LYP values for this quantity.

For pure gold species considered, a general feature of magic clusters Au_6, Au_{14} and Au_{20} is their particularly low CO adsorption energies (<0.50 eV), while the most unstable Au_3system induces the highest value (~1.34 eV). Another striking result is that the CO adsorption energies of the smallest clusters sizes, i.e. Au_2, Au_3 and Au_4, turn out to be much higher than those of the larger sizes. The tetrahedron Au_{16}, i.e. structure **Au$_{16}$-II** shown in Fig. 4.7, also has an exceptionally high E_{ads} value (0.84 eV), though this isomer is relatively less stable than the C_s isomer **Au$_{16}$-I**. The former is computed to stay around 0.33 and 0.77 eV above the latter at both BB95 and B3LYP levels of theory, respectively.

Concerning the bimetallic Au$_n$V clusters, the largest binding energy is predicted for the dimer AuV (1.14 eV), then a significant drop occurs in going to larger sizes. From Au_3V to $Au_{10}V$, the E_{ads} values exhibit the even-odd oscillations with distinct maxima at n = 4, 6 and n = 8. The remaining species with only Au sites exposed, namely the $Au_{12}V$–$Au_{20}V$ cages, have a much smaller CO absorption energy, in which the highest value (0.71 eV) is found for $Au_{14}V$.

As seen in Fig. 4.16, introduction of V leads to a significant change in the trend of CO absorption on gold clusters. Particular differences are observed for clusters in the range of $n = 1-10$ as CO directly attaches to V atom and become less pronounced for larger clusters. Except for small sizes of $n = 1-3$, such a presence is accompanied with an increase of CO adsorption affinity. After replacing one Au by V, the CO binding energies of Au_3 and Au_4 are reduced from 1.34 (Au_3) and 1.28 (Au_4) to 0.85 (Au_2V) and 0.83 eV (Au_3V), respectively.

The larger clusters, on the contrary, generally have the higher E_{ads} values than their isoatomic pure counterparts Au_n, with an exception of Au_{16}. The $Au_{11}V$ cluster is quite exceptional as BB95 predicts it has a higher E_{ads} (1.30 eV) than Au_{12} (0.90 eV), while B3LYP yields a reversed result.

Interactions between metals and CO moiety in carbonyl complexes are basically characterized by donor and acceptor factors, in such a way that the charge transfer constitutes a key element for understanding their mechanism. The natural charges distributed on the adsorbed CO in Au_nVCO adducts are summarized in Table 4.6. The results obtained within BB95 level clearly show that adsorbed CO is negatively charged on Au_{1-4} V clusters, indicating that an amount of electron has been transferred back from metal to CO. AuVCO is a special case, as it possesses a π-back bonding between V and CO that is more typical than other Au_nVCO adducts. Indeed, the adsorbed CO in **1n-I** is much more negatively charged as compared to the adsorbed CO in other complexes (see Table 4.6).

From Au_5VCO to $Au_{14}VCO$, we find a minor electron density flow from CO to metals as the full charge of CO in these complexes is slightly positive. However, there also exist some exceptions. Indeed, the CO in$Au_{11}VCO$ and $Au_{12}VCO$is likely to play a role of electron acceptor. In other words, the negative charge has been transferred from clusters to CO in these species, though it is not significant as in AuVCO. A similar process is also observed for systems with n from 15 to 20. The full charge of CO now is consistently negative, indicating a charge transfer from metals to the adsorbed CO.

Bonding Lengths and Vibrational Frequencies. In Table 4.6 we also list the computed results for CO adsorption process on binary Au_nV clusters, including bond lengths, CO and M–CO stretching frequencies, NBO charges of the CO in complexes, and CO binding energies. As a consequence of the stronger relativistic effects, the Au–CO bond lengths in gold-carbonyl complexes are consistently shorter than those of the V–CO counterparts in the vanadium-carbonyl complexes (Table 4.6). However, the latter is characterized by a much larger E_{ads} value than the former. Especially, the smallest vanadium-carbonyl complex AuVCO has the longest V–CO and C–O bond distances. In this case, an amount of electron has effectively been transferred from metal to CO, as discussed above. Thus the existence of a π back-donation enhances the metal–CO bond in weakening the C–O bond.

The interaction between AuV and CO is the strongest one which is consistent with the fact that the V–CO bond has a partial triple character. Therefore, in spite of having a longer M–CO distance, AuVCO has the largest CO binding energy.

Table 4.6 Calculated data for the CO adsorption on bimetallic Au_nV clusters. The values are computed at BB95/cc-pVDZ-(PP) level of theory[a]

	Bond distance (Å)			ν_{CO}	ν_{M-CO}	NBO Charge		
Species	V–CO	Au–CO	C–O	(cm^{-1})	(cm^{-1})	C	O	E_{ads} (eV)
1n-I	2.005		1.159	1,961	418	0.22	−0.40	1.50
1n-II		2.044	1.153	1,988	331	0.27	−0.41	0.43
2n-I	1.985		1.152	1,997	411	0.32	−0.38	1.26
3n-I	1.959		1.151	1,996	427	0.37	−0.38	1.33
4n-I	1.984		1.151	1,997	410	0.38	−0.38	1.28
5n-I	1.973		1.150	2,006	417	0.42	−0.38	1.36
6n-I	1.985		1.149	2,011	407	0.44	−0.38	1.42
7n-I	1.982		1.148	2,014	408	0.45	−0.38	1.36
8n-I	1.998		1.151	1,991	419	0.45	−0.37	1.41
8n-II	1.985		1.148	2,015	406	0.44	−0.37	1.46
9n-I	1.999		1.150	1,997	414	0.49	−0.37	1.40
9n-II	2.000		1.149	2,004	410	0.49	−0.37	1.41
9n-V	1.976		1.147	2,024	412	0.46	−0.37	1.35
10n-I	2.006		1.158	1,940	407	0.43	−0.41	1.45
10n-II	1.986		1.154	1,968	422	0.46	−0.40	1.52
11n-I	2.023		1.166	1,892	402	0.34	−0.46	1.31
12n-I		1.946	1.147	2,049	389	0.35	−0.37	1.07
13n-I		1.943	1.147	2,047	391	0.46	−0.36	1.02
13n-II		1.968	1.147	2,047	369	0.53	−0.36	0.95
14n-I		1.946	1.147	2,048	387	0.48	−0.36	0.93
14n-II		1.982	1.146	2,049	353	0.54	−0.36	0.82
15n-I		1.946	1.147	2,047	384	0.33	−0.35	0.90
15n-II		1.948	1.147	2,046	384	0.32	−0.37	0.90
16n-I		1.958	1.147	2,045	372	0.32	−0.37	0.79
16n-II		1.954	1.146	2,049	376	0.33	−0.37	0.84
17n-I		1.941	1.148	2,041	393	0.32	−0.37	1.05
17n-III		1.971	1.147	2,042	357	0.33	−0.37	0.70
18n-I		1.954	1.146	2,055	379	0.33	−0.36	0.90
18n-II		1.973	1.145	2,054	359	0.33	−0.36	0.74
19n-I		1.952	1.146	2,052	379	0.33	−0.37	0.86
19n-II		1.961	1.146	2,054	369	0.33	−0.37	0.81
20n-I		1.948	1.147	2,046	385	0.32	−0.36	0.94
20n-II		1.952	1.146	2,051	280	0.33	−0.37	0.89

[a]For free CO, the BB95 values of bond length, vibrational frequency and the natural charge distributed on carbon are 1.138 Å, 2,117 cm^{-1} and +0.45 electron, respectively

Nevertheless, this kind of bond significantly weakens the C–O bond since the $\pi_{2p}{}^{*}$ – anti-bonding orbital (LUMO of CO) does not accommodate electron occupancy. This is clearly reflected in the CO vibrational frequency, as shown in Table 4.6. The AuVCO **1n-I** is computed to have the lowest-energy C–O stretching mode among small complexes considered. The CO stretching frequency in AuVCO is ~1,961 cm^{-1} (BB95 value), while in other complexes such vibrations occur at generally above 2,000 cm^{-1}.

Fig. 4.17 Variation of CO stretching frequencies in the optimal Au_nCO and Au_nVCO complexes as a function of cluster size

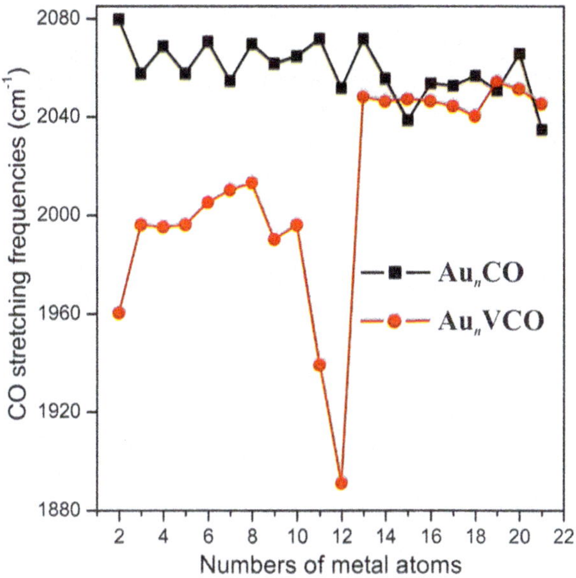

The presence of V as dopant results in a larger C–O frequency red shift. The evolution of CO stretching frequencies of Au_nCO and Au_nVCO as a function of cluster size is illustrated in Fig. 4.17. In agreement with previous reports [29, 33, 39, 50], the CO frequencies in Au_nCO adducts exhibit a clear odd–even oscillation up to Au_{10}. However, following replacement of one Au atom by one V, such a pattern is destroyed. Instead, the CO frequencies of Au_nVCO complexes are predicted to increase monotonically with respect to cluster size up to Au_7VCO. Then an abrupt drop takes place at Au_8VCO, reaching the bottom at $Au_{11}VCO$, and followed by a sudden increase. The large C–O frequency red-shift in $Au_{11}VCO$ is found to correlate with the significant charge transfer from metals to CO.

In general, we find that CO stretching frequencies are shifted to a much distinguishably lower frequency range in the species that CO directly attaches to V atom. For instance, CO frequencies in **1n-I** and **2n-I** amount to 1,961 and 1,997 cm^{-1}, respectively, as compared to ~2,050 cm^{-1} of **12n-I** and **13n-I**. However, no linear correlation can be established between CO vibrational frequency (ν_{CO}) and adsorption energy (E_{ads}). For example, while the E_{ads} values follow the ordering **6n-I** > **5n-I** ≈ **7n-I**, the CO frequency sequence is **7n-I** > **6n-I** > **5n-I**. Nonetheless, an interesting point is that the gold-carbonyl complexes, i.e. systems larger than$Au_{11}VCO$, consistently have higher ν_{CO} values and lower binding energies than corresponding vanadium-carbonyl complexes. Such a correlation between ν_{CO} and E_{ads} will be discussed in a following Section.

4.3.3.3 Chemical Bonding in Au_nVCO Complexes

As stated above, the interaction between CO and transition metals in carbonyl complexes is classically characterized by both σ and π electron components. An overlap of the nonbonding C electron pair with the partially filling *d*-orbitals gives rise to the σ bond. On the other hand, the π bonds arise from an overlap of filled *d*-orbitals on the metal with a pair of $\pi_{2p}{}^{*}$- anti-bonding orbitals of CO.

Gold has no empty *d* orbitals (the *d* shell being fulfilled) and thus an electron donation from the σ_{2p} nonbonding CO orbital to gold atoms is a rather difficult process. On the contrary, vanadium with partially filling *d* orbitals is willing to donate as well to accept electrons. Hence, interaction between CO and V turns out to be stronger than that of CO and Au. This is in line with our calculated results predicting that for small Au_nV clusters, the dopant V binds more favorably to CO than the Au core. For larger clusters, Au atoms become the favored sites simply due to a steric effect, as V is now completely confined inside a gold cage.

Information from frontier orbitals further provides us with some insights into the bonding mechanism between CO and clusters. The most important combinations are related to the highest occupied (HOMO) and the lowest unoccupied (LUMO) molecular orbitals of CO and frontier orbitals of clusters. Depending on their symmetry and energy gap, the forward or backward donation bond will be prevailing. The problem becomes more complicated in open-shell systems because the singly occupied molecular orbitals (SOMO) can equally donate or accept electrons. Nevertheless, the orbital symmetry and energy are also the key factors deciding the bonding formation.

Let us take a look at AuVCO as a simple example. The GS of AuV is a quintet state with the valence orbital configuration $^5\Delta$: $\ldots(2\sigma)^2(3\delta)^1(2\pi)^1(2\pi)^1(3\sigma)^1$. Shapes and symmetries of these MOs are plotted in Fig. 4.1 (see above). In this case, the energy difference between the 3σ orbital (AuV) and the CO σ_{2p}-nonbonding (the HOMO of CO) is much larger than that between 2π orbitals (AuV) and the CO $\pi_{2p}{}^{*}$-anti-bonding (the LUMO of CO). Hence, the charge transfer from 2π orbitals of AuV to LUMO(CO) is more effective than the reversed process, which is in agreement with the NBO analysis above. In addition, the LUMO(CO) can interact with the orbital 2δ(AuV), forming further π-type bonding orbital. Consequently, the back-donation plays a much more important role than the forward-transfer, and in AuVCO a charge amount is actually transferred from AuV to CO. The combination between these orbitals is illustrated in Fig. 4.18.

It is also stressed that the singly occupied orbitals of AuV have large contributions from *d*(V) AOs. As a consequence, CO prefers to bind to V than to Au site. The resulting adduct thus prefers a bent structure over a linear form as a result of the 2δ(V) – LUMO(CO) interaction. A similar interaction disappears in molecules with $C_{\infty v}$ geometry because the symmetry of d_δ(V) orbitals and LUMO(CO) does not overlap with each other.

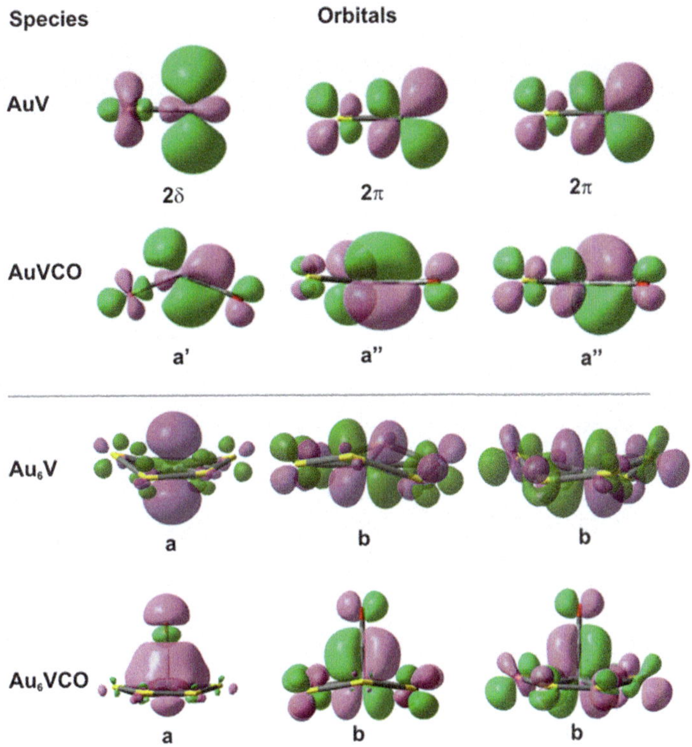

Fig. 4.18 Shapes and symmetries of frontier MOs for AuV (*upper*) and Au_6V (*lower*) before and after interactions with CO

Figure 4.18 in addition shows the shapes and symmetries of orbitals generated from combinations between frontier MOs of CO and Au_6V. In its GS, Au_6V (C_2) has an orbital configuration of ^4B: ...$(1a^2 1b^1 2b^1 3b^1 2a)$.

Due to a difference in symmetry, neither the LUMO nor the HOMO of CO does combine with the SOMO 3b. In contrast, while combinations between lower-energy SOMOs 1b and 2b with LUMO(CO) yield the π-bonding orbitals, the LUMO 2a can interact with HOMO(CO) giving rise to the σ-bonding orbital. These bonding orbitals are fully occupied. This demonstrates the existence of both forward and backward donations but the former (charge transfer from CO to metal) is somewhat dominant over the latter in **6n-I** adduct, as mentioned above, based on NBO charges.

A direct correlation between the CO frequency and CO adsorption energy is not clearly established because both parameters depend on two different bonding environments. While the former directly depends on the C and O bonding, the latter is the consequence of the C to metal attachment. However, in the whole system, one type of bonding may to some extent affect the other, and we find that gold-carbonyl complexes with smaller binding energies constantly possess higher CO frequencies than vanadium-carbonyl complexes. This observation can be

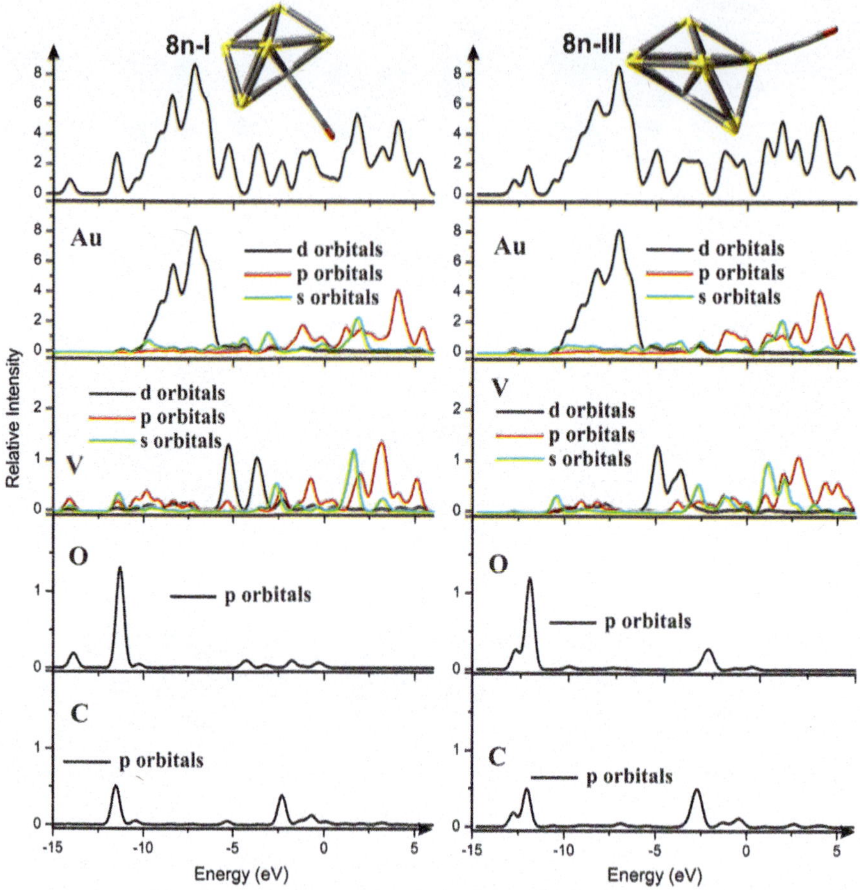

Fig. 4.19 Total and partial densities of states for two Au_8VCO isomers

understood on the basis of an analysis of densities of states (DOS) for the atoms in a complex. The DOS plots are here simply the graphical distributions of the contributions of the atomic orbitals to the different molecular orbitals.

As for an illustration, Fig. 4.19 displays the DOS of two isomers of Au_8VCO, namely **8n-I** and **8n-III**. Both isomers have the same Au_9V backbone but the site bound to CO and the binding energy differ from each other.

In the stronger adsorption complex **8n-I**, CO is attached to V whereas in the weaker adsorption complex **8n-III**, CO absorbs on a low-coordinated Au atom. In **8n-III**, two distinct bonding peaks between the C and O atoms are visibly observed in the DOS graphic. One peak is located at around −12.0 eV and the other is above −2.5 eV. However, the peak corresponding to the CO bonding at around −2.5 eV does not appear on the DOS of **8n-I**. Therefore, the CO bond in **8n-I** is getting weaker compared to that in **8n-III**, which is manifested in the CO frequency shift. In **8n-III**, the CO frequency is actually shifted from 2,117 cm^{-1} of free CO to 2,052 cm^{-1}, as compared to 1,991 cm^{-1} in **8n-I**.

In general, vanadium-carbonyl complexes have a weaker V-CO bond and a larger CO frequency shift than gold-carbonyl complexes (see Table 4.6) as a result of a shrink of the bonding peak between C and O at around –2.5 eV. The DOS plots also emphasize that the interaction between V and CO in **8n-I** is more effective than that involving Au and CO in **8n-III**. The resulting binding energy of **8n-I** (1.41 eV) is in fact much larger than that of **8n-III** (0.97 eV).

4.4 Concluding Remarks

In this theoretical study, we performed a systematic investigation on the electronic, geometric and energetic properties and the CO adsorption of bimetallic clusters Au_nV in the range of $n = 1{-}20$, using the BB95 and B3LYP functionals in conjunction with the pseudo-potential cc-pVDZ-PP basis set for metals and the full-electron cc-pVTZ basis set for non-metals. The effects of the dopant atom are analyzed in detail. A number of interesting results emerge as follows:

(i) All neutral clusters with an even number of electrons prefer a high spin state, namely a quintet state for AuV and triplet state for others with an exception of the singlet GS for $Au_{11}V$ and $Au_{13}V$. For odd-electron systems, a quartet state is invariably favored as the ground state (GS) up to Au_8V, while systems with $n = 10{-}14$ are likely to prefer a low spin (doublet) GS. However, larger systems again exhibit a high-spin quartet GS.
(ii) Concerning the growth pattern, we can now draw some key points. The clusters prefer 2D geometries up to Au_6V involving a weak charge transfer. On the contrary, the larger systems bear 3D conformations with a more effective electron transfer from gold atoms to vanadium. In general, the lowest-energy isomer of an Au_nV species is built upon the most stable form of Au_{n-1} V.
(iii) Another important observation is that during the growth, vanadium is always endohedrally doped as a way of maximizing its coordination numbers and augmenting the charge transfer.
(iv) Analyses of energetic properties including the bond energies (BEs), fragmentation energies (E_fs), and second-order energy difference (Δ^2E) demonstrate that the presence of vanadium atom considerably enhances the thermodynamic stability of odd-numbered Au clusters but tends to reduce that of even-numbered systems. The atomic shapes have an apparently more important effect on the clusters stability than the electronic structure. Even though Au_2V, Au_6V, $Au_{12}V$, $Au_{14}V$ possess an odd number of electrons, they are rather stable due to their ability to form compact symmetric structures. Especially, if both atomic shape and electronic condition are satisfied, the resulting cluster becomes particularly stable such as the anion $Au_{12}V^-$, which

can thus combine with the cation Au^+ to form the superatomic molecule of the type $[Au_{12}V]Au$.

(v) The calculated properties of the systems with $n = 8-12$ turn out to be extremely sensitive with the functionals employed. The hybrid functional B3LYP is significantly biased towards 2D conformations for metal frameworks. Moreover, while the meta-GGA BB95 predicts an icosahedral growth pattern in these species, the former yields a cuboctahedral evolution. In addition, several lower-lying isomers with very small energy gaps (<0.2 eV) are located for a certain system, in such a way that DFT computations cannot clearly establish their ground electronic states. Thus further information from experiment, for example the far-IR spectra of binary metallic complexes, is necessary for more definitive assignment of their equilibrium structures.

(vi) For the adsorption of CO on these binary clusters, we find that when both sites Au and V are exposed, CO adsorption is energetically favorable on V atom because with partially filling *d* orbitals vanadium is more willing to undergo an interaction with either empty or fulfilled orbitals of CO. However, for larger systems, when vanadium is now completely doped inside the gold cage, low-coordinated Au atoms become, as in the pure clusters, the preferred sites of adsorption.

(vii) The binding energies of Au_nVCO adducts are also computed and compared with those of $Au_{n+1}CO$ species to evaluate the effects V on the CO adsorption affinity. In the smallest systems such as Au_2V, Au_3V and Au_4V, the interaction of CO with V-doped clusters is weaker than that with the pure hosts. For the larger systems, *the presence of V consistently results in a reinforcement of CO adsorption*. Among the doped clusters considered, AuV has the highest CO adsorption affinity as it can form a typical π-back donation bond between V and CO. Further examination of densities of states clearly provides support for the view that vanadium-carbonyl complexes have the larger CO binding energies and the larger CO frequency shift than iso-atomic gold-carbonyl counterparts. We should stress that no correlation could be found for the calculated BEs with the recently measured rates of dissociation of the Au_nVCO adducts [52].

Acknowledgements PVN thanks the Can Tho University for support. MTN is indebted to the KU Leuven Research Council (GOA and IDO programs) and thanks ICST and the Department of Science and Technology of Ho Chi Minh City, Vietnam for supporting his stays in Vietnam.

References

1. Haruta M (1997) Catal Today 36:153
2. Valden M, Lai X, Goodman DW (1998) Science 281:1647
3. Sanchez A, Abbet S, Heiz U, Schneider WD, Haekkinen H, Barnett RN, Landman U (1999) J Phys Chem A 103:9573
4. Wallace WT, Whetten RL (2000) J Phys Chem B 104:10964

5. Schwerdtfeger P (2003) Angew Chem Int Ed 42:1892
6. Riboh JC, Haes AJ, McFarland AD, Ranjit C, Van Duyne RP (2003) J Phys Chem B 107:1772
7. Taylor KJ, Pettiette-Hall CL, Cheshnovsky O, Smalley RE (1992) J Chem Phys 96:3319
8. Ho J, Ervin KM, Lineberger WC (1990) J Chem Phys 93:6987
9. Handschuh H, Ganteför G, Bechthold PS, Eberhardt W (1994) J Chem Phys 100:7093
10. Hansen K, Herlert A, Schweikhard L, Vogel M (2000) Phys Rev A73:063202
11. Gruene P, Rayner DM, Redlich B, van der Meer AFG, Lyon JT, Meijer G, Fielicke A (2008) Science 321:674
12. Pyykkö P (2004) Angew Chem Int Ed 43:4412
13. Assadollahzadeh B, Schwerdtfeger P (2009) J Chem Phys 13:064306
14. Majumder C (2007) Phys Rev B 75:235409
15. Majumder C, Kandalam AK, Jena P (2006) Phys Rev B 74:205437
16. Pyykkö P, Runeberg N (2002) Angew Chem Int Ed 41:2174
17. Autschbach J, Hess BA, Johansson MP, Neugebauer J, Patzschke M, Pyykkö P, Reiher M, Sundholm D (2004) Phys Chem Chem Phys 6:11
18. Nhat PV, Nguyen MT (2011) Phys Chem Chem Phys 13:16254
19. Häkkinen H (2008) Chem Soc Rev 37:1847
20. Zhai HJ, Li J, Wang LS (2004) J Chem Phys 121:8369
21. Li X, Kiran B, Li J, Zhai HJ, Wang LS (2002) Angew Chem Int Ed 41:4786
22. Gao Y, Bulusu S, Zeng XC (2005) J Am Chem Soc 127:15680
23. Wang LM, Bulusu S, Zhai HJ, Zeng XC, Wang LS (2007) Angew Chem Int Ed 46:2915
24. Wang LM, Bai J, Lechtken A, Huang W, Schooss D, Kappes MM, Zeng XC, Wang LS (2009) Phys Rev B 79:033413
25. Wang LM, Pal R, Huang W, Zeng XC, Wang LS (2009) J Chem Phys 130:051101
26. Lin L, Claes P, Gruene P, Meijer G, Fielicke A, Nguyen MT, Lievens P (2010) ChemPhysChem 11:1932
27. Haruta M (2003) Chem Rec 3:75
28. Hagen J, Socaciu LD, Heiz U, Bernhardt TM, Wöste L (2003) Eur Phys J D 24:327
29. Wu X, Senapati L, Nayak SK, Selloni A, Hajaligol M (2002) J Chem Phys 117:4010
30. Socaciu LD, Hagen J, Bernhardt TM, Wöste L, Heiz U, Häkkinen H, Landman U (2003) J Am Chem Soc 125:10437
31. Phala NS, Klatt G, van Steen E (2004) Chem Phys Lett 395:33
32. Yuan DW, Zeng Z (2004) J Chem Phys 120:6574
33. Fernández EM, Ordejón P, Balbás LC (2005) Chem Phys Lett 408:252
34. Wang YL, Zhai HJ, Xu L, Li J, Wang LS (2010) J Phys Chem A 114:1247
35. Davran-Candan T, Günay ME, Yildirim R (2010) J Chem Phys 132:174113
36. Blyholder G (1964) J Phys Chem B68:2772
37. Zhao Y, Li Z, Yang J (2009) Phys Chem Chem Phys 11:2329
38. Bernhardt TM, Socaciu-Siebert LD, Hagen J, Wöste L (2005) Appl Catal A 291:170
39. Bernhardt TM, Hagen J, Lang SM, Popolan DM, Socaciu-Siebert LD, Wöste L (2009) J Phys Chem A 113:2724
40. Neumaier M, Weigend F, Hampe O, Kappes MM (2006) J Chem Phys 125:104308
41. Neumaier M, Weigend F, Hampe O, Kappes MM (2008) Faraday Discuss 138:393
42. Popolan DM, Nössler M, Mitric R, Bernhardt TM, Koutecky VB (2011) J Phys Chem A 115:951
43. Joshi AM, Tucker MH, Delgass WN, Thomson KT (2006) J Chem Phys 125:194707
44. Sadek MM, Wang L (2006) J Phys Chem A 110:14036
45. Song C, Ge Q, Wang L (2005) J Phys Chem B 109:22341
46. Graciani J, Oviedo J, Sanz JF (2006) J Phys Chem B 110:11600
47. Fu Y, Li J, Wang SG (2010) J Mol Model 16:9
48. Johansson MP, Pyykkö P (2010) Chem Commun 46:3762
49. Jena NK, Chandrakumar KRS, Ghosh SK (2009) J Phys Chem C 113:17885

50. Lin L, Lievens P, Nguyen MT (2010) Chem Phys Lett 498:296
51. Fernández EM, Torres MB, Balbás LC (2009) Eur Phys J D 52:135
52. Le HT, Lang SM, De Haeck J, Lievens P, Janssens E (2012) Phys Chem Chem Phys 14:9350
53. Frisch MJ et al (2009) Gaussian 09 Revision B.01. Gaussian Inc., Wallingford
54. Becke AD (1988) Phys Rev A 38:3098
55. Becke AD (1996) J Chem Phys 104:1040
56. Peterson KA (2003) J Chem Phys 119:11099
57. Peterson KA, Puzzarini C (2005) Theor Chem Accounts 114:283
58. Harvey JN (2006) Annu Rep Prog Chem Sect C Phys Chem 102:203
59. Cramer CJ, Truhlar DG (2009) Phys Chem Chem Phys 11:10757
60. Johansson MP, Lechtken A, Schooss D, Kappes MM, Furche F (2008) Phys Rev A 77:053202
61. Perdew JP (1986) Phys Rev B 33:8822; 34: 7406
62. Perdew JP (1991) In: Ziesche P, Eschrig H (eds) Electronic structure of solids '91. Akademie Verlag, Berlin, p 11
63. Perdew JP, Burke K, Enzerhof M (1996) Phys Rev Lett 77:3865
64. Becke AD (1993) J Chem Phys 98:5648
65. Lee C, Yang W, Parr RG (1988) Phys Rev B Conden Matter Mater Phys 37:785
66. Stephens PJ, Devlin FJ, Chabalowski CF, Frisch MJ (1994) J Phys Chem 98:11623
67. Schwabe T, Grimme S (2006) Phys Chem Chem Phys 8:4398
68. Adamo C, Barone V (1997) Chem Phys Lett 274:242
69. Ernzerhof M, Scuseria GE (1999) J Chem Phys 110:5029
70. Schmider HL, Becke AD (1998) J Chem Phys 108:9624
71. Tao J, Perdew JP, Staroverov VN, Scuseria GE (2003) Phys Rev Lett 91:146401
72. Staroverov VN, Scuseria GE, Tao J, Perdew JP (2003) J Chem Phys 119:12129; (2004) J Chem Phys 121: 11507
73. Zhao Y, Truhlar DG (2008) Theor Chem Accounts 120:215
74. Raghavachari K, Trucks GW, Pople JA, Gordon MH (1989) Chem Phys Lett 157:479
75. Bartlett RJ, Watts JD, Kucharski SA, Noga J (1990) Chem Phys Lett 165:513
76. Morse MD (1986) Chem Rev 86:1079
77. Koutecký VB, Burda J, Mitrić R, Ge M, Zampella G, Fantucci P (2002) J Chem Phys 117:3120
78. Häkkinen H, Yoon B, Landman U, Li X, Zhai HJ, Wang LS (2003) J Phys Chem A 107:6168
79. Neumaier M, Weigend F, Hampe O, Kappe MM (2005) J Chem Phys 122:104702
80. Glendening ED, Badenhoop JK, Reed AE, Carpenter JE, Bohmann JA, Morales CM, Weinhold F (2001) NBO 5.0 Program, Madison, Wisconsin, USA
81. Iseda M, Nishio T, Han SY, Yoshida H, Terasaki A, Kondow T (1997) J Chem Phys 106:2182
82. Tenderholt A (2005) PyMOlyze-2.0. Stanford University, Stanford
83. Lee HM, Ge M, Sahu BR, Tarakeshwar P, Kim KS (2003) J Phys Chem B 107:9994
84. Xiao L, Tollberg B, Hu X, Wang L (2006) J Chem Phys 124:114309
85. Idrobo JC, Walkosz W, Yip SF, Ögut S, Wang J, Jellinek J (2007) Phys Rev B76:205422
86. Han Y (2006) J Chem Phys 124:024316
87. Li X, Kiran B, Cui LF, Wang LS (2005) Phys Rev Lett 95:253401
88. Zhang M, He LM, Zhao LX, Feng XJ, Luo YH (2009) J Phys Chem C 113:6491
89. Höltzl T, Lievens P, Veszprémi T, Nguyen MT (2009) J Phys Chem C 113:21016
90. Chen MX, Yan XH (2008) J Chem Phys 128:174305
91. Barreteau C, Desjonqueres MC, Spanjaard D (2000) Eur Phys J D 11:395
92. Stener M, Nardelli A, Fronzoni G (2008) Chem Phys Lett 462:358
93. Höltzl T, Veldeman N, Haeck JD, Veszprémi T, Lievens P, Nguyen MT (2009) Chem Eur J 15:3970
94. Lechtken A, Schooss D, Stairs JR, Blom MN, Furche F, Morgner N, Kostko O, von Issendorff B, Kappes MM (2007) Angew Chem Int Ed 46:2944
95. Jug K, Zimmermann B, Calaminizi P, Kçster A (2002) J Chem Phys 116:4497
96. Li J, Li X, Zhai HJ, Wang LS (2003) Science 299:864

97. Li XB, Wang HY, Yang XD, Zhu ZH, Tang YJ (2007) J Chem Phys 126:084505
98. Fernández EM, Soler JM, Garzón IL, Balbás LC (2004) Phys Rev B 70:165403
99. Zhao HY et al (2010) Phys Lett A 374:1033
100. Wang J, Wang G, Zhao J (2002) Phys Rev B 66:035418
101. Schwerdtfeger P, Dolg M, Schwarz WHE, Bowmaker GA, Boyd PDW (1989) J Chem Phys 91:1762
102. Schwerdtfeger P, Dolg M (1991) Phys Rev A 43:1644
103. Die D, Kuang XY, Guo JJ, Zheng BX (2009) J Mol Struct Theochem 54:902
104. Die D, Kuang XY, Guo JJ, Zheng BX (2010) J Phys Chem Solids 71:770
105. Tanaka H, Neukermans S, Janssens E, Silverans RE (2003) J Chem Phys 119:14
106. Die D, Kuang XY, Guo JJ, Zheng BX (2010) Physica A 389:5216
107. Martin TP (1996) Phys Rep 273:199
108. Baletto F, Ferrando R (2005) Rev Mod Phys 77:371
109. Knickelbein MB (1999) Annu Rev Phys Chem 50:79
110. Pakiari AH, Jamshidi Z (2007) J Phys Chem A 111:4391
111. Shafer-Peltier KE, Haynes CL, Glucksberg MR, Van Duyne RP (2003) J Am Chem Soc 125:588

Chapter 5
Structure and Energetics of Polyhedral Oligomeric Silsesquioxane (T_8,T_{10},T_{12}-POSS) Cages with Atomic and Ionic Lithium Species

Habib U. Rehman and Steven R. Gwaltney

Abstract The structures of endohedral complexes of polyhedral oligomeric silsesquioxane (POSS) cage molecules $(HSiO_{3/2})_8$, $(HSiO_{3/2})_{10}$, and $(HSiO_{3/2})_{12}$, containing either atomic or ionic lithium species are determined using density functional theory with the B3LYP functional and the 6-311G(d,p) and 6-311+G(d,p) basis sets. The structures and stabilities of these nanostructures depend on the cage size and the number and charge of the Li species encapsulated in the $(HSiO_{3/2})_8$, $(HSiO_{3/2})_{10}$, and $(HSiO_{3/2})_{12}$ host cages. Li cation encapsulation shows attractive interactions with cage oxygen atoms leading to cage shrinkage. Li anion encapsulation breaks the $(HSiO_{3/2})_8$ host cage. Stable endohedral POSS cages with varying number of neutral and ionic lithium were identified by calculating their inclusion energies and adiabatic and vertical ionization potentials.

5.1 Introduction

Polyhedral oligomeric silsesquioxanes (POSS) are hybrid inorganic-organic cage-like molecules with diameters in the range of 0.7–3 nm. POSS cages have a silicon and oxygen core with the general formula $(RSiO_{3/2})_n$, where R could be a hydrogen or an organic or inorganic ligand [1]. A hydrosilesquioxane cage made of n$(HSiO_{3/2})_n$ units is denoted by the notation "T_n–POSS". Thus, octahidridosilsesquioxane, T_8-POSS consists of eight silicon atoms that are each connected with three cage oxygen atoms and with a single hydrogen atom. Similarly, $(HSiO_{3/2})_{10}$, T_{10}-POSS contains 10 silicon and 15 oxygen atoms with each silicon attached to one hydrogen outside the cage. Dodecahedral silsesquioxane, $(HSiO_{3/2})_{12}$, T_{12}-POSS has 12 cage silicon atoms and 18 cage oxygen atoms along with 12 hydrogen atoms attached to the cage silicons.

H.U. Rehman • S.R. Gwaltney (✉)
Department of Chemistry, Center for Environmental Health Sciences, and HPC² Center for Computational Sciences, Mississippi State University, MS 39762 Starkville, MS, USA
e-mail: gwaltney@ccs.msstate.edu

J. Leszczynski and M.K. Shukla (eds.), *Practical Aspects of Computational Chemistry III*, DOI 10.1007/978-1-4899-7445-7_5,

POSS molecules and their derivatives have been widely combined with organic polymers to generate hybrid inorganic-organic complexes with applications in materials science and catalysis [1–12]. POSS compounds have found potential uses as commercial flow improvers for thermoplastics, as fillers in high performance nanocomposites [2], and as hardening agents for polymer nanocomposites [13]. POSS-polymer nanocomposites have increased glass transition temperatures (T_g), decomposition temperatures, and mechanical strength [2, 14–16]. Cationic POSS derivatives have been shown to serve as drug carriers and potential delivery agents [17]. POSS cages also have a variety of applications in chemical reactions. The best known examples are their uses as supports for Ziegler-Natta catalysts [18], in the synthesis of polymer-derived ceramics [19], for both metathesis and alkene epoxidation catalysts [20–24], as cores for dendrimer catalysts [25], and for in gas separations [26–28].

The structural and electronic properties of polyhedral H-silsesquioxanes (T_n-POSS, where n = 8, 10, 12) [29] have been studied, and it has been found that POSS cages have a void in the center. This space can be filled with endohedral atomic neutral or ionic species. Significant efforts have been made to utilize the cavity of POSS cages for various purposes [30–42]. Päch and Stösser [43] have suggested that ϒ-irradiation of octaalkyloctasilsesquioxane $(RSiO_{3/2})_8$ or octatrialkylsiloxy-octasilsesquioxane $(R_3SiO\text{-}SiO_{3/2})_8$ leads to encapsulation of an H atom within the T_8-POSS cage. Hossain et al. [44–46] have reported the structures, stabilities, and electronic properties of T_n-POSS complexes with different endohedral transition metals and their ions. The predicted POSS cage properties can change significantly when alkali metals, their ions, halides, noble gases, or transition metals are encapsulated. Very recently, Skelton et al. [42] have used density functional theory (DFT) calculations to study the insertion of different gas molecules into the center of POSS cages. All these studies provide useful information for the future investigations of cavity sizes of the POSS cages.

Different internal cavity sizes of POSS cages make these nanocomplexes interesting candidates for theoretical studies. At this point no detailed investigation of the maximum number of atoms or ions that can be incorporated endrohedrally into the T_n-POSS cages has been carried out. Therefore, we have investigated clusters of compositions $nLi^{+,0,-}$(where n = 1,2,3, ... depending on the size of host cage) and T_n-POSS cages (with n = 8, 10, 12). Inserting a large number of Li species inside POSS-cages may serve as an alternative for lithium storage in lithium ion batteries, provided the energy barriers for these guest species to enter and leave the parent cage are small. Hence, we will explore the following possibilities in this article: Are these POSS host cages stable enough to hold varying amounts of Li atoms, cations, or anions endohedrally? What kind of interactions exist between the encapsulated ions (or atoms) and the atoms of the T_n-POSS host cage? And, can we use these POSS cages as a source for Li storage? Computations on the structures, stabilities, and energetics of endohedral T_8, T_{10} and T_{12}-POSS complexes with different atomic and ionic Li species are carried out in this article to address these questions. For notation we will represent endohedral complexes of neutral, cationic or anionic lithium-POSS systems as $Li^{0,+,-}$@T_n-POSS throughout the paper.

5.2 Computational Details

All calculations were performed using the Q-Chem 3.2 *ab initio* quantum chemistry package [47]. Initially, the geometries of the empty POSS cages were optimized. For single lithium species, the Li atom was placed at the center of the cage. Starting geometries for the cages with two or more Li atoms or ions were generated by placing the Li equally inside the POSS cage. We have used density functional theory (DFT/B3LYP) [48, 49] combined with the 6-311G(d,p) and 6-311+G(d,p) basis sets [50] to perform the geometry optimization and subsequent vibrational analysis for the empty T_n-POSS cages and the endohedral $Li^{0,+,-}$@T_n-POSS complexes (where n = 8,10,12-POSS and endohedral lithium species are $Li^{0,+,-}$, $2Li^{0,+2}$, $3Li^{0,+3}$ depending on type of the host POSS-cage). The geometrical differences/similarities between all the optimized geometries and their respective empty host cages were obtained through root-mean square deviation (RMSd) calculations on the cartesian coordinates of all the POSS atoms. RMSd calculations were performed with Pymol and VMD [51]. For all optimized geometries harmonic vibrational frequencies were calculated to verify that the optimized structures were minima. The inclusion of diffuse functions did not change the respective results significantly, but they were used for the cations and anions in order to allow flexibility for the molecular density. The inclusion energy, E_{inc}, for the endohedral complexes was calculated using the equation

$$E_{inc} = E_{endo} - \left(E_{cage} + E_x\right) + \Delta E_{zp} \tag{5.1}$$

where, E_x is the isolated total energy of the guest species X ($X = Li^{0,+,-}$); E_{cage} is the total energy of the empty T_n-POSS cage, and E_{endo} is the total energy of the endohedral species. The symbol ΔE_{zp} represents the difference in zero point vibrational energies, which were obtained from the harmonic vibrational frequencies.

Adiabatic ionization potentials (IP's) were computed for the X@$(HSiO_{3/2})_n$ complexes (where n = 8,10,12-POSS and X = $Li^{0,+}$, $2Li^{0,+2}$, $3Li^{0,+3}$) as the difference between the total energies of the optimized cationic and the optimized neutral complexes of $(HSiO_{3/2})_n$. Similarly, vertical IP's were computed for the X@ $(HSiO_{3/2})_n$ complexes (where n = 8,10,12-POSS and X = $Li^{0,+}$, $2Li^{0,+2}$, $3Li^{0,+3}$) as the difference between the total energies of the optimized neutral species and single-point energies of the cations at the neutral geometries.

Throughout this paper total energies are given in hartrees, relative energies in kcal mol^{-1}, bond lengths in ångströms (Å), and bond angles in degrees.

5.3 Results and Discussion

This section is divided into five subsections. The first three subsections discuss the geometries of the three POSS host cages and their corresponding endohedral complexes. Section 5.3.4 discusses the energetics of these endohedral POSS-cages. Natural bond order charges are given in Sect. 5.3.5.

Table 5.1 Selected bond lengths (in Å) and point groups of the $Li^{0,+,-}$@T_8-POSS endohedral complexes

System	$R_{Si\text{-}O}$ (min)	$R_{Si\text{-}O}$ (max)	$R_{Si\text{-}H}$ (min)	$R_{Si\text{-}H}$ (max)	$R_{Li\text{-}Si}$ (min)	$R_{Li\text{-}Si}$ (max)	$R_{Li\text{-}O}$ (min)	$R_{Li\text{-}O}$ (max)
T_8-POSS	1.64	1.64	1.46	1.46				
T_8-Li	1.63	1.68	1.45	1.46	2.35	4.08	1.75	4.11
T_8-2Li	1.58	1.84	1.46	1.48	2.65	5.63	1.76	6.41
T_8-Li^-	1.57	1.82	1.47	1.55	2.57	5.41	2.04	5.37
T_8-Li^+	1.63	1.65	1.45	1.45	2.72	2.72	2.03	2.90
T_8-$2Li^{+2}$	1.64	1.70	1.45	1.45	2.69	5.56	2.08	5.8

Here R (min) represents the shortest whereas R (max) represents the longest distance (in Å) between two atoms in subscript in $Li^{0,+,-}$@T_8-POSS endohedral complexes

5.3.1 *T_8-POSS Host Cage*

The O_h form of T_8-POSS is slightly more stable than the T_h form [52]. Therefore we have used the O_h form of T_8-POSS to host different numbers of neutral and ionic Li species. The geometrical parameters of the host T_8-POSS cage are given in Table 5.1, where $R_{X\text{-}Y}$ (min) and $R_{X\text{-}Y}$ (max) are the minimum and maximum distances between the atoms X and Y in $nLi^{0,+,-}$@$(HSiO_{3/2})_8$ endohedral POSS-cages, respectively. In case of the host T_8-POSS cage, the distances between Si and O atoms and between Si and H atoms are 1.641 Å and 1.461 Å, respectively, which are in good agreement with the earlier studies [29, 53–56].

5.3.1.1 Endohedral Neutral, Anionic, and Cationic Complexes of Li@$(SiHO_{3/2})_8$

Table 5.1 summarizes the selected optimized bond lengths for the O_h isomer of T_8-POSS and its endohedral $nLi^{0,+,-}$@T_8-POSS complexes, calculated with B3LYP/6-311(d,p) and B3LYP/6-311+G(d,p). Figure 5.1 includes optimized geometries for all the neutral endohedral nLi^0@$(HSiO_{3/2})_8$ and charged endohedral $nLi^{+,-}$@$(HSiO_{3/2})_8$ complexes. Table 5.2 includes RMSd's calculated for each endohedral complex by comparing it with the host T_8-POSS cage. This comparison is made for the individual atom types of the host and the endohedral complex as well as the overall RMSd's of the two systems. Table 5.3 contains total energies, point groups, inclusion energies, and adiabatic and vertical ionization potentials for the endohedral $nLi^{+,0,-}$@$(HSiO_{3/2})_8$ complexes.

All of these structures are energy minima on the PES, even those with positive inclusion energies. Stable endohedral geometries were obtained for a variety of neutral and ionic Li complexes. However, all attempts to find endohedral Li^{-1} @T_8-POSS resulted in breaking the T_8-POSS cage. Geometric deformations of the cage induced by the embedded atoms increased with increasing number of Li neutrals but decreased with increasing number of Li cations, as suggested by the

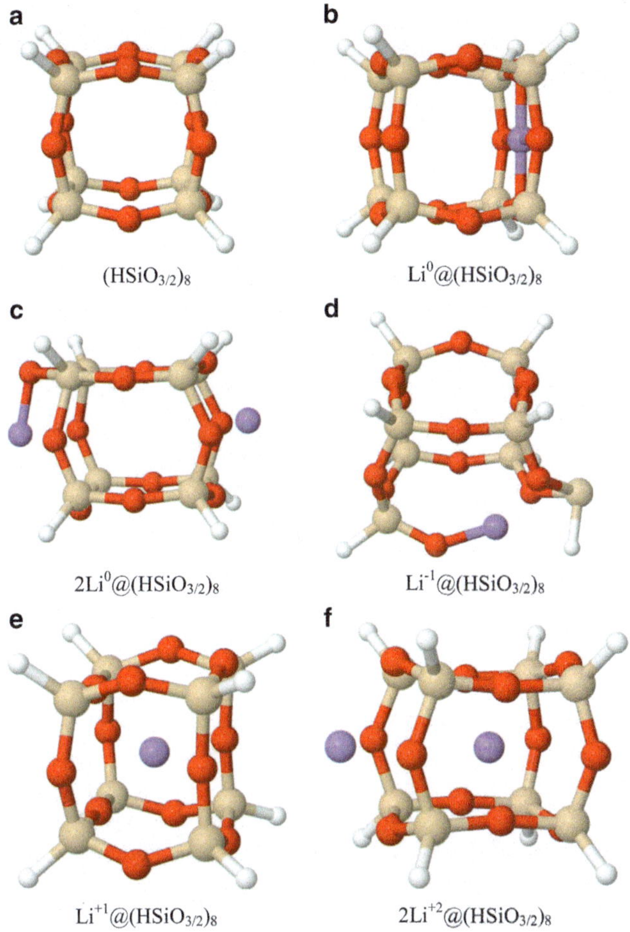

Fig. 5.1 Optimized geometries of T_8-POSS along with endohedral $nLi^{0,-,+}@(HSiO_{3/2})_8$ complexes

RMSd values (Table 5.2). All of these endohedral structures have low symmetries. The minimum and maximum Si-O distances, when compared to the Si-O distances in pure T_8-POSS cage, indicate that the bond lengths increase as the number of Li^0 increases inside the cage. The Si-O bond elongation in the $Li^-@T_8$-POSS is substantial. The bond elongation for the cationic Li species inside the T_8-POSS cage is significantly smaller than for the anionic and neutral Li cages. The Si-H bond elongates in $Li^-@T_8$-POSS, whereas it remains constant at 1.45 in all the cationic Li endohedral complexes. In case of the neutral endohedral cages, the Si-H bond increases with the number of Li neutrals.

A neutral Li atom placed at the center of the T_8-POSS cage moves out towards face of the T_8-POSS cage. The oxygen atoms present at that particular face move

Table 5.2 RMSD data for $Li^{0,+,-}$@T_8-POSS endohedral complexes

System	RMSd (O)	RMSd (Si)	RMSd (H)	RMSd (all cage atoms)
T_8-Li	0.106	0.073	0.105	0.101
T_8-2Li	0.424	0.240	0.186	0.339
T_8-Li^-	0.295	0.661	0.434	0.498
T_8-Li^+	0.424	0.108	0.305	0.329
T_8-$2Li^{+2}$	0.399	0.084	0.279	0.305

Table 5.3 Total Energies (in Hartrees), molecular point groups, zero-point energies (ZPE), inclusion energies (kcal/mol), adiabatic IP's (eV), and vertical IP's (eV) for $Li^{0,+,-}$@T_8-POSS endohedral complexes

System	Point groups	Total energies	ZPE	E_{incl}	Adiabatic IP (eV)	Vertical IP (eV)
T_8-POSS	O_h	−3,225.15	87.899		10.41	11.28
T_8-Li	C_1	−3,232.71	94.764	−40.903	6.64	7.37
T_8-2Li	C_1	−3,240.16	88.484	−13.34	5.78	6.57
T_8-Li^-	C_1	−3,232.75	86.890	−57.178		
T_8-Li^+	C_1	−3,232.46	88.951	−17.17		
T_8-$2Li^{+2}$	C_1	−3,239.71	88.808	4.36		

towards the Li atom, whereas oxygen atoms present on the opposite faces move away from the Li atom. This effect gets more prominent when the number of the Li neutrals increases inside the cage and is obvious in the Li-O distances shown in the Table 5.1. For the cationic nLi^+@POSS complexes, the maximum Li-O distances are shorter than in the neutral Li^0@T_8-POSS cages.

We have compared all the endohedral complexes with the host T_8-POSS cage by calculating RMSd values between the two structures and also between individual types of atoms [Table 5.2]. There are minor differences in the basic cage structures except for the Li^-@T_8-POSS. RMSd values for the individual atoms show that oxygen atoms have more deviation from the parent cage than the Si and H atoms, indicating strength of the interactions of the oxygen atoms with the Li.

5.3.2 *T_{10}-POSS Host Cage*

We have chosen a stable D_{5h} form of T_{10}-POSS as the host structure for our calculations on this system. This structure has been reported in previous theoretical studies [29, 54–56]. The T_{10}-POSS cage consists of two $(–SiH–O–)_5$ rings bridged by five oxygen atoms as shown in the Fig. 5.2(a). Our geometrical parameters are in good agreement with the previous theoretical [54–56] and experimental [53] values. Our typical Si-O bond length is 1.64 Å, whereas previous experimental [53] and theoretical [54–56] calculations have reported Si-O bond lengths between 1.62 Å and 1.64 Å.

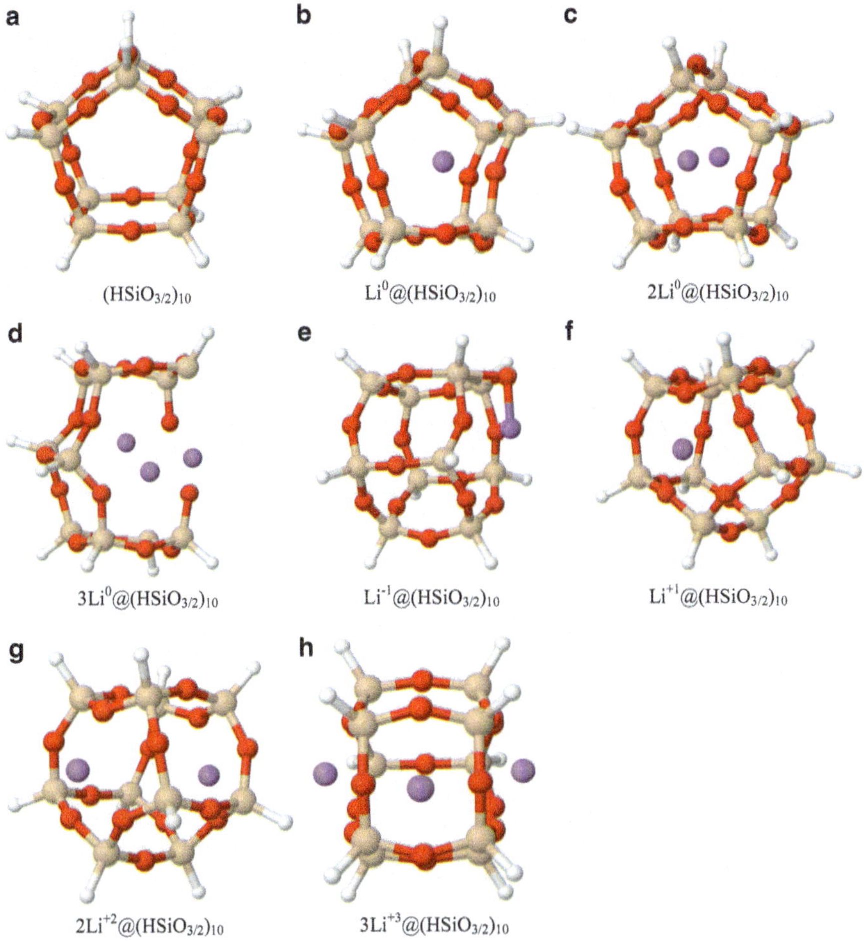

Fig. 5.2 Optimized geometries of T_{10}-POSS (a) along with endohedral $nLi^{0,+,-}@(HSiO_{3/2})_{10}$ complexes

5.3.2.1 Endohedral Neutral, Anionic, and Cationic Complexes of $Li@(SiHO_{3/2})_{10}$

The geometrical parameters for these clusters are given in Table 5.4. Figure 5.2 shows the optimized geometries of T_{10}-POSS and different endohedral $nLi^{0,+,-}@T_{10}$-POSS species calculated with B3LYP/6-311G (d,p) and B3LYP/6-311G+(d,p). Table 5.5 shows the calculated RMSd values for the endohedral $nLi^{0,+,-}@(HSiO_{3/2})_{10}$ complexes. Table 5.6 contains total energies, point groups, inclusion energies, and adiabatic and vertical ionization potentials for the endohedral $nLi^{0,+,-}@(HSiO_{3/2})_{10}$ complexes.

All of the endohedral $Li^{0,+,-}@T_{10}$-POSS cage structures are energy minima. The host cage's D_{5h} symmetry was not conserved for any of the endohedral Li

Table 5.4 Selected bond lengths (in Å) for $Li^{0,+,-}$@T_{10}-POSS endohedral complexes

System	$R_{Si\text{-}O}$ (min)	$R_{Si\text{-}O}$ (max)	$R_{Si\text{-}H}$ (min)	$R_{Si\text{-}H}$ (max)	$R_{Li\text{-}Si}$ (min)	$R_{Li\text{-}Si}$ (max)	$R_{Li\text{-}O}$ (min)	$R_{Li\text{-}O}$ (max)
T_{10}-POSS	1.64	1.64	1.46	1.46				
T_{10}-Li	1.63	1.67	1.46	1.46	2.74	4.59	2.12	4.71
T_{10}-2Li	1.64	1.67	1.46	1.46	2.66	4.18	2.11	4.22
T_{10}-3Li	1.62	1.87	1.46	1.47	2.65	5.20	2.01	5.61
T_{10}-Li^-	1.61	1.82	1.47	1.49	2.64	5.76	1.79	5.79
T_{10}-Li^+	1.63	1.71	1.46	1.46	2.72	5.16	1.99	5.17
T_{10}-$2Li^{+2}$	1.63	1.69	1.45	1.46	2.74	5.31	1.99	5.47
T_{10}-$3Li^{+3}$	1.63	1.71	1.45	1.45	2.79	7.16	2.07	7.12

Here R (min) represents the shortest whereas R (max) represents the longest distance in (in Å) between two atoms in subscript in $Li^{0,+,-}$@T_{10}-POSS endohedral complexes

Table 5.5 RMSD values for $Li^{0,+,-}$@T_{10}-POSS endohedral complexes

System	RMSd (O)	RMSd (Si)	RMSd (H)	RMSd (all cage atoms)
T_{10}-Li	0.0885	0.03	0.048	0.067
T_{10}-2Li	0.153	0.56	0.075	0.112
T_{10}-3Li	0.374	0.127	0.289	0.302
T_{10}-Li^-	0.378	0.249	0.194	0.313
T_{10}-Li^+	0.190	0.065	0.133	0.161
T_{10}-$2Li^{+2}$	0.302	0.096	0.18	0.226
T_{10}-$3Li^{+3}$	0.404	0.126	0.24	0.302

Table 5.6 Total energies (in Hartrees), molecular point groups, zero-point energies (ZPE), inclusion energies (kcal/mol), adiabatic IP's (eV), and vertical IP's (eV) for $Li^{0,+,-}$@T_{10}-POSS endohedral complexes

System	Point groups	Total energies	ZPE	E_{incl}	Adiabatic IP (eV)	Vertical IP (eV)
T_{10}-POSS	D_{5h}	−4,031.43	110.62		9.96	10.15
T_{10}-Li	C_1	−4,038.91	110.21	4.65	2.07	2.60
T_{10}-2Li	C_1	−4,046.41	109.67	0.83	2.68	3.59
T_{10}-3Li	C_s	−4,053.96	109.39	25.77	1.88	7.08
T_{10}-Li^-	C_s	−4,038.97	109.19	−25.39		
T_{10}-Li^+	C_1	−4,038.83	111.35	−76.05		
T_{10}-$2Li^{+2}$	C_1	−4,046.11	112.39	−72.05		
T_{10}-$3Li^{+3}$	C_1	−4,053.24	112.49	27.07		

complexes. Neutral Li atoms, when placed inside the T_{10}-POSS host cage, tend to move towards the face of the cage, except for the $3Li^0$@T_{10}-POSS cage, where the T_{10}-POSS cavity is not large enough to hold three Li neutrals, resulting in breaking of the T_{10}-POSS cage. In Li^-@T_{10}-POSS, the Li anion moves out of the cage, and in this process it takes one of the bridging oxygens, leaving behind two silicons with a very short distance of 2.28 Å. Interestingly, all of the Li cations move towards the face of the T_{10}-POSS cage. The oxygens present on the face of T_{10}-POSS were closer to the cation than the five bridging oxygens, and the bridging oxygens moved slightly outward compared to the empty cage. Overall distortions of the endohedral T_{10}@POSS cage increased with increasing numbers of Li cations and neutrals, as shown from the RMSd data in Table 5.5.

5.3.3 *T_{12}-POSS Host Cage*

We have selected the D_{2h} form of the T_{12}-POSS cage as the host structure for different Li species. This structure has been reported in theoretical calculations [57] and prepared experimentally [58]. The T_{12}@POSS cage structure consists of two $(–SiH–O–)_6$ rings bridged by six oxygen atoms as shown in Fig. 5.3(a). The Si-O bond lengths are between 1.63 and 1.64 Å, whereas the Si-H bond lengths are the same as in the T_8-POSS and T_{10}-POSS host cages. The T_{12}-POSS cage has the largest cavity of the three host POSS cages that we have considered.

5.3.3.1 Endohedral Neutral, Anionic, and Cationic Complexes of Li@$(SiHO3/2)_{12}$

Figure 5.3 shows the optimized geometries of the T_{12}-POSS host cage and different endohedral $nLi^{+,0,-}$@T_{12}-POSS species calculated with B3LYP/6-311G(d,p) and B3LYP/6-311G+(d,p). The geometrical parameters for these clusters are given in Table 5.7. Table 5.8 gives the calculated RMSD values for the endohedral $Li^{+,0,-}$@$(HSiO_{3/2})_{12}$ complexes. Table 5.9 contains total energies, point groups, inclusion energies, and adiabatic and vertical ionization potentials for the endohedral $Li^{+,0,-}$@$(HSiO_{3/2})_{12}$ complexes.

Insertion of the first and second Li neutral atom into the T_{12}-POSS cage changes the symmetry of the host cage to C_{2v}, with both Li atoms present inside the cage near the T_{12}@POSS face. Surprisingly, the third and fourth lithium neutral atoms distort the host cage considerably and move out of the cage. This effect is obvious through the RMSd calculations, which show large RMSd values and which describe how different these structures are from the T_{12}-POSS host cage. The Li cations' trend is usually the same that we have seen in T_{10}@POSS. Cage distortion increases with an increase in the number of Li cations. Generally, oxygen atoms are attracted more towards the guest species than the Si and H atoms, as can be seen from the RMSd data.

5.3.4 *Energetics and Ionization Potentials*

Tables 5.3, 5.6, and 5.9 summarize the total energies, point groups, zero point energies, endohedral inclusion energies, and adiabatic and vertical ionization potentials for the $nLi^{0,+,-}$@T_n-POSS complexes with B3LYP/6-311G(d,p) and B3LYP/6-311+G(d,p). The inclusion energy, E_{incl}, is defined as the difference

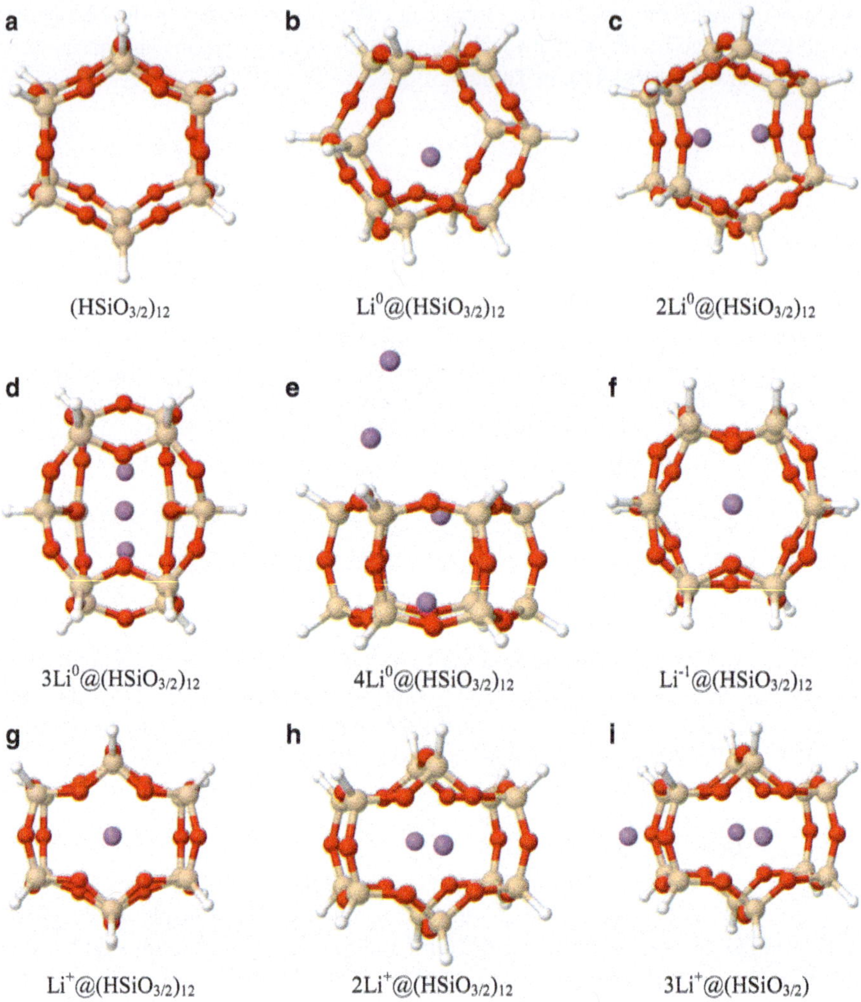

Fig. 5.3 Optimized geometries of T_{12}-POSS (a) along with endohedral $nLi^{0,+,-}@(HSiO_{3/2})_{12}$ complexes

between the energy of the endohedral complex and the sum of energies of the isolated empty cage and the isolated species X. The sign of E_{incl} tells us whether or not formation of this particular endohedral complex is energetically favorable. A negative sign means that the endohedral complex is more stable than the isolated components. Calculated E_{incl} energies for T_8-POSS complexes indicate that $nLi^0@T_8$-POSS and $Li^+@T_8$-POSS are stable when compared to the energies of the isolated species. As the number of lithium atoms increases, it gets more difficult to incorporate those species into the T_8-POSS cage (data not shown). For T_{10}-POSS, only Li^+ and $2Li^+@T_{10}$-POSS are energetically favorable, whereas no T_{10}-POSS complex with neutral Li inside is stable. For T_{12}-POSS we have similar

Table 5.7 Selected bond lengths (in Å) and point groups of the $Li^{0,+,-}$@T_{12}-POSS endohedral complexes

System	$R_{Si\text{-}O}$ (min)	$R_{Si\text{-}O}$ (max)	$R_{Si\text{-}H}$ (min)	$R_{Si\text{-}H}$ (max)	$R_{Li\text{-}Si}$ (min)	$R_{Li\text{-}Si}$ (max)	$R_{Li\text{-}O}$ (min)	$R_{Li\text{-}O}$ (max)
T_{12}-POSS	1.63	1.64	1.46	1.46				
T_{12}-Li	1.62	1.67	1.46	1.47	2.97	4.07	2.10	4.11
T_{12}-2Li	1.63	1.66	1.46	1.47	2.74	4.56	2.03	4.74
T_{12}-3Li	1.63	1.91	1.46	1.49	2.56	6.35	1.90	6.17
T_{12}-4Li	1.64	1.69	1.46	1.46	2.86	9.62	2.03	9.8
T_{12}-Li^-	1.60	1.79	1.47	1.48	2.92	4.87	2.11	4.95
T_{12}-Li^+	1.62	1.69	1.46	1.46	2.78	5.01	2.08	4.97
T_{12}-$2Li^{+2}$	1.63	1.69	1.46	1.46	2.81	5.14	2.12	5.35
T_{12}-$3Li^{+3}$	1.63	1.71	1.45	1.46	2.82	7.85	2.04	8.07

Here R (min) represents the shortest whereas R (max) represents the longest distance in (in Å) between two atoms in subscript in $Li^{0,+,-}$@T_{12}-POSS endohedral complexes

Table 5.8 RMSD values for $Li^{0,+,-}$@T_{12}-POSS endohedral complexes

System	RMSd (O)	RMSd (Si)	RMSd (H)	RMSd (for all atoms)
T_{12}-Li	0.313	0.142	0.312	0.275
T_{12}-2Li	0.296	0.088	0.244	0.238
T_{12}-3Li	0.899	0.710	1.08	0.907
T_{12}-4Li	0.160	0.06	0.15	0.130
T_{12}-Li^-	0.408	0.165	0.48	0.382
T_{12}-Li^+	0.339	0.181	0.285	0.288
T_{12}-$2Li^+$	0.472	0.255	0.332	0.382
T_{12}-$3Li^+$	0.48	0.25	0.34	0.390

Table 5.9 Total Energies (in Hartrees), molecular point groups, zero-point energies (ZPE), inclusion energies (kcal/mol), adiabatic IP's (eV), and vertical IP's (eV) for $Li^{0,+,-}$@T_{12}-POSS endohedral complexes

System	Point groups	Total energies	ZPE	E_{incl}	Adiabatic IP (eV)	Vertical IP (eV)
T_{12}-POSS	D_{2h}	−4,837.73	132.74		10.40	10.55
T_{12}-Li	C_{2v}	−4,845.21	132.91	10.22	3.73	4.56
T_{12}-2Li	C_{2v}	−4,852.69	133.26	14.31	3.60	4.22
T_{12}-3Li	C_s	−4,860.35	133.19	−88.61	5.46	5.70
T_{12}-4Li	C_s	−4,867.75	133.45	−31.97	2.99	3.22
T_{12}-Li^-	C_1	−4,845.23	130.90	5.30		
T_{12}-Li^+	C_1	−4,845.17	133.10	−94.74		
T_{12}-$2Li^{+2}$	C_1	−4,852.43	134.52	−79.39		
T_{12}-$3Li^{+3}$	C_1	−4,859.57	134.966	14.268		

results for the lithium cations inside the cage. Li^+ and $2Li^+$@T_{12}-POSS are stable compared to the isolated cage and species. For Li neutrals inside the T_{12} cage, we found contrasting results in that $3Li^0$ and $4Li^0$ are more favorable inside the T_{12}-POSS cage than Li^0 and $2Li^0$. This suggests favorable Li-Li interactions inside the cage which lower the overall energy of these Li-encapsulated @T_{12}-POSS complexes. In the case of the 4Li^0 @T_{12}-POSS endohedral cage, two lithium neutral atoms move outside the cage, forming a lithium dimer with a bond length of 2.74 Å. This agrees with the literature value for the bond length of the Li_2 dimer [59].

The adiabatic IP is defined as the difference between the total energy of a cationic endohedral complex at its optimized geometry and the energy of the corresponding neutral endohedral complex at its optimized geometry. The vertical IP is defined as the difference in the total energy between a cationic endohedral complex and its neutral endohedral cluster at the neutral geometry. Vertical IP's are always larger than the corresponding adiabatic IP's. In Tables 5.3, 5.6, and 5.9 we have computed the adiabatic and vertical IPs for empty and Li-encapsulated T_8-POSS, T_{10}-POSS, and T_{12}-POSS complexes, respectively. The ionization potentials of the T_{10}-POSS and T_{12}-POSS encapsulated Li^0 are considerably lower (Tables 5.6 and 5.9) than that of the free metal (which was found to be 5.61 eV at the same level of theory), whereas for T_8-POSS encapsulated Li^0 and 2Li^0, the ionization potentials are higher than the free Lithium atom (Table 5.3). In case of T_{10}-POSS encapsulated 3Li^0 (Table 5.6), there is a significant structural rearrangement upon formation of cation as evident from the difference between the adiabatic and vertical IP values. Also, for T_{12}-POSS encapsulated 3Li^0 (Table 5.9), IP values are higher compared with the T_{12}-POSS@Li^0 and 2Li^0 endohedral complexes, suggesting that interactions between the Li atoms in T_{12}-POSS@3Li^0 stabilize this endohedral complex.

5.3.5 *Natural Bond Order Charges*

The Natural Bond Order (NBO) charges [60] for the host T_n-POSS cages and the endohedral guest species were obtained and are given in Table 5.10. Information about the interactions between the Li species and the host POSS cages can be gleaned by examining the charges on the atoms. All silicon atoms are equivalent in the three bare host cages, with atomic charges of +2.12, +2.14 and +2.14 each for the Si atoms in T_8-POSS, T_{10}-POSS, and T_{12}-POSS cages, respectively. The positive charge of the cage silicons decreases as the number of endohedral lithium cations increases. This decrease in silicon atomic charges is −0.065 for T_{12}-POSS cage upon the insertion of the maximum number of lithium atoms, as compared to −0.044 for T_8-POSS and −0.053 for T_{10}-POSS cage. All the oxygens have same atomic charge in the T_8-POSS host cage, which is −1.262 each. On the other hand it is slightly different between the oxygens in the T_{10}-POSS and T_{12}-POSS host cages. The atomic charges on oxygen upon interaction with the lithium species are given in Table 5.10. We see that the negative charge on the oxygens which are close to the lithium cations decreases – owing to the electrostatic interaction between positive charge of Li cations and negative charge of the interacting oxygen atoms – whereas this charge increases for the oxygens at the opposite faces of the cages when Li cations are present. A sharp decrease in the negative charge of the hydrogens in endohedral Li-POSS cages also occurs. Lithium endohedral species transfer a small

Table 5.10 Natural charge analysis with B3-LYP/6-311G(d,p) for $nLi^{+,-}$@T_n-POSS cages

X	Q_{Li}	$Q_{Si(min)}$	$Q_{Si(max)}$	$Q_{O(min)}$	$Q_{O(max)}$	Q_H
T_8-POSS		2.12	2.12	−1.262	−1.262	−0.230
T_8-Li^+	0.83	2.105	2.106	−1.235	−1.308	−0.187
T_8-$2Li^{+2}$	0.836/0.947	2.076	2.093	−1.239	−1.300	−0.14
T_8-Li^-	0.900	0.796	2.129	−1.258	−1.345	−0.255/−0.324
T_{10}-POSS		2.135	2.135	−1.265	−1.266	−0.237
T_{10}-Li^+	0.925	2.104	2.133	−1.259	−1.320	−0.19/−0.21
T_{10}-$2Li^{+2}$	0.93	2.105	2.112	−1.256	−1.316	−.168/−.174
T_{10}-$3Li^{+3}$	0.94/0.96	2.082	2.108	−1.265	−1.311	−0.14/−0.148
T_{10}-Li^-	0.881	1.372	2.139	−1.253	−1.304	−0.23/−0.275
T_{12}-POSS		2.14	2.14	−1.263	−1.267	−0.24
T_{12}-Li^+	0.923	2.091	2.134	−1.250	−1.306	−0.20/−.21
T_{12}-$2Li^{+2}$	0.93	2.090	2.111	−1.250	−1.303	−0.17
T_{12}-$3Li^{+3}$	.937/0.95	2.08	2.107	−1.251	−1.311	−0.15/−0.16
T_{12}-Li^-	0.453	1.06	1.07	−.571	−.649	−0.12/−0.13

Here, the symbols Q_{li}, Q_{Si}, Q_H and Q_O represent the charges on selected types of Li, Si, O and H atoms of the all T_n@$(HSiO_{3/2})_n$ cages

amount of their electron density to the host cages (see Table 5.10). In the case of the anionic Li species, the decrease of the positive charges of the silicons in all three of the host cages is larger compared to the lithium cationic species, effectively localizing the negative charge on the cage. The oxygens and hydrogens present in the Li^-@T_{12}-POSS complex lose much of their negative charge.

5.4 Conclusions

This work predicts that a large variety of endohedral guest T_n-POSS clusters could be prepared experimentally and predicts some of the salient structural features of these POSS cages. The following conclusions can be drawn from these calculations. The cavities of these POSS cages are large enough to accommodate varying amounts of neutral and charged Li species. As the size of the cage increases, a greater chance exists for more Li species to reside inside these cages. Despite the varying amounts of neutral and ionic Li species, stable endohedral geometries were predicted for all the POSS cages. No relation has been found between the size of T_n-POSS host cages and the stability of endohedral complexes. Geometric deformations induced by the varying amounts of Li species depend on number of Li species and also on the size of host cage. In principle, the larger POSS cages could serve as media for lithium cation storage.

Acknowledgment This work was supported in part by the National Science Foundation under awards number EPS 0903787 and EPS 1006883

References

1. Li G, Wang L, Ni H, Pittman CU Jr (2002) J Inorg Organomet Polym 11:123
2. Fu BX, Hsiao BS, Pagola S, Stephens P, White H, Rafailovich M, Sokolov J, Mather PT, Jeon H, Philips S, Lichtenhan J, Schwab J (2001) Polymer 42:599
3. Li H, Eddaouddi M, Richardson DA, Yaghi OM (1998) J Am Chem Soc 120:8567
4. Brown JF Jr, Vogt LH Jr (1965) J Am Chem Soc 87:4313
5. Dance B (2001) Semiconduct Int 24:46
6. Feher FJ, Wyndham KD (1998) Chem Commun :323
7. Li GZ, Wang L, Toghiani H, Daulton TL, Koyama K, Pittman CU Jr (2001) Macromolecules 34:8686
8. Li GZ, Wang L, Toghiani H, Daulton TL, Pittman CU Jr (2002) Polymer 43:4167
9. Maxim N, Abbenhuis HCL, Stobbelaar PJ, Mojet BL, Van Santen RR (1999) Phys Chem Chem Phys 18:4473
10. Maxim N, Magusin PCMM, Kooyman PJ, Van Wolput JHMC, Van Santen RA, Abbenhuis HCL (2001) Chem Mater 13:2958
11. Wada K, Yamada K, Kondo T, Mitsudo TA (2001) Chem Lett 30:12
12. Lamm MH, Chen T, Glotzer SC (2003) Nano Lett 3:989
13. Romo-Uribe A, Mather PT, Haddad TS, Lichtenhan JD (1998) J Polym Sci Polym Phys 36:1857
14. Lee A, Lichtenhan JDJ (1999) J Appl Polym Sci 73:1993
15. Haddad TS, Lichtenhan JD (1996) Macromolecules 29:7302
16. Schwab JJ, Lichtenhan JD (1998) Appl Organomet Chem 12:707
17. McCusker C, Carroll JB, Rotello VM (2005) Chem Commun :996
18. Liu JC (1999) Appl Organomet Chem 13:295
19. Raj R, Riedel R, Soraru GD (2001) J Am Ceram Soc 84:2158
20. Abbenhuis HCL (2000) Chem Eur J 6:25
21. Ducateau R, Cremer U, Harmsen RJ, Mohamud SI, Abbenhuis CL, Saten RA, Meetsma A, Thiele SKH, van Tol FFH, Kranenburg M (1999) Organometallics 18:5447
22. Klunduk MC, Maschmeyer T, Thomas JM, Johnson FG (1999) Chem Eur J 5:1481
23. Pescarmona PP, Waal JCD, Maxwell IE, Maschmeyer T (2001) Angew Chem Int Ed 40:740
24. Xiao FS, Han Y, Yu Y, Meng X, Yang M, Wu S (2002) J Am Chem Soc 124:888
25. Coupar PI, Jaffres PA, Morris RE (1999) J Chem Soc, Dalton Trans :2183
26. de Vos RM, Verweij HJ (1998) J Membr Sci 143:37
27. Suzuki F, Nakane K, Yasuo H (1995) J Membr Sci 104:183
28. Feng X, Shao P, Huang YR, Jiang G, Xu X (2002) Sep Purif Tech 27:211
29. Xiang KH, Pandey R, Pernisz UC, Freeman J (1998) J Phys Chem B 102:8704
30. Villaescusa LA, Lightfoot P, Morris RE (2002) Chem Commun 19:2220
31. Bassindale AR, Pourny M, Taylor PG, Hursthouse MB, Light ME (2003) Angew Chem 42:3488
32. Bassindale AR, Parker DJ, Pourny M, Taylor PG, Horton PN, Hursthouse MB (2004) Organometallics 23:4400
33. Park SS, Xiao C, Hagelberg F, Hossain D, Pittman CU Jr, Saebø S (2004) J Phys Chem A 108:11260
34. Satre G, Pulido A, Corma A (2005) Chem Commun 18:2357
35. Päch M, Macrae RM, Carmichael I (2006) J Am Chem Soc 128:6111
36. Tossel JA (2007) J Phys Chem C 111:3584
37. Anderson SE, Bodzin DJ, Haddad TS (2008) Chem Mater 20:4299
38. Geoge AR, Catlow CRA (1995) Chem Phys Lett 247:408
39. Kudo T, Akasaka M, Gordon MS (2008) Theo Chem Accounts 120:155
40. Kudo T (2009) J Phys Chem A 113:12311
41. Kudo T, Taketsugu T, Gordon MS (2011) J Phys Chem A 115:2679
42. Skelton AA, Fried JR (2013) Phys Chem Chem Phys 15:4341

43. Päch M, Stösser R (1997) J Phys Chem A 101:8360
44. Hossain D, Pittman CU Jr, Saebø S, Hagelberg F (2007) J Phys Chem C 111:6199
45. Hossain D, Pittman CU Jr, Hagelberg F, Saebø S (2008) J Phys Chem C 112:16070
46. Hossain D, Gwaltney SR, Pittman CU Jr, Saebø S (2009) Chem Phys Lett 467:348
47. Shao Y, Fusti-Molnar L, Jung Y, Kussmann J, Ochsenfeld C, Brown ST, Gilbert ATB, Slipchenko LV, Levchenko SV, O'Neill DP, Distasio RA Jr, Lochan RC, Wang T, Beran GJO, Besley NA, Herbert JM, Lin CY, Voorhis TV, Chien SH, Sodt A, Steele RP, Rassolov VA, Maslen PE, Korambath PP, Adamson RD, Austin B, Baker J, Byrd EFC, Dachsel H, Doerksen RJ, Dreuw A, Dunietz BD, Dutoi AD, Furlani TR, Gwaltney SR, Heyden A, Hirata S, Hsu CP, Kedziora G, Khalliulin RZ, Klunzinger P, Lee AM, Lee MS, Liang W, Lotan I, Nair N, Peters B, Proynov EI, Pieniazek PA, Rhee YM, Ritchie J, Rosta E, Sherrill CD, Simmonett AC, Subotnik JE, Woodcock HL III, Zhang W, Bell AT, Chakraborty AK, Chipman DM, Keil FJ, Warshel A, Hehre WJ, Schaefer HF III, Kong J, Krylov AI, Gill PMW, Head-Gordon M (2006) Phys Chem Chem Phys 8:3172
48. Lee C, Yang W, Parr RG (1998) Phys Rev B: Condens Matter Mater Phys 37:785
49. Becke AD (1993) J Chem Phys 98:5648
50. Raghavachari K, Trucks GW (1989) J Chem Phys 91:1062
51. Humphrey W, Dalke A, Schulten K (1996) VMD – visual molecular dynamics. J Molec Graphics 14:33
52. Agaskar PA, Klemperer WG (1995) Inorg Chim Acta 229:355
53. Auf der Heyde TPE, Bürgi HB, Bürgy H, Törnroos KW (1991) Chimia 45:38
54. de Man AJM, Sauer J (1996) J Phys Chem 100:5025
55. Tossell JA (1996) J Phys Chem 100:14828
56. Pasquarello A, Hybertsen MS, Car R (1996) Phys Rev B 54:R2339
57. Tejerina B, Gordon MS (2002) Phys Chem B 106:11764
58. Cho H, Liang K, Chatterjee S, Pittman CU Jr (2005) J Inorg Organometal Polym, Mater 15:541
59. Lee SY, Boo BH, Kang HK, Kang D, Judai K, Nishijo J, Nishi N (2005) Chem Phys Lett 411:484
60. Glendening ED, Reed AE, Carpenter JE, Weinhold F (1993) NBO 4.M. University of Wisconsin, Madison

Chapter 6
Intrinsic Magnetism in Single-Walled Carbon Nanotubes of Finite Length

Frank Hagelberg, Jianhua Wu, Anahita Ayasoufi, and Jerzy Leszczynski

Abstract The magnetic properties of axially confined hydrogenated single-walled carbon nanotubes (SWCNTs) of the (n, 0) type, as well as cross-linking architectures based on these units, are systematically explored by use of density functional theory. Emphasis is placed on the relation between the ground state magnetic moments of SWCNTs and zigzag graphene nanoribbons (ZGNRs). Comparison between SWCNTs with n = 5–24 and ZGNRs of equal length gives rise to two basic questions: (1) how does the nanotube curvature affect the antiferromagnetic order known to prevail in ZGNRs, and (2) to what extent do the magnetic moments localized at the SWCNT edges deviate from the zero-curvature limit n/3 μ_B? The studies on single SWCNTs are extended to cross-linking carbon nanotubes (CLCNTs) composed of three axially confined single-walled carbon nanotubes (SWCNTs) of the (10,0) type. Three CLCNT models, differing from each other by the structure of the contact regions of the three SWCNT constituents, are explored in terms of their geometric, electronic, and magnetic properties. Various magnetic phases, as obtained by combining finite SWCNTs in ferromagnetic (FM) or antiferromagnetic (AFM) coordination, are distinguished. The characteristics of these phases are shown to depend on the contact region geometry which

F. Hagelberg (✉)
Department of Physics and Astronomy, East Tennessee State University,
37614 Johnson City, TN, USA
e-mail: hagx@icnanotox.org

J. Wu
Department of Physics, Atmospheric Sciences, and Geoscience, Jackson State University,
39217 Jackson, MS, USA

A. Ayasoufi
Department of Mathematics and Statistics, East Tennessee State University,
37614 Johnson City, TN, USA

J. Leszczynski
Department of Chemistry and Biochemistry, Interdisciplinary Center for Nanotoxicity,
Jackson State University, Jackson, MS 39217, USA

J. Leszczynski and M.K. Shukla (eds.), *Practical Aspects of Computational Chemistry III*, DOI 10.1007/978-1-4899-7445-7_6,

plays an essential role in defining the order of their stabilities. Prospects of applying either of the two systems analyzed here, SWCNTs and CLCNTs, as transmission elements in spintronics are discussed.

6.1 Introduction

The study of magnetism in carbon nanostructures is a rapidly expanding segment of current materials science. This progress is driven by the realization that carbon-based magnetic materials can usually be fabricated at lower cost than their metallic analogs. Further, these materials are lighter and tend to be more efficient in terms of power consumption. In particular, during the past two decades, many actual and potential applications of carbon nanostructure magnetism have been identified, ranging from fullerene-based magnetic resonance imaging (MRI) contrast agents [1, 2] to zigzag graphene nanoribbon (ZGNR) transmission elements in nano-spintronics [3, 4].

6.1.1 Focus on Nanotubes

Magnetic carbon nanostructures hold great promise for potential applications in nanodevices based on spintronics [5–7], operating with spin currents rather than currents of electric charge, as employed in conventional electronics [8]. While basic carbon nanostructures, such as graphene, periodic carbon nanotubes, or fullerenes are spin zero systems in their ground states, magnetism may be induced in these structures by encapsulating magnetic components into the interior of the nanosystem [9].

Also, radiation damage [10, 11] and defect formation [12, 13], as achieved by irradiating the nanosystem with high-energy particles, have been shown to give rise to magnetic ordering. However, extended carbon structures can also acquire magnetic ground states as a consequence of dimensional reduction. As is well known, finite-sized graphene strips with zigzag-structured edges, zigzag graphene nanoribbons (ZGNRs), display ground-state magnetism [14, 15] as the dimension of graphene is lowered from two to one or zero, and an analogous observation has been made for SWCNTs of the zigzag type as their dimension drops from one to zero [16–18].

This behavior is due to a truncation effect, originating from unpaired π electrons localized at the edges of the carbon nanostructure. Termination of the nanostructure truncation sites diminishes this effect but does not necessarily eliminate it. In the case of hydrogenated ZGNRs, this can readily be seen by use of a heuristic argument. Each carbon atom at the ZGNR edge contributes to each C–C bond 4/3 electrons, while one electron is saturated by the adjacent hydrogen atom, leaving one-third of an electron unsaturated per edge carbon atom. This remnant is localized at the end of the ZGNR. Each edge atom of an infinitely long ZNGR of finite width is thus associated with a magnetic moment of one-third of a Bohr magneton. In the limit of long ZGNRs, the asymptotic total magnetic moment at each of their two zigzag edges is therefore expected to approach $n_r/3\ \mu_B$, with n_r as the number carbon atoms located at the edge.

For SWCNTs, tube curvature makes any qualitative estimate of magnetic moment magnitudes at the tube edges speculative. In Sect. 6.3 of this chapter, we will pursue the question for the quantitative relation between the ground-state magnetic moments of ZGNRs and SWCNTs in a systematic fashion, extending previous calculations for selected species. Among the latter are density functional theory (DFT) computations involving (10,0) tubes of different lengths as reference systems [19] which clearly established the magnetic phases of these systems as more stable than their non-magnetic counterparts. This pattern was confirmed for three termination prototypes, namely hydrogenation, truncation with edge reconstruction, and capping by fullerene hemispheres (see Fig. 6.1). In comparison with the two former systems, the latter case was found to be associated with a diminished magnetic moment magnitude and a more delocalized spin density distribution along the tube axis.

In each investigated case, both ferromagnetic (FM) and antiferromagnetic (AFM) solutions were identified, related to parallel and antiparallel orientation of the spins at the two edges. For nearly all systems explored, AFM was seen to prevail unanimously over FM order. This conclusion has been reached by several authors, e.g. for SWCNTs of the (14,0) type [20] and for short systems of the form (n,0) with n ranging from 4 to 14 [16].

The dominance of the AFM phase has been rationalized by reference to Lieb's theorem [8, 16, 21] which states that the total spin S of a half-filled bipartite system, as realized by graphene, is found from

$$S = \tfrac{1}{2}\,|N_a - N_b|\hbar, \tag{6.1}$$

with N_a, N_b as the numbers of sites in the two sublattices, labeled a and b. Lieb's theorem relies on the Hubbard model and assumes a positive on-site interaction term U, corresponding to repulsive electron–electron interaction. However, predictions based on Lieb's theorem have displayed a certain resistance with respect to the computational method, as they have been confirmed also by use of plane wave–DFT [15].

Recent DFT calculations suggest that both ZGNRs and zigzag SWCNTs display the phenomenon of half-metallicity [16, 18, 22]. Half-metallic materials show a zero band gap for electrons with one spin orientation and a finite band gap for the other as a suitably tuned electric field is applied to these structures. This mechanism results in a spin polarized current and so provides a spin filter.

Further, *spin valves* based on the magnetic properties of carbon nanostructures have been proposed. These potential applications consist in designing magnetoresistive devices by exploiting electron edge localization. As demonstrated by DFT computation [23], ferromagnetically ordered ZGNRs are expected to display a

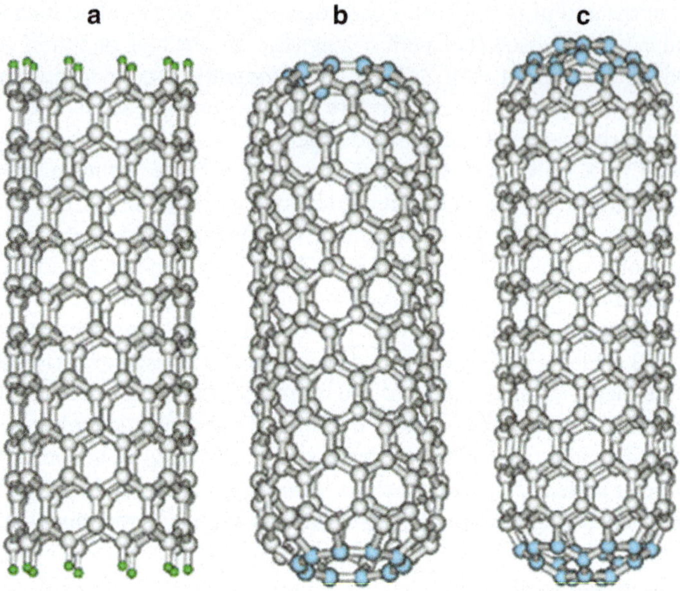

Fig. 6.1 The equilibrium geometries of a (10,0) SWNT with (**a**) H-termination ($C_{200}H_{20}$) (**b**) truncation with reconstructed edge layers (C_{200+20}), and (**c**) capping with fullerene hemispheres (C_{200+40}). *Green symbols* represent H, and *grey symbols* C atoms of intermediate layers, while light *blue symbols* indicate C atoms in the terminating decagons (**b**) or fullerene hemispheres (**c**) (Reprinted with permission from [19]. Copyright 2009, American Physical Society)

substantially lower electric resistance for spin-polarized currents than their antiferromagnetic counterparts. The strength of the magnetic interaction that favors antiferromagnetic coordination for ZGNRs falls off rapidly with the ZGNR width w. More specifically, DFT calculation arrives at a power law of the form $w^{-1.82}$ for the difference between the total ZGNR energy in the FM and the AFM state [24]. This suggests that, for ZGNRs of sufficient width, an external magnetic field can easily switch between both configurations.

As implied by these properties, single SWCNTs of the zigzag type as well as single ZGNRs may be of interest as transmission elements in spintronics applications. More complex networks call for more complex nanostructures, or for extended architectures that involve the SWCNT or ZGNR geometries as basic structural motives.

Architectures consisting of nanotube building blocks are relevant for both nanoelectronics and nanomechanics. Thus, a multi-terminal heterojunction, as realized on the most elementary level by T or Y shaped units, can be used as a transistor [25–28]. Constructing superstructures from joining several SWCNTs, one may create fibers that benefit from the extraordinary mechanic properties of carbon nanotubes, in terms of Young's modulus and the tensile strength [29].

Various computational studies on both the electronic and the mechanic properties of periodic SWCNT superstructures have been performed [30–33].

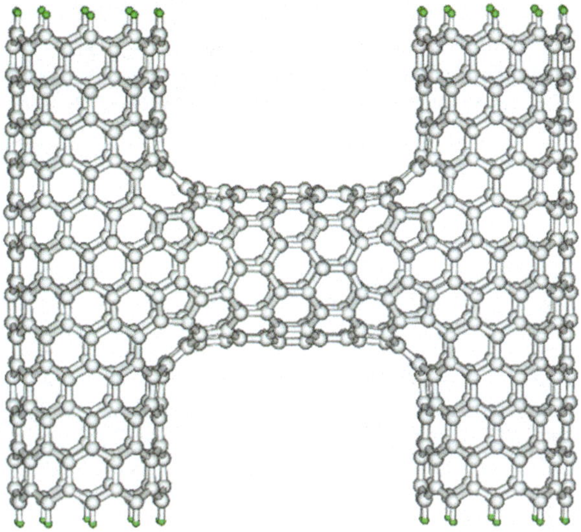

Fig. 6.2 Representation of a hydrogen-terminated cross-linking carbon nanotube (CLCNT) (Reprinted with permission from [34]. Copyright (2013) American Chemical Society)

Coluci et al. carried out classical molecular dynamics simulations of super nanotubes, replacing the C–C bonds of regular carbon tubes by SWCNTs and the C atoms by Y-shaped junctions between SWCNTs [30]. Crucial mechanical parameters of these super-architectures, such as the in-plane stiffness and the shear stiffness, were shown to be independent of the chirality of both the super-nanotube and the constituent nanotubes [31]. Further, cubic crystals with T-shaped junctions as basic elements have been explored using the density-functional based tight binding method (DFTTB) [32]. All of these superstructures share common features that make them extremely attractive for technological applications. Thus, they all are characterized by low mass density, continuous porosity, and high surface area [33].

In Sect. 6.4 of this contribution we analyze H-shaped carbon nanotube composites, or cross-linking carbon nanotubes (CLCNT) architectures (see Fig. 6.2). The particular interest in these units is motivated by the realization that the strongest connection between two nanotubes is provided by another nanotube [29]. At the same time, these architectures are of relevance for nanoelectronics, as they provide a model for a nanotube connecting two carbon electrodes and so eliminate the problem of simulating the contact between a metal electrode surface and the carbon transmission element. Alternatively, they may be viewed as segments of an electronic network built entirely out of SWCNTs.

From the foregoing, the aim of this chapter is twofold. We first inquire about SWCNT magnetism, specifically asking for the relation between the magnetic structure of SWCNTs and ZGNRs (Sect. 6.3). Dealing with this topic, we will place emphasis on examining the onset and the persistence of AFM order in SWCNTs in terms of tube length and diameter. Secondly, we will study the geometric, energetic, and magnetic features of supramolecular H-shaped composites consisting of three (10,0) SWCNTs (Sect. 6.4). This segment of our work is motivated by both the interest in a basic understanding of magnetic effects in cross-linking carbon nanotubes and their potential relevance for the future design of magnetic micro-networks. A summary of the computational methods applied here, and an outlook at the continuation of this research are provided in Sects. 6.2 and 6.5, respectively.

6.2 Computational Details

The results presented in this contribution were obtained by DFT employing a plane wave basis set as implemented by the Vienna ab initio simulation package (VASP) [35]. The finite temperature version of Local Density Functional (LDF) Theory [36, 37] was used. The generalized Kohn-Sham equations were solved utilizing a residual minimization scheme, namely the direct inversion in the iterative subspace (RMM-DIIS) method [38, 39]. The optimization of the atomic geometry was performed via conjugate-gradient minimization of the total energy with respect to the atomic coordinates. Instead of Fermi-Dirac broadening of the one-electron energies, it may be computationally convenient to choose Gaussian broadening which was employed in this work. The width of the Gaussian distribution was selected as 0.001 eV. The total energy of the system refers to the limit of vanishing width.

The interaction of valence electrons and core ions is described by the projector-augmented wave (PAW) method [40]. All DFT calculations involved the gradient correction (GGA) for the exchange-correlation functional as prescribed by Perdew et al. [41].

For our research on SWCNTs, periodic boundary conditions were imposed on a unit cell with of dimension $d_x \times d_y \times d_z$ Å^3. The parameter d_z depends on the length of the chosen nanotube which varies from 20 to 80 Å, while d_x and d_y depend on the diameter of the nanotube which varies from 20 to 35 Å. In each individual case, the unit cell was chosen sufficiently large to minimize the interaction between neighboring systems. From inspection of the converged equilibrium geometries, the nearest-neighbor distance between atoms in adjacent cells is larger than 12 Å, making the interaction between them negligible. The length L of an SWCNT is given by its number of transpolyene rings.

The computations involving CLCNTs were done imposing periodic boundary conditions on a unit cell with of dimension 20 × 40 × 40 Å^3 for all the calculations. The sidewall tube axis was oriented along the z direction.

In both cases, only the Γ point of the Brillouin zone was included, and the geometry optimizations were performed enforcing a total energy difference between two subsequent steps of less than 1 meV as convergence criterion.

Spin density distributions for SWCNTs as well as CLCNTs were evaluated by computing the difference between the alpha and the beta spin density.

6.3 Magnetism in Axially Confined Single-Walled Carbon Nanotubes of the Zigzag Type

In what follows, we present our findings on hydrogenated (n,0) systems with n = 5–24 as a function of the tube length. We comment first on the magnetic coupling in selected SWCNTs (Sect. 6.3.1) and then discuss the relation between the magnetic moments obtained for ZGNRs and SWCNTs (Sect. 6.3.2).

6.3.1 Magnetic Coupling in Selected SWCNTs

An important tool for analyzing the magnetic effects associated with both SWCNTs and ZGNRs is the linear spin density [19] as a function of their length or width coordinates, respectively. As demonstrated in Fig. 6.3, SWCNTs and ZGNRs display very similar linear spin polarization profiles. Instead of being confined to the edges of the respective nanostructure, the spin polarization distribution shows an oscillatory pattern, fading out towards the middle of the structure, in all cases compared. This behavior is analogous to that of the spin polarization in a non-magnetic material sandwiched by layers with itinerant magnetism [19, 43].

The results for all systems shown in Fig. 6.3 refer to the FM phase. Inspecting the magnetic interaction between the SWCNT ends as a function of length, we calculated the energy difference between various (n, 0) systems in FM and AFM coordination, $E_{FM} - E_{AFM}$. Figure 6.4 shows the result for n = 6 (9, 10) at SWCNT lengths up to L = 30 (11). For n = 6, $E_{FM} - E_{AFM}$ displays sign changes as a function of SWCNT length, corresponding to an alternation between different magnetic phases. Where AFM order prevails, the energy difference is small. Larger magnitudes are found in the FM regime, but here the difference decreases markedly as the length increases. In the case of n = 9, a monotonic decrease with length reflects the reduction of the magnetic coupling as the separation between the SWCNT edges increases. The same observation is made for n = 10, except for a pronounced increase from L = 3 to L = 4. This behavior is associated with an alteration of the magnetic moment of the (10,0) system which changes from 2 μ_B at L = 3 to 6 μ_B at L = 4, related to a change of the electronic state of the SWCNT.

We mention that the magnetic coupling of the two cases displayed in Fig. 6.4b has been investigated previously for tube lengths beyond L = 3 [18] and by use of a non-local exchange-correlation functional, involving short-range exact exchange [44, 45]. While this computation yielded consistently larger values for the difference $E_{FM} - E_{AFM}$, it arrived at the same trends as those visible in Fig. 6.4b.

AFM coordination turns out to be more stable than the FM and NM alternatives if n > 6. This finding answers the question for the smallest tube index n that still guarantees AFM ordering of (n,0) systems, independent of their lengths, in accordance with the expectation based on Lieb's theorem, and thus ZGNR-analogous magnetic behavior.

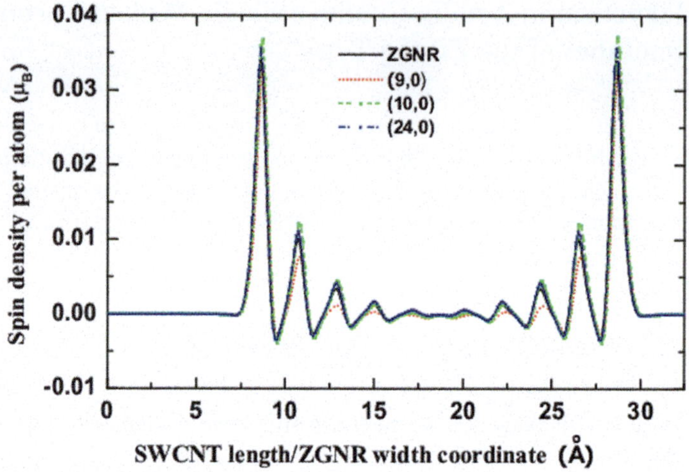

Fig. 6.3 Linear spin density distribution as a function of the SWCNT length (ZGNR width) coordinate for the case of FM coordination. SWCNTs of the (n,0) type with n = 9, 10, and 24 are included. The full length (width) of all systems is L = 10. Shown is the spin polarization average over the C atoms of an atomic layer. From [42]

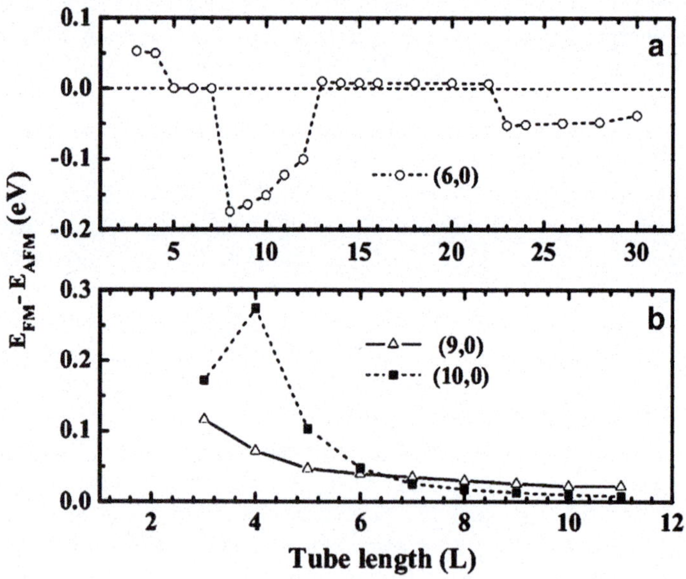

Fig. 6.4 Energy difference $E_{FM} - E_{AFM}$ between (n,0) systems in FM and AFM coordination with n = 6 (**a**), and n = 9, 10 (**b**)

Table 6.1 Magnetic phases for SWCNTs with n = 5, 6, as a function of the tube length, where FM (AFM, NM) stands for ferromagnetic (antiferromagnetic, non-magnetic). For n = 5, m_{FM} = 2 μ_B when L = 3–6 while no FM could be identified for L = 7. For n = 6, m_{FM} = 2 μ_B (4 μ_B) if the ground state is antiferromagnetic (ferromagnetic)

	L													
N	3	4	5	6	7	8	9	10	11	12	13	14	15	16
5	FM	FM	AFM	AFM	AFM	NM	NM	NM	NM	NM	NM	NM	NM	NM
6	AFM	AFM	NM	NM	NM	FM	FM	FM	FM	FM	AFM	AFM	AFM	AFM
	L													
N	17	18	19	20	21	22	23	24	25	26	27	28	29	30
5	NM	NM	NM	NM	NM	NM	NM	NM	NM	NM	NM	NM	NM	NM
6	AFM	AFM	AFM	AFM	AFM	AFM	FM	FM	FM	FM	FM	FM	FM	FM

For the most strongly curved systems considered here, those with n = 5, 6, more than one phase is seen to be adopted as the tube length is varied (see Table 6.1). Specifically, for n = 5, the NM phase turns out to be most stable asymptotically, while for n = 6, the FM phase, with a magnetic moment of magnitude 4 μ_B, prevails for sufficiently long SWCNTs. These results were ascertained for SWCNTs at the asymptotic ZGNR length, i.e. L = 30.

We emphasize that the variation of the magnetic phase with the length of the system, as exhibited by the SWCNTs with n = 5 and 6, draws a parallel between these structures and layered transition-metal compounds that consist of two ferromagnetic segments enclosing a spacer unit, such as Fe/Cr/Fe or Co/Ru/Co [46]. By experimental observation, changing the spacer length was shown to induce oscillations of the magnetic exchange coupling between the two boundary layers with respect to both the sign and the magnitude, with periods between about 12 to about 21 Å. Aspects of this phenomenon were explained by use of quantum-well theory [47] and Ruderman-Kittel-Kasuya-Yosida (RKKY) theory [48], quantitative modeling of the oscillation periods succeeded in the framework of a more generalized approach [49]. While the qualitative difference between the complexes compared here, i.e. layered transition-metal systems involving a spacer unit that differs from the ferromagnetic boundary segments, and homogeneous zigzag SWCNTs with sufficiently high curvature, must be kept in mind, both materials allow for controlled variation of the magnetic phase by manipulating the length of the system.

6.3.2 *Magnetic Moments of Zigzag SWCNTs with n = 7–24*

As SWCNTs of the zigzag type may be viewed as ZGNRs with lengths equal to the SWCNT circonferences rolled up to form a seamless cylinder, a unique relation may be expected to hold between the magnetic moments of these two carbon nanostructures. To establish this relation, we carried out a comparative study of SWCNT versus ZGNR magnetism, including (n,0) systems with n = 7–24. In an

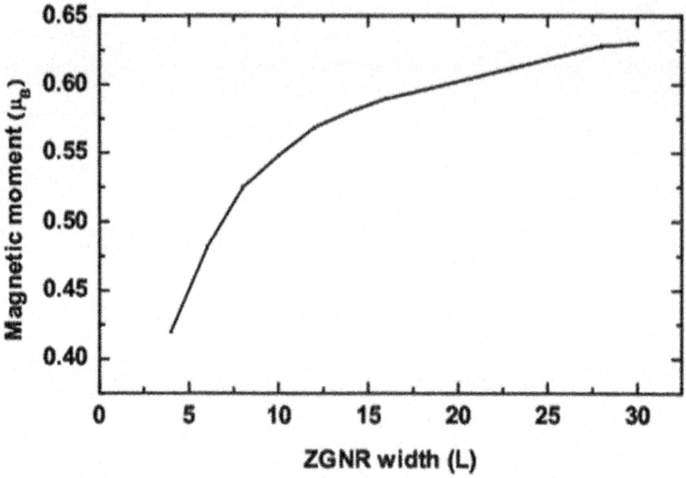

Fig. 6.5 The ZGNR magnetic moment as function of the ZGNR width. From [42]

initial step, we evaluated the magnetic moment for ZGNRs with a periodic length coordinate as a function of their width. The result is displayed in Fig. 6.5 for the case of ferromagnetic coordination. The limit of 2/3 μ_B is closely approximated at a ZGNR width of L = 30, in agreement with results obtained earlier by use of a quantum chemical approach [24].

As a ZGNR in FM coordination transforms into an SWCNT system, the spin-polarized bands become discrete energy levels. Those levels that match the energy of the spin-polarized bands may be expected to maintain the original ZGNR polarization. This expectation has been examined by comparing the Kohn-Sham eigenvalues of the unpaired spin states of (n,0) FM systems with the energy interval spanned by the corresponding ZGNR spin-up bands in the spin-polarized regime.

As documented in Table 6.2, the variation of SWCNT magnetic moments in the FM phase with the tube index n obeys a simple pattern if sufficiently long tubes are considered (specifically, for L $\geq$ 9). For clarification, we distinguish three sets of tube indices by selecting the series n = 3 N−2, n = 3 N−1, and n = 3 N, with N = 3,. . .,8. In order to understand the progressive change of the magnetic moment as a function of the tube index n, we focus first on two of these three series, 3 N−2 and 3 N. For these two, the variation of the magnetic moment is rationalized by observing that, as one goes from N to N + 1, a total of two unpaired electrons is added to the corresponding idealized ZGNR fragments that consist of n parallel ZGNR unit cells, and whose widths equal the lengths of their SWCNT counterparts. These two electrons are located at opposing ends of the ZGNR, enhancing the number of spin-polarized electrons at the ZGNR edges. Thus, within both series, the magnetic moment in FM coordination increases by 2 μ_B as N increases by 1.

Table 6.2 Magnetic moments (in μ_B) in the FM phase for zigzag SWCNTs with n = 7–24, and L ≤ 11 (for n = 10, L ≤ 15). The symbol m_{FM} stands for the tentative assessment of the asymptotic magnetic moment in the FM phase. Also included is the limiting magnetic moment per ZGNR unit for n = 3 N (N = 2–8)

	N																	
L	7	8	9	10	11	12	13	14	15	16	17	18	19	20	21	22	23	24
3	4	2	4	2	4	6	4	6	4	6	4	6	8	6	8	6	8	10
4	4	2	4	6	4	6	4	6	8	6	8	6	8	10	8	10	8	10
5	4	2	4	6	4	6	4	6	8	6	8	10	8	10	8	10	12	10
6	4	6	4	6	4	6	8	6	8	6	8	10	8	10	12	10	12	10
7	4	6	4	6	4	6	8	6	8	10	8	10	8	10	12	10	12	14
8	4	6	4	6	4	6	8	6	8	10	8	10	12	10	12	10	12	14
9	4	6	4	6	8	6	8	6	8	10	8	10	12	10	12	14	12	14
10	4	6	4	6	8	6	8	6	8	10	8	10	12	10	12	14	12	14
11	4	6	4	6	8	6	8	6	8	10	8	10	12	10	12	14	12	14
m_{FM}	4	6	4	6	8	6	8	6	8	10	8	10	12	10	12	14	12	14
2n/3			6			8			10			12			14			16

To motivate the magnitude of the obtained magnetic moments m_{FM} for the series 3 N−2 and 3 N, we inspect systems of length L = 10. Since the change of that magnitude as a function of N is related to the increasing population of spin-polarized electrons at the ZGNR edges as N is increased, the remaining task lies in understanding the magnitude of the initial values of m_{FM} for both series. Examining the ZGNR of width L = 10, we find an overall spin-polarized population of 0.55, yielding an effective number of 3.85 (4.95) unpaired electrons at the edges of the idealized finite ZGNR counterparts of the SWCNTs with n = 7 (n = 9). These electrons occupy the flat band of the spin-alpha set. In both cases, n = 7 and n = 9, SWCNT formation turns the flat band of spin alpha polarization into four degenerate alpha-polarized orbitals, giving rise to the observed magnetic moment of 4 μ_B. The energy of the spin-polarized band segment of the ZGNR (defined by $2\pi/3a \leq |k| \leq \pi/a$) ranges from −3.68 to −3.44 eV, matching the energies of the respective spin-polarized orbitals of the SWCNTs which are −3.46 (−3.54) eV for the n = 7 (n = 9) system.

Similar arguments seem to suggest a magnetic moment of 4 μ_B also for the initial system of the 3 N−1 series (the case N = 3, i.e. the SWCNT with n = 8 and L = 10). This, however, is not the case (see Table 6.2). A computation performed with the magnetic constraint of $\mu = 4\ \mu_B$ yields equal energy eigenvalues for the highest occupied and the lowest unoccupied alpha KS orbital, both values being −3.372 eV, and thus a vanishing energy gap. Comparing the alternative cases $\mu = 2\ \mu_B$ and $\mu = 6\ \mu_B$, we find that the total energy is lower by a margin of 0.512 eV for the latter than the former alternative. This result is in accordance with a widening of the spin-alpha energy gap from 0.101 to 0.376 eV as one goes from $\mu = 2\ \mu_B$ to $\mu = 6\ \mu_B$. Indeed, a calculation for n = 8 at L = 10 where no magnetic constraint was defined, yielded 6 μ_B as ground state magnetic moment.

The single deviation from the simple progression scheme indicated above, involving the increase of the magnetic moment by two units as one goes from N to N + 1 within each of the three series, is found in the 3 N−1 series, and, more specifically, for the transition from N = 4 to N = 5 (or n = 11 to n = 14). Here a decrease of the magnetic moment from 8 to 6 μ_B is noticed, while one would expect a result of 10 μ_B when using the systems with N = 3 and 4 as basis for extrapolation. From our calculations, the solution with $\mu = 6\ \mu_B$ is preferred over the $\mu = 10\ \mu_B$ alternative by a margin of 147 meV. To rationalize this behavior, we note a strong mismatch between the number of alpha-polarized electrons in the $\mu = 10\ \mu_B$ SWCNT system and the corresponding ZGNR fragment (10 versus 6.97), while much better matching is found between the latter and the $\mu = 6\ \mu_B$ SWCNT. Inquiring about the alternative n = 14, L = 10, $\mu = 8\ \mu_B$, we find that for this case, a similar configuration is realized as for n = 8, L = 10, $\mu = 8\ \mu_B$. The highest occupied and the lowest unoccupied alpha KS orbitals turn out to be near-degenerate with energy eigenvalues of −2.972 and −2.928 eV, resulting in an extremely small energy gap of 44 meV.

Addressing the dependence of the SWCNT magnetic moment on the tube length, as documented in Table 6.2, we observe that the magnetic moment increases in steps of four units. This pattern reflects the twofold degeneracy of the KS orbitals adjacent to the SWCNT energy gap.

Comparing the above findings on SWCNT magnetism with those on ZGNR prototypes, i.e. ZGNRs at asymptotic width, a rather simple pattern emerges. For the three series n = 3 N, n = 3 N−1, and n = 3 N−2, with N = 3,..,8, the FM magnetic moments at the maximum SWCNT lengths investigated here may be described as 2(N−1), 2 N, 2(N−1) μ_B, respectively, where it should be noted that 2 N μ_B is the magnetic moment for an idealized finite ZGNR that consists of n = 3 N unit cells, with $n > 7$. A deviation from this rule, as observed for the 3 N−1 series at N = 3 and 4, has been related to the degeneracies found for the electronic structure close to the SWCNT energy gap, and to differences in the gap sizes for different choices of magnetic moments in the FM phase.

6.4 Magnetic Properties of Cross-Linking Carbon Nanotubes

In this section, we extend our analysis of SWCNT magnetism, as presented in Sect. 6.3, to CLCNTs (cross-linking carbon nanotubes, see Sect. 6.1), or H-shaped SWCNT assemblies. For all CLCNTs considered in this work, a (10,0) SWCNT provides the basic structural motif. Thus, each CLCNT described here is composed of two (10,0) tubes with L = 12 as sidewalls which are bridged by a tube with L = 4. Both sidewall edges are terminated by hydrogen atoms. Three CLCNT types were included in this research, as sketched in Fig. 6.6. The three types differ from each other with respect to the geometry of the contact regions between the

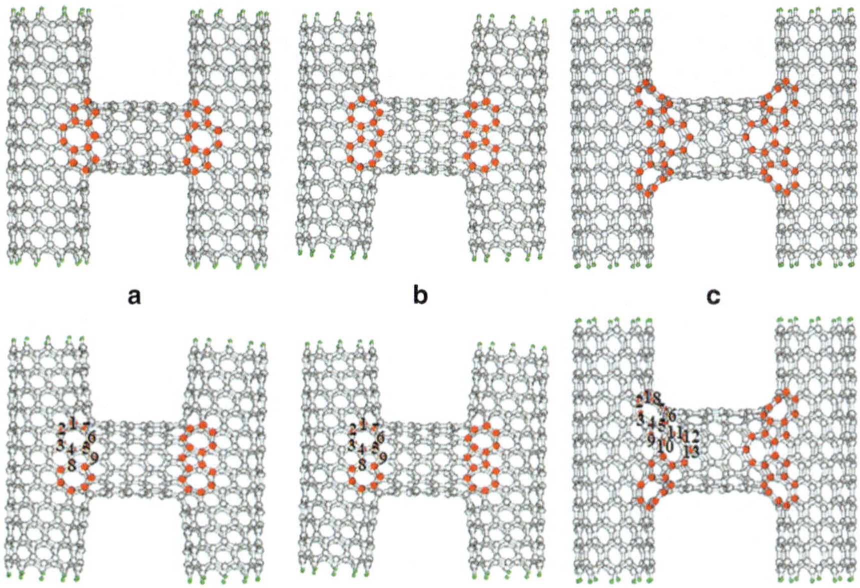

Fig. 6.6 The three prototypes for cross-linking architectures investigated in this work. For clear representation of these structures, they are shown both with (*lower panel*) and without (*upper panel*) labels for characteristic carbon atom sites (Reprinted with permission from [34]. Copyright (2013) American Chemical Society)

sidewall and the bridge segments. In terms of different combinations of non-hexagonal rings, the three types are characterized in the following way:

(a) a heptagon at the center of the sidewall is surrounded symmetrically by two pentagons,
(b) a pentagon at the center of the sidewall is surrounded symmetrically by two heptagons,
(c) four octagons, four pentagons, and two heptagons form the contact region. This is the connection described in [33].

The total number of carbon atoms is different in these three cases, being 560, 556, and 580 in connections (a), (b), and (c), respectively.

In the following, we focus on geometric and energetic properties of the CLCNTs under study (Sect. 6.4.1), preceding a discussion of their electronic and magnetic features (Sect. 6.4.2).

6.4.1 *Geometric and Energetic Properties*

Since the three CLCNT types included in this work differ with respect to the contact regions interfacing between the sidewalls and the bridge segment, comparing the three structures in terms of their geometry can be confined to these regions. The

Table 6.3 Characteristic C–C bond lengths for the three CLCNT types considered in this work. The bond length labels refer to the carbon atom sites indicated in the lower panel of Fig. 6.6

	Connection (**a**)	Connection (**b**)	Connection (**c**)
d_{12}	1.532	1.456	1.439
d_{23}	1.402	1.463	1.472
d_{34}	1.445	1.401	1.434
d_{45}	1.411	1.440	1.484
d_{56}	d_{36} = 1.426	1.399	1.416
d_{67}	1.435	1.433	1.471
d_{78}	d_{48} = 1.431	d_{48} = 1.427	1.407
d_{89}		d_{59} = 1.410	d_{49} = 1.420
d_{9-10}			1.424
d_{10-11}			1.433
d_{11-12}			1.416
d_{12-13}			1.469

lower panel of Fig. 6.6 indicates some characteristic carbon atom sites. The amount of geometric deformation induced by the topological defects joining the CLCNT segments can be assessed by recording the C–C bond lengths in the interface region. The standard of comparison is here defined by the equilibrium C–C bond length in a regular SWCNT, 1.420 Å. The deviation from this reference value provides a measure for the strain associated with merging the three SWCNT constituents into a CLCNT.

Table 6.3 summarizes the bond lengths most relevant for a comparison between the three structures. For connection (a), the bonds within the pentagon range between 1.40 and 1.53 Å while those within the center heptagon are close to 1.430 Å, excepting two elongated bonds with lengths of 1.445 Å. Likewise, connections (b) and (c) exhibit significant deviations from the bond length equilibrium value. Thus, for connection (c), the maximum elongation (contraction) of any bond in the contact region amounts to 1.484 Å (1.407 Å).

Determining for each structure the maximum enhancement of any bond length over the reference value of 1.420 Å, we find elongations of 8.0 %, 3.0 % and 4.5 % in connections (a), (b), and (c), respectively.

We mention that these values are in accord with E_C, the magnitude of the ground state stability (i.e. the difference between the energy of the ground state and that of N carbon atoms in separation from each other, with N as the number of carbons atoms contained in the system) per carbon atom. This quantity favors, by a small margin, connection (b) with $E_C = 9.300$ eV, followed by connection (c) ($E_C = 9.290$ eV), and lastly connection (a) ($E_C = 9.255$ eV), thus reflecting the sequence of the maximally strained C–C bond lengths among the three structures. The variations between the values of E_C may therefore be ascribed to the structural differences between the three models and so to the different strengths with which the three segments cohere. As the three prototypes differ with respect to the number of carbon atoms, a direct comparison between them in terms of stability is not possible.

Table 6.4 Relative energies (ΔE), energy gaps (E_{gap}), and total magnetic moments of the most stable states identified for connections (a), (b), and (c) in the AFM, FM, and mixed states (in units of eV and μ_B)

	Connection (**a**)			Connection (**b**)			Connection (**c**)		
	ΔE[a]	E_{gap}[b]	m[c]	ΔE	E_{gap}	m	ΔE	E_{gap}	m
AFM-AFM	0	0.182	0	0.008	0.103	0	0	0.440	0
Mixed	0.006	0.069	4	0	0.137	6	0.005	0.408	6
FM-FM	0.020	0.140	10	0.012	0.080	12	0.010	0.393	12

[a]Relative energy
[b]Energy gap size
[c]Magnetic moment

6.4.2 Magnetic and Electronic Properties

The magnetic properties of a CLCNT may be described in terms of SWCNT magnetism. The axially confined (10,0) SWCNTs that form the systems considered here may display AFM or FM coordination, or may be non-magnetic, with AFM order as the preferred alternative for single nanotubes. This suggests that the two CLCNT sidewalls likewise display magnetic order. Three basic combinations are possible: the sidewall tubes may both be in FM or AFM states, or one sidewall may display FM, and the other AFM order. The relative energies and energy gaps for these three configurations are listed in Table 6.4.

As AFM coordination is known to prevail in the (10,0) constituent, it is expected to be the most stable magnetic order for the CLCNTs considered here as well. From Table 6.4, this expectation is borne out for connections (a) and (c). The entries for both reflect the stability of the AFM versus the FM phase of the isolated (10,0) tube for which a magnetic coupling $E_{FM} - E_{AFM}$ of 7 meV was reported at L = 12 [19]. The configuration with both sidewalls in the FM-FM state is highest in energy, and the mixed FM/AFM solution is intermediate. We point out that all three schemes admit different realizations. With reference to the FM-FM and AFM-AFM cases, the directions of the magnetic moments at the ends of the two sidewalls may be identical or opposite. From test calculations involving the AFM-AFM phase, the energy difference between these two variants is extremely small, not surpassing 0.5 meV.

The stability sequence found for connection (b) deviates from the standard set by connection (**c**), as the stability order of the mixed and the AFM-AFM solution is reversed.

Commenting on the magnetic moments of the three considered connections, as indicated in Table 6.4, we note that respective entries for (b) and (c) conform with the expectation based on the isolated (10,0) system. For the latter, a magnetic moment of 6 μ_B in the FM phase was reported. This agrees with the finding of 12 μ_B (6 μ_B) in the FM-FM (mixed) phase of both connections (b) and (c). The result for connection (a), however, deviates from this scheme. Here the magnetic

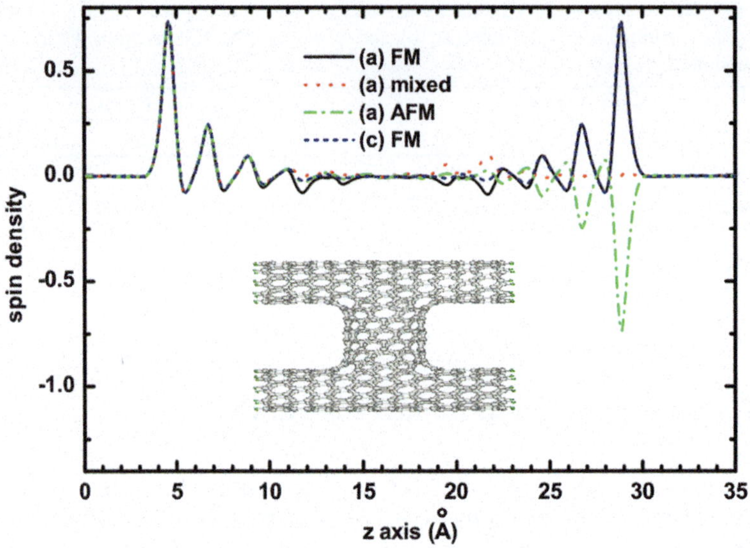

Fig. 6.7 Linear spin density distribution for connections (**a**) and (**c**). The linear spin density was evaluated by integrating the three-dimensional spin density over the directions perpendicular to the length coordinate z which is indicated in the insert by the example of connection (**c**) (Reprinted with permission from [34]. Copyright (2013) American Chemical Society)

moment of both the mixed solution and the FM-FM solution is reduced by 2 μ_B as compared with the two other cases.

To interpret these findings we add local information on CLCNT magnetism, as provided by linear spin density distributions. In Fig. 6.7, the linear spin density profile of all three magnetic phases of connection (a) is compared with that of connection (c) in the FM phase. In accordance with the behavior of individual finite zigzag SWCNTs ([19], see also Fig. 6.3), the spin polarization displays an oscillatory pattern, fading out towards the middle of the structure, instead of being confined to the CLCNT edges. Comparing the FM distributions for connections (a) and (c), one finds stronger damping in the latter case. This feature is consistent with additional magnetic moments, beyond those present at the tube edges, localizing in the contact region of (a).

As revealed by closer analysis of the spin density distributions of all three connections, the emergence of finite magnetic moments in the contact region is a common feature of (a) and (b), in contrast to (c). In accordance with the general preference for AFM coordination, these magnetic moments tend towards antiparallel orientation in the most stable states of connections (a) and (b). For the mixed and the FM-FM cases of connection (a), however, they adopt FM order, thus affecting the total magnetic moment of the CLCNT which is found to be smaller than that of connection (c). The spin density distributions for the two most stable states of connection (a) are represented in Fig. 6.8(i, ii) which shows the dominance of AFM (FM) coordination between the two contact regions in the most stable (the

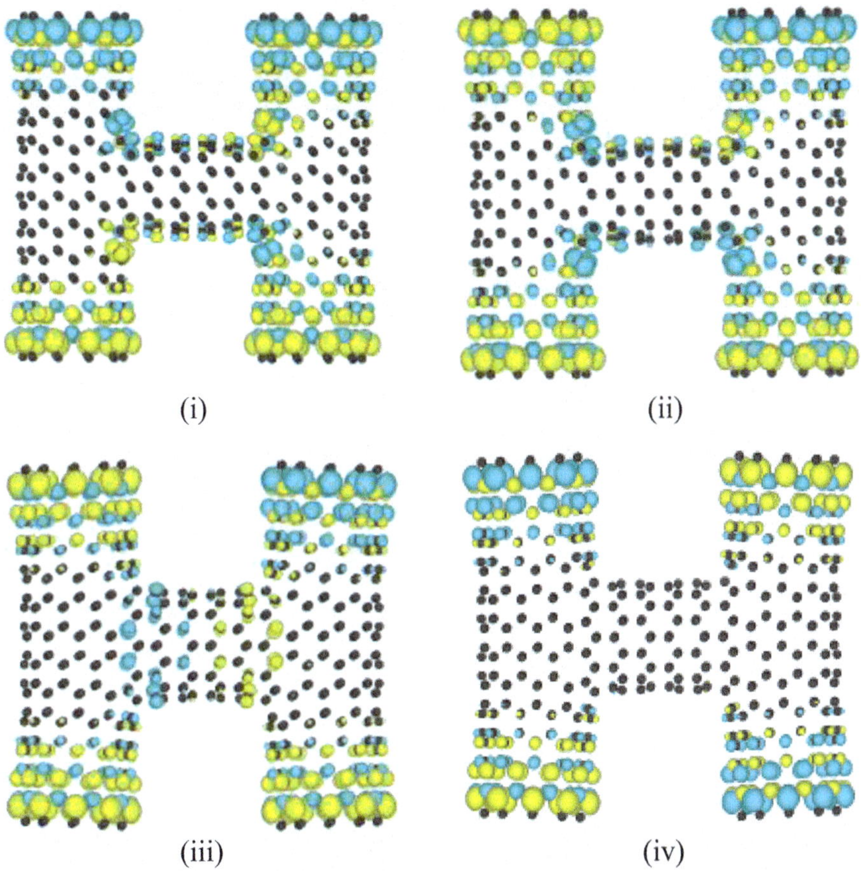

Fig. 6.8 (*i–ii*): Representation of the three-dimensional spin density distribution for the two most stable states of connection (**a**), namely the AFM-AFM (*i*) and the AFM-FM (or mixed) solution (*ii*). (*iii–iv*): Same for the two most stable states of connection (**b**), namely the AFM-FM (or mixed) (*iii*) and the AFM-AFM (*iv*) solution. Spin alpha (beta) contributions to the total spin density are indicated in *blue* (*yellow*) (Reprinted with permission from [34]. Copyright (2013) American Chemical Society)

second most stable) state. Figure 6.8(iii, iv) refers to connection (b). We observe that the AFM-FM solution (iii) exhibits finite magnetic moments in the contact region while this is not the case for the AFM-AFM solution (iv), a feature that may account for the higher stability of the former as compared with the latter. From inspection of the ground state configurations for (a) and (b), as displayed in Fig. 6.8 (i, iii), respectively, it can be seen that magnetic moments localized in the contact region tend to adopt AFM order across the bridge, in accordance with the behavior of finite SWCNTs of the zigzag type.

We point out that for three connections, the relative stability of the AFM-AFM, FM-FM, and mixed cases tends to correlate with the size of the energy gap. As compared with the energy gap of the hydrogenated component SWCNT with L = 12, which amounts to 0.59 eV, all energy gaps listed in Table 6.4 are reduced. This reduction is markedly lower in the case of connection (c) than for connections (a) and (b). Thus, the ground state solution for connection (c), i.e. the only one that does not display finite magnetic moments in the contact region, is characterized by a substantially wider energy gap than the two rivaling systems.

6.5 Summary and Outlook

Magnetic SWCNTs as well as CLCNTs were designed and analyzed by plane-wave density functional theory (PL-DFT) modeling as potential transmission elements in nano-spintronic circuits.

Among the axially confined hydrogenated zigzag SWCNTs examined in this contribution, with tube indices n ranging from 5 to 24, and with lengths up to 11 transpolyene rings, AFM coordination has been found to prevail for n = 7–24. Thus, beyond a threshold index of n = 6, the systems investigated here display the magnetic behavior expected on the basis of Lieb's theorem. For SWCNTs of higher curvature, as represented in this work by the systems with n = 5 and 6, distinct variations of the magnetic phase with the tube length were observed, in analogy to layered transition metal composites consisting of a spacer unit surrounded by two ferromagnetic segments.

With respect to the size of the magnetic moment generated in the FM phase by unpaired electrons at the SWCNT edges, and its variation with n, a universal relation if found to hold between the magnetic moments of these finite ZGNRs in the asymptotic range and those of the longest SWCNTs analyzed in this work. Focusing on systems with n as a multiple of 3, this pattern simplifies to a linear increase of the magnetic moment. Its value is consistently below the result for the corresponding nanoribbons by exactly 2 μ_B. As the tube is flattened, its magnetic moment increases by the same amount, irrespective of its diameter.

More complex units, namely CLCNTs consisting of three (10,0) constituents, were analyzed in terms of their geometric, energetic and magnetic characteristics. Three CLCNT prototypes were included in this study, differing from each other by the structure of the contact region between the sidewalls and the bridge. In particular, a previously established structure (connection (c)) with a complex arrangement of topological defects in the contact region, involving pentagons, heptagons, and octagons, is contrasted with two alternative structures with pentagons and heptagons only (connections (a) and (b)). In terms of magnetism, three configurations were distinguished: both sidewalls in AFM (FM) coordination, and a mixed case where one sidewall realizes AFM, the other FM coordination. Connections (a) and (c) were shown to display the expected hierarchy of magnetic phases, with the AFM-AFM (FM-FM) ordering at highest (lowest) stability. Different stability sequences were found for connection (b), demonstrating that the structure of the contact region exerts a strong influence on the relative energies of the three magnetic phases. Inspection of the corresponding linear spin density distribution profiles in conjunction with the associated total magnetic moments suggests that these differences are correlated with the presence of finite magnetic moments in the contact regions of connections (a) and (b) which may exhibit AFM or FM order across the bridge.

In continuation of this project, it will be interesting to explore how to integrate the systems considered here into spintronics circuits. Embedding them into a nano-electronic device will require designing junctions between them and electrode units. Identifying conditions that preserve the SWCNT edge magnetism to the highest possible extent is crucial for the success of this effort. Composites consisting of the SWCNTs, nano-electrodes as well as anchor molecules connecting these two components are to be tested with respect to their overall magnetism.

With reference to CLCNTs, the stabilities of selected magnetic configurations, involving AFM-AFM, FM-FM, and mixed junctions, can be sensitively manipulated by adjusting the structure of the contact region joining the bridge and the sidewalls. This effect might be exploited to generate a wide variety of magnetic networks of interest for spintronics applications. For quantitative assessment of these units as transmission elements, their transport properties will have to be determined. In the light of the results on the CLCNTs examined here, exploring extended arrays composed of these systems with respect to their magnetism as well as their conductivity will be a worthwhile topic for future studies.

Acknowledgment This work was performed in collaboration with Jackson State University and supported by the DoD through the US Army/Engineer Research and Development Center (ERDC) Vicksburg, MS Contract #W912HZ-10-C-0107. F. H. acknowledges support from the Tennessee NSF-EPSCoR grant TN-SCORE (NSF EPS 1004083). J.L. would like to thank the ONR grant number 08PRO2615-00/N00014-08-1-0324 and NSF RISE program for support.

References

1. Bolskar AD, Alford JM, Benedetto RF, Husebo LO, Price RE, Jackson EF, Wallace S, Wilson IJ (2003) J Am Chem Soc 125:5471
2. Iezzi EB, Duchamp JC, Fletcher KR, Glas TE, Dorn HC (2002) Nano Lett 2:1187
3. Maassen J, Ji W, Guo H (2011) Nano Lett 11:151
4. Kan E, Li Z, Yang J, Hou JG (2008) J Am Chem Soc 130:4224
5. Kuemmeth F, Churchill HOH, Herring PK, Marcus CM (2009) Mater Today 13:18
6. Schneider CM, Kohzuharova R, Groudeva-Zutova S, Zao B, Muehl T, Moenchj I, Vinschelberg H, Leonhardt RA, Fink J (2005) Nanotube spintronics: magnetic systems based on carbon nanotubes. In: Buzaneva E, Scharff P (eds) Frontiers of multifunctional integrated nanosystems, vol 152, Nato science series. Springer Science + Business Media, Inc, Dordrecht, p 359, 2005
7. Jhang SH, Marganska M, Skoursi Y, Preusche D, Witkamp B, Grifoni M, van der Zant H, Strunk C, Wosnitza J (2010) Phys Rev B 82:041404
8. Yazyev OV (2011) ChemInform abstract: emergence of magnetism in graphene materials and nanostructures. Chem Inform 42. doi:10.1002/chin.201101219
9. Wu J, Hagelberg F (2011) J Phys Chem C 115:4571

10. Banhart F (1999) Rep Prog Phys 62:1181
11. Krasheninnikov AV, Banhart F (2007) Nat Mater 6:723
12. Lehtinen P, Foster AS, Ma Y, Krasheninnikov AV, Nieminen RM (2004) Phys Rev Lett 93:187202
13. Telling RH, Ewels CP, El-Barbary AA, Heggie MI (2009) Nat Mater 2:333
14. Fujita M, Wakabayashi K, Nakada K, Kusakabe K (1996) J Phys Soc Jp 65:1920
15. Nakada K, Fujita M, Dresselhaus G, Dresselhaus M (1996) Phys Rev B 54:17954
16. Du AJ, Chen Y, Lu GQ, Smith SC (2008) Appl Phys Lett 93:073101
17. Kim YH, Choi J, Chang KJ (2003) Phys Rev B 68:125420
18. Hod O, Scuseria GE (2008) ACS Nano 2:2243
19. Wu J, Hagelberg F (2009) Phys Rev B 79:115436
20. Mananes A, Duque F, Ayuela A, Lopez MJ, Alonso JA (2008) Phys Rev B 78:035432
21. Lieb EH (1989) Phys Rev Lett 62:1201
22. Son YW, Cohen ML, Louie SG (2006) Nature 444:347
23. Kim WY, Kim KS (2008) Nat Nanotechnol 3:408
24. Pisani L, Chan JA, Montanari B, Harrison NM (2007) Phys Rev B 75:064418
25. Kong J, Cao J, Dai HJ, Anderson E (2002) Appl Phys Lett 80:73
26. Kong J, Dai HJ (2001) J Phys Chem B 105:2890
27. Andriotis AN, Menon M, Srivastava D, Chernozatonskii L (2001) Phys Rev Lett 87:066802
28. Menon M, Srivastava D (1997) Phys Rev Lett 79:4453
29. Qian D, Wagner DJ, Liu WK, Yu MF, Ruoff RS (2002) Appl Mech Rev 55:495
30. Coluci VR, Pugno NM, Dantas SO, Galv˜ao DS, Jorio A (2007) Nanotechnology 18:335702
31. Li Y, Qiu XM, Yang F, Wang XS, Yin Y, Fan Q (2008) J Phys D Appl Phys 41:155423
32. Enyashin AN, Ivanovskii AL (2008) JETP Lett 87:372
33. Romo-Herrera JM, Terrones M, Terrones H, Dag S, Meunier V (2007) Nano Lett 7:570
34. Wu J, Ayasoufi A, Leszczynski J, Hagelberg F (2013) J Phys Chem C 117:3646
35. (a) Kresse G, Hafner J (1993) Phys Rev B 47:558; (b) Kresse G, Hafner J (1994) Phys Rev B 49:14251; (c) Kresse G, Furthmüller J (1996) J Comput Mater Sci 6:15
36. Mermin NM (1965) Phys Rev 137:A1441
37. Methfessel M, Paxton AT (1981) Phys Rev B 40:3616
38. Wood DM, Zunger A (1984) J Phys A 18:1343
39. Pulay P (1080) Chem Phys Lett 73:393
40. Blöchl PE (1994) Phys Rev B 50:17953
41. Perdew JP, Burke K, Ernzerhof M (1996) Phys Rev Lett 77:3865
42. Wu J, Hagelberg F (2013) Chem Phys Chem 14:1696
43. Wu J, Herrmann T, Nolting W (1999) Phys Rev B 60:12226
44. Kudin KN, Scuseria GE, Martin RL (2002) Phys Rev Lett 89:266402
45. Prodan ID, Sordo JA, Kudin KN, Scuseria GE (2005) J Chem Phys 123:014703
46. Parkin SSP, More N, Roche KP (1990) Phys Rev Lett 64:2304
47. Edwards DM, Mathon J, Muniz RB, Phan MS (1991) Phys Rev Lett 67:493
48. Bruno P, Chappert C (1991) Phys Rev Lett 67:1602
49. Ferreira MS, d'Albuquerque e Castro J, Edwards DM, Mathon J (1996) J Phys Condens Matter 8:11259

Chapter 7
Modelability Criteria: Statistical Characteristics Estimating Feasibility to Build Predictive QSAR Models for a Dataset

Alexander Golbraikh, Denis Fourches, Alexander Sedykh, Eugene Muratov, Inta Liepina, and Alexander Tropsha

Abstract It is not always possible to build predictive Quantitative Structure-Activity Relationships (*QSAR*) models for a given chemical dataset. In this work, we propose several statistical criteria, which can with high confidence answer a question, whether it is possible to build a predictive model for a dataset *prior* to actual modeling, i.e. to establish, whether the dataset is ***modelable***. Calculation of these criteria is fast, and using them in *QSAR* studies could dramatically reduce modelers' time and efforts, as well as computational resources necessary to build *QSAR* models for at least some datasets, especially for those which are not modelable. The calculation of modelability criteria is based on the k-nearest neighbors approach. For all datasets, as modelability criteria we have proposed dataset diversity (*MODI_DIV*) and new activity cliff indices (*MODI_ACI*). For datasets with binary end points, as modelability criteria we have proposed the correct classification rate (*MODI_CCR*) *CCR* = 0.5(sensitivity + specificity) for leave-one-out (*LOO*) cross-validation in the entire descriptor space, and correct classification rate for similarity search (*MODI_ssCCR*) in the entire descriptor space with leave 20 %-out (five-fold) cross-validation. For binary datasets, all these modelability criteria were tested on 42 datasets with previously generated *QSAR* models. Two latter criteria (*MODI_CCR* and *MODI_ssCCR*) were found to have high correlation with the predictivity of *QSAR* models (*QSAR_CCR*) and were additionally tested on 60 ToxCast end points with *QSAR* modeling results published recently (Thomas RS, Black MB, Li L, Healy E, Chu T-M, Bao W, Andersen MD, Wolfinger RD. Toxicol Sci: Off JSoc

A. Golbraikh • D. Fourches • A. Sedykh • A. Tropsha (✉)
University of North Carolina at Chapel Hill, Chapel Hill, NC, USA
e-mail: alex_tropsha@unc.edu

E. Muratov
University of North Carolina at Chapel Hill, Chapel Hill, NC, USA

A.V. Bogatsky Physical-Chemical Institute NAS of Ukraine, Odessa, Ukraine

I. Liepina
Latvian Institute of Organic Synthesis, Riga, Latvia

J. Leszczynski and M.K. Shukla (eds.), *Practical Aspects of Computational Chemistry III*, DOI 10.1007/978-1-4899-7445-7_7,

Toxicol 128(2):398–417, 2012). These modelability criteria can be used to classify many datasets as modelable or non-modelable. These criteria can be generalized to datasets with compounds belonging to more than two categories or classes. Additionally, criteria which take into account errors of prediction $MODI\_CAT_i$ and $MODI\_CLASS_i$ were proposed for datasets with compounds belonging to more than two ($i > 2$) categories or classes and continuous end points, divided into $i > 2$ bins. For continuous end points, *LOO* cross-validation q^2 for similarity search with different numbers of nearest neighbors in the entire descriptor space ($MODI\_q^2$), and similarity search coefficient of determination ($MODI\_ssR^2$) in the entire descriptor space were proposed as modelability criteria. Our preliminary studies demonstrated high correlation between the external predictivity of QSAR models ($QSAR\_R^2$) and each of the $MODI\_q^2$ and $MODI\_ssR^2$. On the other hand, for datasets with any binary or continuous response variable, *MODI_DIV*s and *MODI_ACI*s were found to be less useful to establish dataset modelability.

7.1 Introduction

Quantitative Structure-Activity Relationships analysis (*QSAR*) is a set of mathematical and statistical methods aimed at the development of models predicting a given molecular property (typically, biological activity) for compounds similar to those in a given dataset. In other words, *QSAR* analysis can be thought of deriving a mathematical equation: predicted property = function(structure). During the development of a *QSAR* model, the cumulative error of prediction for the dataset is minimized. In *QSAR* analysis, the compounds are represented by independent variables (descriptors), which are quantitative characteristics of molecular structures, and a dependent variable *Y*, or activity, or end point, that is a property, for which a predictive model is to be built. Let *M* and *N* be the number of compounds and the number of descriptors, respectively. Elements X_{ij} ($i = 1, \ldots, M$; $j = 1, \ldots, N$) of $M \times N$ descriptor matrix $\mathbf{X}$ are *N* descriptor values for each compound (so $M \times N$ elements in total). Elements of activity vector $\mathbf{Y}$ are activity values Y_i ($i = 1, \ldots, M$) for each compound. Elements of vector $\boldsymbol{\varepsilon}$ are errors of prediction ε_i ($i = 1, \ldots, M$) for each compound. The goal of *QSAR* analysis is to obtain the Equation $\mathbf{Y} = f(\mathbf{X}) + \boldsymbol{\varepsilon}$, where the length of vector $\boldsymbol{\varepsilon}$ is minimized. To obtain this equation different *QSAR* methodologies (supervised machine learning techniques) are used. Examples of descriptors include Dragon [1], MOE [2], Molconn-Z [3], Mold2 [4], CDK [5], Volsurf [6], Adriana [7], USEPA [8], SiRMS [9], different types of fingerprints [2, 10], etc. Examples of QSAR methodologies include multiple linear regression (MLR) [11], principal component regression (PCR) [11], partial least squares (PLS) [12], *k* nearest neighbors (*k*NN) [13], support vector machines (SVM) [14], decision tree (DT) [15, 16], random forest (RF) [17, 18], Hierarchical Clustering (HC) [8], etc. Typically, *QSAR* models are validated using external sets of compounds (those, which were not included in modeling). If a *QSAR* model demonstrates high predictivity for training, test and external sets in terms of different statistical criteria [19–22], it is considered as a predictive model.

One of the problems in *QSAR* analysis is related to the dataset modelability, i.e. the question is, whether and when it is *at all possible* to build a predictive *QSAR* model for a given dataset. In many cases, multiple attempts to build a predictive model are undertaken (like in Combi-*QSAR* modeling workflow developed in our laboratory [23, 24]): different sets of chemical descriptors are used, and different *QSAR* methodologies are applied, but all the models developed appeared to be not predictive. Is it possible to predict in advance (prior to modeling), whether a dataset is modelable or not? There are some examples of non-modelable datasets like ToxCast™ phase I datasets [25] with ToxRefDB [26] in vivo endpoints [27], a Bioavailability dataset [28] of 1,110 compounds, and a 91 anticonvulsant dataset [29, 30] compiled in our laboratory. For example, for Bioavailability dataset, two descriptor sets were used, Dragon [1] and SiRMS [9], and for each set of descriptors, two *QSAR* methods were used, PLS [12] and RF [17]. In all cases, predictive R^2 was about 0.5. Should we continue to work with this dataset? For ToxCast dataset, many calculations have been conducted by us and other authors [27], but no truly predictive models were obtained, except that for rat cholinesterase inhibitors. We tried to build models for Growth Inhibition assay for C. Elegans [31]. The dataset included 212 compounds after curation of the initial dataset of 320 compounds: 9 duplicates, 9 salts and mixtures, 3 large molecules, 13 elementoorganic and inorganic compounds, 26 metabolites and 20 bad structures according to EPA were removed. Two different binary definitions of toxicity were used separately as end points (see [31] and a method described in [32]). Dragon [1] descriptors were used, 615 binary in vitro assays were used as descriptors, and both Dragon and in vitro assays were used together. Here we have to mention that in vitro assays are biological descriptors. We have used *k*NN *QSAR* method. The best *QSAR* model for one toxicity definition appeared to have correct classification rate [$CCR = 0.5$ (Sensitivity + Specificity)] of about 0.7, (but Sensitivity was only 0.69), what was marginally acceptable. All other models had unacceptable statistics. For 91 anticonvulsant datasets, Molconn-Z [3] and Dragon [1] descriptors were used. Multiple *k*NN QSAR models were built, and all of them had $R^2 < 0.5$ for external test sets [30]. Evidently, much more unsuccessful attempts to build predictive QSAR models are not published than published.

In this paper, we are trying to establish statistical criteria, which could be used to distinguish modelable and non-modelable datasets. These criteria include dataset diversity (independent of activity values), activity cliff indices, modelability indices based on leave-one-(three, five)-out cross-validation in the entire descriptor space (for category, classification and continuous end points [33]), and leave-group-out cross-validation similarity search in the entire descriptor space, i.e. using all descriptors (without building *k*NN models by descriptor selection). In result, best criterion was obtained by leave-group-out similarity search followed by modelability index and activity cliff index. Dataset diversity indices, which are calculated without taking activities of compounds into account, were the weakest modelability criteria. The latter result was expected, because formally it is possible to label the same data points in the same descriptor space as active or inactive or by numbers from a continuous interval so that the similarity search would give better or worse results. In fact, it happened in our experiments with *C. Elegans* data, where one definition of toxicities gave better model than the other, and other modelability criteria for these two toxicities were different and corroborated our modeling results. Another example was the ToxCast [25] data, for which excellent modelability criteria (and modeling results) were obtained for rat cholinesterase inhibitors, while for the same compounds but different end points modelability criteria were poor and no predictive models were possible to build.

In this paper, to establish the modelability of datasets we have used Dragon 5.5 descriptors [1] in most cases. For *C. Elegans* dataset we have also used in vitro assays data as descriptors, as well as Dragon [1] descriptors along with in vitro assays data. Dragon software calculates most of the widely used chemical descriptors (for the list of Dragon descriptors see [1]). In most calculations Euclidean distances between representative points were used as dissimilarity measure.

7.2 Methods

7.2.1 Modelability Criteria

Several modelability criteria have been defined in this work for datasets with binary and continuous end points. We have also theoretically considered datasets with compounds belonging to more than two categories or classes. The goal of this study was to establish fast and reliable approach to evaluate, whether for a given dataset it is possible to build predictive *QSAR* models. High correlation coefficients R^2 between some modelability criteria and prediction power of *QSAR* models were obtained. Below, *CCR* is the correct classification rate for the discrete (mainly binary) end point, and R^2 is the square of the correlation coefficient (coefficient of determination) between predicted and observed activities for the continuous end point. To distinguish between modelability criteria and predictive power of *QSAR* models, the prefixes *MODI_* and *QSAR_* are used, for example, *MODI_CCR* and $QSAR\_R^2$.

The following *QSAR modelability* criteria have been introduced for a molecular dataset.

1. **Dataset diversity indices (*MODI_DIVs*).** These criteria are independent of the activities of molecules. The idea was to find, whether the datasets with higher diversity are less modelable than datasets with lower diversity. Suppose compounds are represented by points in the multidimensional descriptor space. Each coordinate of this space represents a descriptor; all coordinates are orthogonal to each other. Then each compound in this space can be represented as a point according to its descriptor values. The distance between two compounds can be defined as Euclidean distance between the corresponding representative points. The Euclidean distance between compounds *i* and *j* (*i* and *j* are equal to 1, ..., *M*; *M* is the number of compounds in the dataset) with descriptors X_{ik} and X_{jk}, ($k = 1, \ldots, N$; *N* is the number of descriptors) is defined as

$$D_{ij} = \sqrt{\sum_{k=1}^{N} \left(X_{ik} - X_{jk}\right)^2}. \quad (7.1)$$

Descriptors are considered to be range-scaled, so that their values are within the interval [0, 1]. The following dataset diversity indices based on distance threshold (or dissimilarity level) were introduced by Golbraikh [34]. For each distance threshold, D_{thresh}, a hypersphere in the descriptor space with the center at each compound and radius $R = D_{thresh}$ is built. Let M_b be the number of compounds, for which there are no other compounds within these spheres. Then the ratio

$$I(D_{thresh}) = M_b/M \tag{7.2}$$

is the diversity index of the dataset corresponding to threshold D_{thresh}. These kind of indices depend on the distance threshold D_{thresh}: the larger the threshold is, the lower the value of $I(D_{thresh})$ is. The diversity indices considered in this work, are independent of distance threshold and the number of descriptors. So, it is possible to assert that the new diversity indices are true characteristics of the dataset, and that the corresponding indices for two datasets with the same descriptors can be compared with each other. However, since these indices do not depend on the distribution of activities, we did not expect high correlation of them with *QSAR_CCR*. Nevertheless, we expected that for the more diverse datasets there are lower chances to build predictive *QSAR* models.

Let K be the number of nearest neighbors of each of the M compounds, and D_{ij} ($i = 1, \ldots, M; j = 1, \ldots, K$) are the Euclidean distances between compounds and their nearest neighbors. Taking into account that the descriptors are range-scaled and take values from 0 to 1, the maximum theoretical value of the distance will be $\sqrt{N}$, where N is the number of descriptors. Thus, the normalized distances between compounds can be defined as follows:

$$D_{ij}^n(1) = \frac{D_{ij}}{\sqrt{N}}. \tag{7.3}$$

"1" in parentheses means that it is definition 1 of the normalized distances. These normalized distances are independent of the number of descriptors. While the maximum theoretical value of $D_{ij}^n(1)$ is one, in fact, in almost all cases it will be much lower than one. $D_{ij}^n(1)$ could be equal to one, only if compound i had all descriptor values of one (zero), for which compound j had all descriptor values of zero (one). The minimum value of $D_{ij}^n(1)$ will be higher than zero, unless two compounds are identical within the descriptor space considered (i.e. two compounds have identical descriptor values). Typically, one of the pair of identical compounds is (or sometimes both are) excluded from the dataset prior to QSAR studies [35]. The second way to normalize distances is as follows:

$$D_{ij}^n(2) = \frac{D_{ij}}{D_{ij}^{\max}}. \tag{7.4}$$

"2" in parentheses means that it is definition 2 of the normalized distances, and $D_{ij}^{max} = \max D_{ij}$ is the maximum over all D_{ij} values. The theoretical and real maximum value of $D_{ij}^{n}(2)$ is one, while its minimum value is higher than zero. At the same time, $D_{ij}^{n}(2)$ is independent of the number of descriptors, because the application to its numerator and denominator formula (7.3), will not change its value. It could be possible also to consider range-scaled distances:

$$D_{ij}^{n}(3) = \frac{D_{ij} - D_{ij}^{min}}{D_{ij}^{max} - D_{ij}^{min}}, \tag{7.5}$$

"3" in parentheses means that it is definition 3 of the normalized distances, and $D_{ij}^{min} = \min D_{ij}$ is the minimum over all D_{ij} values.

Dataset diversity indices are defined as follows:

$$MODI\_DIV(k) = \frac{1}{MK} \sum_{i=1}^{M} \sum_{j=1}^{K} D_{ij}^{n}(k), \tag{7.6}$$

where k = 1,2,3 according to definitions of normalized distances between compounds and their nearest neighbors. According to the above definitions,

$$MODI\_DIV(2) = \frac{MODI\_DIV(1)\sqrt{N}}{D_{ij}^{max}}; \tag{7.7}$$

$$\begin{aligned} MODI\_DIV(3) &= \frac{MODI\_DIV(1)\sqrt{N}}{D_{ij}^{max} - D_{ij}^{min}} - \frac{D_{ij}^{min}}{D_{ij}^{max} - D_{ij}^{min}} \\ &= \frac{MODI\_DIV(2)D_{ij}^{max}}{D_{ij}^{max} - D_{ij}^{min}} - \frac{D_{ij}^{min}}{D_{ij}^{max} - D_{ij}^{min}}, \end{aligned} \tag{7.8}$$

i.e. all three definitions are linearly dependent of each other, and thus they are, in fact, equivalent. In this work we have used definition 1 [see formula (7.3)]. Definition 1 is also preferable, if it is necessary to compare diversities of different datasets, as in this work. In fact, using definition 1, the comparison will be made between mean differences of values of one descriptor. There is a drawback of such a comparison. Ideally, for all the datasets under comparison the same descriptors should be used. However, even in this case there might be problems, because some descriptors could have higher variance for one of the datasets than for the other. In these studies we used descriptors which had non-zero variance. So for different datasets different descriptors were used. Further, the index of the type of the diversity will be omitted.

It is also possible to calculate the standard deviation of *DIV*. It is defined by the standard formula

$$SDDIV = \sqrt{\frac{\sum_{i=1}^{M}\sum_{j=1}^{K}\left(D_{ij}^{n} - MODI\_DIV\right)^{2}}{MK - 1}}. \tag{7.9}$$

Alternative definitions of distances between compounds are possible. Instead of formula (7.1), it is possible to use the Minkowski's distance of any order, for example, the Manhattan distance

$$D_{ij}^{'} = \sum_{k=1}^{N} \left|X_{ik} - X_{jk}\right|. \tag{7.1a}$$

In this case, the normalized distances should be defined as

$$D_{ij}^{'n} = \frac{D_{ij}^{'}}{N} \tag{7.3a}$$

and the diversity as

$$MODI\_DIV = \frac{1}{MK}\sum_{i=1}^{M}\sum_{j=1}^{K} D_{ij}^{'n}. \tag{7.6a}$$

Other diversity indices considered were based on the Tanimoto Coefficient (TC). The Tanimoto coefficient between two compounds i and j is defined as

$$TC_{ij} = \frac{\sum_{k=1}^{N} X_{ik}X_{ij}}{\sum_{k=1}^{N} X_{ik}^{2} + \sum_{k=1}^{N} X_{jk}^{2} - \sum_{k=1}^{N} X_{ik}X_{jk}}, \tag{7.1b}$$

where N is the number of descriptors. In case of binary descriptors like *MACCS* keys [2], it can be calculated as

$$TC_{ij} = \frac{c}{a + b - c}, \tag{7.1c}$$

where a and b are the counts of descriptors that are ones either for compound i or j, and c is the count of descriptors that are ones for both compounds i and j. If the number of nearest neighbors is K, then the modelability index can be defined as

$$MODI\_TC = \frac{1}{MK}\sum_{i=1}^{M}\sum_{j=1}^{K} TC_{ij}. \tag{7.6b}$$

MODI_TC(Dragon) were calculated using Dragon descriptors and *MODI_TC* (*MACCS*) were calculated using *MACCS* keys. Since *MODI_TC* criteria are independent of activity distribution, we did not expect high correlation between them and *QSAR_CCR* and *QSAR_*R^2, however some correlation was expected, because for many datasets with higher diversity there is a lower chance that query compounds have the same activities as their nearest neighbors.

2. **Activity cliff indices (*MODI_ACIs*).** Several similar *MODI_ACIs* have been introduced in this work. To calculate these indices, activities of compounds should be range-scaled, so that their values are between zero and one. This makes the ACIs independent of the ranges of activities in different datasets, and thus the *MODI_ACIs* of different datasets become comparable. The first *MODI_ACI* is based on activities of nearest neighbors and is independent of distances of compounds to their nearest neighbors. Again, let M be the number of compounds and K the number of nearest neighbors of each compound. Activities of compounds are A_i ($i = 1,\ldots, M$) and activities of nearest neighbors of compound i are A_{ij} ($j = 1, \ldots, K$). Then the activity cliff index for compound i is defined as

$$MODI\_ACI_i(1) = \frac{1}{K}\sum_{j=1}^{k} |A_i - A_{ij}|, \tag{7.10}$$

where "1" in parentheses denotes that this is definition 1 of the *MODI_ACI*$_i$. The *MODI_ACI* for the entire dataset is define

$$\begin{aligned} MODI\_ACI(1) &= \frac{1}{M}\sum_{i=1}^{M} MODI\_ACI_i(1) \\ &= \frac{1}{MK}\sum_{i=1}^{M}\sum_{j=1}^{K}|A_i - A_{ij}|. \end{aligned} \tag{7.11}$$

The other three definitions of *ACIs* depend not only on activities, but also on distances between compounds, which are defined and normalized according to formulas (7.1) and (7.3), respectively. Of course, the corresponding formulas (7.1a) and (7.3a) can also be used. In fact, non-normalized distances can also be used in the formulas (7.12), (7.13), (7.14), and (7.15) below, and the results will be the same. All these *ACIs* make use of distance-dependent weights, and for each compound they are defined by the following formula:

$$MODI\_ACI_i(k) = \frac{\sum_{j=1}^{K} w_{ij}(k)|A_i - A_{ij}|}{\sum_{j=1}^{K} w_{ij}}, \tag{7.12}$$

where $k = 2{,}3$, or 4. The weights w_{ij} are defined as follows:

$$w_{ij}(2) = \exp\left(-\frac{D_{ij}^{n}}{\sum_{j=1}^{K} D_{ij}^{n}}\right); \tag{7.13}$$

$$w_{ij}(3) = \left(1 - \frac{D_{ij}^{n}}{\sum_{j=1}^{K} D_{ij}^{n}}\right); \tag{7.14}$$

$$w_{ij}(4) = \left(1 + \frac{D_{ij}^{n}}{\sum_{j=1}^{K} D_{ij}^{n}}\right)^{-1}. \tag{7.15}$$

The *ACI*s for the entire dataset are defined as

$$MODI\_ACI(k) = \frac{1}{M}\sum_{i=1}^{M} MODI\_ACI_i(k) \tag{7.16}$$

where $k = 2, 3, 4$.

Properties of *MODI_ACI*s are as follows. (i) *MODI_ACI*s are functions of the range-scaled activities of compounds and their distribution in the descriptor space. This property allows comparing of *MODI_ACI*s for different datasets. (ii) If $K = 1$, all *MODI_ACIs* are equal to each other. (iii) Even if $K > 1$, *ACI*s are close to each other, if $x_{ij} = \frac{D_{ij}}{\sum_{j=1}^{K} D_{ij}}$ are small. This can be shown by the direct expansion of the right sides of Eqs. (7.13), (7.14), and (7.15) in the Taylor series by values of x_{ij}.

Our approach is slightly different than that of *SALI* [36]. (i) Unlike our approach, *SALI* was not used for the entire dataset to calculate a global dataset characteristic like our *MODI_ACI*s. (ii) Besides, the original *SALI* value (without additional correction) can become infinitely high, if the similarity between two compounds even with small but finite activity difference becomes smaller and smaller. In our *MODI_ACI*s, the maximum $MODI\_ACI_i$ for a compound cannot exceed the maximum of $|A_i - A_{ij}|$. (iii) Range-scaling of activities used in our approach allows comparison of our global *ACI*s obtained for different datasets.

The questions that will be addressed in this paper are as follows. How good are *MODI_ACI*s as modelability criteria? Is it possible to establish a threshold (or two thresholds) of *MODI_ACI*s such that the datasets with the corresponding *MODI_ACI* lower than the lower threshold are modelable, and conversely, the datasets with the corresponding *MODI_ACI* higher than the higher threshold are not modelable.

3. **Modelability indices for datasets with binary and continuous response variables.** For a binary response variable (where compounds belong to classes 0 and 1, or actives and inactives, etc.), a modelability criterion *MODI_CCR* was proposed. It is the correct classification rate (*CCR*) for leave-one-out cross-validation with 1-nearest neighbor (1-NN) in the entire descriptor space. For binary datasets, for each class of compounds it is easy to count how many of their nearest neighbors belong to each class. There are

N_0 compounds of class 0.
N_1 compounds of class 1.
N_{00} compounds of class 0 with the nearest neighbor of class 0 (True negatives – TN).
N_{11} compounds of class 1 with the nearest neighbor of class 1 (True positives – TP).
N_{10} compounds of class 0 with the nearest neighbor of class 1 (False positives –FP).
N_{01} compounds of class 1 with the nearest neighbor of class 0 (False negatives – FN).
$N_0 = N_{00} + N_{10}$
$N_1 = N_{11} + N_{01}$

Predictions by nearest neighbors can be defined as follows.

$MODI\_SE = N_{11}/N_1$ as sensitivity
$MODI\_SP = N_{00}/N_0$ as specificity
MODI-False Positive Rate: $MODI\_FPR = N_{10}/N_0 = 1 - MODI\_SP$
MODI-False Negative Rate: $MODI\_FNR = N_{01}/N_1 = 1 - MODI\_SE$

$$MODI\_CCR = 0.5(MODI\_SE + MODI\_SP) \tag{7.17}$$

These data can be represented by a confusion matrix (Tables 7.1a and 7.1b). It means that if the nearest neighbors were used for prediction of classes of compounds, then the results could be represented in a form of the confusion matrix (Table 7.1a), where *TP*, *TN*, *FP* and *FN* are counts of true positives, true negatives, false positives and false negatives, respectively. Modelability index *MODI_CCR* is defined by formula (7.17), where *CCR* is the correct classification rate in the corresponding leave-one-out cross-validation (*LOO-CV*) procedure. Unfortunately, formula (7.17) can be used only for binary datasets.

A slightly more complex approach can be implemented, if the number of classes is higher than two, or if the response variable is continuous, and the

Table 7.1a 2 × 2 confusion matrix

		Observed		
		Active	Inactive	Total
Predicted	Active	$N_{11} = TP$	$N_{10} = FP$	$TP + FP$
	Inactive	$N_{01} = FN$	$N_{00} = TN$	$FN + TN$
	Total	$N_1 = TP + FN$	$N_0 = FP + TN$	$N = N_0 + N_1$

Table 7.1b Normalized 2 × 2 confusion matrix

		Observed		
		Active	Inactive	Total
Predicted	Active	$MODI\_SE = TP/N_1$	$MODI\_FPR = FP/N_0$	$MODI\_SE + MODI\_FPR$
	Inactive	$MODI\_FNR = FN/N_1$	$MODI\_SP = TN/N_0$	$MODI\_FNR + MODI\_SP$
	Total	1	1	2

dataset is divided in more than two bins by activity values. The classes will be referred to as categories, if the compounds can be ordered by them (for example, very active compounds, active compounds, moderately active compounds and inactive compounds). Otherwise, if the compounds cannot be ordered by classes (for example, ligands to different receptors), they will be referred to as true classes or just classes [22].

Thus, we have to distinguish datasets, the compounds of which belong to different categories or different classes. In the first case, the compounds can be ordered by the category they belong. In the second case, compounds cannot be ordered. In the case of classes, the same approach as above can be used, and the modelability criterion can be defined as

$$MODI\_CCR = \frac{1}{C}\sum_{i=1}^{C}\frac{N_{ii}}{N_i}, \tag{7.18}$$

where C is the number of classes, N_{ii} is the count of compounds of class i predicted correctly, and N_i is the total number of compounds of class i.

Formula (7.18) is incorrect for categories, if we want to take into account the values of the errors of prediction. For example, prediction of a compound of category 1 as a compound of category 2 has lower error than its prediction as a compound of category 4. The same problem arises, when a dataset with continuous response variable is divided into bins. In both these cases, confusion matrices (non-normalized and normalized) are as follows (Tables 7.2a and 7.2b).

Thus, compounds in ordered adjacent categories have closer activity values than non-adjacent ones. This evident observation is used as the basis of modelability index definition. For compounds belonging to each bin or category,

Table 7.2a Confusion matrix for *C* categories or bins

		Observed categories or bins				
		1	2	...	C	Totals
Predicted categories or bins	1	N_{11}	N_{12}	...	N_{1C}	$N_{1\cdot}$
	2	N_{21}	N_{22}	...	N_{2C}	$N_{2\cdot}$
	...	...	...	...	...	...
	C	N_{C1}	N_{C2}	...	N_{CC}	$N_{C\cdot}$
	Totals	$N_{\cdot 1}$	$N_{\cdot 2}$	...	$N_{\cdot C}$	$N_{\cdot\cdot}$

Table 7.2b Normalized confusion matrix for *C* categories or bins

		Observed categories or bins				
		1	2	...	C	Totals
Predicted categories or bins	1	$D_{11} = N_{11}/N_{\cdot 1}$	$D_{12} = N_{12}/N_{\cdot 2}$	...	$D_{1C} = N_{1C}/N_{\cdot C}$	$D_{1\cdot}$
	2	$D_{21} = N_{21}/N_{\cdot 1}$	$D_{22} = N_{22}/N_{\cdot 2}$	...	$D_{2C} = N_{2C}/N_{\cdot C}$	$D_{2\cdot}$
	...	...	...	...	...	...
	C	$D_{C1} = N_{11}/N_{\cdot 1}$	$D_{C2} = N_{C2}/N_{\cdot 2}$	...	$D_{CC} = N_{CC}/N_{\cdot C}$	$D_{C\cdot}$
	Totals	1	1	...	1	C

the counts of their nearest neighbors belonging to each bin or category are obtained.

Results can be represented in a form of a confusion matrix (see Tables 7.2a and 7.2b). In Table 7.2a, C is the number of categories or bins, N_{ij} is the number of compounds of category or bin j which are predicted as compounds of category or bin i. $N_{i\cdot}$ is the sum for one (i-th) predicted category or bin (one row) over all observed categories or bins (all columns), i.e. sum of all matrix elements of the i-th row. Similarly, $N_{\cdot j}$ is the sum for one (j-th) observed category or bin (one column) over all predicted categories or bins (all rows), i.e. sum of all matrix elements of the j-th column. $N_{\cdot\cdot} = N$ is the sum of all elements of the matrix. Table 7.2b is the normalized confusion matrix 0.2.a. Each element D_{ij} can be considered as a conditional probability of a compound of category j to be predicted as that of category i. $D_{i\cdot}$, $D_{\cdot j}$, and $D_{\cdot\cdot}$ are sums defined similarly to the corresponding sums of matrix 0.2.a. Evidently that $D_{\cdot j} = 1$ and $D_{\cdot\cdot} = C$. The diagonal elements of the normalized confusion matrix represent the generalization of sensitivity and specificity in 2×2 confusion matrix, and non-diagonal elements correspond to the generalization of false positive and false negative rates. The error of prediction can be calculated as the sum of non-diagonal elements of matrix 0.2.b, multiplied by a monotonically increasing function f of the error of prediction $|i-j|$ for element D_{ij}:

$$E = \sum_{i=1}^{C} \sum_{\substack{j=1 \\ j \neq i}}^{C} D_{ij} f(|i-j|), \tag{7.19}$$

In this work,

$$f(|i-j|) = |i-j|. \tag{7.20}$$

For calculation of the expected error E_{exp}, we assume that the expected value of each matrix element in column i of the normalized confusion matrix is $1/C$. It is correct, if we assume that the number of compounds of each category (or bin) is unknown (or equal). It is true for external test sets, and in our opinion this assumption should also be used for calculation of E_{exp} of test sets. The same formulas (7.19) and (7.20) can be used for the calculation of E_{exp}. It can be proven that

$$E_{exp}(\text{categories or bins}) = \frac{C^2-1}{3}. \tag{7.21}$$

The modelability index is defined by formula (7.22):

$$MODI\_CAT_C = 1 - \frac{E}{E_{exp}}. \tag{7.22}$$

For binary response variable, $E_{exp} = 1$ and $MODI\_CAT_2 = 1-E = 2\times MODI\_CCR-1$. We note the evident fact: according to the definitions, the lower the error of prediction is, the higher the $MODI\_CAT_2$ and $MODI\_CCR$ (see above) are.

In case of more than one nearest neighbor, each nearest neighbor is used for prediction separately, i.e., if $K = 3$, for each compound three predictions by nearest neighbors are considered. The total number of predictions is equal to NK, and N_0 and N_1 are replaced by N_0K and N_1K. So the procedure becomes different from that of *LOO-CV*.

For true classes, the modelability index is also defined by formula (7.22). However, in this case, any error of prediction cannot take into account the differences between predicted and observed classes, since they cannot be ordered. The expected error for each class j will be (see above) $E_{exp}(\text{class } j) = (C-1) \times 1/C = (C-1)/C$. $E_{exp}(\text{all classes}) = E_{exp}(\text{class } j) \times C = C-1$:

$$E_{exp}(\text{classes}) = C-1. \tag{7.23}$$

For calculation of true errors, the differences between predicted and observed classes are not taken into account, i.e.

$$E(\text{classes}) = \sum_{i=1}^{C} \sum_{\substack{j=1 \\ j \neq i}}^{C} D_{ij}. \tag{7.23a}$$

For modelability index, $MODI\_CLASS_C$, instead of formula (7.18) the same formula (7.22) can also be used:

$$MODI\_CLASS_C = 1 - \frac{E}{E_{\text{exp}}} \tag{7.22a}$$

Alternative definitions of modelability indices can be based on the maximum expected error E_{max}:

$$MODI\_CAT^* = 1 - E/E_{\text{max}} \tag{7.24}$$

For categories or bins, E_{max} is defined for the prediction, when all compounds are predicted as belonging to the category or bin which maximizes $|i-j|$. For observed category or bin $j = 1$, $\max|i-j| = C-1$, i.e. $i = C$. For observed category or bin $j \le C/2$, $\max|i-j| = C-j$ is achieved when $i =$ C. For observed category or bin $j > C/2$, $\max|i-j| = j-1$ is achieved when $i = 1$. The general formulas for E_{max} for C categories or bins is as follows:

$$E_{\text{max}}(\text{categories or bins}) = \frac{3C^2 - 2C}{4} \text{ for even } C, \tag{7.25a}$$

and

$$E_{\text{max}}(\text{categories or bins}) = \frac{3C^2 - 2C - 1}{4} \text{ for odd } C. \tag{7.25b}$$

For classes, E_{max} is equal to the number of classes:

$$E_{\text{max}}(\text{classes}) = C, \tag{7.26}$$

Further, in this paper, we will not consider cases with more than two classes or categories. However, we will consider division of datasets with continuous endpoints into more than two bins. Evidently, for two classes, categories, and bins the same formula (7.18) [or (7.17)] can be used.

If number of nearest neighbors $K = 1$, our approach is similar to *LOO-CV*. As in the case of binary response variable, if K is higher than 1, each nearest neighbor is used for prediction separately, and the Totals in the Table 7.2a are multiplied by K. It is also possible to use *LOO-CV*, if $K > 1$.

4. **Similarity Search.** The similarity search procedure is a k-nearest neighbors approach without a variable selection, i.e. it is carried out in the entire descriptor space, and no optimization of the descriptor set is carried out like in the standard *kNN QSAR* developed in our laboratory [13]. As in the regular *kNN*, leave-one-out (*LOO*) and leave-group-out (*LGO*) cross-validation (*CV*) were used. *LOO-CV* can also be used to calculate *MODI_CCR* (see above) for binary datasets and $MODI\_q^2$ for continuous datasets. In this work, five-fold *LGO*

was also used: each dataset was divided into five equal parts. One part, in turn, was used as the prediction set, and the remaining four parts together were used as the modeling set. In case of the continuous response variable, one most active and one most inactive compound as well as three compounds with activities uniformly distributed within the entire range of activities, were always included in the modeling set. Modelability indices *MODI_ssCCR* and *MODI_ssR*^{2} were calculated for binary and continuous datasets, respectively.

In general, the Euclidean distances in the descriptor space between a compound and each of its k nearest neighbors are not the same. Thus, the neighbor with the smaller distance from a compound is given a higher weight in calculating the predicted activity as follows:

$$\hat{y}_i = \frac{\sum_{j=1}^{k} y_{ij} w_{ij}}{\sum_{j=1}^{k} w_{ij}} \tag{7.27}$$

where y_{ij} is the observed activity value for the nearest neighbor j of compound i, $\hat{y}_i$ is the predicted activity value for compound i; weights w_{ij} are defined as

$$w_{ij} = \left(1 + \frac{d_{ij}^2}{\sum_{j'=1}^{k} d_{ij'}^2} \right)^{-1} \tag{7.28}$$

where d_{ij} are the Euclidean distances between compound i and each of its k nearest neighbors. In the case of the binary response variable, predicted activities are rounded to the nearest integer, or according to the specified threshold between categories.

For similarity search criteria with 20 %-out (five-fold) cross-validation, we have also used the applicability domain (*AD*). The *AD* for prediction was defined by the average distance between k nearest neighbors d_{av} and its standard deviation s in the modeling set. The threshold distance

$$D_{thresh} = d_{av} + Zs \tag{7.29}$$

was used to find, whether a compound from the prediction set was within the *AD*. Z in formula (7.29) is a parameter (so called Z-cutoff) defined by a user. If a compound of the prediction set was out of the *AD* (the average distance to all k nearest neighbors within the modeling set), it was not predicted. Output statistics was calculated for all compounds within the *AD*.

The output from the similarity search was as follows. In the case of the binary response variable, the correct classification rate (*MODI_ssCCR*), as well as the sensitivity (*MODI_ssSE*) and the specificity (*MODI_ssSP*) were calculated for each hold-out prediction set, and then the results were combined for the entire dataset, which consisted of five prediction sets. Similarly, for datasets with continuous response variable, predictions of all five prediction sets were combined and coefficient of determination between predicted and observed activities R^2 along with coefficients of determination R_0^2 and R'^2_0 and slopes k and k' for trend lines through the origin for predicted versus observed and observed versus predicted activities were calculated [22]. The formulas to calculate these statistics were as follows.

$$q^2 = 1 - \frac{\sum_{i=1}^{N} (y_i - \tilde{y}_i)^2}{\sum_{i=1}^{N} (y_i - \bar{y})^2} \tag{7.30}$$

$$R = \frac{\sum (y_i - \bar{y})(\tilde{y}_i - \bar{\tilde{y}})}{\sqrt{\sum (y_i - \bar{y})^2 \sum (\tilde{y}_i - \bar{\tilde{y}})^2}} \tag{7.31}$$

$$R_0^2 = 1 - \frac{\sum (\tilde{y}_i - \tilde{y}_i^{r_0})^2}{\sum (\tilde{y}_i - \bar{\tilde{y}})^2} (QSAR\_R^2 \text{ vs. } MODI\_ssR^2) \tag{7.32a}$$

$$R_0'^2 = 1 - \frac{\sum (y_i - y_i^{r_0})^2}{\sum (y_i - \bar{y})^2} (MODI\_ssR^2 \text{ vs. } QSAR\_R^2) \tag{7.32b}$$

$$k = \frac{\sum y_i \tilde{y}_i}{\sum y_i^2} (QSAR\_R^2 \text{ vs. } MODI\_ssR^2) \tag{7.33a}$$

$$k' = \frac{\sum y_i \tilde{y}_i}{\sum \tilde{y}_i^2} (MODI\_ssR^2 \text{ vs. } QSAR\_R^2), \tag{7.33b}$$

where y_i and $\tilde{y}_i$ are the corresponding *MODI* and *QSAR* criteria, R_0^2 and R'^2_0 are the coefficients of determination for trend lines through the origin for *QSAR* versus *MODI*, and *MODI* versus *QSAR* criteria, respectively, k and k' are the corresponding slopes, and $\tilde{y}^{r_0} = ky$ and $y^{r_0} = k'\tilde{y}$ are the regressions through the origin for *QSAR* versus *MODI* and *MODI* versus *QSAR* criteria.

7.2.2 Datasets

The datasets used to evaluate the modelability criteria described above, can be divided into two parts: datasets with binary and continuous end points.

7.2.2.1 Datasets with Binary End Points (See Also Table 7.3a)

1–2. Multidrug resistance protein 1 datasets: MDR1 (572 compounds: 217 (class 0) and 355 (class 1) and MDR1i (1,541 compounds: 828 (class 0) + 743 (class 1)) [37]. MDR1 inhibitors and non-inhibitors were defined based on a potency threshold set to 10 μM. MDR1 substrates and non-substrates were defined based on the evidence of their transport by MDR1 across the membrane. For calculations of modelability criteria, Dragon descriptors were used.

3–8. C. Elegans toxicity datasets for ToxCast: (212 compounds; for calculations of modelability criteria and QSAR studies 2 toxicity scales; 3 combinations of descriptors: Dragon 5.5, ToxCast in vitro assays, and Dragon with in vitro assays) [31] were used.

9–42. GPCRome of Receptorome. Compounds to the most studied 34 GPCR targets were extracted from the databases of ChEMBL [38] and PDSP [39]. The targets included 10 serotonin receptors (5-HT$_{1A}$, 5-HT$_{1B}$, 5-HT$_{1D}$, 5-HT$_{1E}$, 5-HT$_{2A}$, 5-HT$_{2C}$, 5-HT$_3$, 5-HT$_5$, 5-HT$_6$, and 5-HT$_7$), 7 adrenoceptors (α_{1A}, α_{1B}, α_{2A}, α_{2B}, α_{2C}, β_1, and β_2), 5 dopamine receptors (D_1, D_2, D_3, D_4, and D_5), 5 muscarinic receptors (M_1, M_2, M_3, M_4, and M_5), 4 histamine receptors (H_1, H_2, H_3, and H_4), and 3 neurotransmitter transporters [serotonin (SERT), norepinephrine (NET), and dopamine (DAT)]. The datasets were curated using the protocol described elsewhere [35]. The datasets included compounds with K_i values (in mol/L) tested on the human tissues. The pK_i cutoffs dividing the datasets into active and inactive classes were defined in such a way that the datasets were balanced, i.e. there had to be not too much difference between counts of actives and inactives (the ratio between these counts had to be between 1/3 and 3. For most datasets, $pK_i = 7$ was used as the cutoff value. The exceptions were: 5-HT$_{1E}$, 5-HT$_3$ and H_2 for which the pK_i cutoff was equal to 5, 5-HT$_5$, D_5, M_4 and M_5 for which the pK_i cutoff was equal to 6, and H_3 for which it was equal to 8. Counts of compounds in each class for all 34 datasets are given in Table 7.3a. For calculations of modelability criteria Dragon descriptors were used.

43–102. ToxCast datasets for 60 end points including 229–241 compounds [27]. Dragon 5.5 [1] descriptors were used for both *QSAR* studies and modelability criteria. For each dataset, models using 84 different *QSAR* methodologies were built [27].

The first 42 datasets were the initial datasets, for which the method was developed. The remaining 60 datasets were used to further validate and to extend the results to lower modelability indices.

Table 7.3a Modelability indices **MODI_CCR** and **MODI_ssCCR** and the corresponding **QSAR_CCR** for all the datasets under study: 34 GPCRome datasets (for all datasets, AD in **MODI_ssCCR** calculations included 80–90 % of compounds)

	Dataset statistics					MODI_CCR			MODI_ssCCR			SVM_QSAR_CCR		
Dataset	Comps	Class 1	Class 0	DIV	ACI	SENS	SPEC	CCR	SENS	SPEC	CCR	SENS	SPEC	CCR
5-HT1A	1,275	750	525	**0.063**	**0.22**	**0.83**	**0.71**	**0.77**	**0.84**	**0.70**	**0.77**	**0.82**	**0.64**	**0.73**
5-HT1B	508	272	236	**0.069**	**0.16**	**0.87**	**0.81**	**0.84**	**0.89**	**0.81**	**0.85**	**0.84**	**0.77**	**0.80**
5-HT1D	504	291	213	**0.072**	**0.19**	**0.86**	**0.73**	**0.8**	**0.87**	**0.76**	**0.82**	**0.88**	**0.72**	**0.80**
5-HT1E	108	37	71	**0.106**	**0.19**	**0.73**	**0.85**	**0.79**	**0.74**	**0.85**	**0.79**	**0.54**	**0.82**	**0.68**
5-HT2A	1,036	455	581	**0.072**	**0.17**	**0.84**	**0.82**	**0.83**	**0.86**	**0.84**	**0.85**	**0.76**	**0.85**	**0.80**
5-HT2C	892	417	475	**0.072**	**0.22**	**0.82**	**0.75**	**0.78**	**0.81**	**0.78**	**0.79**	**0.76**	**0.76**	**0.76**
5-HT3	153	55	98	**0.094**	**0.16**	**0.76**	**0.88**	**0.82**	**0.78**	**0.88**	**0.83**	**0.76**	**0.90**	**0.83**
5-HT5	181	87	94	**0.083**	**0.20**	**0.83**	**0.78**	**0.8**	**0.87**	**0.76**	**0.82**	**0.84**	**0.85**	**0.85**
5-HT6	1,107	804	303	**0.07**	**0.15**	**0.9**	**0.69**	**0.8**	**0.91**	**0.68**	**0.79**	**0.93**	**0.66**	**0.79**
5-HT7	464	222	242	**0.08**	**0.22**	**0.81**	**0.76**	**0.79**	**0.82**	**0.79**	**0.80**	**0.81**	**0.76**	**0.79**
alpha1A	495	295	200	**0.077**	**0.14**	**0.91**	**0.79**	**0.85**	**0.91**	**0.82**	**0.86**	**0.88**	**0.73**	**0.81**
alpha1B	420	117	303	**0.077**	**0.13**	**0.78**	**0.9**	**0.84**	**0.78**	**0.91**	**0.84**	**0.70**	**0.90**	**0.8**
alpha2A	412	180	232	**0.078**	**0.21**	**0.82**	**0.77**	**0.8**	**0.84**	**0.81**	**0.82**	**0.74**	**0.81**	**0.78**
alpha2B	283	72	211	**0.08**	**0.19**	**0.65**	**0.87**	**0.76**	**0.66**	**0.88**	**0.77**	**0.58**	**0.87**	**0.73**
alpha2C	332	153	179	**0.08**	**0.25**	**0.78**	**0.73**	**0.76**	**0.77**	**0.77**	**0.77**	**0.72**	**0.73**	**0.73**

Beta1	194	57	137	**0.076**	**0.27**	**0.6**	**0.79**	**0.69**	**0.60**	**0.76**	**0.68**	**0.68**	**0.89**	**0.79**
Beta2	239	110	129	**0.075**	**0.23**	**0.84**	**0.72**	**0.78**	**0.83**	**0.77**	**0.80**	**0.76**	**0.74**	**0.75**
D1	488	168	320	**0.076**	**0.16**	**0.82**	**0.85**	**0.84**	**0.83**	**0.85**	**0.84**	**0.73**	**0.84**	**0.78**
D2	2,063	688	1,375	**0.068**	**0.2**	**0.73**	**0.84**	**0.79**	**0.74**	**0.84**	**0.79**	**0.63**	**0.85**	**0.74**
D3	1,498	849	649	**0.065**	**0.21**	**0.82**	**0.75**	**0.79**	**0.81**	**0.72**	**0.77**	**0.81**	**0.73**	**0.77**
D4	749	380	369	**0.075**	**0.2**	**0.84**	**0.77**	**0.8**	**0.84**	**0.79**	**0.81**	**0.81**	**0.77**	**0.79**
D5	252	110	142	**0.077**	**0.22**	**0.76**	**0.79**	**0.78**	**0.79**	**0.79**	**0.79**	**0.74**	**0.78**	**0.76**
H1	398	151	247	**0.08**	**0.21**	**0.77**	**0.8**	**0.79**	**0.79**	**0.81**	**0.80**	**0.70**	**0.84**	**0.77**
H2	94	38	56	**0.105**	**0.27**	**0.71**	**0.75**	**0.73**	**0.72**	**0.79**	**0.75**	**0.68**	**0.79**	**0.73**
H3	1,389	1,094	295	**0.06**	**0.18**	**0.9**	**0.5**	**0.7**	**0.90**	**0.54**	**0.72**	**0.94**	**0.44**	**0.69**
H4	445	172	273	**0.07**	**0.21**	**0.77**	**0.8**	**0.78**	**0.69**	**0.79**	**0.74**	**0.69**	**0.81**	**0.75**
M1	384	140	244	**0.072**	**0.2**	**0.72**	**0.84**	**0.78**	**0.69**	**0.84**	**0.76**	**0.66**	**0.83**	**0.74**
M2	431	179	252	**0.071**	**0.23**	**0.76**	**0.78**	**0.77**	**0.73**	**0.78**	**0.75**	**0.65**	**0.78**	**0.71**
M3	434	219	215	**0.069**	**0.23**	**0.79**	**0.76**	**0.77**	**0.80**	**0.77**	**0.79**	**0.83**	**0.74**	**0.79**
M4	234	114	120	**0.077**	**0.18**	**0.81**	**0.84**	**0.82**	**0.77**	**0.85**	**0.81**	**0.78**	**0.85**	**0.82**
M5	253	86	167	**0.082**	**0.18**	**0.76**	**0.86**	**0.81**	**0.74**	**0.87**	**0.81**	**0.65**	**0.88**	**0.77**
NET	970	377	593	**0.07**	**0.21**	**0.77**	**0.8**	**0.79**	**0.78**	**0.79**	**0.78**	**0.77**	**0.82**	**0.79**
SRT	1,262	681	581	**0.067**	**0.17**	**0.87**	**0.78**	**0.83**	**0.88**	**0.79**	**0.83**	**0.84**	**0.81**	**0.83**
DAT	876	198	678	**0.071**	**0.16**	**0.67**	**0.89**	**0.78**	**0.67**	**0.89**	**0.78**	**0.67**	**0.91**	**0.79**

Table 7.3b Modelability indices **MODI_CCR** and **MODI_ssCCR** and the corresponding **QSAR_CCR** for all the datasets under study: Multi-drug resistance datasets (in **MODI_ssCCR** calculations, AD for MDR1 and MDR1i included 90 % and 81 % of compounds, respectively)

	Dataset statistics					MODI_CCR			MODI_ssCCR			QSAR_CCR		
Dataset [23]	Comps	Class 1	Class 0	DIV	ACI	SENS	SPEC	CCR	SENS	SPEC	CCR	kNN	RF	SVM
MDR1	572	317	255	**0.090**	**0.17**	**0.55**	**0.48**	**0.52**	**0.70**	**0.67**	**0.68**	**0.73**	**0.72**	**0.71**
MDR1i	1,571	743	828	**0.067**	**0.12**	**0.9**	**0.86**	**0.88**	**0.92**	**0.86**	**0.89**	**0.90**	**0.91**	**0.90**

Table 7.3c Modelability indices **MODI_CCR** and **MODI_ssCCR** and the corresponding **QSAR_CCR** for all the datasets under study: ToxCast datasets including 212 compounds with different descriptors and toxicity values for C. Elegans

	Dataset statistics					MODI_CCR			MODI_ssCCR			kNN_QSAR_CCR		
Dataset[a]	Comps	Class 1	Class 0	DIV	ACI	SENS	SPEC	CCR	SENS	SPEC	CCR	SENS	SPEC	CCR
615-ind[b]	212	129	83	**0.264**	**0.37**	**0.67**	**0.53**	**0.6**	**0.62**	**0.52**	**0.57**	**0.87**	**0.36**	**0.61**
615-scr[b]	212	71	141	**0.264**	**0.40**	**0.45**	**0.72**	**0.58**	**0.42**	**0.75**	**0.59**	**0.79**	**0.52**	**0.66**
1,224-ind[c]	212	129	83	**0.153**	**0.42**	**0.62**	**0.52**	**0.57**	**0.59**	**0.57**	**0.58**	**0.61**	**0.55**	**0.58**
1,224-scr[c]	212	71	141	**0.153**	**0.33**	**0.46**	**0.77**	**0.62**	**0.48**	**0.75**	**0.62**	**0.51**	**0.85**	**0.68**
1,839-ind[d]	212	129	83	**0.215**	**0.47**	**0.57**	**0.47**	**0.52**	**0.60**	**0.50**	**0.55**	**0.67**	**0.57**	**0.62**
1,839-scr[d]	212	71	141	**0.215**	**0.30**	**0.51**	**0.8**	**0.65**	**0.50**	**0.81**	**0.66**	**0.52**	**0.87**	**0.70**

[a]ind [31] and scr(method is described elsewhere [32]) are two toxicity scales: toxicity indices and toxicity scores we developed
[b]615 in vitro assays were used as descriptors
[c]1,224 Dragon 5.5 descriptors were used
[d]615 In vitro assays and 1,224 Dragon descriptors were used

Table 7.3d Modelability indices **MODI_CCR** and **MODI_ssCCR** and the corresponding **QSAR_CCR** for all the datasets under study: 60 ToxCast datasets with the **QSAR** model best out of 84 (for all datasets, AD in **MODI_ssCCR** calculations included 80–90 % of compounds)

	Dataset statistics			MODI_CCR			MODI_ssCCR			QSAR_CCR[a]		
Dataset [21]	Comps	Class 1	Class 0	SENS	SPEC	CCR	SENS	SPEC	CCR	SENS	SPEC	CCR
C_Mus_Kidney_AnyLes	229	44	185	**0.23**	**0.81**	**0.52**	0.27	0.81	0.54	**0.13**	**0.91**	**0.52**
C_Mus_Kidney_Pathology	229	44	185	**0.23**	**0.81**	**0.52**	0.35	0.80	0.58	**0.13**	**0.91**	**0.52**
C_Mus_Liver_AnyLes	229	124	105	**0.62**	**0.62**	**0.62**	0.61	0.56	0.59	**0.62**	**0.56**	**0.59**
C_Mus_Liver_Hypertrophy	229	89	140	**0.47**	**0.76**	**0.61**	0.47	0.78	0.62	**0.18**	**0.89**	**0.54**
C_Mus_Liver_Necrosis	229	68	161	**0.38**	**0.78**	**0.58**	0.37	0.77	0.57	**0.09**	**0.96**	**0.53**
C_Mus_Liver_NeoplastLes	229	61	168	**0.33**	**0.78**	**0.55**	0.35	0.79	0.57	**0.18**	**0.89**	**0.54**
C_Mus_Liver_PreneoplastLes	229	38	191	**0.26**	**0.86**	**0.56**	0.25	0.87	0.56	**0.39**	**0.80**	**0.60**
C_Mus_Liver_ProliferatLes	229	87	142	**0.46**	**0.76**	**0.61**	0.44	0.76	0.60	**0.33**	**0.79**	**0.56**
C_Mus_Liver_Tumors	229	68	161	**0.38**	**0.78**	**0.58**	0.42	0.77	0.60	**0.18**	**0.89**	**0.54**
C_Mus_Spleen_AnyLes	229	37	192	**0.14**	**0.83**	**0.48**	0.13	0.83	0.48	**0.00**	**1.00**	**0.50**
C_Mus_Tumorigen	229	88	141	**0.49**	**0.73**	**0.61**	0.49	0.74	0.62	**0.41**	**0.70**	**0.56**
C_Rat_Cholinester_Inhibit	238	42	196	**0.81**	**0.94**	**0.87**	0.79	0.93	0.86	**0.72**	**0.97**	**0.85**
C_Rat_Eye_AnyLes	238	37	201	**0.19**	**0.81**	**0.5**	0.27	0.83	0.55	**0.15**	**0.92**	**0.54**
C_Rat_Kidney_AnyLes	238	83	155	**0.49**	**0.65**	**0.57**	0.49	0.62	0.56	**0.24**	**0.86**	**0.55**
C_Rat_Kidney_Nephropathy	238	35	203	**0.17**	**0.84**	**0.51**	0.19	0.83	0.51	**0.11**	**0.95**	**0.53**
C_Rat_Liver_AnyLes	238	124	114	**0.62**	**0.56**	**0.59**	0.63	0.57	0.60	**0.72**	**0.60**	**0.66**
C_Rat_Liver_Hypertrophy	238	57	181	**0.33**	**0.79**	**0.56**	0.33	0.79	0.56	**0.28**	**0.89**	**0.58**
C_Rat_Liver_PreneoplastLes	238	62	176	**0.34**	**0.74**	**0.54**	0.28	0.77	0.53	**0.05**	**0.92**	**0.48**
C_Rat_Liver_ProliferatLes	238	60	178	**0.33**	**0.78**	**0.56**	0.35	0.78	0.56	**0.09**	**0.90**	**0.49**
C_Rat_Liver_Tumors	238	21	217	**0.1**	**0.91**	**0.5**	0.05	0.91	0.48	**0.01**	**0.97**	**0.49**
C_Rat_Testes_AnyLes	238	40	198	**0.3**	**0.81**	**0.56**	0.32	0.79	0.55	**0.08**	**0.93**	**0.51**
C_Rat_ThyrGl_PreneoplastLes	238	49	189	**0.29**	**0.84**	**0.56**	0.23	0.84	0.53	**0.13**	**0.96**	**0.54**
C_Rat_Thyroid_ProliferatLes	238	38	200	**0.29**	**0.88**	**0.58**	0.27	0.86	0.57	**0.13**	**0.96**	**0.54**

C_Rat_ThyroidGlnd_AnyLes	238	38	200	**0.29**	**0.88**	**0.58**	0.16	0.87	0.52	**0.01**	**1.00**	**0.50**
C_Rat_Tumorigen	238	93	145	**0.45**	**0.58**	**0.52**	0.39	0.55	0.47	**0.35**	**0.61**	**0.48**
D_Rab_Dev	228	104	124	**0.48**	**0.56**	**0.52**	0.55	0.50	0.52	**0.47**	**0.68**	**0.57**
D_Rab_Dev_GenFetal	228	44	184	**0.11**	**0.77**	**0.44**	0.18	0.77	0.47	**0.17**	**0.92**	**0.55**
D_Rab_Dev_GenFetal_WghtRed	228	55	173	**0.2**	**0.74**	**0.47**	0.27	0.74	0.51	**0.02**	**0.99**	**0.50**
D_Rab_Dev_Skel	228	53	175	**0.32**	**0.83**	**0.58**	0.30	0.85	0.58	**0.20**	**0.89**	**0.54**
D_Rab_Dev_Skel_Axial	228	66	162	**0.44**	**0.8**	**0.62**	0.47	0.82	0.65	**0.25**	**0.89**	**0.57**
D_Rab_Mat	228	208	20	**0.9**	**0.2**	**0.55**	0.90	0.24	0.57	**0.94**	**0.15**	**0.54**
D_Rab_Mat_GenMat	228	199	29	**0.83**	**0.14**	**0.49**	0.87	0.16	0.52	**0.94**	**0.13**	**0.53**
D_Rab_Mat_GenMat_Systemic	228	199	29	**0.83**	**0.14**	**0.49**	0.87	0.04	0.46	**0.93**	**0.12**	**0.52**
D_Rab_Mat_PregRel	228	131	97	**0.54**	**0.44**	**0.49**	0.56	0.46	0.51	**0.82**	**0.21**	**0.52**
D_Rab_Mat_PregRel_PregLoss	228	131	97	**0.54**	**0.44**	**0.49**	0.57	0.44	0.51	**0.82**	**0.21**	**0.52**
D_Rab_Prenatal_Loss	228	131	97	**0.54**	**0.44**	**0.49**	0.54	0.42	0.48	**0.82**	**0.21**	**0.52**
D_Rat_Dev	241	138	103	**0.7**	**0.52**	**0.61**	0.67	0.49	0.58	**0.70**	**0.53**	**0.61**
D_Rat_Dev_GenFetal	241	80	161	**0.46**	**0.65**	**0.56**	0.46	0.64	0.55	**0.15**	**0.89**	**0.52**
D_Rat_Dev_GenFetal_WghtRed	241	88	153	**0.52**	**0.64**	**0.58**	0.51	0.66	0.59	**0.26**	**0.79**	**0.52**
D_Rat_Dev_Skel	241	106	135	**0.56**	**0.6**	**0.58**	0.59	0.59	0.59	**0.57**	**0.63**	**0.60**
D_Rat_Dev_Skel_Appendicular	241	115	126	**0.58**	**0.56**	**0.57**	0.58	0.51	0.55	**0.01**	**0.98**	**0.49**
D_Rat_Dev_Skel_Axial	241	45	196	**0.24**	**0.81**	**0.53**	0.24	0.81	0.52	**0.48**	**0.67**	**0.58**
D_Rat_Dev_Skel_Cranial	241	37	204	**0.27**	**0.81**	**0.54**	0.22	0.81	0.51	**0.09**	**0.96**	**0.52**
D_Rat_Mat	241	211	30	**0.9**	**0.33**	**0.61**	0.90	0.36	0.63	**0.99**	**0.00**	**0.50**
D_Rat_Mat_GenMat	241	209	32	**0.89**	**0.34**	**0.62**	0.91	0.31	0.61	**0.99**	**0.01**	**0.50**
D_Rat_Mat_GenMat_Systemic	241	209	32	**0.89**	**0.34**	**0.62**	0.91	0.28	0.60	**0.99**	**0.01**	**0.50**
D_Rat_Mat_PregRel	241	79	162	**0.38**	**0.72**	**0.55**	0.36	0.68	0.52	**0.11**	**0.90**	**0.51**
D_Rat_Mat_PregRel_PregLoss	241	79	162	**0.38**	**0.72**	**0.55**	0.38	0.73	0.56	**0.11**	**0.90**	**0.51**
D_Rat_Prenatal_Loss	241	80	161	**0.39**	**0.67**	**0.53**	0.39	0.63	0.51	**0.00**	**1.00**	**0.50**
M_Rat_Fem_ReproductTract	235	42	193	**0.19**	**0.82**	**0.51**	0.20	0.80	0.50	**0.00**	**1.00**	**0.50**
M_Rat_Kidney	235	69	166	**0.38**	**0.72**	**0.55**	0.38	0.75	0.56	**0.34**	**0.91**	**0.62**
M_Rat_LactationPND21	235	42	193	**0.26**	**0.82**	**0.54**	0.31	0.80	0.55	**0.11**	**0.94**	**0.53**
M_Rat_LitterSize	235	43	192	**0.28**	**0.84**	**0.56**	0.29	0.85	0.57	**0.07**	**0.95**	**0.51**

(continued)

Table 7.3d (continued)

Dataset [21]	Dataset statistics			MODI_CCR			MODI_ssCCR			QSAR_CCR[a]		
	Comps	Class 1	Class 0	SENS	SPEC	CCR	SENS	SPEC	CCR	SENS	SPEC	CCR
M_Rat_Liver	235	100	135	**0.51**	**0.69**	**0.6**	0.55	0.70	0.63	**0.54**	**0.75**	**0.65**
M_Rat_Male_ReproductTract	235	48	187	**0.17**	**0.79**	**0.48**	0.14	0.77	0.46	**0.20**	**0.93**	**0.56**
M_Rat_OffspringSurvival	235	85	150	**0.45**	**0.69**	**0.57**	0.47	0.73	0.60	**0.25**	**0.87**	**0.56**
M_Rat_Reproduct_Outcome	235	64	171	**0.36**	**0.78**	**0.57**	0.34	0.79	0.56	**0.22**	**0.88**	**0.55**
M_Rat_Reproduct_Perform	235	48	187	**0.27**	**0.8**	**0.54**	0.23	0.80	0.51	**0.01**	**0.98**	**0.50**
M_Rat_Testis	235	38	197	**0.05**	**0.83**	**0.44**	0.07	0.85	0.46	**0.13**	**0.92**	**0.53**
M_Rat_ViabilityPND4	235	66	169	**0.35**	**0.72**	**0.54**	0.37	0.75	0.56	**0.17**	**0.88**	**0.52**

[a]**84 QSAR models were built for each dataset using different QSAR methods. Highest QSAR_CCR value is given for each dataset**

7.2.2.2 Datasets with Continuous End Points

1. *P. promelas* LC_{50} dataset (809 compounds) compiled from the ECOTOX database [40, 41]. The toxicity is expressed as the concentration (in mol/L)) which kills 50 % of fathead minnow after 96 h. Dragon descriptors were used in calculations of modelability criteria. Multiple *QSAR* models were built using USEPA [8], Dragon 5.5 [1], SiRMS [9], and CDK descriptors [5, 42], and HC [8], *k*NN [13], and RF [18] methods. Two RF implementations were used: available in R [43] and developed by Polishchuk et al. [44].
2. *T. pyriformis* IGC_{50} dataset (1,093 compounds) compiled by Zhu and coworkers [45] from the work of Schultz and coworkers [46–48]. The toxicity is expressed as the 50 % growth inhibitory concentration (in mmol/L) of the *T. pyriformis* organism (a protozoan ciliate) after 40 h. Dragon 5.5 descriptors [1] were used in calculations of modelability criteria. *QSAR* models were built using USEPA [8], Dragon [1], SiRMS [9] and CDK [5, 42] descriptors, and HC [8], *k*NN [13], RF [43, 44] and SVM [14] methods.
3. An acute toxicity oral rat LD_{50} dataset (7,285 compounds) compiled from the ChemIDplus database [49, 50]. The oral rat LD_{50} endpoint represents the amount of the chemical (mass of the chemical per body weight of the rat) which when orally ingested kills half of rats. This dataset was recently modeled by a variety of *QSAR* techniques [51, 52]. Dragon [1] descriptors were used in calculations of modelability criteria. *QSAR* models were built using USEPA [8], Dragon [1], SiRMS [9] and CDK [5, 42] descriptors, and HC [8], RF [43, 44] and SVM [14] methods.
4. A bioaccumulation factor (BCF) dataset (643 compounds) [53]. The bioconcentration factor BCF is defined as the ratio of the chemical concentration in biota as a result of absorption via the respiratory surface to that in water at steady state [54]. The final dataset consists of 610 chemicals (after removing salts, mixtures, and ambiguous compounds). The modeled endpoint was the log(BCF). Dragon descriptors were used in calculations of modelability criteria. *QSAR* models were built using USEPA [8], Dragon [1], SiRMS [9] and CDK [5, 42] descriptors, and HC [8], kNN [13], RF [43, 44] and SVM [14] methods.

5–14. 10 Regulatory Information datasets with activity data (see Table 7.4). Counts of compounds and ranges of activities are also given in Table 7.4. Dragon 5.5 [1] descriptors were used in calculations of modelability criteria. *QSAR* models were built using RF [44] and SVM [14] methods with Dragon [1], CDK [5, 42] and SiRMS [9] descriptors.

15. Multiple Recommended Therapeutic Dose (MRTD) database (1,169 compounds). 1,220 pharmaceuticals listed in *Martindale* [59]*: The Extra Pharmacopoeia* (1973, 1983, and 1993) and *The Physicians' Desk Reference* (1995 and 1999) [60]. Dragon [1] descriptors were used in calculations of modelability criteria. *QSAR* models were built using SVM [14] and RF [44] methods with Dragon [1], CDK [5, 42] and SiRMS [9] descriptors.

Table 7.4 10 Regulatory Information (RI1-RI10) datasets

No	Activity	Compounds after curation	Activity range	Ref
5	Cancer potency value_CalEPA	214	2.48–9.44	[55]
6	Oral slope factor regional screening level – superfund	189	2.12–7.97	[56]
7	Reference concentration_regional screening level – superfund	125	3.11–9.71	[56]
8	Reference dose point of departure EPA IRIS	269	1.81–7.88	[57]
9	Reference dose dog office of pesticide programs	139	4.64–9.95	[58]
10	Reference dose EPA IRIS	269	4.48–10.59	[57]
11	Reference dose point of departure dog office of pesticide programs	139	2.64–7.87	[58]
12	Reference dose point of departure rat office of pesticide programs	190	2.79–7.36	[58]
13	Reference dose rat office of pesticide programs	190	4.79–9.86	[58]
14	Reference dose regional screening level – superfund	424	4.29–10.51	[56]

16. Bioavailability dataset of 1,110 compounds [28]. *QSAR* models were built using Dragon [1] and SiRMS [9] descriptors, and for each set of descriptors, PLS [12] and RF [17] methods were used. Modelability criteria were built using Dragon [1] descriptors.
17. Alpha 1a (α_{1a}) adrenergic receptor agonists and antagonists (415 compounds). Dragon [1] descriptors were used in calculations of modelability criteria. *QSAR* models were built using SVM [14] and RF [44] methods with Dragon [1], CDK [5, 42] and SiRMS [9] descriptors.
18. 59 Histone Deacetylase Inhibitors (HDAC) [61], for which predictive *QSAR* models were built using *k*NN [13] and SVM [14] with Molconn-Z [3] descriptors. Dragon [1] and Molconn-Z [3] descriptors were used in calculations of modelability criteria.

QSAR models were built for datasets 1–14, and for datasets 1–4 [53] they were published. Modelability criteria were calculated for datasets 1–14, and linear regressions of predictivity of best *QSAR* models (in terms of R^2 between predicted and observed activities) versus these criteria were built. Datasets 15–18 were used to evaluate predictivity of *QSAR* models for these datasets. For datasets 15 and 16 no predictive *QSAR* models were built (see Introduction). Predictive *QSAR* models were built for a dataset of adrenergic receptor agonists and antagonists (dataset 17; see below), and HDAC inhibitors [61] (dataset 18). Modelability criteria for datasets 15–18 were used to evaluate, if it was possible at all to build predictive *QSAR* models for these datasets, as if there were no attempts building them. So datasets 15–18 were the external test set.

7.3 Results

7.3.1 Binary Response Variable

In Tables 7.3a, 7.3b, and 7.3c modelability criteria as well as some other statistics for the initial 42 binary datasets are given. As expected, in most cases *MODI_CCR* and *MODI_ssCCR* are relatively close to each other. A small difference between them is explained by the different number of compounds used in prediction in each case: in *LOO-CV*, a nearest neighbor is selected out of N-1 compounds, while in five-fold cross-validation it is selected from 0.8 × *N* compounds. *MODI_CCR* and *MODI_ssCCR* have highest correlations with *QSAR_CCR*; they appeared to be the best modelability criteria (Figs. 7.1 and 7.2). Interestingly, for some datasets *MODI_CCR* and/or *MODI_ssCCR* appeared to be very high, sometimes even higher than *QSAR_CCR*, so *LOO-CV* or similarity search in the entire descriptor space could be used instead of *QSAR* modeling at least in some cases, what would dramatically save the modeling time and computational resources. If for a dataset several models were built using different methods, the highest *QSAR_CCR* value was used on the plots (Figs. 7.1 and 7.2). Coefficient of determination R^2 between *MODI_CCR* and *QSAR_CCR* was equal to 0.67 (Fig. 7.1). R^2 between *MODI_ssCCR* and *QSAR_CCR* was 0.75 (Fig. 7.2). It was expected that *MODI_ACI* should also correlate with the *QSAR_CCR*. Indeed, the R^2 between

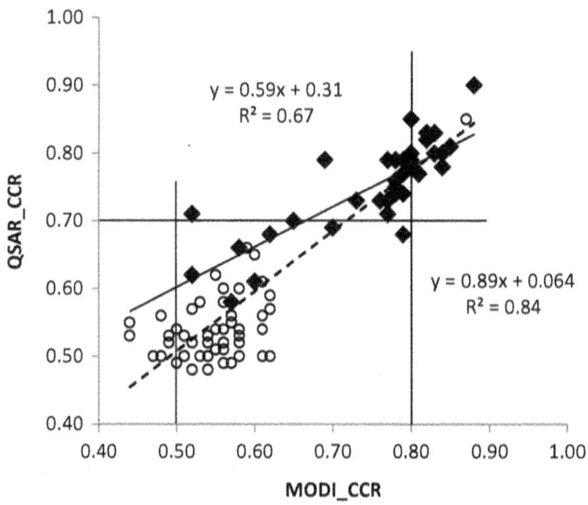

Fig. 7.1 QSAR_CCR versus MODI_CCR for initial 42 datasets (*black diamonds*) and 60 ToxCast datasets (*hollow circles*). Regression lines and the corresponding equations are shown for 42 datasets (*solid line*) and all 102 datasets (*dashed line*). Datasets with MODI_CCR ≤ 0.5 are not modelable (all QSAR_CCR values < 0.7). Datasets with MODI_CCR ≥ 0.8 are modelable (all QSAR_CCR values > 0.7). For datasets with 0.5 < MODI_CCR < 0.8 it is impossible to say, whether they are modelable or not, and QSAR modeling is required

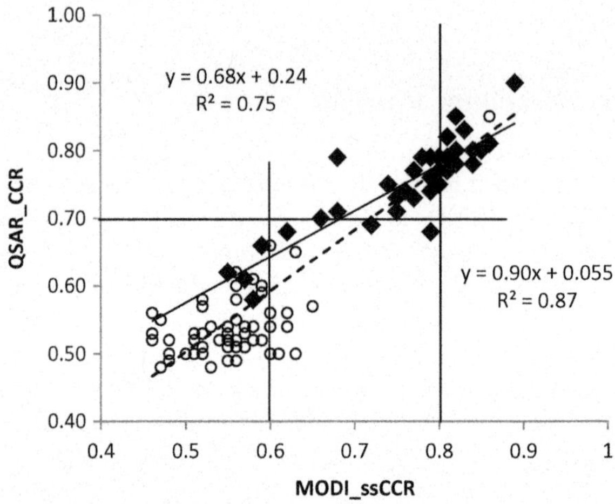

Fig. 7.2 QSAR_CCR versus MODI_ssCCR for initial 42 datasets (*black diamonds*) and 60 ToxCast datasets (*hollow circles*). Regression lines and the corresponding equations are shown for 42 datasets (*solid line*) and all 102 datasets (*dashed line*). Datasets with MODI_ssCCR ≤ 0.6 are not modelable (all QSAR_CCR values < 0.7). Datasets with MODI_ssCCR ≥ 0.8 are modelable (all QSAR_CCR values > 0.7). For datasets with $0.6 <$ MODI_ssCCR < 0.8 it is impossible to say, whether they are modelable or not, and QSAR modeling is required

Fig. 7.3 QSAR_CCR versus MODI_ACI (with one nearest neighbor) for 42 initial datasets

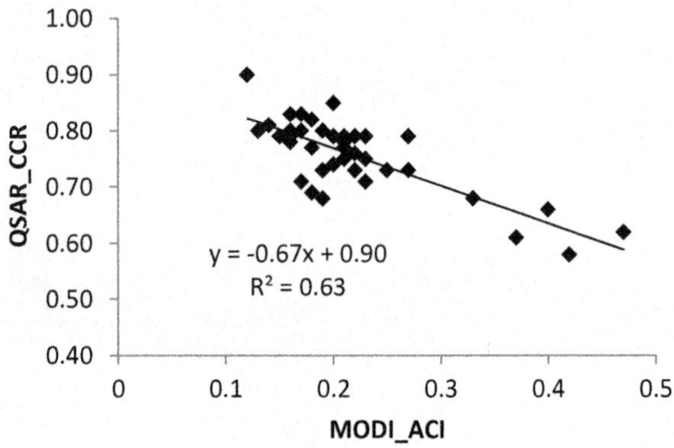

them is 0.63 (Fig. 7.3), but it is lower than for *MODI_ssCCR* and *MODI_CCR*. *MODI_ACI* was calculated for one nearest neighbor, so the distances between nearest neighbors were not taken into account. So the quantity based only on activities of nearest neighbors, which does not take into account the distances between them, only partially explains the dataset modelability. Besides, the sign of the difference between activities of a compound and its nearest neighbor is also important, but it is not taken into account by our *ACI*s. The weakest criteria of modelability appeared to be the *MODI_DIV*, *MODI_TC* (calculated using Dragon [1] descriptors) and *MODI_TC* (calculated using MACCS keys [2]); the R^2 between *MODI_DIV* and the *QSAR_CCR* was only 0.48 (Fig. 7.4), despite high

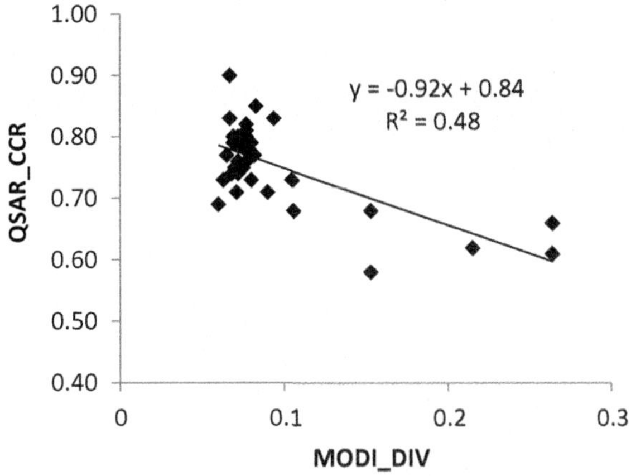

Fig. 7.4 QSAR_CCR versus MODI_DIV for 42 initial datasets

density of points in one part of the plot and lower density in the other, which is far away from the first part. R^2 between *MODI_TC*(Dragon [1]) and *QSAR_CCR* was 0.47, and between *MODI_TC*(MACCS [2]) and *QSAR_CCR* it was only 0.42. All these results were obtained for one nearest neighbor; lower R^2 values were obtained with 3 and 5 nearest neighbors. These low correlations mean that the contribution of dataset diversity to the explanation of dataset modelability is limited. As expected, both factors diversity and activity distribution, define the dataset modelability.

Results for the 60 ToxCast datasets [27] are shown for the best modelability criteria only: *MODI_CCR* (Fig. 7.1) and *MODI_ssCCR* (Fig. 7.2). If all 102 datasets are considered, the R^2 value between *MODI_CCR* and *QSAR_CCR* amounts to 0.84, and between *MODI_ssCCR* and *QSAR_CCR* amounts to 0.87. Regression lines for 102 compounds on Figs. 7.1 and 7.2 are almost identical, and any *MODI_CCR* or *MODI_ssCCR* can be used as a good modelability criterion. Here we should mention one important problem we did not take into account to the full extent in this work: imbalance of datasets (it means that one class contains significantly larger number of compounds than the other class). For these datasets, it is typical that sensitivity can be much higher (or lower) than specificity. This is the case with most of 60 ToxCast datasets (see Table 7.3d). However, despite rat cholinesterase inhibitors dataset is highly biased, excellent *QSAR* models as well as modelability criteria were obtained for this dataset. The problem of imbalance was not taken into account to the full extent in this work.

We consider a model predictive, if the expected *QSAR_CCR* ≥ 0.7. From Fig. 7.1, we can see that if *MODI_CCR* is higher than 0.8, the dataset is most probably modelable (the problem of discrepancy between *MODI_SENS* and *MODI_SPEC*, if it is large, should still be addressed). If *MODI_CCR* is lower than 0.5, the dataset is most probably not modelable. If *MODI_CCR* is between 0.5 and 0.8, it is impossible to say, whether the dataset is modelable or not. In addition, it is possible to use equations for regression lines to estimate the *QSAR_CCR*, if *MODI_CCR* is known. Thus, from the regression line for all 102 compounds, we obtain that the expected *QSAR_CCR* ≥ 0.7, if *MODI_CCR* ≥ 0.71.

From Fig. 7.2, we also can see that if *MODI_ssCCR* is higher than 0.8, the dataset is most probably modelable (the problem of discrepancy between *MODI_SENS* and *MODI_SPEC*, if it is large, should still be addressed). If *MODI_ssCCR* is lower than 0.6, the dataset is most probably not modelable. If *MODI_ssCCR* is between 0.6 and 0.8, it is impossible to say, whether the dataset is modelable or not. In addition, it is possible to use equations for regression lines to estimate the *QSAR_CCR*, if *MODI_ssCCR* is known. Thus, from the regression line for all 102 compounds, we obtain that the expected *QSAR_CCR* $\geq$ 0.7, if *MODI_ssCCR* $\geq$ 0.72.

On Fig. 7.1, we can see some outliers (as examples, see also Beta1 in Table 7.3a and MDR1 in Table 7.3b). Outliers can be explained, for example, by descriptor selection procedure, when the orders of distances between compounds in the selected and entire descriptor spaces are very different. However, as it can be seen from high R^2 values between *QSAR_CCR* and *MODI_ssCCR* or *QSAR_CCR* and *MODI_CCR*, large changes of orders of distances are relatively rare. This also partially explains, why it is sufficient to calculate modelability criteria just with one comprehensive set of descriptors like Dragon [1].

7.3.2 Continuous Response Variable

Initially, modelability criteria were calculated for 14 datasets (see in Methods). Some modelability criteria showed high correlation with *QSAR_*R^2 for these datasets. Additionally, four datasets were used as an external validation set to predict *QSAR_*R^2 values for them. These results demonstrated high potential of these modelability criteria to establish, whether it is possible to build predictive models for given datasets. We note that insufficient number of datasets were included in these studies, so the results in this section are preliminary.

Results of modelability studies are presented in Tables 7.5, 7.6, and 7.7 and Figs. 7.5, 7.6, 7.7, 7.8, and 7.9. IGC50 and BCF were the datasets which gave highest *QSAR_*R^2. Table 7.5 shows the highest values of *MODI_ss*R^2 and *MODI_*q^2. Both *MODI_ss*R_0^2 and *MODI_ss*R'^2_0 are close to *MODI_ss*R^2, what also indicates on goodness of these datasets for *QSAR* studies. *MODI_*CAT_3 and *MODI_*CAT_5 (see Sect. 7.2.1) obtained after splitting the datasets into 3 and 5 bins (see also below) are also best for these datasets. Other criteria *MODI_ACI* and *MODI_DIV* are not indicative, whether datasets are modelable or not. For 10 RI datasets, no highly predictive *QSAR* models were built (see Table 7.6). RI4, RI5 and RI7 datasets gave marginally acceptable RF [44] models (R^2 was equal to 0.54, 0.57 and 0.63 for out of bag sets), and other datasets did not give any predictive models at all. Even for these marginally acceptable datasets, modelability criteria seem to be poor; some *MODI_ss*R_0^2 were negative, a large difference was observed between *MODI_ss*R^2 and *MODI_ss*R_0^2.

It is seen that that *MODI_ss*R^2 with 1, 3 and 5 nearest neighbors are good modelability criteria; their coefficients of determination R^2 with highest *QSAR_*R^2 are equal to 0.88, 0.85 and 0.88, respectively (see Figs. 7.5a–c). It is possible to

Table 7.5 Modelability criteria for *P. promelas* (LC50), *T. Pyriformis* (IGC50), the oral rat (LD50) and bioaccumulation factor (BCF) datasets and their comparison with QSAR modeling results

Dataset	MODI_DIV	MODI_ACI	MODI_CAT	MODI_ssR^2 (R_0^2, R'^2_0; k, k′), MODI_q^2	QSAR_R^2 and coverage for external prediction: HC	RF1	kNN	RF2
LC50	0.072	0.088	0.59 (3 bins)	0.51 (R_0^2, R'^2_0: 0.42, 0.45; k, k′: 0.95, 0.99) (K = 1)	0.60–0.67	0.67–0.69	0.61–0.71	0.66–0.69
	0.24		0.61 (5 bins)	0.60 (R_0^2, R'^2_0: 0.43, 0.60; k, k′: 0.95, 1.01) (K = 3)	Coverage: 0.99	Coverage: 0.82–0.85	Coverage: 0.84–1.00	Coverage: 0.62–0.71
				0.55 (R_0^2, R'^2_0: 0.29, 0.55; k, k′: 0.94; 1.02) (K = 5)				
				MODI_q^2: 0.46 (K = 1); 0.61 (K = 3); 0.60 (K = 5)				
IGC50	0.068	0.071	0.69 (3 bins)	0.69 (R_0^2, R'^2_0: 0.66, 0.66; k, k′: 0.98, 0.99) (K = 1)	0.85	0.80–0.81	0.79–0.85	0.84–0.85
	0.32		0.74 (5 bins)	0.76 (R_0^2, R'^2_0: 0.70, 0.76; k, k′: 0.97, 1.01) (K = 3)	Coverage: 0.98	Coverage: 0.79–0.86	Coverage: 0.82–1.00	Coverage: 0.96–0.98
				0.75 (R_0^2, R'^2_0: 0.66, 0.75; k, k′: 0.96, 1.02) (K = 5)				
				MODI_q^2 : 0.68 (K = 1); 0.76 (K = 3); 0.76 (K = 5)				
LD50	0.049	0.071	0.54 (3 bins)	0.511 (R_0^2, R'^2_0: 0.44, 0.44; k, k′: 0.97, 0.96) (K = 1)	0.53–0.55	0.65–0.66	0.59–0.60	0.58–0.65
	0.27		0.58 (5 bins)	0.572 (R_0^2, R'^2_0: 0.44, 0.56; k, k′: 0.96, 0.99) (K = 3)	Coverage: 0.83–0.84	Coverage: 0.79–0.85	Coverage: 0.71–0.81	Coverage: 0.95–0.96
				0.562 (R_0^2, R'^2_0: 0.38, 0.56; k, k′: 0.95, 0.99) (K = 5)				
				MODI_q^2 : 0.36 (K = 1); 0.53 (K = 3); 0.55 (K = 5)				
BCF	0.072	0.092	0.72 (3 bins)	0.71 (R_0^2, R'^2_0: 0.69, 0.70; k, k′: 0.97, 0.92) (K = 1)	0.73–0.81	0.87–0.89	0.79–0.86	0.83–0.84
	0.29		0.67 (5 bins)	0.76 (R_0^2, R'^2_0: 0.72, 0.76; k, k′: 0.95, 0.96) (K = 3)	Coverage: 0.94–0.95	Coverage: 0.77–0.81	Coverage: 0.82–1.00	Coverage: 0.92–0.95
				0.76 (R_0^2, R'^2_0: 0.68, 0.75; k, k′: 0.96, 0.96) (K = 5)				
				MODI_q^2 = 0.66 (K = 1); 0.73 (K = 3); 0.72 (K = 5)				

Notes:
After curation: LC50: 809 compounds; IGC50: 1,085 compounds; BCF: 610 compounds; LD50: 7,285 compounds
Both MODI_DIV(1) and MODI_DIV(2) are given for one nearest neighbor (K = 1) (see Methods)
MODI_ssR^2: for LC50, IGC50 and BCF: Z-cutoff = 1; for LD50: Z-cutoff = 0.5
MODI_ssR^2 are given for 1, 3, and 5 nearest neighbors
Modi_ssR^2, R_0^2, R'^2_0, k, k′ and MODI_q^2 were calculated according to formulas 7.31, 7.32a, 7.32b, 7.33a, 7.33b, and 7.30, respectively (see Methods)
QSAR models for these datasets were reported recently [8]
HC – Hierarchical clustering [8], RF – Random Forest [43, 44], *k*NN – k Nearest Neighbors [13]
HC: USEPA descriptors [8] + Kennard-Stone division into training and test sets[62]
RF1: Dragon [1] descriptors + division into training and test sets using Sphere-Exclusion [63]
*k*NN: Dragon [1] descriptors + Multiple divisions into training and test sets using Sphere-Exclusion [63]
RF2: Simplex [9] descriptors + Minimal Test Set Dissimilarity [64] used for divisions into training and test sets
Different applicability domain (AD) definitions were used with different methods [8]

Table 7.6 Modelability criteria for Regulatory Information (RI) datasets and their comparison with QSAR modeling results

					QSAR modeling R^2 (best OOB)		
Dataset	MODI_DIV	MODI_ACI	MODI_CAT	MODI_ssR^2 (R_0^2, $R'_0{}^2$; k, k′), MODI_q^2	RF(Dragon)	RF(Simplex)	RF(CDK)
RI1	0.10	0.15	0.53 (3 bins)	0.28 (R_0^2, $R'_0{}^2$: 0.11, 0.15; k, k′: 0.96, 0.98) (K = 1)	0.36	0.31	0.30
	0.36		0.38 (5 bins)	0.29 (R_0^2, $R'_0{}^2$: −0.14, 0.21; k, k′: 0.95, 1.01) (K = 3)			
				0.27 (R_0^2, $R'_0{}^2$: −0.48, 0.23; k, k′: 0.95, 1.01) (K = 5)			
				MODI_q^2: 0.005 (K = 1); 0.15 (K = 3); 0.19 (K = 5)			
RI2	0.019	0.18	0.43 (3 bins)	0.18 (R_0^2, $R'_0{}^2$: −0.094, −0.19; k, k′: 0.99, 0.93) (K = 1)	0.43	0.34	0.31
	0.25		0.33 (5 bins)	0.095 (R_0^2, $R'_0{}^2$: −1.29, −0.021; k, k′: 0.97, 0.96) (K = 3)			
				0.16 (R_0^2, $R'_0{}^2$: −1.62, 0.14; k, k′: 0.97, 0.98) (K = 5)			
				MODI_q^2: −0.14 (K = 1); 0.071 (K = 3); 0.12 (K = 5)			
RI3	0.10	0.17	0.42 (3 bins)	0.34 (R_0^2, $R'_0{}^2$: 0.17, 0.16; k, k′: 0.94, 1.03) (K = 1)	0.46	0.38	0.36
	0.29		0.51 (5 bins)	0.46 (R_0^2, $R'_0{}^2$: 0.14, 0.45; k, k′: 0.94, 1.03) (K = 3)			
				0.36 (R_0^2, $R'_0{}^2$: −0.39, 0.36; k, k′: 0.94, 1.03) (K = 5)			
				MODI_q^2: 0.24 (K = 1); 0.42 (K = 3); 0.44 (K = 5)			
RI4	0.11	0.14	0.53 (3 bins)	0.38 (R_0^2, $R'_0{}^2$: 0.29, 0.16; k, k′: 0.93, 1.03) (K = 1)	0.53	0.54	0.45
	0.34		0.54 (5 bins)	0.41 (R_0^2, $R'_0{}^2$: 0.22, 0.34; k, k′: 0.93, 1.03) (K = 3)			
				0.44 (R_0^2, $R'_0{}^2$: 0.088, 0.42; k, k′: 0.92, 1.05) (K = 5)			
				MODI_q^2 : 0.18 (K = 1); 0.38 (K = 3); 0.37 (K = 5)			
RI5	0.14	0.15	0.61 (3 bins)	0.18 (R_0^2, $R'_0{}^2$: −0.072, −0.23; k, k′: 0.97, 1.00) (K = 1)	0.51	0.48	0.57
	0.59		0.48 (5 bins)	0.35 (R_0^2, $R'_0{}^2$: −0.20, 0.32; k, k′: 0.97, 1.02) (K = 3)			
				0.39 (R_0^2, $R'_0{}^2$: −0.54, 0.39; k, k′: 0.97, 1.02) (K = 5)			
				MODI_q^2 : 0.041 (K = 1); 0.34 (K = 3); 0.36 (K = 5)			
RI6	0.11	0.14	0.42 (3 bins)	0.22 (R_0^2, $R'_0{}^2$: 0.027, −0.15; k, k′: 0.98, 1.00) (K = 1)	0.37	0.37	0.30
	0.34		0.48 (5 bins)	0.26 (R_0^2, $R'_0{}^2$: −0.16, 0.15; k, k′: 0.97, 1.02) (K = 3)			
				0.31 (R_0^2, $R'_0{}^2$: −0.36, 0.28; k, k′: 0.96, 1.03) (K = 5)			
				MODI_q^2 : 0.18 (K = 1); 0.19 (K = 3); 0.19 (K = 5)			

RI7	0.14 0.59	0.14	0.58 (3 bins) 0.55 (5 bins)	0.37 (R_0^2, $R'_0{}^2$: 0.28, 0.16; k, k′: 0.97, 0.99) (K = 1) 0.40 (R_0^2, $R'_0{}^2$: −0.005, 0.37; k, k′: 0.95, 1.03) (K = 3) 0.32 (R_0^2, $R'_0{}^2$: −0.93, 0.32; k, k′: 0.94, 1.04) (K = 5) MODI_q^2 : 0.18 (K = 1); 0.39 (K = 3); 0.42 (K = 5)	0.58	0.57	0.63
RI8	0.14 0.47	0.17	0.43 (3 bins) 0.39 (5 bins)	0.21 (R_0^2, $R'_0{}^2$: −0.058, −0.087; k, k′: 0.97, 1.00) (K = 1) 0.23 (R_0^2, $R'_0{}^2$: −0.42, 0.15; k, k′: 0.95, 1.02) (K = 3) 0.22 (R_0^2, $R'_0{}^2$: −1.01, 0.20; k, k′: 0.95, 1.02) (K = 5) MODI_q^2 : 0.12 (K = 1); 0.16 (K = 3); 0.19 (K = 5)	0.39	0.40	0.37
RI9	0.14 0.46	0.17	0.31 (3 bins) 0.41 (5 bins)	0.18 (R_0^2, $R'_0{}^2$: −0.11, −0.21; k, k′: 0.98, 1.00) (K = 1) 0.10 (R_0^2, $R'_0{}^2$: −1.02, −0.058; k, k′: 0.98, 1.00) (K = 3) 0.19 (R_0^2, $R'_0{}^2$: −1.38, 0.16; k, k′: 0.97,1.01) (K = 5) MODI_q^2: 0.28 (K = 1); 0.066 (K = 3); 0.11 (K = 5)	0.29	0.28	0.23
RI10	0.090 0.30	0.12	0.52 (3 bins) 0.52 (5 bins)	0.37 (R_0^2, $R'_0{}^2$: 0.26, 0.17; k, k′: 0.98, 1.00) (K = 1) 0.49 (R_0^2, $R'_0{}^2$: 0.34, 0.45; k, k′: 0.97, 1.01) (K = 3) 0.44 (R_0^2, $R'_0{}^2$: 0.18, 0.41; k, k′: 0.97, 1.02) (K = 5) MODI_q^2: 0.18 (K = 1); 0.41 (K = 3); 0.42 (K = 5)	0.48	0.46	0.48

OOB – Out Of Bag

Table 7.7 Modelability studies of other datasets

Dataset	MODI_DIV	MODI_ACI	MODI_CAT	MODI_ssR^2 (R$_0^2$, R$'_0^2$; k, k'), MODI_q^2
Alpha1a	0.070	0.12	0.57 (3 bins)	0.47 (R$_0^2$, R$'_0^2$: 0.33, 0.39; k, k': 1.01 0.98) (K = 1)
	0.25		0.57 (5 bins)	0.55 (R$_0^2$, R$'_0^2$: 0.37, 0.54; k, k': 1.00, 0.99) (K = 3)
				0.56 (R$_0^2$, R$'_0^2$: 0.30, 0.56; k, k': 1.00, 0.99) (K = 5)
				MODI_q^2 = 0.40 (K = 1); 0.52 (K = 3); 0.52 (K = 5)
MRTD	0.092	0.11	0.45 (3 bins)	0.36 (R$_0^2$, R$'_0^2$: 0.23, 0.24; k, k': 0.90, 0.91) (K = 1)
	0.37		0.45 (5 bins)	0.45 (R$_0^2$, R$'_0^2$: 0.22, 0.42; k, k': 0.90, 0.96) (K = 3)
				0.42 R$_0^2$, R$'_0^2$: 0.088, 0.40; k, k': 0.88, 0.98) (K = 5)
				MODI_q^2 = 0.19 (K = 1); 0.40 (K = 3); 0.41 (K = 5)
Bioavailability	0.080	0.22	0.42 (3 bins)	0.26 (R$_0^2$, R$'_0^2$: 0.087, 0.11; k, k': 0.86, 0.84) (K = 1)
	0.39		0.38 (5 bins)	0.29 (R$_0^2$, R$'_0^2$: −0.19, 0.255; k, k': 0.82, 0.94) (K = 3)
				0.22 (R$_0^2$, R$'_0^2$: −0.63, 0.20; k, k': 0.80, 0.95) (K = 5)
				MODI_q^2 = 0.051 (K = 1); 0.24 (K = 3); 0.26 (K = 5)

Notes

Alpha1a dataset was shown to be modelable (as expected from the results). RF with simplex descriptors gave QSAR_R^2 = 0.80

MRTD dataset: based on the results, modelability of this dataset is inconclusive. In fact, RF with simplex descriptors gave QSAR_R^2 = 0.81, i.e. this dataset appeared to be modelable

Bioavailability dataset: most probably, it is not modelable as shown by similarity search. QSAR_R^2 obtained with four combinations of descriptors and methods was lower than 0.5 (see in Introduction)

obtain the conditions for $MODI\_ssR^2$ to estimate, whether a *QSAR* model will be predictive. Usually, we consider a *QSAR* model predictive, if R^2 between predicted and observed activities of external test set is at least 0.6. Thus, from regression equations (see Fig. 7.5a–c) we obtain: $MODI\_ssR^2 \geq 0.44$ for one nearest neighbor, $MODI\_ssR^2 \geq 0.49$ for 3 nearest neighbors, and $MODI\_ssR^2 \geq 0.48$ for 5 nearest neighbors.

There are different ways of splitting a dataset into bins. In this work, all the datasets with continuous end points were divided into 3 and 5 bins by activity, namely, the entire range of activities was divided into 3 and 5 equal intervals, and compounds with activities within each interval were considered as those belonging to the same bin. $MODI\_CAT_3$ and $MODI\_CAT_5$ in Fig. 7.6a and b are modelability criteria for datasets divided into 3 and 5 bins, respectively. It is seen that $MODI\_CAT_3$ and $MODI\_CAT_5$ are not as good modelability criteria as $MODI\_ssR^2$; however they are still acceptable: the R^2 between them and $QSAR\_R^2$ are equal to 0.78 and 0.81, respectively. Since for predictive *QSAR* models R^2 between predicted and observed activities of external test set should be not lower than 0.6, from regression equations (see Fig. 7.6a, b) we obtain: $MODI\_CAT_3 \geq 0.57$ and $MODI\_CAT_5 \geq 0.56$.

$MODI\_q^2$ criterion is leave-one-out cross-validation q^2 calculated in the entire descriptor space. $MODI\_q^2$ with 3 and 5 nearest neighbors appeared to have the highest $R^2 = 0.90$ with $QSAR\_R^2$ (Fig. 7.7b, c). $MODI\_q^2$ with 1 nearest neighbor has $R^2 = 0.65$ with $QSAR\_R^2$ (Fig. 7.7a), which is much worse. Taking into account that acceptable models have $R^2 \geq 0.6$, we find that the corresponding modelability criteria should satisfy the following inequalities: $MODI\_q^2 \geq 0.46$ for 3 nearest neighbors, and $MODI\_q^2 \geq 0.47$ for 5 nearest neighbors (see regression equations on Fig. 7.7b, c).

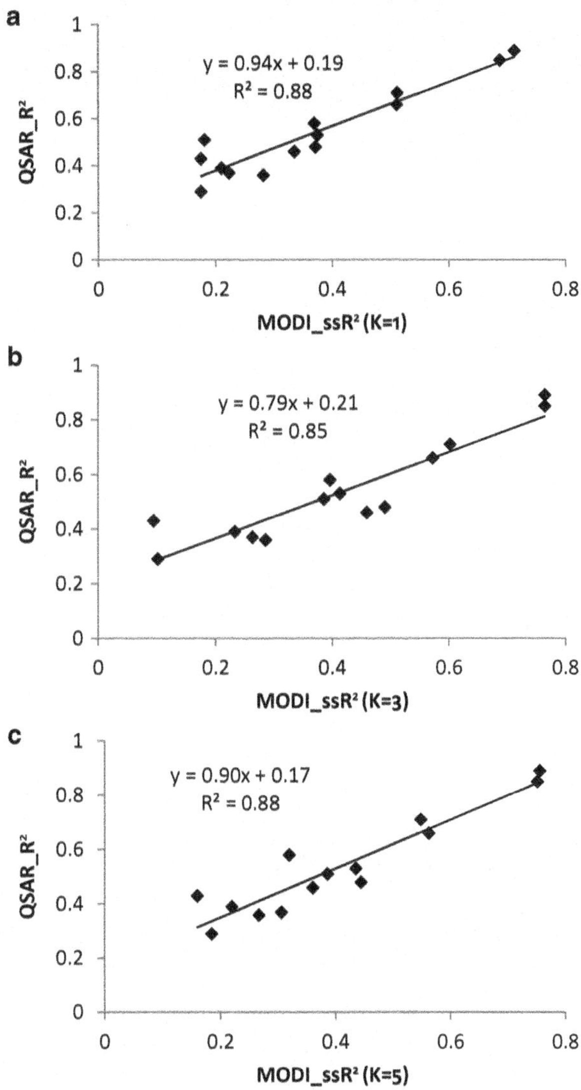

Fig. 7.5 QSAR_R^2 versus MODI_ssR^2 for 14 datasets with continuous end points. K is the number of nearest neighbors. (**a**) K = 1; (**b**) K = 3; (**c**) K = 5. High R^2 values suggest that all the MODI_ssR^2 are good modelability criteria

Modelability activity cliff index *MODI_ACI* was calculated for one nearest neighbor. It was noticed (see in Methods section), that in the case of one nearest neighbor all four *MODI_ACI* defined in this work have equal values. *MODI_ACI* has the $R^2 = 0.72$ with *QSAR_R*^{2} (Fig. 7.8), what is much lower than that for the best modelability criteria *MODI_q*^{2} and *MODI_ssR*^{2} (see above).

Diversity of a dataset independent of the distribution of activities *MODI_DIV* (see Methods, *MODI_DIV*(1)) is the weakest modelability criterion (Fig. 7.9) with $R^2 = 0.17$ between it and the *QSAR_R*^{2}. If to remove one obvious outlier (RI2, Tables 7.4 and 7.6), R^2 becomes equal to 0.43, what is still unacceptable.

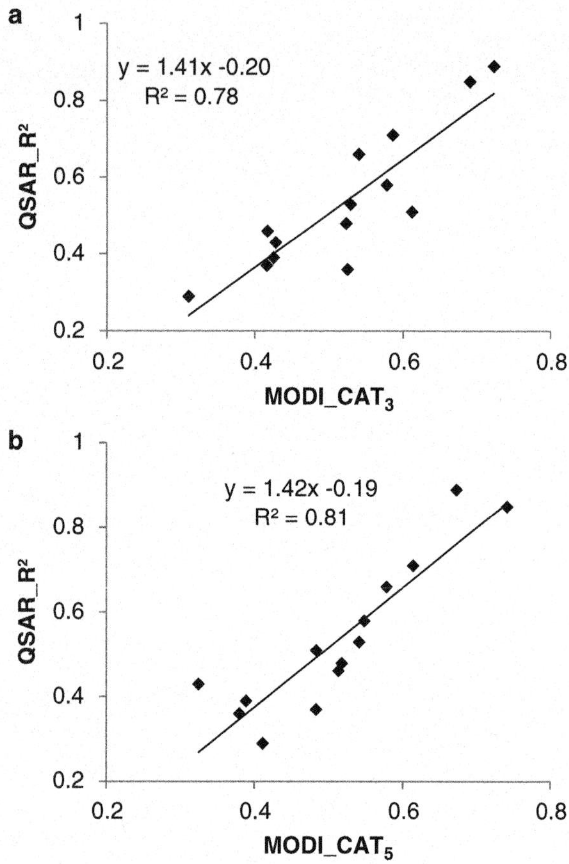

Fig. 7.6 (**a**) QSAR_R^2 versus MODI_CAT_3 for 14 datasets with continuous end points (see the main text). The datasets were divided into 3 bins by activity values. (**b**) QSAR_R^2 versus MODI_CAT_5 for the same 14 datasets. The datasets were divided into 5 bins by activity values. High R^2 values indicate that both MODI_CAT_3 and MODI_CAT_5 are good modelability criteria

As for the three datasets of the external validation set (see Table 7.7), Alpha 1a, MRTD and Bioavailability, modelability criteria demonstrate that the expected $QSAR\_R^2$ values for Alpha 1A dataset should be the highest, while for Bioavailability dataset it should be the lowest. Result for MRTD dataset was inconclusive. QSAR models built for these datasets were highly predictive for Alpha 1a and MRTD datasets, while for Bioavailability dataset no predictive *QSAR* models were obtained (see also notes below Table 7.7).

Relatively high predictivity of models built for HDAC inhibitors dataset was also corroborated by modelability studies. Multiple *k*NN and SVM *QSAR* models were built for HDAC dataset of 59 compounds; 163 Molconn-Z descriptors were used [61]. The dataset was divided into external validation set of 9 compounds, and multiple pairs of training and test sets. Training sets were used to build models and test sets were used to validate them. Then the models with best statistics (q^2(train) > 0.75 and R^2(test) > 0.75) were used for consensus prediction of external validation set. The applicability domain with $Z = 0.5$ was applied (see Sect. 7.2.1). Consensus prediction gave coefficient of determination $R^2 = 0.87$ for *k*NN models

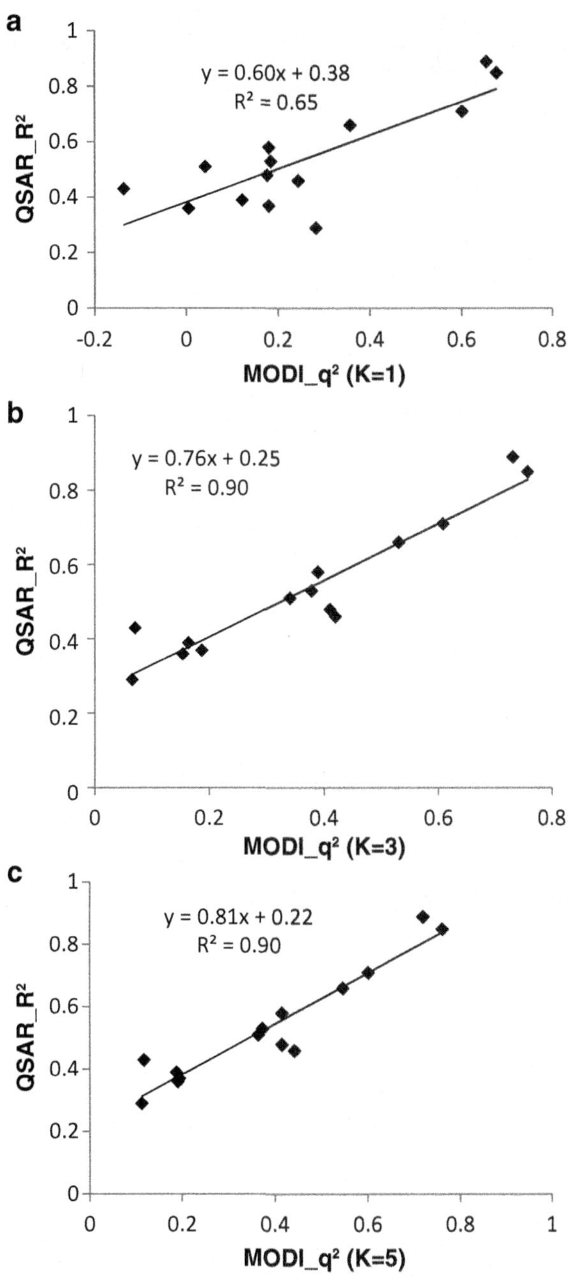

Fig. 7.7 QSAR_R^2 versus leave-one-out cross-validation (LOO-CV) MODI_q^2 for 14 datasets with continuous end points. In (**a**), (**b**) and (**c**) K is the number of nearest neighbors used to calculate MODI_q^2. High R^2 values indicate that MODI_q^2 (K = 3) and MODI_q^2 (K = 5) are good modelability criteria

Fig. 7.8 QSAR_R^2 versus Activity Cliff Index (ACI) for 14 datasets with continuous end points. ACI was calculated for one nearest neighbor (K = 1). In case K = 1, all four ACIs are equal to each other, so only one plot is shown. ACIs for K > 1 are slightly higher than ACIs for K = 1, as well as different from each other. (see Sect. 7.2.1)

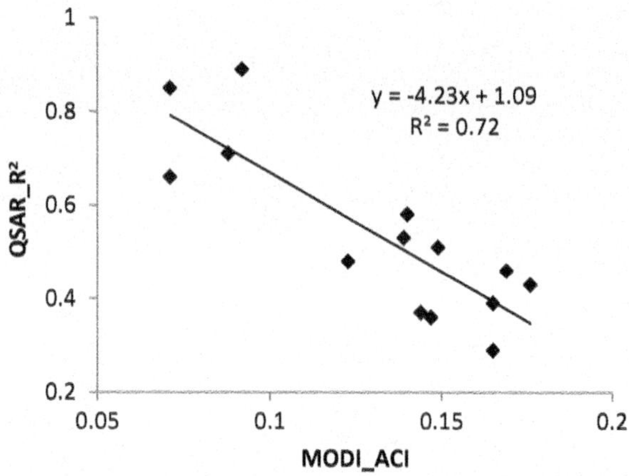

Fig. 7.9 QSAR_R^2 versus MODI_DIV for 14 datasets with continuous end points. Very low R^2 indicates that MODI_DIV is very poor modelability criterion. If one obvious outlier (RI2, Tables 7.4 and 7.6) is removed, R^2 becomes equal to 0.43, what is still very low

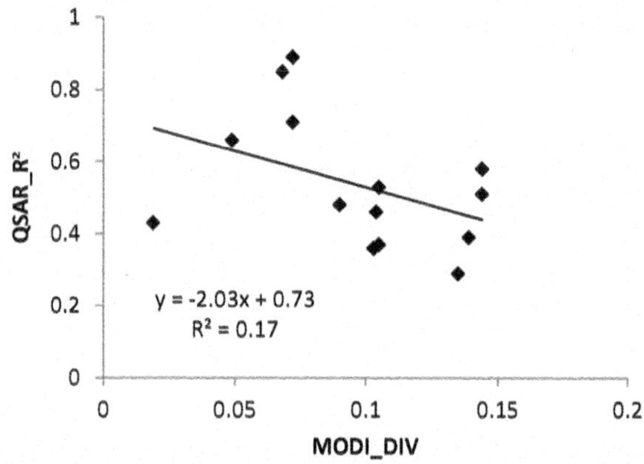

and 0.71 for SVM models. All 9 compounds were predicted. Additionally, *k*NN *QSAR* studies were conducted using leave five-fold external cross validation (each external validation set contained about 20 % of compounds), so that all external test sets together included 54 compounds, for which R^2 was 0.77 (Molconn-Z [3] descriptors were used). In this work, we have studied the modelability of HDAC dataset using 163 Molconn-Z [3] and 1,283 Dragon [1] descriptors. One of the best modelability criteria $MODI\_ssR^2$ (see Figs. 7.5a–c) was calculated for this dataset. The number of Dragon [1] descriptors was reduced using the pairwise correlation analysis: the number of remaining descriptors was 409 and 229 for threshold correlation coefficient between descriptors of 0.99 and 0.95, respectively. Results of calculations are shown in Table 7.8. From Table 7.8, we see that the highest predicted $QSAR\_R^2$ is 0.75, which is relatively close to our result $QSAR\_R^2$ = 0.77 and $QSAR\_R^2$ = 0.87 obtained elsewhere [61] (see above).

Table 7.8 MODI_ssR^2 and the corresponding predicted QSAR_R^2 for HDAC dataset (see regression equations in Fig. 7.5a–c)

	Dragon descriptors						Molconn-Z descriptors	
K	1289[a] MODI_ssR^2	Pred QSAR_R^2	409[a] MODI_ssR^2	Pred QSAR_R^2	229[a] MODI_ssR^2	Pred QSAR_R^2	163[a] MODI_ssR^2	Pred QSAR_R^2
1	0.45	0.61	0.71	0.68	0.60	0.75	0.60	0.75
3	0.45	0.57	0.49	0.60	0.56	0.65	0.54	0.64
5	0.20	0.35	0.38	0.51	0.34	0.48	0.44	0.57

[a]MODI_ssR^2 and predicted QSAR_R^2 values are given for this number of descriptors (see the main text)

7.4 Conclusions and Future Studies

In this work, we tried to answer the question, whether it is possible to evaluate predictivity of QSAR models for a given dataset prior to actual QSAR modeling (i.e. to establish the dataset modelability). The approach can save a lot of time and efforts by excluding unnecessary QSAR studies of not modelable datasets. The procedure we propose is fast and reliable. It is based on calculations in the entire descriptor space, without variable selection. So it corresponds to just one step of the *k*NN QSAR [13] approach. Several modelability criteria have been suggested and validated for datasets with binary and continuous end points. Some of the criteria were successfully used to establish dataset modelability, while others seemed to be ineffective for this goal. Moreover, we propose to use the best criteria *instead of* much more tedious and time consuming QSAR modeling. Generalizations of the criteria on the datasets with compounds belonging to more than two classes or categories have been suggested. Below are summaries of the results for datasets with binary and continuous response variables.

7.4.1 Conclusions for Datasets with Binary Response Variable

1. Similarity Search Correct Classification Rate (*MODI_ssCCR*) is the best criterion of dataset modelability in terms of its coefficient of determination with predictive power of *QSAR* models (*QSAR_CCR*).
2. Modelability Index based on activities of nearest neighbors (*MODI_CCR*) is slightly worse criterion of dataset modelability.
3. Activity Cliff Indices (*MODI_ACI*) have lower correlation with *QSAR_CCR* than *MODI_ssCCR* and *MODI_CCR*.
4. Activity-independent Dataset Diversity indices (*MODI_DIV* and *MODI_TC*) have the lowest coefficient of determination with predictive power of QSAR models (*QSAR_CCR*).
5. Both dataset diversity and activity distribution of compounds are important for evaluation of dataset modelability.

We suggest that *MODI_ssCCR* should be used as modelability criterion for binary datasets. If for some dataset it is higher than 0.8, then the dataset is modelable. If it is lower than 0.6, the dataset is not modelable. In these cases, no extensive QSAR modeling is necessary. *MODI_CCR* is also a very good modelability criterion. In comparison with *MODI_ssCCR*, *MODI_CCR* has lower discriminative power between modelable and not modelable datasets, however, it gives almost the same prediction of *QSAR_CCR*, and its calculation is faster.

7.4.2 Conclusions for Datasets with Continuous Response Variable

These results are preliminary, because insufficient number of datasets was involved in these studies. However, we think that most of these results will hold after more datasets will be included.

1. *LOO-CV* q^2 in the entire descriptor space (*MODI_*q^2) with $K = 3$ and $K = 5$ nearest neighbors (*K* other than 1, 3 or 5 were not considered) are the best modelability criteria.
2. R^2 for Similarity Search performed with the external five-fold cross validation (*MODI_ss*R^2) is almost as good as *MODI_*q^2.
3. Bin-based modelability indices (*MODI_CAT*s) and Activity Cliff Indices (*MODI_ACI*s) have lower correlation with model predictive *QSAR_*R^2 than *MODI_*q^2 and *MODI_ss*R^2.
4. Activity-independent Dataset Diversity (*MODI_DIV*s) indices have the lowest R^2 with *QSAR_*R^2.
5. Both dataset diversity and activity distribution of compounds are important for evaluation of dataset modelability.

We suggest that *MODI_*q^2 with 3 and 5 nearest neighbors and *MODI_ss*R^2 should be used as modelability criteria for continuous datasets. Other criteria have lower correlation with *QSAR_*R^2 for external test sets and should not be used as modelability criteria.

7.4.3 Future Studies

1. Influence of descriptor sets on modelability criteria. Correlations between descriptor sets. Stability of modelability criteria across different descriptor sets. For preliminary results see [65].
2. Modelability criteria for datasets with category end points, when the number of categories is higher than 2.
3. Influence of bias in the category (not only binary) datasets on modelability criteria.
4. Datasets with continuous response variables. In this preliminary study, we evaluated modelability criteria on only 14 datasets, what is insufficient. More studies are necessary.
5. Influence of the method of dataset division into bins on the corresponding modelability criteria.

Acknowledgements The authors are thankful to Dr. Guiyu Zhao (AstraZeneca, Shanghai, China) for providing QSAR modeling results for 34 GPCRome datasets, Dr. Jessica Wignall (University of North Carolina at Chapel Hill) for providing the QSAR modeling results for 10 regulatory information datasets, and Dr. Jonathan Freedman (National Institute of Environmental Health

Science, Research Triangle Park, NC) and Dr. Ruchir Shah (SciOme, LLC, Research Triangle Park, NC) for providing experimental results for *C. Elegans* datasets.

References

1. Dragon Descriptors. http://www.talete.mi.it/products/dragon_description.htm. Accessed 21 Aug 2012
2. Molecular Operating Environment (MOE). http://www.chemcomp.com/software.htm. Accessed 21 Aug 2012
3. Molconn-Z descriptors. http://www.edusoft-lc.com/molconn. Accessed 08 Sept 2013
4. Mold2 descriptors. http://www.fda.gov/ScienceResearch/BioinformaticsTools/Mold2/default.htm. Accessed 08 Sept 2013
5. CDK Descriptor Calculator. http://pele.farmbio.uu.se/nightly/dnames.html. Accessed 08 Sept 2013
6. Volsurf Descriptors. http://www.moldiscovery.com/soft_volsurf.php. Accessed 08 Sept 2013
7. Adriana Descriptors. http://molecular-networks.com/node/45. Accessed 08 Sept 2013
8. Martin TM, Harten P, Venkatapathy R, Das S, Young DM (2008) A hierarchical clustering methodology for the estimation of toxicity. Toxicol Mech Method 18(2–3):251–266
9. Kuz'min VE, Artemenko AG, Muratov EN (2008) Hierarchical QSAR technology based on the simplex representation of molecular structure. J Comput Aided Mol Des 22(6–7):403–421
10. Isida Fragments. http://infochim.u-strasbg.fr/recherche/Download/FragmentorNomenclature_of_ISIDA_fragments_2011.pdf. Accessed 08 Sept 2013
11. Adams MJ (2004) Chemometrics in analytical spectroscopy. Royal Society of Chemistry, Cambridge, UK
12. Wold S, Sjöström M, Eriksson L (2001) PLS-regression: a basic tool of chemometrics. Chemometrics Intel Lab Syst 58(2):109–130
13. Zheng W, Tropsha A (2000) Novel variable selection quantitative structure–property relationship approach based on the k-nearest-neighbor principle. J Chem Inf Comput Sci 40 (1):185–194
14. Vapnik VN (1995) The nature of statistical learning theory. Springer, New York
15. Quinlan JR (1986) Induction of decision trees. Mach Learn 1(1):81–106
16. Breiman L, Friedman JH, Olshen RA, Stone CJ (1998) Classification and regression trees. Chapman & Hall/CRC, New York
17. Breiman L (2001) Random forests. Mach Learn 45(1):5–32
18. Breiman L, Cutler A. Random forests. http://www.stat.berkeley.edu/~breiman/andomForests/. Accessed 08 Sept 2013
19. Chirico N, Gramatica P (2011) Real external predictivity of QSAR models: how to evaluate it? Comparison of different validation criteria and proposal of using the concordance correlation coefficient. J Chem Inf Model 51(9):2320–2335
20. Chirico N, Gramatica P (2012) Real external predictivity of QSAR models. Part 2. New intercomparable thresholds for different validation criteria and the need for scatter plot inspection. J Chem Inf Model 52(8):2044–2058
21. Kovatcheva A, Golbraikh A, Oloff S, Feng J, Zheng W, Tropsha A (2005) QSAR modeling of datasets with enantioselective compounds using chirality sensitive molecular descriptors. SAR QSAR Environ Res 16(1–2):93–102
22. Tropsha A, Golbraikh A (2010) Predictive quantitative structure–activity relationships modeling: development and validation of QSAR models. In: Faulon J-L, Bender A (eds) Handbook of chemoinformatics algorithms. Chapman & Hall/CRC, London, pp 213–233

23. Kovatcheva A, Golbraikh A, Oloff S, Xiao Y-D, Zheng W, Wolschann P, Buchbauer G, Tropsha A (2004) Combinatorial QSAR of ambergris fragrance compounds. J Chem Inf Comput Sci 44(2):582–595
24. de Cerqueira Lima P, Golbraikh A, Oloff S, Xiao Y-D, Tropsha A (2006) Combinatorial QSAR modeling of P-glycoprotein substrates. J Chem Inf Model 46(3):1245–1254
25. ToxCastTM. http://epa.gov/ncct/toxcast. Accessed 11 Jan 2012
26. U.E.-N.C. for C. Toxicology, Computational Toxicology Research Program (CompTox). http://www.epa.gov/ncct/toxrefdb/. Accessed 08 Sept 2013
27. Thomas RS, Black MB, Li L, Healy E, Chu T-M, Bao W, Andersen MD, Wolfinger RD (2012) A comprehensive statistical analysis of predicting in vivo hazard using high-throughput in vitro screening. Toxicol Sci: Off J Soc Toxicol 128(2):398–417
28. Veber DF, Johnson SR, Cheng H-Y, Smith BR, Ward KW, Kopple KD (2002) Molecular properties that influence the oral bioavailability of drug candidates. J Med Chem 45 (12):2615–2623
29. Shen M, LeTiran A, Xiao Y-D, Golbraikh A, Kohn H, Tropsha A (2002) Quantitative structure-activity relationship analysis of functionalized amino acid anticonvulsant agents using k nearest neighbor and simulated annealing PLS methods. J Med Chem 45 (13):2811–2823
30. Goret M, Wang-Bell M, Golbraikh A, Tropsha A (2006) QSAR analysis of a dataset of 91 functionalized amino acids anticonvulsant agents using k nearest neighbor. Unpublished results
31. Boyd WA, McBride SJ, Rice JR, Snyder DW, Freedman JH (2010) A high-throughput method for assessing chemical toxicity using a Caenorhabditis elegans reproduction assay. Toxicol Appl Pharmacol 245(2):153–159
32. Sedykh A, Zhu H, Tang H, Zhang L, Richard A, Rusyn I, Tropsha A (2011) Use of in vitro HTS-derived concentration-response data as biological descriptors improves the accuracy of QSAR models of in vivo toxicity. Environ Health Persp 119(3):364–370
33. Tropsha A, Golbraikh A (2007) Predictive QSAR modeling workflow, model applicability domains, and virtual screening. Curr Pharm Des 13(34):3494–3504
34. Golbraikh A (2000) Molecular dataset diversity indices and their applications to comparison of chemical databases and QSAR analysis. J Chem Inf Comput Sci 40(2):414–425
35. Fourches D, Muratov E, Tropsha A (2010) Trust, but verify: on the importance of chemical structure curation in cheminformatics and QSAR modeling research. J Chem Inf Model 50(7):1189–1204
36. Guha R, Van Drie JH (2008) Structure–activity landscape index: identifying and quantifying activity cliffs. J Chem Inf Model 48(3):646–658
37. Zhang L, Sedykh A, Tripathi A, Zhu H, Afantitis A, Mouchlis VD, Melagraki G, Rusyn I, Tropsha A (2013) Identification of putative estrogen receptor-mediated endocrine disrupting chemicals using qsar- and structure-based virtual screening approaches. Toxicol Appl Pharmacol 23. doi:pii: S0041-008X(13)00216-0. 10.1016/j.taap.2013.04.032. Epub ahead of print
38. ChemBL databases. https://www.ebi.ac.uk/chembl/. Accessed 08 Sept 2013
39. PDSP database. http://pdsp.med.unc.edu/indexR.html. Accessed 08 Sept 2013
40. USEPA ECOTOX database (2008). http://cfpub.epa.gov/ecotox. Accessed 08 Sept 2013
41. U.E.-N.C. for C. Toxicology, Computational Toxicology Research Program (CompTox). http://www.epa.gov/ncct/toxcast/. Accessed 08 Sept 2013
42. CDK Descriptor Names. http://pele.farmbio.uu.se/nightly/dnames.html. Accessed 08 Sept 2013
43. R: Classification and Regression with Random Forest. http://rss.acs.unt.edu/Rdoc/library/randomForest/html/randomForest.html. 08 Sept 2013
44. Polishchuk PG, Muratov EN, Artemenko AG, Kolumbin OG, Muratov NN, Kuz'min VE (2009) Application of random forest approach to QSAR prediction of aquatic toxicity. J Chem Inf Model 49(11):2481–2488
45. Zhu H, Tropsha A, Fourches D, Varnek A, Papa E, Gramatica P, Öberg T, Phuong D, Cherkasov A, Tetko IV (2008) Combinatorial QSAR modeling of chemical toxicants tested against Tetrahymena pyriformis. J Chem Inf Model 48(4):766–784

46. Schultz TW (1997) TETRATOX: Tetrahymena pyriformis population growth impairment endpoint – a surrogate for fish lethality read more. http://informahealthcare.com/doi/abs/10.1080/105172397243079. Toxicol Mech Method 7(4):289–309
47. Schultz TW, Netzeva TI (2004) Development and evaluation of QSARs for ecotoxic endpoints: the benzene response-surface model for Tetrahymena toxicity. In: Cronin MTD, Livingstone DJ (eds) Modeling environmental fate and toxicity. CRC Press, Boca Raton
48. Schultz TW, TETRATOX. http://www.vet.utk.edu/TETRATOX/index.php. Accessed 08 June 2013
49. ChemiDplus Advanced Database National Library of Medicine 2011 (NLM). http://chem.sis.nlm.nih.gov/chemidplus/. Accessed 24 Feb 2011
50. USEPA, User's Guide for T.E.S.T. (Toxicity Estimation Software Tool). http://www.epa.gov/ORD/NRMRL/std/cppb/qsar/testuserguide.pdf. Accessed 27 Oct 2009
51. Zhu H, Martin TM, Ye L, Sedykh A, Young DM, Tropsha A (2009) Quantitative structure-activity relationship modeling of rat acute toxicity by oral exposure. Chem Res Toxicol 22 (12):1913–1921
52. Zhu H, Ye L, Richard A, Golbraikh A, Wright FA, Rusyn I, Tropsha A (2009) A novel two-step hierarchical quantitative structure-activity relationship modeling work flow for predicting acute toxicity of chemicals in rodents. Environ Health Persp 117(8):1257–1264
53. Martin TM, Harten P, Young DM, Muratov EN, Golbraikh A, Zhu H, Tropsha A (2012) Does rational selection of training and test sets improve the outcome of QSAR modeling? J Chem Inf Model 52(10):2570–2578
54. Hamelink JL (1977) Current bioconcentration test methods and theory. In: Mayer FL, Hamelink JL (eds) Aquatic toxicology and hazard evaluation. ASTM STP 634, American Society for Testing and Materials, Baltimore, pp 149–161
55. OEHHA Toxicity Criteria Database. http://www.oehha.ca.gov/risk/ChemicalDB/index.asp. Accessed 08 May 2013
56. Regional Screening Levels | Region 9: Superfund | US EPA. http://www.epa.gov/region9/superfund/prg/. Accessed 08 May 2013
57. O. US EPA, Integrated Risk Information System (IRIS). http://www.epa.gov/iris/. Accessed 08 May 2013
58. O. of P.P. US EPA, Pesticide Reregistration Status | Pesticides | US EPA. http://www.epa.gov/oppsrrd1/reregistration/status.htm. Accessed 08 May 2013
59. Pharmaceutical Press. Martindale: the complete drug reference, 37th edn. http://www.pharmpress.com/product/9780853699330/martindale. Accessed 08 July 2013
60. U.E.-N.C. for C. Toxicology, Computational Toxicology Research Program (CompTox). http://www.epa.gov/ncct/dsstox/sdf_fdamdd.html. Accessed 08 Sept 2013
61. Tang H, Wang XS, Huang X-P, Roth X-P, Butler KV, Kozikowski AP, Jung M, Tropsha A (2009) Novel inhibitors of human histone deacetylase (HDAC) identified by QSAR modeling of known inhibitors, virtual screening, and experimental validation. J Chem Inf Model 49(2):461–476
62. Kennard RW, Stone L (1969) Computer aided design of experiments. Technometrics 11(1):137–148
63. Golbraikh A, Shen M, Xiao Z, Xiao Y-D, Lee K-H, Tropsha A (2003) Rational selection of training and test sets for the development of validated QSAR models. J Comput Aided Mol Des 17(2–4):241–253
64. Kuz'min VE, Artemenko AG, Muratov EN, Volineckaya IL, Makarov VA, Riabova OB, Wutzler P, Schmidtke M (2007) Quantitative structure-activity relationship studies of [(biphenyloxy)propyl]isoxazole derivatives. Inhibitors of human rhinovirus 2 replication. J Med Chem 50(17):4205–4213
65. Golbraikh A, Muratov E, Fourches D, Tropsha A. Data set modelability by QSAR. J Chem Inf Model. 8 Jan 2014 [Epub ahead of print]

Chapter 8
A Combination of the Sequential QM/MM and the Free Energy Gradient Methodologies with Applications

H.C. Georg, T.S. Fernandes, S. Canuto, N. Takenaka, Y. Kitamura, and M. Nagaoka

Abstract To obtain stable states (SS) and transition states (TS) in chemical reactions in condensed phase, the free energy gradient (FEG) method was proposed in 1998 as an optimization method on a multidimensional free energy surface (FES). Analogous to the method for the Born Oppenheimer potential energy surface (PES) using ab initio molecular orbital calculation, the FEG method utilizes the force and Hessian on the FES, which can be adiabatically calculated by molecular dynamics (MD) or Monte Carlo (MC) methods, and, originally, the free energy (FE) perturbation theory. In fact, since then, a number of excellent approximate methods have been developed, e.g., the averaged solvent electrostatic potential (ASEP)/MD method and the average solvent electrostatic configuration (ASEC) method. In this chapter, the FEG methodology is reviewed in general and a future perspective to explore the FE landscape is introduced together with several applications of these methods. Based on computational demands and on the numerical accuracy, we believe that a family of the FEG methodologies should become more efficient as one strategic setting and will play promising and important roles to survey condensed state chemistry on the basis of recent supercomputing technology.

H.C. Georg
Instituto de Física, Universidade Federal de Goiás, CP 131, 74001-970 Goiânia, GO, Brazil

T.S. Fernandes
Instituto de Química, Universidade Federal de Goiás, CP 131, 74001-970 Goiânia, GO, Brazil

S. Canuto (✉)
Instituto de Física, Universidade de São Paulo, CP 66318, 05314-970 São Paulo, SP, Brazil
e-mail: canuto@if.usp.br

N. Takenaka • Y. Kitamura • M. Nagaoka
Graduate School of Information Science, Nagoya University, Chikusa Ku, Nagoya, Aichi 4648601, Japan

J. Leszczynski and M.K. Shukla (eds.), *Practical Aspects of Computational Chemistry III*, DOI 10.1007/978-1-4899-7445-7_8,

8.1 Introduction

Atomistic computer simulation methods can provide microscopic information such as the transition state (TS) structure of a chemically reacting molecule that is not able to be observed easily by conventional experimental approaches. So far, quantum mechanical (QM) approaches using the ab initio molecular orbital (MO) methods have been carried out for many reaction systems in solutions and biological environment. However, for the solution chemical reaction system, it is still difficult to obtain the stable state (SS) and TS geometries because such reactions are involved in both solute species and a vast number of solvent molecules. In such cases, the hybrid quantum mechanical/molecular mechanical (QM/MM), using molecular dynamics (MD) or Monte Carlo (MC), methods are very useful to treat the whole solution reaction system. Only the reactive parts in the whole system are treated quantum mechanically while the other parts are treated molecular mechanically. Nevertheless, the potential energy surfaces (PESs) for the solution reaction system are too high dimensional, and then it is almost impossible to search even TS with respect to all degrees of freedom for the whole system. Therefore, the free energy surface (FES) has been often estimated approximately along only a certain reaction coordinate since FES makes it possible to consider effectively a vast number of solvent molecules. However, the SS and TS geometries should be optimized with respect to all degrees of freedom of the solute in solution for the purpose to identify these accurate structures on the FES. Under the circumstances, we have been developing the FEG methodology, utilizing the forces and Hessian on the FES [1–4]. In this chapter we briefly review the essential aspects of the FEG methodology and its implementation in different possibilities along with some applications on the solvent effects on the structure and spectroscopy of molecular systems.

8.2 The Iterative Sequential QM/MM Methodology

Hybrid QM/MM methodologies are largely used for simulations of systems with hundreds of molecules, as fully QM simulations are, still nowadays, a computational challenge for large systems. The Sequential QM/MM methodology [5, 6] consists in a variant in which the simulations are performed (with either MC or MD) using a purely classical Hamiltonian, and after the sampling of the ensemble, uncorrelated configurations are submitted to electronic structure calculations of the reference, solute molecule, perturbed by the electrostatic field of a sufficiently thick shell of solvent molecules surrounding it. The main advantages of the sequential variant is that it is more economical in the amount of QM calculations necessary to converge the average of any electronic property and that it is possible to select the size of the QM (and MM) region a posteriori and according to the necessity.

However, in the sequential procedure, the solute and the solvent molecules are uncoupled because one is not permitted to be electronically polarized by the other during the simulation. The solvent force-field is usually parameterized to reflect the bulk situation, including the electronic polarization. Hence most of the attention is devoted to the reference solute that must be polarized according to the environment in which it is embedded.

Using the Sequential QM/MM iteratively it is possible to polarize the solute molecule in the solvent environment [6–8]. At each step of the iteration, it is calculated the (ensemble) average charge distribution of the solute molecule in the presence of the specific solvent in which it is immersed. The solute interaction with the solvent is updated in the charge distribution until convergence [6–8].

8.3 The Free Energy Gradient (FEG) Methodology

According to the analogy of the FES to the Born-Oppenheimer PES obtained by ab initio MO calculation, if we know the gradients on the FES, i.e., the force and Hessian, the "reaction ergodography" [9] would become possible in solution [1–4]. For this purpose, the analytic gradient method (AGM) on the PES could be utilized appropriately to the optimization problems on the FES, i.e., the searching of equilibrium structures of solute molecules in solution. From the viewpoint of the efficiency of calculation in the standard FEG method, it is true that the active use of AGM on the PES obtained by ab initio MO method must be practically the best in the QM/MM framework. However, the calculation methods of the derivatives, or analogs to find any optimum states on the FES is discussed in this section, especially, paying attention to the efficiency of the calculation.

8.3.1 The Basics and the Standard FEG Method

In the MD method, the forces acting on each atom of the solute molecule by all solvent molecules are always calculated every time step. Moreover, by time averaging the forces, the force on the FES is obtained as follows

$$\mathbf{F}^{\mathrm{FE}}(\mathbf{q}) = -\frac{\partial G(\mathbf{q})}{\partial \mathbf{q}} = -\left\langle \frac{\partial V_{\mathrm{SB}}(\mathbf{q})}{\partial \mathbf{q}} \right\rangle \tag{8.1}$$

where $\mathbf{F}^{\mathrm{FE}}$ $(\mathbf{q})$ is a force vector on the FES, $\mathbf{q}$ denotes a vector of the solute mass-weighted coordinates, $G(\mathbf{q})$ is a free energy (FE) function, V_{SB} is the sum of both the potential energy of the solute and the interaction energy between the solute and solvent molecules. The brackets denote the time average that is equal to the equilibrium ensemble average [1]. Therefore, in the MD methodology, we can evaluate the force on the FES by Eq. 8.1 without such additional efforts as the numerical differentiation. Furthermore, the Hessian on the FES is obtained by

$$\mathbf{H}^{\mathrm{FE}}(\mathbf{q}) = \frac{\partial^2 G(\mathbf{q})}{\partial \mathbf{q} \partial \mathbf{q}} \tag{8.2}$$

$$= \left\langle \frac{\partial^2 V_{\mathrm{SB}}(\mathbf{q})}{\partial \mathbf{q} \partial \mathbf{q}} \right\rangle - \beta \left[\left\langle \frac{\partial V_{\mathrm{SB}}(\mathbf{q})}{\partial \mathbf{q}} \frac{\partial V_{\mathrm{SB}}^{T}(\mathbf{q})}{\partial \mathbf{q}} \right\rangle - \left\langle \frac{\partial V_{\mathrm{SB}}(\mathbf{q})}{\partial \mathbf{q}} \right\rangle \left\langle \frac{\partial V_{\mathrm{SB}}(\mathbf{q})}{\partial \mathbf{q}} \right\rangle^{T} \right] \tag{8.3}$$

where the superscript T denotes the transposition and $\beta = 1/\mathrm{kT}$. While the second and third terms in the RHS of Eq. 8.3 can be calculated out of those quantities usually evaluated in MD simulations, the first term evaluation needs an additional procedure where a number of the second derivatives on "instantaneous" PESs should be calculated including the solvent influence [1].

8.3.2 *The Hybrid Quantum Mechanical/Molecular Mechanical (QM/MM) Method*

In the hybrid QM/MM method, only the reactive parts in the whole solution reaction system are treated quantum mechanically, while the other parts are treated molecular mechanically [10]. The Hamiltonian $\hat{H}$ consists of three terms:

$$H = H_{\mathrm{QM}} + H_{\mathrm{MM}} + H_{\mathrm{QM/MM}}, \tag{8.4}$$

where the first two terms H_{QM} and H_{MM} stand for the standard Hamiltonian of the QM and MM systems, respectively, while the last QM/MM term $H_{\mathrm{QM/MM}}$ represents the interaction between the QM and MM system, and is expressed as a sum of electrostatic (est) and van der Waals (vdW) terms:

$$H_{\mathrm{QM/MM}} = H_{\mathrm{QM/MM}}^{\mathrm{est}} + H_{\mathrm{QM/MM}}^{\mathrm{vdW}}. \tag{8.5}$$

The first term $H_{\mathrm{QM/MM}}^{\mathrm{est}}$ is given by

$$H_{\mathrm{QM/MM}}^{\mathrm{est}} = \sum_{M} q_M V_{\mathrm{QM}}(\mathbf{R}_{\mathrm{M}}), \tag{8.6}$$

with the potential

$$V_{\mathrm{QM}}(\mathbf{R}_{\mathrm{M}}) = -\sum_{s} \frac{1}{R'_{sM}} + \sum_{A} \frac{Z_A}{R_{AM}}, \tag{8.7}$$

which is represented as the sum of coulomb interactions created by the electrons and cores of QM atoms [11]. In these expressions, q_M is the atomic point electric charge on the M-th MM atom in solvent molecules, R'_{sM} is the distance between the s-th QM electron of the solute molecule and the M-th MM atom, Z_A is the core charge of the A-th QM atom, and R_{AM} is the distance between the A-th QM atom and the M-th MM one. The vdW term $H^{\mathrm{vdW}}_{\mathrm{QM/MM}}$ is often expressed using a number of 6–12 Lennard-Jones (LJ) functions:

$$H^{\mathrm{vdW}}_{\mathrm{QM/MM}} = \sum_A \sum_M \varepsilon_{AM} \left[\left(\frac{r_{AM}}{R_{AM}} \right)^{12} - 2 \left(\frac{r_{AM}}{R_{AM}} \right)^{6} \right], \tag{8.8}$$

where ε_{AM} and r_{AM} are a couple of LJ parameters for the A-th QM atom interacting with the M-th MM atom. The potential energy V of the total system is then given by

$$V = \langle \psi | H_{\mathrm{QM}} + H_{\mathrm{QM/MM}} | \psi \rangle + V_{\mathrm{MM}} \tag{8.9a}$$

$$= V_{\mathrm{SB}} + V_{\mathrm{MM}}. \tag{8.9b}$$

A minimal implementation combining AMBER with Gaussian or PAICS for the purpose of performing ab initio QM/MM-MD simulation has been done [12, 13].

8.3.3 *The Average Solvent Electrostatic Potential (ASEP): Mean Field Approximation – ASEP/MD-FEG Method*

Since 1997, Aguilar and his colleagues have utilized the mean field theory to estimate solvent effects by combining QM and MD calculations [14]. The mean field approximation (MFA) reduces drastically the number of quantum calculations and permits a better description of the quantum subsystem, introducing no significant inaccuracies. The MD method with their MFA is called the average solvent electrostatic potential (ASEP)/MD method [15]. Then, they developed, to obtain geometries and properties in solution, such an efficient FEG method that the formalism of the FEG method is reinforced with the ASEP method [16]. In the ASEP/MD-FEG method, the free energetic force, i.e., the negative of the FEG (Eq. 8.1), is approximated by the force of the mean configuration:

$$\mathbf{F}^{\mathrm{FE}}(\mathbf{q}) \approx -\frac{\partial \langle V_{\mathrm{SB}}(\mathbf{q}) \rangle}{\partial \mathbf{q}}. \tag{8.10}$$

In Eq. 8.10, the FEG is replaced by the gradient of the average of energy V_{SB} (Eq. 8.9b) in the spirit of the MFA. In such cases, the energy and wave functions of the solvated solute molecules are obtained by solving the effective Schrödinger equation,

$$\left(\hat{H}_{\mathrm{QM}} + \hat{H}_{\mathrm{QM/MM}} \right)_{\mathbf{q}} | \psi \rangle = E(\mathbf{q}) | \psi \rangle. \tag{8.11}$$

The electrostatic QM-MM interaction term, $\hat{H}^{\text{est}}_{\text{QM/MM}}$ takes the following form:

$$\hat{H}^{\text{est}}_{\text{QM/MM}} = \int d\mathbf{r} \cdot \hat{\rho} \cdot \langle \hat{V}_{\text{S}}(\mathbf{r};\rho) \rangle, \tag{8.12}$$

where $\hat{\rho}$ is the solute charge density operator,

$$\hat{\rho} = -\sum_{s} \delta(\mathbf{r} - \mathbf{r}_s) + \sum_{A} Z_A \delta(\mathbf{r} - \mathbf{R}_A), \tag{8.13}$$

and the quantity $\langle \hat{V}_{\text{S}}(\mathbf{r};\rho) \rangle$ is the average electrostatic potential generated by the solvent (i.e., ASEP) at the position **r**, and is obtained from MD calculations where the solute molecule is represented by the charge distribution ρ associated with a given electronic state $|\psi\rangle$ at a geometry **q** fixed during the MD simulation. The angle brackets here denote a statistical average of configurations. Since the solvent structure, and hence the ASEP, is a function of the solute charge density, Eqs. 8.11 and 8.12 have to be solved iteratively for convergence.

8.3.4 The Average Solvent Electrostatic Configuration (ASEC) and Sequential QM/MM Methodology: ASEC-FEG Method

The Brazilian group introduced a simpler implementation inspired by the idea of ASEP. To include the solvent effects into the electronic calculation of the solute within MFA, they make use of a statistical analysis using an autocorrelation function to prepare a few hundreds of uncorrelated configurations obtained via the MC simulation, and generate what is called the average solvent electrostatic configuration (ASEC) [17]. With a significantly smaller sample of configurations one is able to assure statistical convergence and at the same time avoid generating a fitting of the average potential.

Hence, the ASEC could be regarded as a configurational variant of the ASEP devised by Aguilar's group. The relaxation, namely, optimization process is done by combining the FEG method with the ASEC approach, in a similar way of ASEP, i.e., the ASEC-FEG method [18].

Figure 8.1 illustrates the iterative relaxation process combining the Sequential QM/MM, the ASEC and the FEG methods. In the first QM calculation of the method, a new geometry is obtained with the average of the gradient and Hessian of the central solute in the electrostatic field of the solvent molecules. Then, the electron distribution of the new geometry is estimated by using MFA. The updated molecular geometry and charge distribution are used in a new MC simulation, initiating the next step of the iterative relaxation process. Examples of the applicability of the method are presented in Sect. 8.4.

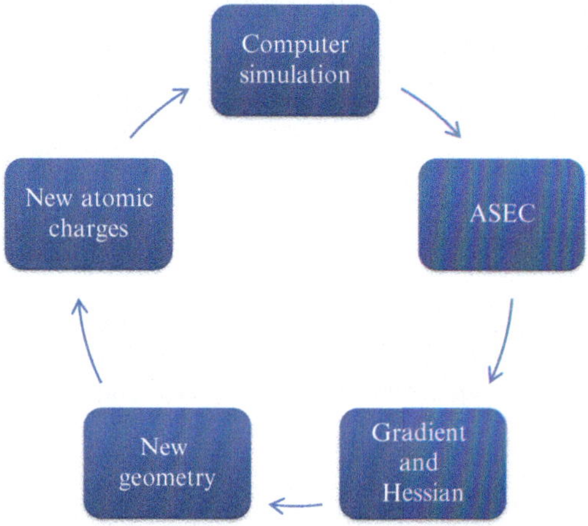

Fig. 8.1 Illustration of the iterative process combining the Sequential QM/MM, the ASEC and the FEG method

8.3.5 Free Energy in Energy Representation

Matubayasi and Nakahara proposed the energy representation (ER) method to estimate the solvation free energy of a solute molecule using an approximate functional [19, 20]. In this method, the solvation free energy is expressed in terms of energy distribution operator $\hat{\rho}^{\mathrm{e}}$, which is defined in the energy representation as

$$\hat{\rho}^{\mathrm{e}}(\varepsilon) = \sum_i \delta\left(v^{\mathrm{f}}(\mathbf{x}_i) - \varepsilon\right), \tag{8.14}$$

where $\mathbf{x}_i$ is the full coordinate of the ith solvent molecule and $v^{\mathrm{f}}(\mathbf{x}_i)$ is the interaction potential between the solute and the ith solvent molecule and the sum is taken over all the solvent molecules. The superscripts f and e are attached to emphasize that a function is represented over the full coordinate $\mathbf{x}$ and over the energy coordinate ε, respectively. Using $\hat{\rho}^{\mathrm{e}}$, the average distribution of the value of the defining potential is expressed in the presence of a certain solute-solvent interaction u as

$$\rho_u^{\mathrm{e}}(\varepsilon) = \langle \hat{\rho}^{\mathrm{e}}(\varepsilon) \rangle_u = \int d\mathbf{x}\delta\left(v^{\mathrm{f}}(\mathbf{x}) - \varepsilon\right)\rho_u^{\mathrm{f}}(\mathbf{x}) \tag{8.15}$$

where $< \ldots >_u$ denotes again the ensemble average taken in the solution with the solute–solvent interaction u. The solvation free energy $\Delta\mu$ is then expressed as

$$\begin{aligned} \Delta\mu = -k_{\mathrm{B}}T\int d\varepsilon\Big[\left(\rho_u^{\mathrm{e}}(\varepsilon) - \rho_0^{\mathrm{e}}(\varepsilon)\right) + \beta w^{\mathrm{e}}(\varepsilon)\rho_u^{\mathrm{e}}(\varepsilon) \\ -\left(\beta\int_0^1 d\lambda w^{\mathrm{e}}(\varepsilon)\right) \times \left(\rho_u^{\mathrm{e}}(\varepsilon) - \rho_0^{\mathrm{e}}(\varepsilon)\right)\Big], \end{aligned} \tag{8.16}$$

with

$$w^{\mathrm{e}}(\varepsilon) = -k_{\mathrm{B}}T\ln\left(\frac{\rho_{u}^{\mathrm{e}}(\varepsilon)}{\rho_{0}^{\mathrm{e}}(\varepsilon)}\right) - u^{\mathrm{e}}(\varepsilon), \tag{8.17}$$

where u^{e} is the solute-solvent interaction, and $\hat{\rho}_{0}^{\mathrm{e}}$ is the distribution function in the pure solvent system ($u^{\mathrm{e}} = 0$) [19, 20]. If we could assume a set of meaningful molecular structures of a solute molecule in solution owing to whatever the standard FEG method or other approximate methods might be used, the ER method would easily provide the free energy difference between those two arbitrary points, e.g., that between two geometries along the reaction path on the FES.

8.3.6 Selecting an Appropriate Theoretical Method: A Protocol of FE Ergodography

Of these several kinds of theoretical methods to study the FE landscapes of complex chemical species in solution there are different and feasible ways to consider. Which method and how we should use for exploring our individual issues is open for proper choice. Today, the continuum model may be looked as a simple choice. In this work we will use and explore the possibilities of the ASEC-FEG in addition to one application by the standard-FEG.

8.4 Applications of Free Energy Gradient (FEG) Methodology

8.4.1 Electronic and Structural Properties of Molecules in Solution

We have applied the FEG method in conjunction with the Iterative Sequential QM/MM methodology to investigate how the solvent affects the electronic properties of a solute molecule in a liquid environment.

An example is liquid ammonia. By using the Iterative Sequential QM/MM, two ensembles of liquid ammonia were simulated (at $P = 1$ atm and $T = 197.2$ K). The first one was obtained by relaxing only the charge distribution of the ammonia molecule and the second by relaxing also its geometry. In both cases, the relaxation was extended to all the molecules in the box. The LJ parameters used are those from the Impey-Klein potential [21] (using another LJ potential for ammonia, from Gao, Xia and George [22], gives the same results) and the charges are obtained from a CHELPG fit of the calculated charge density obtained with MP2/aug-cc-pVTZ at

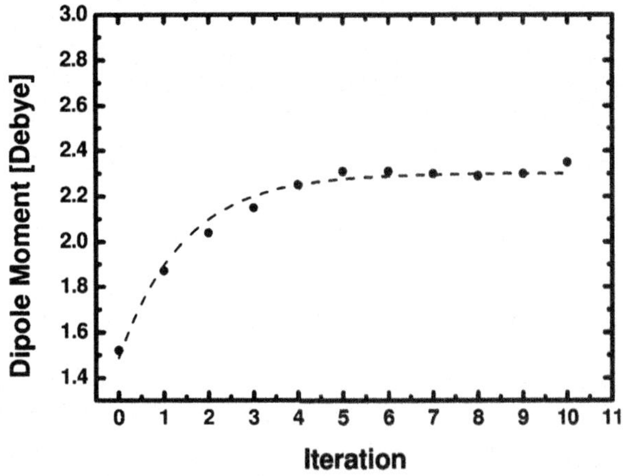

Fig. 8.2 The evolution of the dipole moment of the ammonia molecule, from gas-phase (step 0) to the converged, in-liquid value, including geometry relaxation

each step of the iteration. The dipole moment, shown in Fig. 8.2, increases during the relaxation and with five steps starts to fluctuate around a value of 2.3 D, corresponding to an increase of 50 % from the gas phase, when geometry relaxation is included. A similar value is obtained with fixed geometry.

In the case of geometry relaxation, we found a slight increase of 0.005 Å in the N–H bond length and a small decrease of ~2° in the H–N–H bond angle, corresponding to an increase in the pyramidality of the NH_3.

An important property related to the electronic structure of the molecule and largely dependent on its geometry is the intramolecular indirect spin-spin coupling constants. Because of its high correlation with atomic distances, this property is used in NMR experiments to help elucidating the arrangement of atoms in molecules. Such property can be calculated using, for instance, DFT methods. In particular, the hybrid DFT method B3LYP has been very successful in calculating NMR properties, mainly for oxygen and nitrogen [23]. Hence we used the B3LYP method with the aug-cc-pVTZ-J basis set developed by Provasi et al. [24] for spin-spin coupling calculations. We have also used the aug-pcJ-2 basis developed by Jensen [25]. The gas-phase one-bond coupling ^1J(N,H) is calculated as −64.4 Hz. In the liquid phase, during the iterative relaxation process, the coupling is calculated for a reference ammonia molecule embedded in the field of a 12 Å thick shell of solvent ammonia molecules. Figure 8.3 shows the evolution of the coupling during the inclusion of the polarization effects. It can be seen that the solute polarization influences the spin-spin coupling constant differently depending on whether the geometry relaxation is considered or not. In the first step, including the solvent ammonia molecules as a perturbation in the solute ammonia with rigid, gas-phase geometry, the coupling is changed to approximately −67.3 Hz, with the aug-cc-pVTZ-J basis set. Now, including the geometry relaxation the spin-spin coupling constant is calculated as −63.9 Hz. The converged gas-to-liquid shift is small, calculated as less than 0.5 Hz, consistent with the experimental observations that show a negligible shift [26]. This result shows that even small geometry changes like the one experienced by ammonia, when going from the gas to the

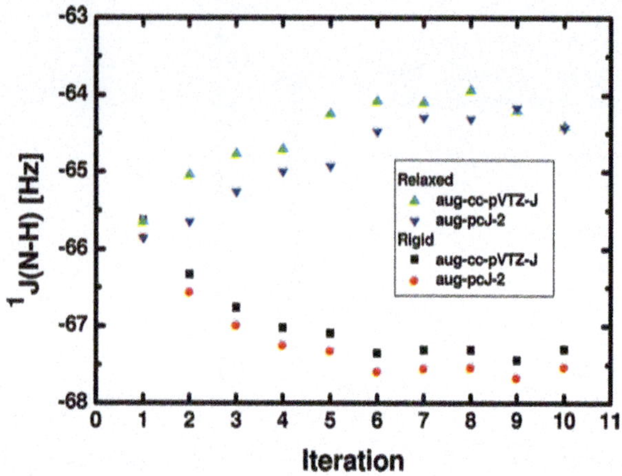

Fig. 8.3 Evolution of the intramolecular, one-bond spin-spin coupling constant parameter in ammonia during the iterative relaxation, with rigid and relaxed geometry

liquid phase, can affect the spin-spin coupling constant and this is well described by the present methodology. Further details and results for the chemical shielding and spin-spin coupling constants in ammonia are discussed in reference [27].

As another interesting application, the electronic properties of water in liquid environment were also investigated using the ASEC-FEG methodology. Water is an essential system because of its importance as a universal solvent in (bio) chemistry. From the theoretical point of view, the understanding of liquid water is still a challenge for the current methodologies [28] and has been subjected to the ASEC-FEG analysis [18].

We performed a full (charge distribution and geometry) relaxation of a central, reference water molecule immersed in solvent water molecules at room pressure and temperature. At each step, four parallel MC Metropolis simulations, starting from different random initial configurations, of a system with 501 water molecules were performed and the ASEC was generated with 400 uncorrelated configurations extracted from these simulations. The gradient and hessian were then calculated using a MP2/aug-cc-pV5Z level and a new geometry was obtained and subsequently submitted to an atomic charge fit (by the CHELPG method) using the electronic density obtained at the same level. The new geometry and atomic charges are used in a new MC simulation, and so on until convergence is observed in the geometric parameters and the dipole moment of the solute molecule.

Figure 8.4 shows the evolution of the average and maximum (over the nine components) atomic forces. We observed that after three steps the average force reaches a plateau in 0.0025 hartree/bohr. This is a sufficiently low value, considering the fluctuations of the liquid situation and the impossibility of a complete representation of the ensemble, which would take an infinite simulation. That plateau corresponds to an in-liquid O–H bond length that fluctuates around an equilibrium value of 0.976 Å (Fig. 8.5), with a fluctuation amplitude of 0.001 Å. This average in-liquid O–H bond length represents an increase of 0.018 Å from the gas-phase value, within the experimental range of 0.012 ± 0.006 Å, obtained from neutron diffraction experiments [29].

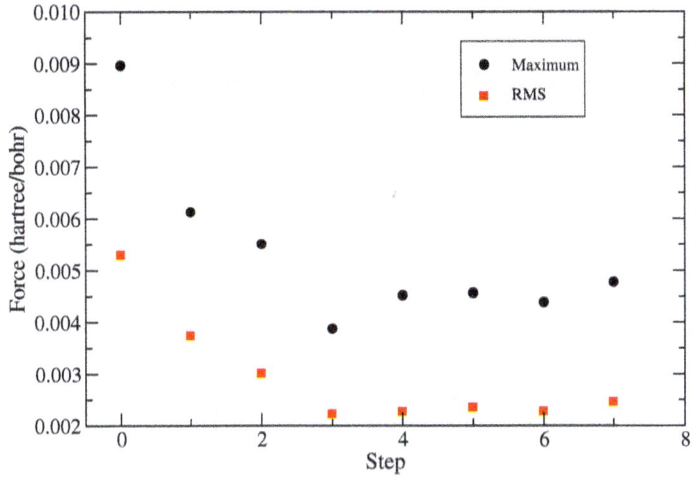

Fig. 8.4 Evolution of the atomic force components of the central H_2O molecule

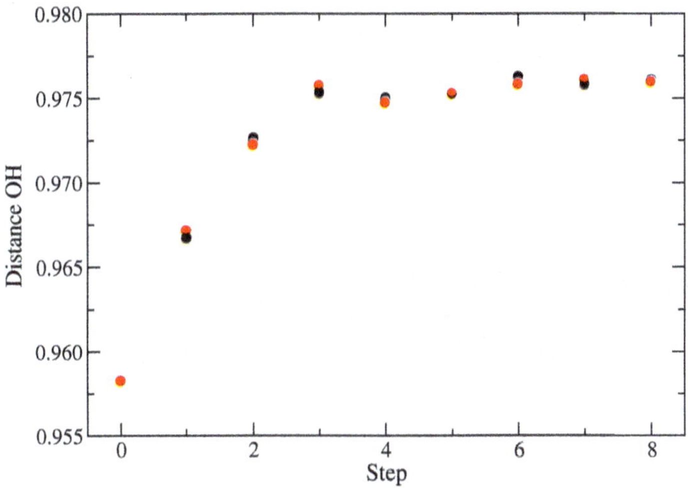

Fig. 8.5 Evolution of the O–H distance during the iterative relaxation process

The main results on the geometric and charge distribution relaxations are shown in Table 8.1. The converged dipole moment obtained for the relaxed central water is around 2.7 D and is consistent with the dielectric constant of liquid water and various theoretical predictions in the literature [30–32].

Again, an important property to analyze is the intramolecular indirect spin-spin coupling constants, mainly the ${}^1J({}^{17}\text{O,H})$ that is very sensitive to geometry relaxation and other solvent effects. Here we used again the hybrid DFT method, B3LYP, with the aug-pcJ-3 basis set designed by Jensen [25] for spin-spin coupling calculations. In gas phase the ${}^1J({}^{17}\text{O,H})$ coupling is calculated (without vibrational

Table 8.1 Equilibrium geometric parameters (distances in Å, angles in degrees) and dipole moment (μ in D) of water in liquid environment

	Calculated	Experimental [29]
R(O-H)	0.976	0.970 ± 0.005
ΔR	0.018	0.012 ± 0.006
θ(H-O-H)	106.6	106.1
Δθ	2.3	1.5
μ	2.72	
Δμ	0.86	

The gas-to-liquid shift, ΔX, of each property is also reported

Table 8.2 Spin-spin coupling constants (Hz) calculated with B3LYP/aug-pcJ-3 for water in liquid environment obtained using ASEC

	$^{1}J(^{17}O,H)$	$^{2}J(H,H)$
Calculated[a]	−88.1	−6.7
Experiment [34]	−89.8 ± 2.3	
Calc. shift	−11.1	0.7
Exp. shift	−11.1 ± 2.3	

[a]Average of the spin-spin coupling constants calculated in steps 3–7

correction) as −77.0 Hz in good agreement with the experimental value of −78.7 Hz [33]. Table 8.2 shows the results for the spin-spin coupling constants in the relaxed central water in the presence of the solvent water molecules treated as an electrostatic embedding constructed from the atomic point charges. The agreement between the calculated and the experimental values for $^{1}J(^{17}O,H)$ is very good, even considering that our value is not vibrationally averaged. The calculated gas-to-liquid shift (in which the vibrational corrections are expected to be mostly cancelled) is not negligible, as in the case of liquid ammonia, and is in sharp agreement with the experimental value, showing that we have an accurate description of the relaxed geometry of the water molecule in liquid environment.

All the results obtained are in very good agreement with the experimental data and they show that the ASEC-FEG method is capable of describing accurately the equilibrium geometry of molecules in solution.

Another example of application of the ASEC-FEG methodology is the investigation of the structure of urea in water. There is some debate in the literature about the structure of the urea molecule in aqueous solution, whether it is planar or the amino groups are pyramidalized [35]. The gas-phase equilibrium conformation of urea is known to have the amino groups in pyramidal conformation [36]. The elucidation of the equilibrium structure of urea in water is important to rationalize the mechanism responsible for protein denaturation.

The ASEC-FEG methodology was applied to find the structure of minimum free energy of the urea molecule in liquid water, at room temperature and pressure, following a similar protocol as in the other previous cases. MC metropolis simulations were performed at each relaxation step and a sampling of the ensemble was generated by combining the configurations of four parallel simulations, each starting from a different random initial configuration. At each step of the iterative procedure the MP2/aug-cc-pVTZ level was used for calculating the electronic

Table 8.3 Analysis of the hydrogen bonds formed between water and urea

	Average number of HBs			Interaction energy of HBs (kcal/mol)		
HB on atom	C_{2V}	C_2	C_S	C_{2V}	C_2	C_S
H	0.88	0.89	0.84	−6.37	−6.18	−5.98
N	0.50	0.52	0.53	−10.05	−8.57	−9.37
O	3.42	3.25	3.31	−10.16	−9.63	−9.75
Total	4.80	4.66	4.68	−26.58	−24.37	−25.10

distribution of the urea molecule immersed in the electrostatic field of a 12 Å thick shell of water, represented by an electrostatic embedding provided by the atomic point charges.

The geometry relaxation was performed starting from the three different gas-phase minimum energy conformations – C_2 (pyramidalized, global minimum) C_S (pyramidalized) C_{2V} (planar). Regardless of the starting conformation, urea was found to relax to a completely planar geometry, with the amino groups preferring the planar conformation over the pyramidalized one in aqueous solution.

The reasons behind this preference were also investigated. We have performed a simple iterative Sequential QM/MM process, relaxing only the charge distribution of the three minimum energy conformers mentioned above, which were kept rigid in their gas-phase geometry. It was found that the amount of hydrogen bonds formed between urea and water is essentially the same for the three conformers, but urea enhances its interaction with water in the planar conformation (see Table 8.3) because of a larger polarization in this conformation. The hydrogen bond analysis shown in Table 8.3 was made using criteria for identifying HBs that are obtained from the distribution functions [5] (in the present case: $R_{X...Y} \leq 3.5$ Å; $\Theta_{HX...Y} \leq 40°$; $E_{bond} \geq 2.0$ kcal/mol).

8.4.2 *Vibrational Frequency Analysis in Solution*

We have demonstrated the vibrational frequency analysis (VFA) in solution within the FEG framework. For the purpose, the approximation to neglect the second term of the force fluctuation in Eq. 8.3, which was often assumed in the ASEP/MD-FEG method [15, 16], was also adopted and then, the approximated Hessian is given in the QM/MM approximation (Eqs. 8.4 – 8.9),

$$\mathbf{H}^{\mathrm{FE}}(\mathbf{q}) \cong \left\langle \frac{\partial^2 V_{\mathrm{SB}}(\mathbf{q})}{\partial \mathbf{q} \partial \mathbf{q}} \right\rangle = \left\langle \frac{\partial^2 \left\langle \Psi \middle| \hat{H}_{\mathrm{QM}} + \hat{H}^{\mathrm{est}}_{\mathrm{QM/MM}} \middle| \Psi \right\rangle}{\partial \mathbf{q} \partial \mathbf{q}} \right\rangle + \left\langle \frac{\partial^2 \hat{H}^{\mathrm{vdW}}_{\mathrm{QM/MM}}}{\partial \mathbf{q} \partial \mathbf{q}} \right\rangle \tag{8.18}$$

with the expression of the matrix element of the second term of RHS in Eq. 8.18,

$$\frac{\partial^2 \hat{H}^{\text{vdW}}_{\text{QM/MM}}}{\partial q_{A\alpha}\partial q_{A\beta}} = \begin{cases} \sum_M \varepsilon_{AM}\left[24\left(7\left(\frac{r_{AM}}{R_{AM}}\right)^{12} - 4\left(\frac{r_{AM}}{R_{AM}}\right)^{6}\right)\cdot\frac{(q_{A\alpha}-q_{M\alpha})^2}{R_{AM}^4}\right. \\ \left. \quad -12\left(\left(\frac{r_{AM}}{R_{AM}}\right)^{12} - \left(\frac{r_{AM}}{R_{AM}}\right)^{6}\right)\cdot\frac{1}{R_{AM}^2}\right] & (\alpha=\beta) \\ \sum_M \varepsilon_{AM}\left[24\left(7\left(\frac{r_{AM}}{R_{AM}}\right)^{12} - 4\left(\frac{r_{AM}}{R_{AM}}\right)^{6}\right)\cdot\right. \\ \left. \quad \frac{(q_{A\alpha}-q_{M\alpha})(q_{A\beta}-q_{M\beta})}{R_{AM}^4}\right] & (\alpha\neq\beta) \end{cases} \tag{8.19}$$

where A and M denote the integral numbers discriminating the QM and MM atoms, and α and β are one among x, y and z.

For an application example using the FE Hessian matrix $\mathbf{H}^{\text{FE}}(\mathbf{q})$ (Eq. 8.18), we considered an H_2O molecule in liquid water. Initially, the structure of the H_2O molecule in water was optimized by the standard FEG method for the H_2O geometry to satisfy the zero-FEG condition [3]

$$\mathbf{F}^{\text{FE}}\left(\mathbf{q}^{\text{FE}}_{\text{eq}}\right) = 0, \tag{8.20}$$

where the optimized structure is denoted by $\mathbf{q}^{\text{FE}}_{\text{eq}}$. A set of three effective normal vibrational frequencies of the OH stretching and the HOH bending motion, $\{\omega^{\text{eff}}_{\text{sym}}, \omega^{\text{eff}}_{\text{asym}}, \omega^{\text{eff}}_{\text{bend}}\}$, have been obtained by diagonalizing $\mathbf{H}^{\text{FE}}(\mathbf{q})$ and are shown as three grey bars in Fig. 8.6, at 3,287 (sym), 3,387 (asym) and 1,676 (bend) cm^{-1}, with the heights in proportion to their infrared (IR) intensities obtained by Gaussian-03 [38]. The gas-to-liquid shifts of the OH stretching show large red-shifts, −354 and −393 cm^{-1}, while that of the HOH bending a slight blue-shift, +77 cm^{-1}. These shifts stem from the electrostatic interactions of a water molecule with ambient water molecules [39] and the tendency is consistent with previous studies of liquid H_2O [18, 39–42].

In addition, we have also evaluated the density of states (DOS) [43] of the IR spectra in liquid water

$$D^{\text{IR}}(\omega) = \left\langle \frac{1}{3N_{H_2O}} \sum_{r\in\{\text{sym,asym,bend}\}} \delta(\omega-\omega_r(t))\cdot\left(\frac{\partial\mu}{\partial Q_r(t)}\right)^2\right\rangle, \tag{8.21}$$

with respect to the "instantaneous" normal vibrational frequencies $\{\omega_{\text{sym}}(t), \omega_{\text{asym}}(t), \omega_{\text{bend}}(t)\}$ and the corresponding "instantaneous" normal coordinates

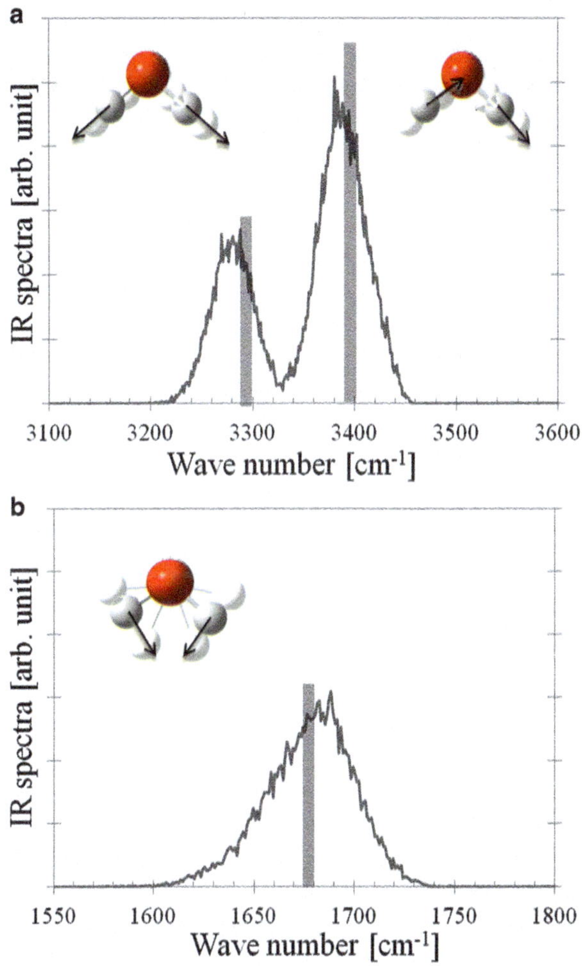

Fig. 8.6 Calculated IR spectra obtained by diagonalizing the FE Hessian (the *grey bars*) and by the density of states (DOS) of IR intensities prepared by diagonalizing the "instantaneous" Hessian matrices along an equilibrium QM/MM-MD trajectory (the *grey jagged curves*). The DOSs are all weighted by the "instantaneous" IR intensities (Eq. 8.21). (**a**) The OH stretching and (**b**) the HOH bending frequencies. (The vibrational frequencies obtained by Gaussian03 were uniformly scaled by the recommended factor of 0.9486 [37])

$\{Q_{\text{sym}}(t), Q_{\text{asym}}(t), Q_{\text{bend}}(t)\}$ obtained by diagonalizing the "instantaneous" mass-weighted Hessian $\partial^2 \hat{V}_{\text{SB}}(\mathbf{q}(t))/\partial q_{A\alpha}(t)\partial q_{A\beta}(t)$ (cf. Eq. 8.18) along an equilibrium QM/MM-MD trajectory. In the frequency region higher than 1,600 cm^{-1}, the IR spectra of liquid water consist of two main bands, which correspond to (a) the stretching motions of the covalent OH bonds and (b) the bending motion with respect to the HOH angle (Fig. 8.6). The bandwidths reflect only the electron density fluctuation dependent on the fluctuations of the "instantaneous" solvent configurations, because the solute H_2O molecule is kept as the structure $\mathbf{q}_{eq}^{FE}$ optimized on FES. In fact, it was understood that such calculated IR spectra correspond more reasonably to the experimental observation in Glycine in aqueous solution [44]. However, it should be noted that the present method of VFA in solution does not necessarily require any clear optimized structures $\mathbf{q}_{eq}^{FE}$. The similar treatment is, therefore, possible for another set of "instantaneous" mass-weighted Hessians, including the intramolecular vibrational motion of the solute, which can be obtained through an equilibrium QM/MM-MD trajectory without fixing the structures of solute molecules. Thus, we believe that this method can reasonably provide us with not only the effective normal vibrational frequencies in solution but also their fluctuations, and must become, in principle, a class of plausible VFA techniques that can be utilized for analyzing unique vibrational characteristics of a great number of functional groups in solution.

8.5 Summary and Conclusions

The study of the influence of the solvent effects on the structure, properties and spectroscopy of molecular systems is one of the leading focuses of quantum chemistry and molecular quantum mechanics. The ability to obtain the free energy surface is important in this direction and essential for the study of chemical reaction in solution. In this chapter the free energy gradient method has been reviewed. Its implementation has been presented along with feasible computational methods such as the average solvent electrostatic potential, or its configuration counterpart, the average solvent electrostatic potential. This very promising combination of these methodologies has been demonstrated by specific applications that included infrared spectroscopy of water in solution. In special, the calculations of the spin-spin coupling constants in water and in ammonia have also been very successful, exemplifying one property that is very dependent on geometry relaxation.

References

1. (a) Okuyama-Yoshida N, Nagaoka M, Yamabe T (1998) Int J Quantum Chem 70:95 (b) Nagaoka M, Okuyama-Yoshida N, Yamabe T (1998) J Phys Chem A 102:8202 (c) Okuyama-Yoshida N, Nagaoka M, Yamabe T (1998) J Phys Chem A 102: 282
2. (a) Okuyama-Yoshida N, Kataoka K, Nagaoka M, Yamabe T (2000) J Chem Phys 113:3519 (b) Hirao H, Nagae Y, Nagaoka M (2001) Chem Phys Lett 348:350
3. (a) Nagae Y, Oishi Y, Naruse N, Nagaoka M (2003) J Chem Phys 110:4555 (b) Nagaoka M, Nagae Y, Koyano Y, Oishi Y (2006) J Phys Chem A 110:4555
4. (a) Koyano Y, Takenaka N, Nakagawa Y, Nagaoka M (2010) Bull Chem Soc Jpn 83:486 (b) Takenaka N, Kitamura Y, Koyano Y, Asada T, Nagaoka M (2011) Theor Chem Acc 130:215
5. Malaspina T, Coutinho K, Canuto S (2002) J Chem Phys 117:1692

6. Coutinho K, Rivelino R, Georg HC, Canuto S (2008) In: Canuto S (ed) Solvation effects in molecules and biomolecules. Challenges and advances in computational chemistry and physics, vol 6. Springer, London, pp 159–189
7. Georg HC, Coutinho K, Canuto S (2006) Chem Phys Lett 429:119
8. Georg HC, Coutinho K, Canuto S (2007) J Chem Phys 126:034507
9. (a) Fukui K (1981) Acc Chem Res 14:363 (b) Fukui K (1982) Angew Chem Int Ed 21:801
10. (a) Warshel A, Levitt M (1976) J Mol Biol 103:227 (b) Singh UC, Kollman PA (1986) J Comput Chem 7:718 (c) Field MJ, Bash PA, Karplus MA (1990) J Comput Chem 11:700
11. Luque FJ, Reuter N, Cartier A, Ruiz-López MF (2000) J Phys Chem A 104:10923
12. (a) Okamoto T, Yamada K, Koyano Y, Asada T, Koga N, Nagaoka M (2011) J Comput Chem 32:932 (b) Okamoto T, Ishikawa T, Koyano Y, Yamamoto N, Kuwata K, Nagaoka M (2013) Bull Chem Soc Jpn 86:210
13. Yamada K, Koyano Y, Okamoto T, Asada T, Koga N, Nagaoka M (2011) J Comput Chem 32:3092
14. Sanchez ML, Aguilar MA, Olivares del Valle FJ (1997) J Comput Chem 18:313
15. Martin ME, Sánchez ML, Olivares del Valle FJ, Aguilar MA (2002) J Chem Phys 116:1613
16. Galván IF, Sánchez ML, Martín ME, Olivares del Valle FJ, Aguilar MA (2003) J Chem Phys 118:255
17. Coutinho K, Georg HC, Fonseca TL, Ludwig V, Canuto S (2007) Chem Phys Lett 437:148
18. Georg HC, Canuto S (2012) J Phys Chem B 116:11247
19. Matubayasi N, Nakahara M (2000) J Chem Phys 113:6070
20. Matubayasi N, Nakahara M (2002) J Chem Phys 117:3605
21. Impey RW, Klein ML (1984) Chem Phys Lett 104:579
22. Gao J, Xia X, George TF (1993) J Phys Chem 97:9241
23. Keal TW, Helgaker T, Salek P, Tozer DJ (2006) Chem Phys Lett 425:163
24. Provasi PF, Aucar GA, Sauer SPA (2001) J Chem Phys 115:1324
25. Jensen F (2006) J Chem Theory Comput 2:1360
26. (a) Bernheim RA, Batiz-Hernandez H (1964) J Chem Phys 40:3446 (b) Alei Jr M, Florin AE, Litchman WM, O'Brien JF (1971) J Phys Chem 75:932 (c) Litchman WM, Alei Jr M, Florin AE (1969) J Chem Phys 50:1897 (d) Jameson CJ, Jameson AK, Cohen SM, Parker H, Oppusunggu D, Burrel PM, Wille S (1981) J Chem Phys 74:1608
27. Gester RM, Georg HC, Canuto S, Caputo MC, Provasi PF (2009) J Phys Chem A 113:14936
28. Moriarty NW, Karlström G (1997) J Chem Phys 106:6470
29. (a) Ichikawa K, Kameda Y, Yamaguchi T, Wakita H, Misawa M (1991) Mol Phys 73:79 (b) Lide DR (ed) (2007) CRC handbook of chemistry and physics, 87th edn. Taylor and Francis, Boca Raton (c) Benedict WS, Gailar N, Plyler EK (1956) J Chem Phys 24:1139
30. Chalmet S, Ruiz-Lopez MS (2001) J Chem Phys 115:5220
31. Tu Y, Laaksonen A (2000) Chem Phys Lett 329:283
32. Guedes RC, Coutinho K, Cabral BJC, Canuto S (2003) Chem Phys Lett 369:345
33. Wasylischen RE, Friedrich JO (1987) Can J Chem 65:2238
34. Burnett LJ, Zeltmann AH (1974) J Chem Phys 60:4636
35. Hermida-Ramón JM, Öhrn A, Karlström G (2007) J Phys Chem B 111:11511
36. Godfrey PD, Brown RD, Hunter AN (1997) J Mol Struct 413–414:405
37. Merrick JP (2007) J Phys Chem A 111:11683
38. Foresman JB, Frisch AE (1996) Exploring chemistry with electronic structure methods, 2nd edn. Gaussian Inc., Pittsburgh
39. Hermansson K (1993) J Chem Phys 99:861
40. Benedict WS, Gailar N, Plyler EK (1956) J Chem Phys 24:1139
41. Ford TA, Falk M (1968) Can J Chem 46:3579
42. Bertie JE, Lan Z (1996) Appl Spectrosc 50:1047
43. Nagaoka M, Yu I, Takayanagi M (2009) In: Leitner DM, Straub JE (eds) Proteins: energy, heat and signal flow, Computation in chemistry. CRC Press, Boca Raton, pp 149–196
44. Kitamura Y, Takenaka N, Koyano Y, Nagaoka M, J Chem Phys, submitted for publication

Chapter 9
A Physics Based Hydrologic Modeling Approach to Simulate Non-point Source Pollution for the Purposes of Calculating TMDLs and Designing Abatement Measures

Nawa Raj Pradhan, Charles W. Downer, and Billy E. Johnson

Abstract Non-point source pollution has become the major source of surface water impairment in the United States. The transport of suspended sediments and nutrients from watersheds to aquatic resources directly influences their environmental quality and ecosystem. The Environmental Protection Agency calculates maximum daily loads from watersheds that allow receiving water bodies to meet water quality standards and mandates load reductions on suspended sediments, nutrients, and other pollutants through the implementation of best management practices. Calculation of these loads and assessment of best management practices is often done with simplified methods, such as spreadsheets and lumped, empirical models that do not account for the spatial heterogeneity or the physical processes occurring in the watershed. The physically based, distributed watershed hydrologic, sediment, constituent transport and fate model Gridded Surface Subsurface Hydrologic Analysis (GSSHA) represents another approach where the spatial heterogeneity, physical and chemical processes in the watershed are explicitly included in the simulation. In this study, GSSHA is used to simulate sediment, nitrogen and phosphorous in Eight-Mile Run, a 264-ha watershed located in the Upper Eau Galle River Basin, west-central Wisconsin. The quality of the GSSHA simulation results demonstrates that the model is capable of quantifying and predicting flow and the transport of sediment and nutrients, nitrogen and phosphorous, in the watershed.

9.1 Introduction

The transport of suspended sediments and nutrients from watersheds to aquatic resources (streams and rivers, lakes and reservoirs, and coastal zones) directly influences their environmental quality and ecosystem. While this is a natural

N.R. Pradhan (✉) • C.W. Downer • B.E. Johnson
U.S. Army Engineering Research and Development Center, Vicksburg, MS 39180, USA
e-mail: Nawa.Pradhan@usace.army.mil

J. Leszczynski and M.K. Shukla (eds.), *Practical Aspects of Computational Chemistry III*, DOI 10.1007/978-1-4899-7445-7_9,

process, excessive inputs from anthropogenic sources (e.g., intensive agriculture) can exceed the assimilative capacity and resilience of aquatic ecosystems. The result is increased turbidity and accelerated rates of eutrophication and destabilized ecological communities susceptible to the establishment of invasive species or to changes in food web structure, loss of valuable habitat, threats to human health, and reduced economic and societal benefits. As point sources of pollutants have become better regulated, non-point source pollution has become the major source of surface water impairment in the United States.

As a result, the National Environmental Policy Act (NEPA) requires all federal agencies to evaluate the environmental implications of their plans, policies, programs, and projects. In addition, Clean Water Act (CWA) regulations concerning water quality and effluent standards have grown exponentially in the past 20 years and require an increased capacity to understand and predict the effects of land use change on our water bodies in order to design best management practices (BMPs) to mitigate their effects.

Under the CWA, receiving water bodies, such as rivers, lakes, and estuaries, have designated uses and corresponding water quality requirements to ensure the receiving water body is suitable for the designated use. Numerical models are used to simulate the receiving body water quality and determine the maximum contribution, or load, of contaminants to the system that will allow the receiving water body to meet its designated uses, expressed as the Total Maximum Daily Load (TMDL). The reduction in load necessary from the contributing watershed may be determined from the existing load and the load that allows the water body to meet its designated uses, the TMDL. Analysis of the contributing watershed is then used to determine the distribution of sources of the watershed and determine appropriate abatement actions through the implementation of BMPs. Total suspended sediment (TSS), and nutrients, nitrogen (N) and phosphorous (P), are typically the targeted pollutants of concern.

The simulation of TSS, N, and P in receiving water bodies typically includes the complex physical and chemical processes recognized as important in the simulation of these NPS pollutants. In the TMDL process, simulations are typically highly simplified, when simulations are conducted at all. Estimation of treatment effects of BMPs are commonly determined from simple spreadsheet calculations where the flows and loads are computed from simple empirical relationships and treatment efficiencies are assigned based on BMP type and/or size. States, such as Virginia, have developed their own spreadsheet to do these computations http://chesapeakestormwater.net/training-library/state-specific-resources/virginia/. The EPA is in the process of developing a similar tool, SUSTAIN, which focuses on providing a user friendly interface to allow easy calculation of loads, treatment by BMPs and optimization of BMPs based on similar methodologies http://www.epa.gov/nrmrl/wswrd/wq/models/sustain/. These types of approaches are meant to ease the burden of computing loads and estimating load reduction through BMPs, as opposed to rigorous analysis. The SUSTAIN model is built on portions of the Hydrological Simulation Program FORTRAN (HSPF) model. HSPF is a lumped parameter, empirical model used for hydrologic, sediment transport, and water quality calculations, and along with the Surface Water Analytical Tool (SWAT) model, it is the most commonly used model for TMDL calculations. While including many of the important components relevant to flow, sediment, and constituent transport, SWAT, HSPF, and similar models, use simplified, empirical relationships, in-lieu of actually simulating the physical and chemical processes that are occurring. When using such a lumped modeling approach, the modeler loses spatial heterogeneity important to the allocation of loads, as well as the assignment of parameters. Utilizing empirical formulas also results in the need to tune the current configuration and conditions during the calibration period. As these parameters include significant implicit information based on the empirical equations and the lumping of information, they may not, and most likely do not, apply to changed physical configurations or meteorological conditions. The process of modifying empirical coefficients to include new information is cumbersome, typically unscientific, and has the common effect of producing non-intuitive results.

9.2 Purpose

The purpose of this chapter is to describe a physics based method for the simulation of watershed non-point source pollution load calculation and abatement. The methods described are suitable for the simulation of non-point source pollution of surface water from sources arising from overland, soils, groundwater, or from point sources for the purposes of calculating TMDLs and abatement with BMPs.

9.3 Background

9.3.1 Regulatory Environment

Two main Federal regulations have driven the TMDL process, NEPA and the CWA.

9.3.1.1 National Environmental Policy Act (NEPA)

NEPA, signed into law by President Richard Nixon on January 1, 1970, acknowledges the decades of environmental neglect that had significantly degraded the nation's landscape and damaged the human environment. The law was established to foster and promote the general welfare, to create and maintain conditions under which man and nature can exist in productive harmony, and fulfill the social, economic, and other requirements of present and future generations of Americans.

NEPA advanced an interdisciplinary approach to Federal project planning and decision-making through environmental impact assessment. This approach requires Federal officials to consider environmental values alongside the technical and economic considerations that are inherent factors in Federal decision making. Environmental impact assessment also calls for the evaluation of reasonable alternatives to a proposed Federal action, solicitation of input from organizations and individuals that could potentially be affected, and the unbiased presentation of direct, indirect, and cumulative environmental impacts. This information is used by a Federal official before a decision is made. Doing so results in informed, and ultimately, improved Federal decision making. Council on Environmental Quality, CEQ, regulations (40 C.F.R. Parts 1500–1508) set the standard for NEPA compliance. EPA also requires agencies to create their own NEPA implementing procedures. These procedures must meet the CEQ standard while reflecting each agency's unique mandate and mission.

9.3.1.2 Clean Water Act (CWA)

The Clean Water Act (CWA) establishes the basic structure for regulating discharges of pollutants into the waters of the United States and regulating quality standards for surface waters. The basis of the CWA was enacted in 1948 and was called the Federal Water Pollution Control Act, but the act was significantly reorganized and expanded in 1972. "Clean Water Act" became the act's common name with amendments in 1972.

Under the CWA, EPA has implemented pollution control programs such as setting wastewater standards for industry. EPA also sets water quality standards for all contaminants in surface waters. The CWA made it unlawful to discharge any pollutant from a point source into navigable waters, unless a permit was obtained. EPA's National Pollutant Discharge Elimination System (NPDES) permit program controls discharges.

To address issues related to non-point source pollution, the Environmental Protection Agency has implemented a process that establishes TMDLs for contaminants of concerns within individual watersheds and enforces these TMDLs on NPDES permits. NPDES permit holders are forced to reduce their non-point contribution through abatement measures called BMPs. BMPs may take many forms, such as actions, no-till farming, or structural, such as installation of storm water retention basins. In any case, the computation of TMDLs and assessment of potential BMPs is typically done with simple techniques, such as hand calculations, spreadsheets, or simple empirical based computer based modeling techniques.

9.3.2 Total Suspended Sediments and Associated Nutrients Dynamics

TSS and nutrients, N and P, are the most common NPS pollution contaminants of concern. TSS, N, and P interact in the soil, on the overland, and in streams, reservoirs and estuaries. The ability to understand and simulate these interactions is key to calculating TMDLs and determining appropriate abatement actions. This section

Table 9.1 List of typical aquatic water quality state variables

No.	Variable name	Symbol	Unit
1	Algae	A_p	[mg-A/L]
	Chlorophyll-a	Chl_p	[ug-Chl/L]
2	Organic Nitrogen	*orgN*	[mg-N/L]
3	Ammonia Nitrogen	*NH4*	[mg-N/L]
4	Nitrate-Nitrite Nitrogen	*NO3*	[mg-N/L]
5	Organic Phosphorous	*orgP*	[mg-P/L]
6	Total Inorganic Phosphorous	*TIP*	[mg-P/L]
7	Carbonaceous BOD	*CBOD*	[mg-O_2/L]
8	Dissolved Oxygen	*DO*	[mg-O_2/L]

describes the general dynamics and simulation considerations of TSS and nutrients in the environment.

9.3.2.1 Aquatic Simulations

A general description of biogeochemical simulation of nutrient dynamics in the aquatic environment is described by [1]. Receiving water aquatic nutrient simulation models are used for waste load allocations, discharge-permit allocations, and other pollution evaluations. Most aquatic simulation models have the capability of simulating water quality variations in a number of state variables such as those listed in Table 9.1.

Models typically include sediment processes, but only as a sink for substances (for example a model may include a settling rate for carbonaceous biochemical oxygen demand, CBOD, but not a re-suspension rate) or as a source of oxygen demand. Figure 9.1 shows a general schematic for aquatic water quality state variables and their interactions. The basic kinetic formulations are developed for computing dissolved oxygen (DO). The major source of dissolved oxygen is algal photosynthesis and atmospheric reaeration. The nitrogen cycle is composed of four compartments: organic nitrogen, ammonia nitrogen, nitrite nitrogen, and nitrate nitrogen. The phosphorus cycle consists of dissolved phosphorus and organic phosphorus. Ultimate carbonaceous biochemical oxygen demand (CBOD) is generally modeled as a first order degradation process.

The symbols shown in are further defined in Table 9.1.

9.3.2.2 Reaction Rates and Physical Constants

Each of the arrows in Fig. 9.1 represents a kinetic process that proceeds at a rate which is controlled by the user through their input of kinetic data. The rate coefficients are all specified by the user for the value at 20 °C and where appropriate the model takes account of the effect of temperature on their value. The kinetic process descriptions require the specification of a number of constants and coefficients related to biogeochemical reactions. It is common practice to use model

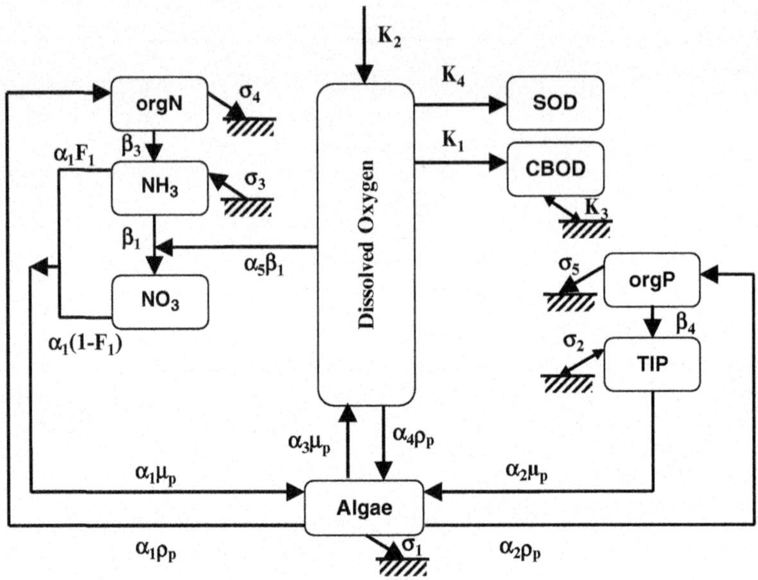

Fig. 9.1 Water quality state variables and interactions in the aquatic environment

coefficients and rate parameters determined from other studies with final selection of the values should be made during model calibration and validation.

9.3.2.3 Overland Flow Nutrient Simulation

While overland flow nutrient simulation can be calculated using the same approach used for aquatic systems with nutrient transformation processes estimated for both particulate and dissolved nitrogen (N) and phosphorous (P) most watershed models do not take this approach since overland flow is typically brief, and chemical decay and transformation may be limited. In addition, techniques used for computing the overland contribution to the receiving water body may compute flow and sediment empirically from large areas without any information about flow path and only crude approximations of transport time. In this context, loads are computed from empirical, or simple first order reactions, without regard to phase, and load reductions due to BMPs are computed from simple treatment efficiencies.

9.3.2.4 Soil Nutrient Simulation

In the simulation of constituents in soil, the modeled subsurface soil is the region from the soil surface to the ground water table [2]. Soil nutrient simulation models are designed to estimate the size of various N and P pools, N and P gains, N and P losses to the environment, and to characterize the interactions among multimedia and multi-pathway processes, Figs. 9.2 and 9.3.

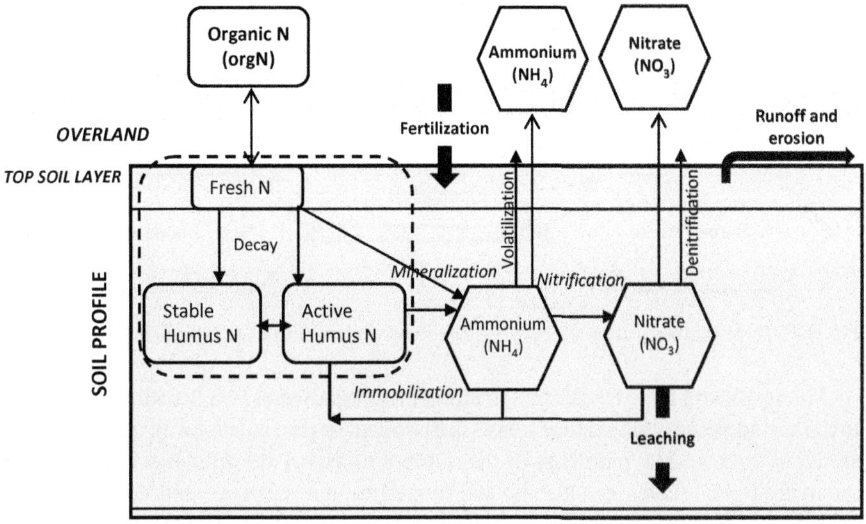

Fig. 9.2 Soil nitrogen cycle and interconnections with nitrogen species in overland flow

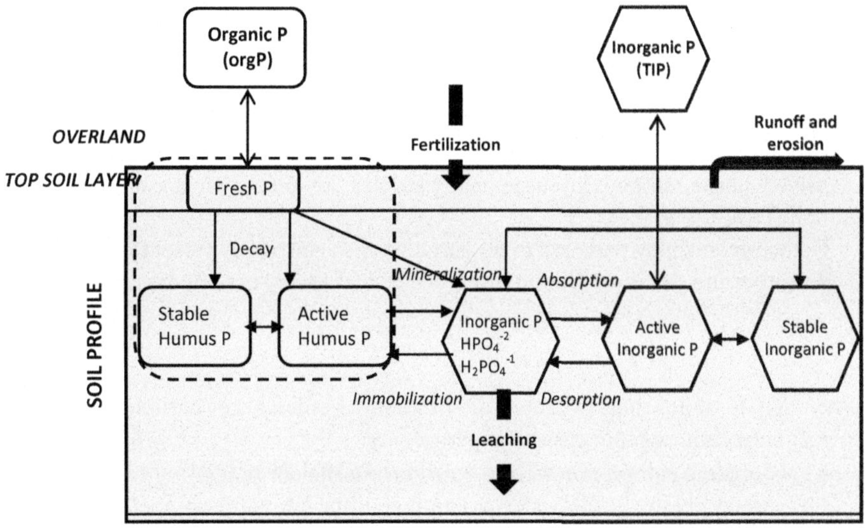

Fig. 9.3 Soil phosphorus cycle and interconnections with phosphorus species in overland flow

Typically one divides the soil profile into multiple layers to simulate N and P dynamics within each layer. The upper soil layer is the only layer that interacts with the overland flow. Soil simulations perform all calculations on a mass basis in the form of storage (pool) accounting procedures.

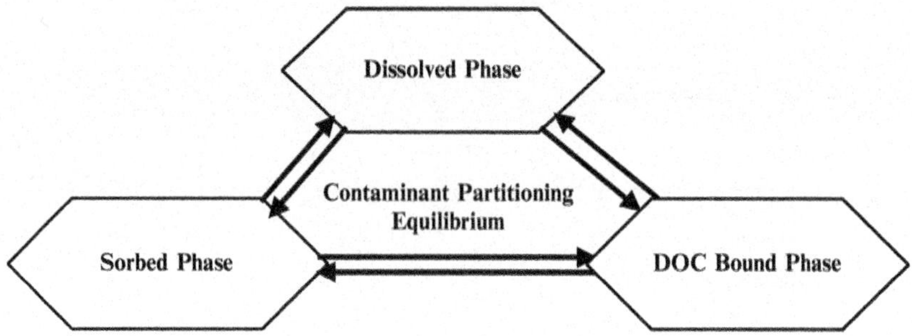

Fig. 9.4 Three phase partitioning between dissolved, sorbed, and DOC bound phases

The soil N and P are transferred to either the groundwater (via leaching) or directly to the surface waters via surface runoff in dissolved or particulate form, or temporarily stored in the soil. The pathways of the nutrient transport are primarily influenced by the hydrological fluxes, but also by soil properties, nutrient availability, and climatic factors such as temperature. The rate of nutrient dynamic processes or transformations rate is described using zero-order, first-order, second-order and Michaelis–Menten kinetics taking into account the effect of soil water content and temperature.

9.3.2.5 Constituent Partitioning

When performing constituent modeling, it is often necessary to partition the contaminant of concern into different phases. Constituents can be found in the dissolved phase, sorbed to suspended particles, or bound to dissolved organic carbon (DOC), Fig. 9.4.

Contaminants may partition to all particle types (sorbents) present in a solution. The equilibrium partition (distribution) coefficient to any particle is defined as [3]:

$$K_{p_n} = f_{oc_n} K_{oc} \tag{9.1}$$

where K_{p_n} is equilibrium partition (distribution) coefficient for particle "n" [L^3/M]

f_{oc_n} is fraction organic carbon of particle "n"

K_{oc} is organic carbon normalized partition coefficient [L^3/M].

For sorbed phases in the water column, equilibrium partition coefficients vary with the concentration of suspended solids as a result of particle interactions, Fig. 9.5. Particle-dependent partition coefficients are described as [4]:

$$\mathrm{K}_{px_n} = \frac{\mathrm{K}_{p_n}}{1 + m_n \mathrm{K}_{p_n}/\nu_x} = \frac{f_{oc_n} K_{oc}}{1 + m_n f_{oc_n} K_{oc}/\nu_x} \tag{9.2}$$

where K_{px_n} is particle-dependent partition coefficient [L^3/M]

n is particle index = 1, 2, 3, etc.

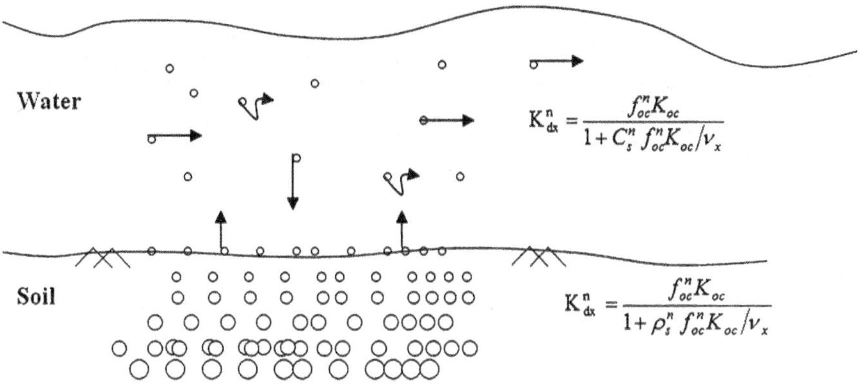

Fig. 9.5 Partitioning between the water column and the bed sediments

m_n is concentration of particle "n" [M/L^3]
ν_x is particle interaction parameter.
For the bound phase, the equilibrium binding coefficient is defined as:

$$\mathrm{K}_b = D_e f_{oc_D} K_{oc} \tag{9.3}$$

where K_b is equilibrium binding coefficient [L^3/M]
f_{oc_D} is fraction organic carbon of DOC
D_e is DOC-binding effectiveness coefficient.
The equilibrium partition coefficient can be used to describe the fraction of the total contaminant that is associated with each phase as follows [3, 5]:

$$f_d = \frac{1}{1 + D_{oc}\mathrm{K}_b + \sum_{n=1}^{N} m_n \mathrm{K}_{px_n}} \tag{9.4}$$

$$f_b = \frac{D_{oc}\mathrm{K}_b}{1 + D_{oc}\mathrm{K}_b + \sum_{n=1}^{N} m_n \mathrm{K}_{px_n}} \tag{9.5}$$

$$f_{p_n} = \frac{m_n \mathrm{K}_{px_n}}{1 + D_{oc}\mathrm{K}_b + \sum_{n=1}^{N} m_n \mathrm{K}_{px_n}} \tag{9.6}$$

$$f_d + f_b + \sum_{n=1}^{N} f_{p_n} = 1 \tag{9.7}$$

where f_d is fraction of the total contaminant in the dissolved phase
f_b is fraction of the total contaminant in the DOC-bound phase

f_{p_n} is fraction of the total contaminant in the sorbed phase associated with particle "*n*"

D_{oc} is dissolved organic carbon concentration [M/L^3].

Equations 9.4, 9.5, 9.6, and 9.7 are presented for the water column. For the soils or bed sediments, the fractions associated with dissolved, DOC, and sorbed, respectively, are derived by using porosity. K_{p_n} is used in place of K_{px_n}, Fig. 9.5.

$$f_d = \frac{1}{\phi + \phi D_{oc} K_b + \sum_{1}^{N} (1 - \phi) \rho_{sn} K_{px_n}} \tag{9.8}$$

$$f_b = \frac{D_{oc} K_b}{\phi + \phi D_{oc} K_b + \sum_{1}^{N} (1 - \phi) \rho_{sn} K_{px_n}} \tag{9.9}$$

$$f_{p_n} = \frac{(1 - \phi) \rho_s K_{p_n}}{\phi + \phi D_{oc} K_b + \sum_{1}^{N} (1 - \phi) \rho_{sn} K_{px_n}} \tag{9.10}$$

where ϕ is porosity of the soils or bed sediments

ρ_{sn} is density of particle "*n*" in the soils or bed sediments [M/L^3].

9.4 Study Area

GSSHA is applied to Eight Mile run a 264-ha sub-watershed located in the Upper Eau Galle River Basin (Fig. 9.6).

The Eight Mile Run watershed is largely agricultural with livestock (dairy) pasture and associated barnyards representing approximately six (6) percent of the watershed, Fig. 9.7. These high TSS, TP, and TN source areas are located immediately adjacent to the tributary in the lower portion of the watershed. The remaining agricultural areas, 49 % of the watershed, are used for alfalfa and corn production. Grass, conservation reserve program (CRP) land, and wooded land uses occupy the remaining 45 % of the watershed. The watershed is bisected into approximately equal areas by a railroad and U.S. Highway Route 12 with the area to the north being primarily natural CRP and wooded areas, 69 % and the area to the south being primarily active agricultural areas, livestock pastures and crops, 61 %. The watershed was selected because of the varied land uses, presence and location of a high source area for NPSP and because of prior work conducted at the site by [6]. Data for rainfall, flow, TSS, N, and P, as well as field measured soil partition coefficients for soluble reactive phosphorous (SRP) where all available at this site.

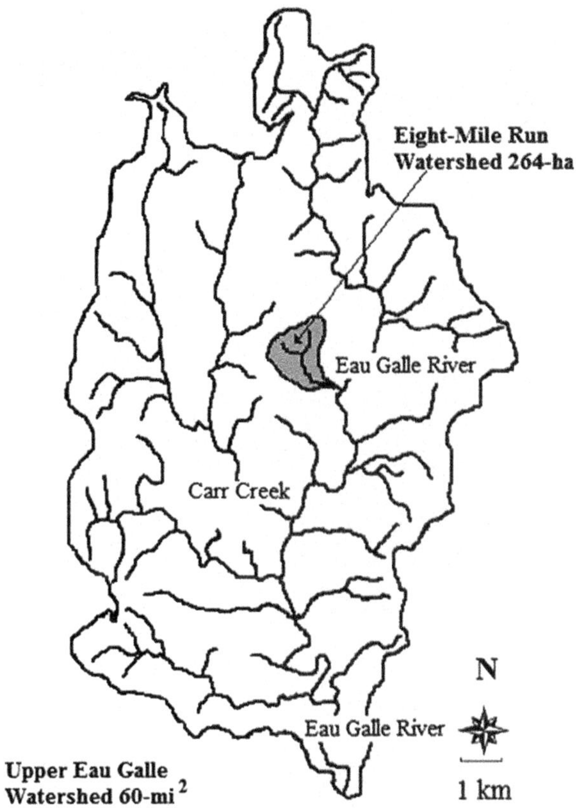

Fig. 9.6 Eight-Mile Run watershed is a sub-watershed of Upper Eau Galle Watershed in northwest Wisconsin

9.5 Methodology

The Gridded Surface Subsurface Hydrologic Analysis (GSSHA) [7, 8] model is used in this study to simulate flow, sediment, nitrogen and phosphorous in the Eight Mile Run watershed and to show how GSSHA solves the major restrictions in nutrients modeling outlined above.

Non-Point Source Pollution (NPSP) modeling in GSSHA involves being able to link overland flow and soil layer processes with aquatic (channel) processes in order to trace the fate and transport of constituents of concern from their source to a decision point downstream. The GSSHA formulation and implementation methodology is presented in this section:

- GSSHA model description
- Key processes included for the GSSHA numerical simulations
- Parameterization

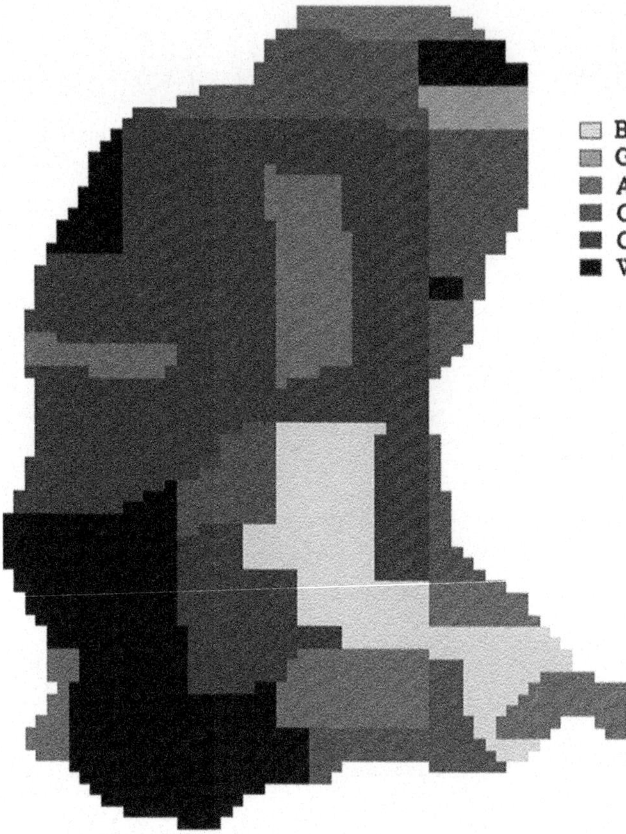

Fig. 9.7 Eight-Mile Run watershed Landuse map

9.5.1 GSSHA Hydrological Model

GSSHA is a physically-based, distributed-parameter, structured grid, hydrologic model that simulates the hydrologic response of a watershed subject to given hydrometeorological inputs. The model is intended to improve simulations of hydrologic, sediments, and nutrient/contaminant transport problems where flow path is important. The GSSHA model is formulated on a structured grid, and uses explicit finite-volume solutions of the diffusive-wave form of the de St-Venant equations of motion for two-dimensional overland and one-dimensional channel flow. Overland flow depth is modified each timestep with the addition of rainfall and subtraction of infiltration, and updated to account for overland fluxes. Flow is then routed in the overland flow plane from regions of higher hydraulic head to those of lower hydraulic head. Infiltration can be simulated in GSSHA using a variety of optional methods including: multilayer Green and Ampt [8], Green and

Ampt with redistribution [9], and the Richards equation [10]. Groundwater can be simulated using a 2-D finite difference scheme [7], allowing simulations of saturation excess overland flow and return flow. The groundwater solution is fully coupled to surface flows using a 1-D implicit finite difference solution of Richards equation [10]. The groundwater solution can also be coupled to surface flows using the Green and Ampt methods, as is done in the example below, as described in [11]. The GSSHA model can operate on individual events, or continuously with calculated evapotranspiration (ET) and soil-moisture accounting between rainfall events [11]. GSSHA also has erosion/deposition simulation capabilities and can be used to calculate the transport of sediments, nutrients, and other aquatic constituents. These processes will be discussed in more detail below. The GSSHA model has been extensively published and the model functioning and numerical solutions are described in detail in several references including [7, 12] and the GSSHA WIKI (www.gsshawiki.com accessed Oct 29, 2013). GSSHA is developed and maintained at the US Army Engineer Research and Development Center.

9.5.1.1 Hydrologic Simulations

GSSHA works on a defined area, typically a watershed bounded by surface water divides. For simulations that include groundwater the study domain may be expanded to incorporate suitable specified head boundary conditions, such as surface water sources or groundwater wells. The study area is divided into square grid cells that comprise a uniform finite difference/finite volume grid. Grids are sized to capture the important spatial scales of the problem and range from 1 to 1,000 m. Because of numerical complexity and computing limitation, GSSHA studies have typically been limited in scope to models with 10^6 nodes, or study areas of about 10^3 km^2, although a MPI parallel version of the model [13] holds promise for larger studies at finer resolution. Both overland flow and groundwater processes are computed on the same grid. Channel routing computations are performed on an independent one dimensional (1-D) grid [14] with typical stream nodes ranging in size from 30 to 200 m.

Processes that occur before, during, and after a precipitation event are calculated for each grid cell and then the responses from individual grid cells are integrated to produce the watershed response. All processes, both point and integrated, are represented by first principle equations, i.e. conversation of mass, momentum, and energy. Many significant simplifications are made to the basic conservation equations as implemented in the *GSSHA* model to make the model both computationally efficient and robust. Point processes include precipitation distribution, snow accumulation and melt, precipitation interception by plants, infiltration, ET, and microtopographical retention. Point process solutions in *GSSHA* are not iterative; they are calculated based on the most current boundary conditions available at the update time. Integrated processes include surface runoff routing, channel flow routing, vadose zone flow, and saturated groundwater flow. A number of watershed structural features are also simulated in the model, including: lakes, reservoirs, wetlands, detention basins, detailed hydraulic structures [14], storm and tile drains [15, 16], frozen soil and permafrost [17] and the influence of roadways, berms, and dykes (GSSHA wiki). Point processes occur at the grid cell level for every cell in the active domain. Processes are updated sequentially; point processes are updated first and the updated cell values are used to simulate integrated processes. Integrated processes are coupled with boundary conditions and fluxes with two way fluxes possible across the vadose zone, overland flow, channel and groundwater domains. The model proceeds at a user specified overall timestep, typically 10–60 s. Individual processes are updated as required. Most processes have internal time step limitations to ensure stability and accuracy.

9.5.1.2 Sediment Simulation

GSSHA simulates erosion and transport of sediments on the overland flow plain and within the 1-D stream channel with flux exchanges between the two domains. Soils on the overland flow plane consist of a user specified number of sediments with specified diameters, specific gravities, and fractional distribution in each overland flow cell. Within a grid cell, sediments are accounted for in three separate storages: parent soil material, suspended sediments, and deposited materials. During a rainfall event, sediment particles are detached by rainfall and overland flow. If available, deposited materials are eroded first. If there are no deposited sediments, or the amount of deposited sediments is less than the transport demand, then parent soil materials are eroded at a rate determined by detachment limits. The need for detachment sets an upper limit on the amount of sediments that can be eroded from a given cell. Transport of sediments on the overland flow plane is constrained by the ability of the water to carry sediments. If the transport capacity is greater than the available materials, then all the suspended materials are transported to downstream cells by advection, using a two-dimensional sequential explicit method, as described below. If the transport capacity cannot accommodate all the suspended materials, then some materials may be deposited according to the calculated trap efficiency. Deposited materials are added to the deposition layer in that cell and are eroded first in subsequent time intervals. Materials not deposited or transported remain in suspension, and may be transported or deposited in subsequent time steps. To save computational time, any suspended sediments remaining in suspension at the end of a specified precipitation event are deposited onto the overland flow cell containing the suspended sediments and sediment computations cease until the next specified event. When constituent transport is simulated the constituents in the soil can be sorbed to soil sizes smaller than sand. These constituents are eroded, transported, deposited and exchanged with the dissolved component along with the sediments.

Upon reaching a model grid cell that contains a stream all suspended and deposited materials in that overland flow cell are added to the stream node. Once in the stream network, particles are transported as either suspended or bed load depending on whether the particles' diameter are larger or smaller than the user-specified value of sand. A unit stream power method of [18] is used to route sand-size total-load in stream channels. Particles smaller than sand size are transported using the advection-dispersion equation as described by [19]. Suspended sediments do not settle in fluvial stream links or structures but may settle in reservoirs, depending on their calculated fall rate and reservoir properties. These particles also may interact with dissolved constituents in the stream network. Only suspended sediments interact with constituents. The bed load sediments are assumed to be quartz and considered inert. Details of the sediment transport methods are given in [20].

9.5.1.3 Transport Processes

In GSSHA the general transport equations are solved in two dimensions on the overland flow plane and in one dimension in the channel network, using a common methodology. The general transport equations describe the fate and transport of suspended soil particles, dissolved constituents and constituents sorbed to suspended soil particles. Each cell is treated as a completely mixed reactor and the concentration in each cell is affected by internal sources and sinks, as well as by advection into and out of the cell and diffusion between surrounding cells. Full details of the transport methods in GSSHA are presented in [19, 21]. Downer and Byrd [19] describe the general transport equations and solutions used for both sediments and constituent transport. Downer [21] describes the details of fate and transport of reactive constituents. Some of the information in these referenced has changed and is updated here. One significant change is that GSSHA now supports transport of both dissolved and sorbed constituents with transfer between the two computed each time interval. Suspended sediment particles, as well as dissolved constituents and constituents sorbed to sediments are simulated with the general methodology described below. The DOC bound phase and transport are not simulated in GSSHA. For the proceeding discussion, the description of nutrient in environment in Sect. 9.2 applies with the exception that the DOC bound phase is not simulated in GSSHA. All the equations in Sect. 9.2 also apply, ignoring the DOC bound component in these equations. In GSSHA, the DOC bound fraction will be distributed between the dissolved and sorbed to sediment fractions.

Precipitation falling on the watershed may contain specified concentrations of contaminants. As rainfall accumulates on the land surface, contaminants on the land surface are dissolved and/or contaminants in the soil column diffuse into the overlying surface waters. Ponded surface water may infiltrate, providing a source of contaminants to the soil column, and/or move as surface runoff to adjacent cells. Additionally, soil particles may erode and be entrained into the water column. Constituents sorbed to these particles may be exchanged with the dissolved component. As water flows across the land surface, it is continuously affected by contaminated precipitation, exchange with contaminants on the land surface or in the soils, infiltration, and mixing with constituents in adjacent cells. Concentrations are also affected by decay. Because overland flow typically occurs for only brief periods of time, decay is generally minor compared to the physical processes affecting the constituents.

As overland flow reaches a stream channel, water, sediment, and dissolved and sorbed contaminants are deposited into the channel and begin to move downstream, mixing with other water, suspended sediments, and dissolved and sorbed constituents already in the channel, and with water, suspended sediments, and constituents entering the channel at downstream locations. Additional point sources may enter the channel at any point. Interaction with the groundwater may represent either a source or a sink. As well as mixing with other sources, constituents in the channel also decay. Decay is generally more important in the stream network than on the overland flow plane. If the channel network includes reservoirs [14], significant decay can be expected to occur due to the long residence times in reservoirs.

At the end of a rainfall event, most or all, of the overland flow plane will be dry. Any standing water will continue to infiltrate and evaporate. Constituent reactions will continue on the overland flow plane in any cells with remaining water. Suspended sediments, and any contaminants sorbed to them, are deposited to the soil depositional layer. Baseflow may occur in channels. Groundwater provides a source of constituents to the channel if baseflow occurs. Reactions continue in the channel network as long as water is in the stream network. If reservoirs are included in the stream network, reactions continue in the reservoirs. In between events, constituents in the soil column may drain into the lower portions of the column and back to the groundwater. As the soil column dries due to drainage and evapotranspiration, the fraction of contaminants dissolved in the pore water versus that attached to the soil matrix changes. Reactions also continue for constituents in the pore water.

Although GSSHA is a fully integrated hydrologic model, water and associated constituents are associated with distinct domains which pass between domains as fluxes. For the current version of GSSHA (6.1), available at www.gsshawiki.com, the domains including or affecting constituent transport are as follows:

Overland flow
Stream network
Reservoirs
Surface soils
Saturated groundwater

Currently the focus of the GSSHA model is the surface water transport of contaminants. Constituent can be tracked in both the dissolved and sorbed phase. When particles are tracked on the sorbed phase the sorbed and dissolved transport are computed in three steps.

- First, computed flows are used to track the erosion, transport, and deposition of soil particles.
- Next, particles sorbed to the eroded particles are tracked, followed by computation of the dissolved phase of any contaminants.
- Exchange between sorbed and dissolved is computed assuming equilibrium kinetics at all points in the process, in the soils, after entrainment into surface waters, after transport, and after deposition. For the suspended sediment in the

water column, the distribution between sorbed and dissolved is based on Eqs. 9.4, 9.5 and 9.6, ignoring the DOC bound phase. The DOC component is not simulated in GSSHA.

- The net exchange between sorbed and dissolved for each overland and channel computational node is represented in the other domain as a source or a sink.

Overland Flow Transport

As described in [19], transport on the overland flow plane is simulated by solving the advection diffusive equation (AD). Advection is the cell to cell movement of contaminants associated with water flow. Strictly speaking, diffusion is the molecular exchange that occurs due to concentration gradients. Practically speaking, the actual diffusion process is small, and the diffusion term is really a mixing term including several processes: physical dispersion, numerical dispersion, and molecular diffusion. The AD equation includes a source term to account for the various additions and subtraction of constituents as they move across the overland flow plane.

$$\frac{\partial(M)}{\partial t}+\frac{\partial(uM)}{\partial x}+\frac{\partial(vM)}{\partial y}-\frac{\partial}{\partial x}\left(D_x\frac{\partial M}{\partial x}\right)-\frac{\partial}{\partial y}\left(D_y\frac{\partial M}{\partial y}\right)+KM=S \quad (9.11)$$

where: C is concentration (M L^{-3}),

M is the mass (M)

D_x is the diffusion coefficient in the x direction for the constituent of concern (L^2 T^{-1}),

D_y is the diffusion coefficient in the y direction for the constituent of concern (L^2 T^{-1}),

K is the decay coefficient (T^{-1}),

u is the flow velocity in the x direction (M T^{-1}),

v is the velocity in the y direction (m s^{-1}), and

S is the term for all sources and sinks (M T^{-1}).

Other than reactions, as described in the following paragraphs, the sources and sinks considered are as follows:

- Addition by rainfall
- Uptake from the soil surface
- Exchange with soil
- Infiltration
- Exchange with channels
- Exchange with groundwater
- Addition by point sources
- Exchange with reservoirs
- Exchange with sorbed materials

Rainfall. The concentration (g m^{-3}) of each constituent being simulated is constant throughout the simulation. Rainfall inputs are added directly to the overland flow plane, where they may either infiltrate, or pond and produce contaminated runoff.

Soil. Contaminants distributed on the overland flow plane at the beginning of the simulation may be mixed into the soil column. When simulating constituents in the soil column, the mass of contaminants is distributed over a specified mixing depth in the soil column, typically on the order of 10 cm. Constituents in the soil partition between the soil matrix and the pore water are based on the chemical properties, the soil properties, and the soil moisture, Eqs. 9.8, 9.9, and 9.10 ignoring the DOC component. Constituent mass transfer into water ponded on the overland flow plane occurs due to the mass transfer coefficient (K) and the concentration difference between the soil pore water volume and the overland flow plane. As the concentration gradient may be in either direction, the flux may also be in either direction, i.e., the dispersive flux may be into the soil, acting as a sink for the overland plane. How constituents are treated in the soil column is discussed in detail later.

Infiltration. Some or all of the water ponded on the land surface may infiltrate, removing contaminants. Water that infiltrates is assumed to contain the same concentration of dissolved contaminants as the ponded water.

Channels. In general, the overland flow plane acts as the primary source of contaminants to the channel, which is a sink for the overland flow plane. For cases where overbank flooding occurs, the stream may overflow and add water, as well as constituents, back to the overland flow plane.

Groundwater. The overland flow plane interacts with the groundwater in two possible ways. If the groundwater table is high enough, water may spill out onto the overland flow plane as exfiltration. Exfiltrated water moves through the soil column and the concentration of the exfiltrated water will be calculated as part of the soil constituent transport routine. Constituents seeping out of the soil column into the groundwater are accounted for but do not affect the static groundwater concentration for the cell.

Point Sources. Point sources may be input into any cell in the watershed. Point sources are defined by discharge rate, Q ($m^3\ s^{-1}$), and concentration, C (g m^{-3}), for each point source.

Reservoirs. Reservoirs are dynamic features that exist in both the channel network and overland flow plane [14]. Water and constituents may be lost to a reservoir by either flowing into the reservoir, or by the reservoir rising and taking over the overland flow cell. Water and constituents may also flow from the reservoir back onto the overland flow plane. This results in a source for the overland cells adjacent to the rising reservoir.

Solution. The 2-D overland transport equations are solved with a five-point explicit alternating direction (ADE) scheme with prediction correction (PC), or ADE-PC [22]. This method is 2nd order accurate in space and time. Details of the numerical methods are discussed in [19].

Stream Network Transport

GSSHA computes values of water flow and depth within a user specified 1-D finite volume stream network. This information is needed for the transport of constituents within the channel network using the general 1-D advection-dispersion equation in terms of the mass of constituent (M) equal to the concentration (C) multiplied by the volume (V) with constant dispersion. The details of the equations are described in [19].

$$\frac{\partial(M)}{\partial t}+\frac{\partial(uM)}{\partial x}-\frac{\partial}{\partial x}\left(D\frac{\partial M}{\partial x}\right)+K(M)=S \tag{9.12}$$

where: u is the flow velocity in the x direction (M T^{-1})
C is concentration (M L^{-3}),
Q_x is discharge in the x direction (L^3 T^{-1}),
A is cross sectional area (L^2),
D_x is the diffusion coefficient in the x direction for the constituent of concern (L^2 T^{-1}),
K is the decay coefficient (T^{-1}),
V is the volume, (L^3), and
S is the term for all sources and sinks (M T^{-1}).

This equation is solved with a three-point explicit finite-volume a predictor-corrector scheme for each constituent of concern. The solution proceeds from the most upstream node and proceeds downstream to the channel outlet. In addition to the predictor-corrector scheme, the solution is iterative and proceeds until the maximum difference in concentrations in the channel network over successive iterations is less than 10^{-6} g m^{-3}, or the maximum number of iterations is reached, currently set at 100 iterations. The predictor-corrector step is employed during each iteration. This scheme is 1st order accurate in space and 2nd order accurate in time. For each node the following mass reactions are accounted for.

1. Exchange with overland flow
2. Exchange with reservoirs
3. Exchange with groundwater
4. Point sources
5. Exchange between sorbed and dissolved phases
6. Chemical reactions

Exchange with overland. As described above.

*Exchange with reservoir*s. When interacting with reservoirs water and constituents are lost to the channel in two ways. Water may flow into the reservoir from one or more upstream tributaries. The reservoir may also expand, taking stream nodes or entire reaches. When this occurs, any water and constituents in the overtaken stream node are removed from the channels and added to the reservoir. Discharges from reservoir outlets act as sources to the channel network.

Exchange with groundwater. Channel losses can be simulated whether or not the saturated groundwater surface is included in the simulation. When water seeps into the channel bottom, subsequent loss of dissolved constituents occurs as well. When the water table is included in the solution, as either static or time varying, exchange can be in either direction. Concentration of constituents is specified for every cell in the groundwater domain. This concentration does not vary temporally throughout the simulation. If flow is from the groundwater domain to the channel, water entering the channel is assumed to have the specified groundwater concentrations of constituents. Seepage from the channel to the groundwater is assumed to have the same concentration as the water in the channel node. Additions and subtractions to the groundwater are accounted for but do not affect the specified groundwater concentrations.

Point sources. Point sources may be input into any node in the stream network. Point sources can be defined as a constant or time varying discharge rate, Q ($m^3\ s^{-1}$), and concentration, C ($mg\ L^{-1}$), for each point source.

Exchange between sorbed and dissolved phases. Both dissolved and sorbed phases of constituents are transported in the channels with equilibrium conditions assumed to partition between the two.

Chemical Reactions. Constituents in the dissolved phase can decay with first order reactions based on dissolved phase concentration.

Soil Column Transport

The simple soil moisture accounting routine [11] allows the user to specify up to two soil layers for computation of soil moisture. Within these layers, downward soil water movement is due to gravity. If groundwater is being simulated, the groundwater may rise into the soil column, causing an upward flow of water in the soil column. Soil movement due to capillary pressure is not considered with these methods. Infiltration is a source to the top layer of the soil model. Leakage from the bottom layer is considered a loss. Loss of water, but not constituents, also occurs due to ET, which is taken from both soil moisture layers. If sediment transport and sorbed constituent transport is being simulated, then erosion of particles results in a loss of and deposition results in a gain of constituents from/to the top soil layer.

Figure 9.8 shows the conceptual model of the soil column transport model. Downward fluxes (infiltration, gravity drainage, groundwater recharge) are shown on the left. Upward fluxes (exfiltration, upward groundwater flux) are shown on the right. Diffusive exchange occurs between the top soil layer and the surface water. Exchange between pore water and soil particles occurs in every layer, as does decay and transformations.

Distribution of constituents in soil column. When simulating constituents, a third constituent transport layer may be considered. Specification of this layer further divides the surface soil moisture layer. Any initial amount of constituents

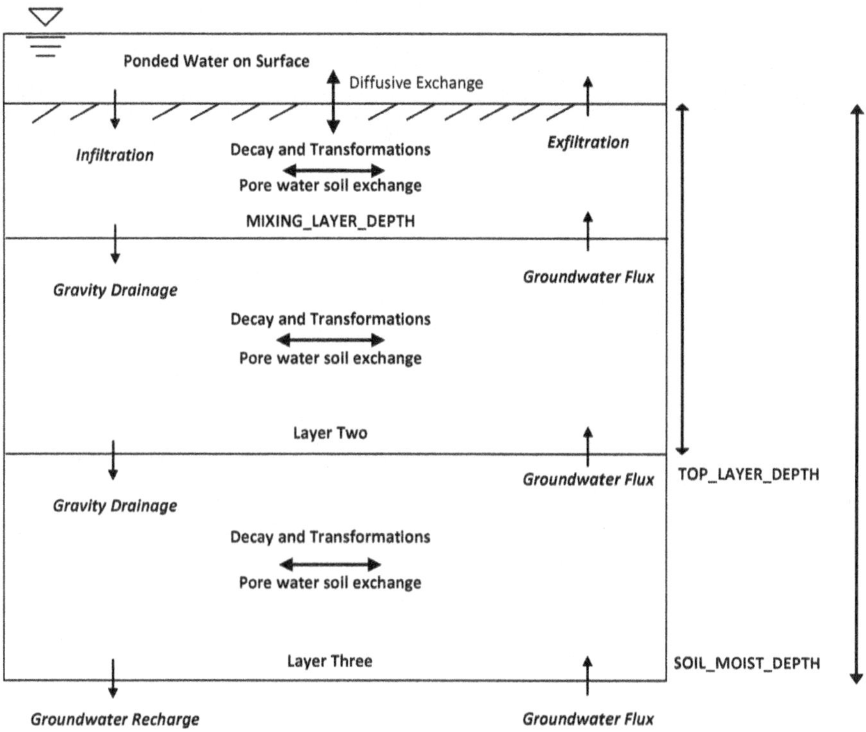

Fig. 9.8 Soil column transport model

distributed on the overland flow plane is assumed to be mixed within this surface mixing layer. If this additional layer is not specified in the project file, the initial amount of contaminants is assumed to be evenly mixed over the top soil layer, and there will be only two layers in the soil transport model. If only one soil moisture layer is specified, then the initial amount of constituents is assumed to be mixed over this single layer, and transport is computed for one soil layer only.

Regardless of the total number of layers, the initial constituent loading is assumed to be mixed over the depth of the top layer. This mass of contaminants is distributed between an amount absorbed to the soil and dissolved in the pore water. The distribution is calculated based on the chemical distribution coefficient (K_d) (m^3 Kg^{-1}) and the soil moisture as described by [23]. For the soils, the distribution between absorbed and dissolved is based on soil moisture (θ), and not porosity, as shown in Eq. 9.13.

$$f_d = \frac{1}{\theta + K_d(1-\theta)\rho_s} \tag{9.13}$$

where: f_d is fraction of the constituent dissolved in the soil layer and ρ_s is dry soil density (Kg m^{-3}). The concentration (g m^3) of constituent in the pore water (C_{pw}) is then

$$C_{pw} = f_d MV \tag{9.14}$$

Where M is the total mass of constituent in the soil layer and V is the volume (m^3) of the soil layer:

$$V = \theta AD \tag{9.15}$$

Where D is the depth of the soil layer (m).

Equation 9.13 is similar to Eq. 9.8 except that Eq. 9.13 is based on soil moisture, and not porosity. Soil moisture is used because on the overland, soils are potentially not saturated. Also, in the GSSHA nutrients simulation in this study, the DOC-bound phase is not simulated. Equilibrium conditions in the soil are described in Eqs. 9.8, 9.9, and 9.10, ignoring the DOC bound phase.

Exchange with surface water. During the simulation, infiltration acts as a source to the top layer. Advection of water transfers water and constituents to lower layers. Leakage from the bottom layer is a sink for that layer. The concentration of constituent in the advected water depends on the mass of contaminant in the layer, the soil moisture, and the distribution coefficient.

As described in [21] mass exchange with water ponded on the land surface occurs due to the concentration gradient between the pore water in the top soil layer and the ponded water.

$$F = \left(C_{ponded} - C_{soil}\right)KA \tag{9.16}$$

As can be seen in the equation, the direction of the flux is dependent of the relationship between the concentration of the surface water to the soil pore water volume. If the water on the land surface has a higher concentration than the soil pore water volume, the flux will be into the soil.

Interaction with groundwater. If the water table is being simulated, the water table may be present in any or all of the soil layers. If the water table is present in a layer, the amount of groundwater in that layer is considered in the calculation of soil moisture in the layer for the purposes of distributing the constituent between dissolved and attached fractions.

If exfiltration occurs, the groundwater is considered to come into the bottom layer, reach equilibrium condition in that layer and then move upward to the next layer, where it reaches equilibrium before being advected upward to the next layer, and ultimately to the land surface. The concentration reaching the land surface may not be the same as that specified for the groundwater in any given grid cell.

Erosion and deposition. If sorbed transport is being simulated, contaminants sorbed to eroded particles are subtracted from the top layer and added due to deposition.

Reservoir Transport

Reservoirs in the stream network are treated separately from the channel network. Each reservoir is considered as a completely mixed reactor. As previously described, reservoirs interact with both the overland flow plane and the channel network. Reservoirs can also interact with the groundwater in the same manner as the channels, where the reservoir water and contaminants can seep to the groundwater, and the groundwater can supply water with static concentrations to reservoirs.

Groundwater Transport

GSSHA does not currently simulate fate and transport in the saturated groundwater. Whenever a water table is simulated in the GSSHA model, a concentration is specified for every grid cell in the watershed. Any flux from the groundwater to any other domain has the static constituent concentration of the groundwater cell that the flux occurs from. Fluxes to and from the groundwater do not affect the groundwater concentrations. This simplified conceptualization of groundwater may not be adequate for simulating conditions where the groundwater exchange is significant and the groundwater concentrations vary with time over the period of the simulation.

9.5.1.4 Kinetic Reactions

As described in [21], all constituent rates and reactions in GSSHA are simulated as first order reactions with specified mass transfer rates from the soil and specified decay rates. The mass decay rate (F) (g s^{-1}) is computed as:

$$F = KCV \tag{9.17}$$

where, K is the decay rate (s^{-1}), C is the concentration (g m^{-3}), and V is the volume (m^3) of water in the computational element.

9.5.2 *Processes and Methods Simulated in the Test Case, Eight Mile Run*

GSSHA is an option driven model with the user selecting both the processes and the solution techniques Table 9.2 lists the key processes taken into account, as a coupled numerical routines in GSSHA, used for simulations of flow, sediment and constituent routing in the Eight Mile Run simulation model.

Table 9.2 Process routines included in Eight-Mile Run numerical model

Process	Representation	Solution
Rainfall distribution	Inverse weighted distance	
Infiltration	Green and Ampt with redistribution	
Evapotranspiration	Penman-Monteith	
Soil moisture	Two layer model	
Overland flow	2-D diffusive wave equation	ADE
Channel flow	1-D diffusive wave equation	Explicit, iterative
Erosion	Kilinc-Richardson	
Overland sediment transport	2-D advection-diffusion	Explicit
Channel sediment transport—fines	1-D advection-diffusion	Explicit
Channel sediment transport—sand	Bedload formula	Yang's method
Constituent loading	Concentration in soils	
Constituent reactions	First order	
Overland transport	2-D advection-diffusion	Explicit, iterative
Channel constituent transport	1-D advection-diffusion	Explicit, iterative
Saturated groundwater	2-D	Implicit scheme

9.5.3 Model Parameterization

Landuse, soil texture data, obtained from satellite imagery, GIS and field survey, were the basis of model parameterization. Parametric values were applied directly inferring physically based values to these maps. Some parameters were left as calibration parameters for which, physically based lower and upper limit are referenced to those maps.

9.5.3.1 Soil Parameters

Initial moisture [24], soil hydraulic properties, and soil particles distributions are needed for GSSHA simulations of waters, sediment, and contaminants. These parameters are assigned based on soil textural classifications. The soils data for the simulations was developed from the Natural Resources Conservation Service (NRCS) STATSGO database. Soils in the watershed consist of sandy and silty loams. The distribution of the soils in the watershed is shown in Fig. 9.9.

9.5.3.2 Landuse Parameters

Land use maps were obtained from field survey and USGS Land Use and Land Cover (LULC) web site at http://edc.usgs.gov/geodata/. Thirty-one individual land use types in the study area were condensed into six basic land uses shown in Fig. 9.7. This land use information was used to derive dominating overland parametric values, soil erosion parametric values, evapotranspiration parametric values and nutrient transport parametric values.

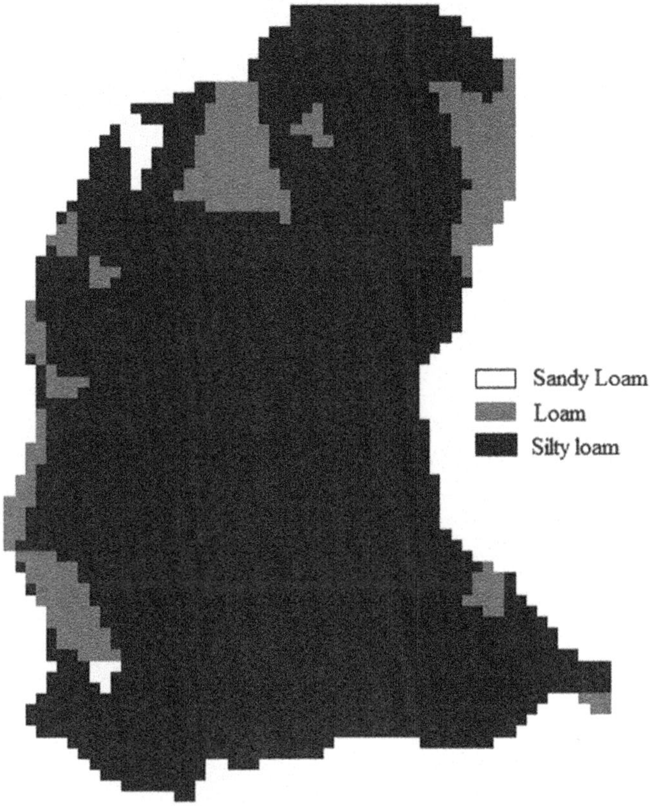

Fig. 9.9 Eight-Mile Run watershed soil map

9.6 Results

Non-point source pollution represents a perfect example of a complex multidisciplinary problem that exists over multiple scales with tremendous spatial and temporal complexity. To address the NPS problem requires the inclusion of soil physics, hydrological processes and sediment erosion and transport process in a watershed level. Solute transports are difficult to infer a priori without a detailed understanding of heterogeneity and its influence on the processes under consideration. This is especially true because the number of interacting processes needs to be considered simultaneously [25] such as in reactive flows where chemical and biological processes interact with hydrological and sediment transport processes. Thus, a model's ability to address the hydrological process and sediment erosion process for acceptable outputs in terms of quality and quantity is required prior to the assessment of the NPS pollution. This section presents the followings:

- the model's simulation response to the included hydrological process,
- the model's simulation response to the combined hydrological and sediment erosion processes, and finally

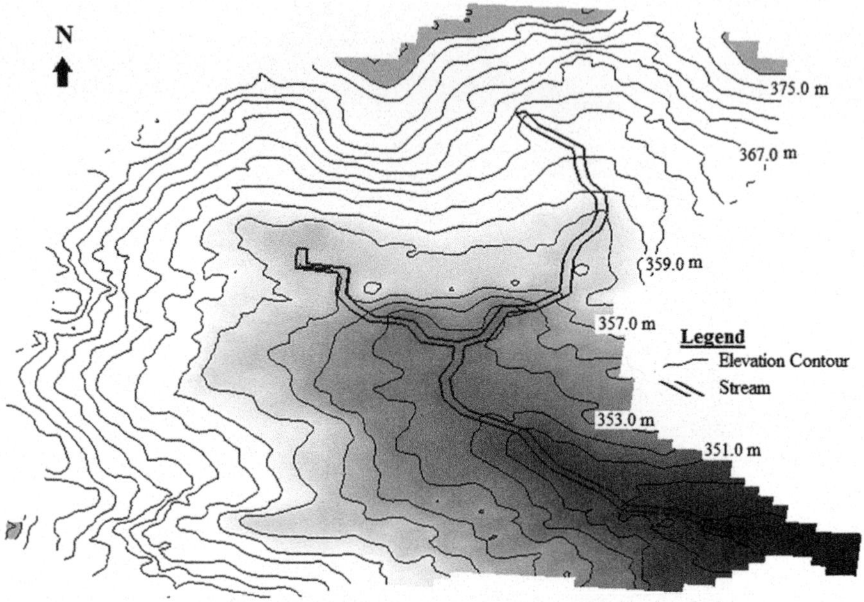

Fig. 9.10 Eight-Mile Run watershed model developed from Shuttle Radar Topography Mission, SRTM, 30 m resolution DEM

- the model's simulation response to the coupled hydrological, sediment erosion and nutrient transport processes.

The Eight-Mile Run experimental watershed data set was used to demonstrate the GSSHA model numerical formulations for simulating hydrologic response, sediment and nutrient fate and transport. Storm event and baseline samples for flow, suspended sediments, total dissolved and particulate components of N and P, as well as rainfall data were collected by the US Army Corps of Engineers for 2002 and 2003 [6]. Rainfall data collected at this site was supplemented with rainfall data from additional sites in the Eau Galle Basin [26]. Data collected at the catchment outlet, see, Fig. 9.10, were used as the input forcing and to calibrate and verify the model simulations.

The primary goal of this exercise is demonstrating the ability of the model to predict the watershed discharge, watershed scale sediment and nutrient yield under a range of precipitation events.

9.6.1 Hydrograph Calibration

Eight-Mile Run data from June 16, 2002 to June 22, 2002 were used to calibrate the model in a continuous simulation mode. The Shuffled Complex Evolution (SCE) [27]

Table 9.3 Hydrograph quality analysis in calibration and verification simulation modes

Simulation event	Observed peak ($m^3 s^{-1}$)	Simulated peak (m^3/s)	Peak error (%)	Observed volume (m^3)	Simulated volume (m^3)	Volume error (%)
Calibration event 2	1.10	1.13	2.0	33,790	32,878	−2.7
Verification event 3	1.40	1.47	4.4	14,680	18,994	29.3
Verification event 4	0.52	0.61	16.9	22,300	11,146	−50.0

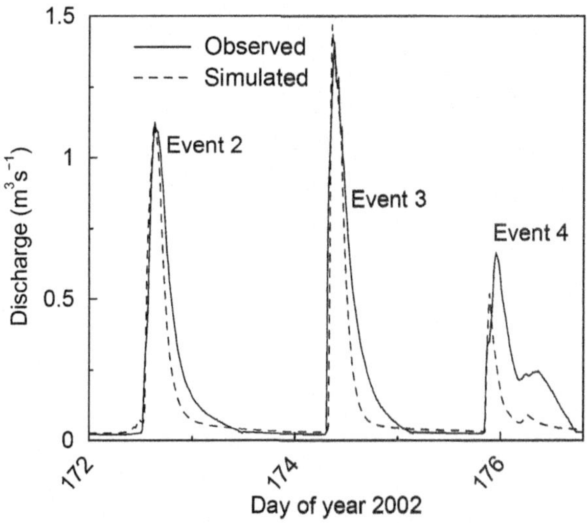

Fig. 9.11 Observed and simulated water discharge from Eight-Mile Run calibration and validation period

method was used to adjust values of retention depth, overland roughness, channel roughness, soil hydraulic conductivity, depth of the soil moisture layer and the top soil layer depth in the soil moisture model. The SCE method is a global optimization routine that converges on an optimal solution based on a user define criteria called a cost function. The calibration period included two rainfall events. The first event occurring between June 16, 2002 and June 17, 2002 was a relatively very small event used to initialize the model, and was not used in the calibration analysis. Senarath et al. [28] discusses the importance of properly initializing hydrologic models when determining parameter sets through calibration. The second event that occurred between June 19, 2002 and June 22, 2002 was used in the cost function analysis where equal weights were placed on both event volume and peak.

As shown in Table 9.3, the GSSHA model was able to reproduce both the storm peak flow and total volume within about 2 %. The calibration hydrograph, indicated by Event 2, is shown in Fig. 9.11.

Table 9.4 Sedograph quality analysis in calibration and verification simulation modes

Simulation event	Observed peak ($m^3 s^{-1}$)	Simulated peak ($m^3 s^{-1}$)	Peak error (%)	Observed volume (m^3)	Simulated volume (m^3)	Volume error (%)
Calibration event 2	0.0002033	0.000199	−2.1	1.75	2.21	26.2
Verification event 4	0.000031	0.000058	87.0	0.65	0.24	−62.7

9.6.2 Hydrograph Validation

Flow data from June 23, 2002 to June 25, 2002 were used to validate the parameter calibration. This validation period included two rainfall events, the event that occurred June 23, 2002, referred to as the third event, and the event that occurred between June 24, 2002 and June 25, 2002, referred as fourth event. The verification simulation period included the calibration events. All calibrated parameter values and initial soil moisture values and groundwater levels remained the same. The hydrologic verification results are listed in Table 9.3. As seen in the table, simulated peak values were within 20 % and simulated event volumes were within 50 % of observed The error was much larger for Event 4, a very small event [29]. Points out the difficulty with simulating very small flows. Still, the model produces very good accuracy on the larger events and acceptable accuracy on the smaller event.

9.6.3 Sedograph Calibration

The sediment calibration and verification periods were the same as those used for hydrology. Calibration of the erosion-related parameters focused on the erodibility factor K_{eff} in the Kilinc-Richardson [30] transport capacity equation for five separate land use/soil complexes. The soil erosivity values for each unique soil texture/land-cover class were identified using SCE. Equal weights were given to the event sediment runoff volume and event sediment peak flux. Results for the sediment calibration are shown in Table 9.4. As seen in Table 9.4, error percentage for peak and discharge in the calibration period was 2.13 % and -26.2 % respectively.

Event 2 in Fig. 9.12 shows the sedograph computed by GSSHA during the calibration period. Sedograph numbers in Fig. 9.12 correspond to the hydrograph numbers in Fig. 9.11.

9.6.4 Sedograph Validation

The same events used for hydrologic verification were used for verification of the sediment parameter values, Event 3 and 4. However TSS data for Event 3 was

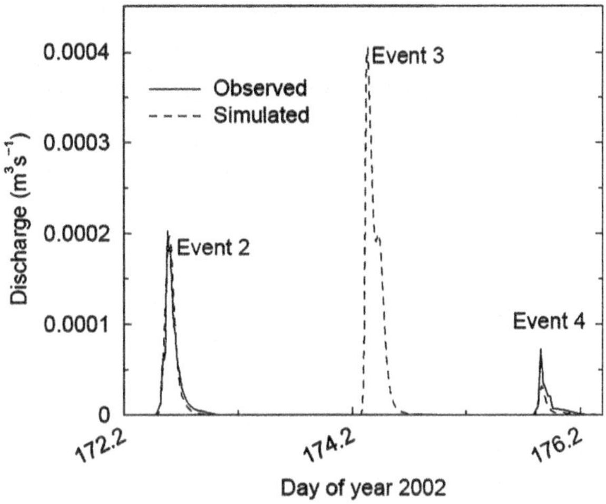

Fig. 9.12 Observed and simulated sediment discharge from Eight-Mile Run calibration and validation period

missing so only data from Event 4 were used in validation. In terms of sediment produced, Event 4 is very small. Even though the event was small and the error for total storm volume was higher than for the other events, the error in terms of sediment prediction was acceptable, with an error of about 60 %. Event 4 in Fig. 9.12 shows the sedograph computed by GSSHA during the verification period. Sedograph event numbers in Fig. 9.12 correspond to the hydrograph event numbers in Fig. 9.11. Figure 9.12 also shows the model sedograph prediction for Event 3. Given the models ability to simulate flow for this event, and the quality of sedograph in calibration and validation periods, the sedograph predicted data through the model numerical formulation, as in case for Event 3, can be used for hydro-environmental quality purpose.

9.6.5 Nutrient Calibration and Verification

Nutrient calibration and verification were done using the same period used for both hydrology and sediments. Unfortunately, only limited observed N and P data were available for the calibration period. Only Event 2 had sufficient data for calibration. No data were available for Event 3 and only limited data were available for Event 4. Figures 9.13 and 9.14 show the total dissolved N and P mass flux for the calibration and verification period. Dominant values, such as soil concentrations (mg Kg^{-1}), concentration in the groundwater, and P partition coefficients in the soil were specified based on observed data. Parameter values for the uptake rate and the partition coefficient used in the water column were calibrated using total observed dissolved N and P fluxes for Event 2. The Root Mean Square Error, RMSE, statistic in the calibration Event 2 was 0.98 g s^{-1} and 0.21 g s^{-1} for N and P, respectively.

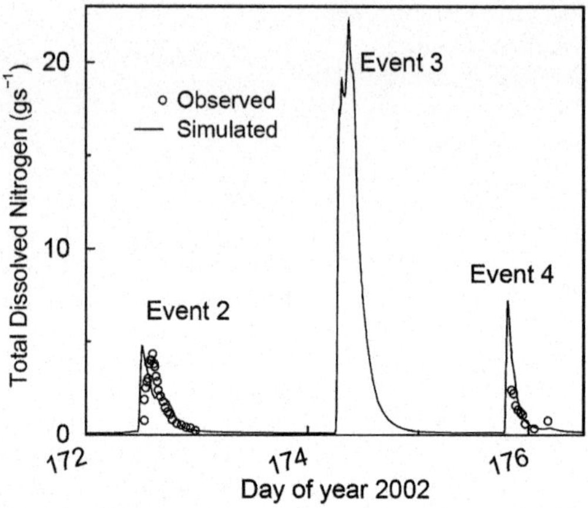

Fig. 9.13 GSSHA simulation of the total dissolved nitrogen load

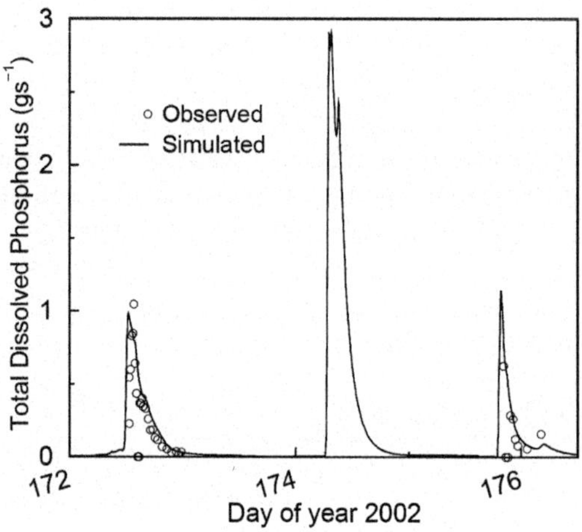

Fig. 9.14 GSSHA simulation of the total dissolved phosphorus load

There was no observed data in Event 3. Thus available observed data in Event 4 was used to validate the model. RMSE for N and P mass flux in the validation Event 4 were 0.88 and 0.27 respectively. While the data set is imperfect, the match between the observed and simulated data for both N and P is quite good, matching both the shape, and the magnitude of the observed values for periods within and outside the calibration period. Given the models ability to simulate total dissolved N and P mass flux in the calibration and validation periods, the predicted data through the model numerical formulation in observed data missing period, as in case for Event 3, could be used for filling data gaps and to calculate the TMDL for monitoring recommendations.

9.7 Discussion

In the preceding example GSSHA demonstrates considerable skill in simulating hydrologic response, event sediment discharge and nutrient event discharge. GSSHA is able to accomplish these results by explicitly including the spatial heterogeneity of the watershed into the simulation as well as explicitly simulating the physical and chemical processes occurring in the watershed, as opposed to simulating only the response from lumped areas with empirical methods. Building this level of detail into the model allows the model to be used for varied conditions, due to changes in the watershed or inputs with confidence. Extending model results beyond the range of calibration to model future conditions requires the use of physically based models that include the important processes that generate stream flow, material transport, uptake, loss, transformation, and recycling. In this study, GSSHA successfully demonstrated the extension of model simulated results beyond the range of calibration to predict future conditions. This demonstration of how GSSHA results could be used to fill the data gaps hints that GSSHA can be a valuable tool for the analysis of non-point source pollutants in ungaged basins.

This approach to including spatial heterogeneity and physical processes and correspondent parameters requires appropriate spatial and temporal input data to populate the model. As described in [8] the GSSHA model has been developed as an application tool to solve real world problems, and has been applied for many practical studies around the world. The model utilizes commonly available input data sets. All the data for the 8 Mile Run watershed, except soil N and P concentrations and partition coefficients were acquired from commonly available data sources, as described above. The availability of soil concentrations of N and P vary. Measured values of N and P partition coefficients are not commonly available. When these required parameter values are not available they may be estimated based on physical properties in the watershed, such as land use, application rates of fertilizers in agricultural areas, vegetation type or crop cover, soil type, soil organic content, etc., and literature values related to these properties and/or local knowledge of conditions in the watershed. Estimated parameters may be calibrated to observed constituent concentration or mass loading data. This process of estimating parameter values from watershed properties and known relationships between the physical property and the parameter value and tuning of these parameters to observed data is common for any watershed modeling approach, including simpler lumped parameter empirical approaches. However, even in a data sparse region, a distributed physics based approach offers advantages to simpler approaches because the physically based parameters in GSSHA can be reasonably estimated from known properties in the watershed, as described above. These parameters are bounded within known physical limits and require less observed data for calibration purposes and even in the complete absence of observed data for calibration, the range of possible results can be reasonably bound. Since empirical models are only applicable within the range of calibration, they require much more calibration data to be accurately applied. Application of such models with no calibration data introduces great uncertainty because the parameter values include significant implicit information that can only be determined by tuning to observed results.

9.8 Summary and Conclusions

The TMDL process has been implemented by EPA to ensure that water quality standards in receiving water bodies are met, requiring NPDES permit holders to implement BMPs to reduce NPSP loads. To date, calculation of TMDLs and estimation of the effects of BMPs has been performed using simplified methods. GSSHA represents a class of numerical hydrologic models that can simulate physical and chemical processes in a watershed at very fine scales. The GSSHA model was used to simulate water, sediment, and nutrient discharge from a small agricultural watershed in the Midwest. The model was able to accurately simulate hydrology, sediment, and dissolved total N and P export from the watershed after calibration of key parameters for each process. Validation shows that the calibrated model parameters are applicable for periods outside the calibration period. The resulting physically based distributed water flow, sediment and nutrient transport and fate modeling system is capable of assessing hydrology and transport and fate of sediments and nutrients within the watershed due to changing land uses and implementation of BMPs. The results show that GSSHA may be useful for a variety of watershed NPSP and TMDL planning activities, including:

1. Watershed characterization—understanding the basic physical, environmental, and human elements of the watershed.
2. Filling data gaps for monitoring recommendations
3. Source assessment—identification of sources of pollutants, and magnitude of sources.
4. Load allocation—determination of background and anthropogenic pollutant load.
5. TMDL implementation plan—a watershed management strategy to attain established targets.

Thus, GSSHA could be a valuable tool for calculating TMDLs, as well as analyzing project alternatives and evaluating best management practices (BMPs) designed to control contaminants, offering significant potential advantages over simpler approaches.

References

1. Brown LC, Barnwell TO Jr (1987) The enhanced stream water quality models QUAL2E and QUAL2E-UNCAS: documentation and user manual. Tufts University/US EPA, Athens
2. Johnson BE, Gerald T (2006) Development of a distributed nutrient sub-model (NSM Version 1.0) for watersheds – kinetic process descriptions. ERDC/EL TR-06–12. U.S. Army Engineer Research and Development Center, Vicksburg. https://swwrp.usace.army.mil/
3. Thomann RV, Meuller JA (1987) Principles of surface water quality modeling and control. Harper and Row Publishers, Inc., New York
4. DiToro DM (1985) A particle interaction model of reversible organic chemical sorption. Chemosphere 14(9–10):1503–1538

5. Chapra SC (1997) Surface water quality modeling. McGraw-Hill, Boston
6. James WF, Johnson BE, Zhang Z, Downer CW, Byrd AR (2009) Soil exchangeable phosphorous pools, equilibrium characteristics and mass distribution coefficients for eight mile run watershed, Wisconsin. ERDC TN-SWWRP-09–6
7. Downer CW, Ogden FL (2004) GSSHA: a model for simulating diverse streamflow generating processes. J Hydrolog Eng 9(3):161–174
8. Downer CW, Ogden FL (2006) GSSHA users' manual. ERDC/CHL SR-06–1. U.S. Army Engineer Research and Development Center, Vicksburg
9. Ogden FL, Saghafian B (1997) Green and Ampt infiltration with redistribution. J Irr Drain Engr 123:386–393
10. Richards LA (1931) Capillary conduction of liquids in porous mediums. Physics 1:318–333
11. Downer CW (2007) Development of a simple soil moisture model in the hydrologic simulator GSSHA, ERDC-TN-SWWRP-07–8. U.S. Army Engineer Research and Development Center, Vicksburg. https://swwrp.usace.army.mil/
12. Downer CW, Ogden FL, Neidzialek J, Liu S (2006) GSSHA: a model for simulating diverse streamflow generating processes. In: Singh, V.P. and D. Frevert (eds) Chapter 6 in. Watershed Models, Taylor and Francis Group, CRC Press, 637 pp
13. Eller RP, Cheng J-RC, Byrd AR, Downer CW, Pradhan NR (2013) Development of parallel GSSHA. ERDC TR-13-8. U. S. Army Engineer Research and Development Center, Vicksburg. http://acwc.sdp.sirsi.net/client/search/asset/1029920
14. Downer CW, Ogden FL, Niedzialek JM, Byrd A (2008) Nonorthogonal channel and reservoir routing in GSSHA. ERDC TN-SWWRP-08–5. U.S. Army Engineer Research and Development Center, Vicksburg. https://swwrp.usace.army.mil/
15. Ogden FL, Pradhan NR, Downer CW, Zahner JA (2011) Relative importance of impervious area, drainage density, width function, and sub-surface storm drainage on flood runoff from an urbanized catchment. Water Resour Res 47, W12503. doi:10.1029/2011WR010550
16. Pradhan NR, Ogden FL, Downer CW (2009) Uncertainty in tile drain locations: implications for hydrologic modelling of agricultural watersheds. Trans Amer Geophys Union 90:52
17. Pradhan NR, Downer CW, Marchenko Sergei, Lijedahl A, Douglas TA, Byrd A (2013) Development of a coupled framework for simulating interactive effects of frozen soil hydrological dynamics in permafrost regions, ERDC TR-13-15. U.S. Army Engineer Research and Development Center, Vicksburg. http://acwc.sdp.sirsi.net/client/search/asset/1032303
18. Yang CT (1973). Incipient motion and sediment transport. J Hydraulics Div, ASCE 99(HY10): Paper 10067
19. Downer CW, Byrd AR (2007) Watershed scale TMDL model: multi-dimensional sediment erosion, transport, and fate. ERDC TN-SWWRP-07–1. U.S. Army Engineer Research and Development Center, Vicksburg. https://swwrp.usace.army.mil/
20. Downer CW, Ogden FL, Pradhan NR, Liu S, Byrd AR (2010) Improved soil erosion and sediment transport in GSSHA. ERDC TN-SWWRP-10–3. U.S. Army Engineer Research and Development Center, Vicksburg. https://swwrp.usace.army.mil/
21. Downer CW (2009) Simulation of reactive constituent fate and transport in hydrologic simulator GSSHA, ERDC TN-SWWRP-09–2. U.S. Army Engineer Research and Development Center, Vicksburg. https://swwrp.usace.army.mil/
22. MacCormick RW (1971) Numerical solution of the interaction of a shock wave with a laminar boundary layer. Proc Second Int Conf Num Methods Fluid Dyn, Lect Notes Phys 8:151–161
23. Johnson, BE, Zhang Z (2007) Development of a distributed source contaminant transport, transformation, and fate (CTT&F) sub-model for military installations. ERDC/EL TR-07–10. U.S. Army Engineer Research and Development Center, Vicksburg. https://swwrp.usace.army.mil/
24. Pradhan NR, Byrd AR, Ogden FL, Hendrickx JMH (2012) SEBAL evapotranspiration estimates for the improvement of distributed hydrologic model runoff and soil moisture predictions. Remote Sensing and Hydrology 2010, IAHS Publ. 352-41, pp 435–439

25. Crawford JW, Baveye P, Grindrod P, Rappoldt C (1999) Application of fractals to soil properties, landscape patterns, and solute transport in porous media In: Corwin DL, Loague K, Ellsworth TR (eds) Assessment of non-point source pollution in the vadose zone. American Geophysical Union, Washington, D.C., pp 151–164
26. Downer CW (2008) Demonstration of GSSHA hydrology and sediment at Eau Galle Watershed near Spring Valley, Wisconsin. ERDC-TN-SWRPP-08–2. U.S. Army Engineer Research and Development Center, Vicksburg. https://swwrp.usace.army.mil/
27. Duan WY, Gupta VK, Sorooshian S (1992) Shuffled complex evolution approach for effective and efficient global minimization. J Optim Theory Appl 76(3):501–521
28. Senarath SUS, Ogden FL, Downer CW, Sharif HO (2000) On the calibration and verification of two-dimensional, distributed, Hortonian, continuous watershed models. Wat Resour Res 36 (6):1495–1510
29. Ogden FL, Senarath SUS, Downer CW, Sharif HO (2002) Reply, Water Resources Research Vol 37. Wat Resour Res 37(12):3397
30. Kilinc, M, Richardson EV (1973) Mechanics of soil erosion from overland flow generated by simulated rainfall: Hydrology Paper No. 63, Colorado State University

Chapter 10
Advances in *In Silico* Research on Nerve Agents

Devashis Majumdar, Szczepan Roszak, Jing Wang, Tandabany C. Dinadayalane, Bakhtiyor Rasulev, Henry Pinto, and Jerzy Leszczynski

Abstract Nerve-agents (NAs) are toxic environment contaminants causing massive health hazard to the plant, animal, and civilian populations. Moreover, these materials have the properties of adsorption on various artificial surfaces which include cement, paints, metal oxides and clay minerals. These adsorption properties also threaten long-lasting toxic after-effects of NA exposure to the environment. Modeling these diverse NA-exposure characteristics through computational techniques has been always of great importance because of the restrictions in using such materials directly in the experiments due to their high toxicity. The present review discusses the recent advancements in the *in silico* research of NA, which include their conformational, biological and surface-occlusion properties. There are some positive sides of NA-adsorption also. The adsorption properties of NAs on oxide surfaces are used as binder to remove and subsequently deactivate them through chemical treatments. Moreover, NA adsorption on various surfaces is also useful to design materials to detect those agents using spectroscopic techniques. The present review also discusses the theoretical advancements in these directions in details. All of these discussions are mostly based on the results of the state of the art quantum-chemical computations. Related experimental results are also discussed to validate the results from such theoretical approaches.

D. Majumdar (✉) • J. Wang • T.C. Dinadayalane • B. Rasulev • H. Pinto
Department of Chemistry and Biochemistry, Jackson State University,
Jackson, MS 39217, USA
e-mail: devashis@icnanotox.org; jingw@icnanotox.org; dina@icnanotox.org; rasulev@icnanotox.org; pavlvs.pinto@gmail.com

S. Roszak (✉)
Institute of Physical and Theoretical Chemistry, Wroclaw University of Technology,
Wybrzeze Wyspianskiego 27, 50-370 Wroclaw, Poland
e-mail: szczepan.roszak@pwr.wroc.pl

J. Leszczynski (✉)
Department of Chemistry and Biochemistry, Interdisciplinary Center for Nanotoxicity,
Jackson State University, Jackson, MS 39217, USA
e-mail: jerzy@icnanotox.org

J. Leszczynski and M.K. Shukla (eds.), *Practical Aspects of Computational Chemistry III*, DOI 10.1007/978-1-4899-7445-7_10,

10.1 Introduction

Nerve-agents (NAs) are chemical warfare with toxic environment contamination properties causing massive health hazard to the plant, animal and civilian populations. Common nerve-agents (NAs) are organophosphorus (OP) compounds and they exhibit their toxicity through achetylcholinesterase (AchE) inhibition [1]. They are clearly differentiated from other chemical warfare (CW) type agents (viz. blister, blood, choking, incapacitating, tear (lachrymator), and vomit agents) due to their phosphorylating mode of action. NAs are phosphorus (V) compounds with a terminal oxide and three singly bonded substituents (two alkoxyl and one alkyl group). Their toxic activity is derived from their organophosphonate structure ($RO(O = P(R_1)OR_2)$. This unique structure is the key to its mammalian toxicity as the group directly participates in AchE inhibition. Several such NAs are presented in Fig. 10.1, together with their technical name. These names are derived from the particular class (G or V series) they belong to.

The NAs are the object of various kinds of research for last 15 years due to their highly toxic behavior and the impending danger of its use as chemical warfare, if fallen into wrong hands. Such research activities involve their structural and conformational properties, binding/adsorption properties on various surfaces, materials to detect and destroy such NAs, and search for the effective antidotes. Several reviews are available [1, 2] elucidating mostly the experimental approaches on such research topics on NAs. The toxic behavior of NAs limited most of such research works (several exceptions apply) to be carried out using non-toxic NA-simulants like dimethyl phosphonate (DMMP), trimethyl phosphate (TMP), and O, O-diisopropylfluorophosphate (DFP). The *in silico* approach in such research is getting importance for last several years, as the computed results could not only be compared with experiments, but the toxic NAs could also be considered directly to assess the experimental data from practical point of view. In many cases such

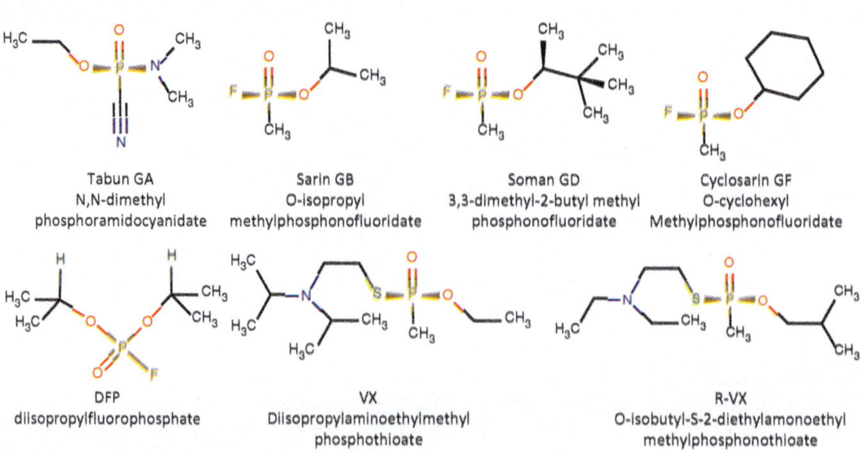

Fig. 10.1 Structures of common nerve agents

approaches provide a vivid insight into the properties and behavior of such molecules. The present review article mostly discusses the use various theoretical approaches used to explore the conformation, reactivity (as a toxic agent), adsorption properties and materials for detection and deactivation of NAs. The available experimental results would be discussed to assess the reliability and utility of such *in silico* techniques in NA-research. The review article is organized under the following three broad aspects of NA-research:

(a) The conformational and biological properties of various NAs. The main objective is to explore how such properties of NAs are related to their toxic properties.
(b) The adsorption/binding of NAs on various surfaces. These properties of NAs are important to assess the long term environmental hazard caused by such toxic materials.
(c) Materials for detection and deactivation of NAs.

10.2 The Conformational and Biological Properties of Various NAs

It is well known that the physiological reaction catalyzed by ACHE is the hydrolysis of acetylcholine (Ach). This reaction proceeds, after reversible substrate binding, through successive acylation and deacylation steps [3].

$$\mathbf{AchE + S \rightleftharpoons AchE\text{-}S \rightarrow F + P \rightarrow AchE + Q}$$

AchE, S, AchE-S and F refer to free enzyme, free substrate, Michaelis complex and the enzyme intermediate respectively. P and Q are successive choline and acetate products. The reaction takes place inside the active cavity of AchE and crystallographic studies have shown this cavity as a deep and narrow gorge (~20 Å) [4]. It penetrates halfway into the enzyme [4, 5] and 40 % of the inner surface of the inner gorge is lined by 14 aromatic residues [6]. It facilitates the two dimensional diffusion of Ach to the active site [4], accommodates the substrate in the AcheE-S complex and is involved in the allosteric modulation of catalysis [6].

The active site of AchE contains two main sub-sites. (a) The esteric sub-site [4, 5]: It is composed of serine, histidine and glutamic acid and lies about 4 Å from the bottom of the cleft. This site is responsible for the acylation and deacylation catalysis steps. (b) The oxyanion hole: It contains two glycine and one alanine moiety and interacts with the tetrahedral intermediate (AchE-S) through their –NH groups. There are two other sub-sites of AchE [4, 5, 7, 8], one of which is highly aromatic in amino acid contents (glutamic acid, tryptophan and phenylalanine). The other sub-site, known as acyl pocket, contains glycine, tryptophan and phenylalanine residues and is responsible for the tight fit of the acetoxy methyl group. Finally, there is a peripheral sub-site, located at the opening of the aromatic gorge. The main constituents are tryptophan, tyrosine and asparate and this part helps the passage of Ach inside the cavity [4, 9].

The inhibition of AchE by NAs involves phosphonylation of the active site at serine residue and the formation of stable phosphonyl adducts [10]. Subsequently, an irreversibly inactivated enzyme is generated by spontaneous dephosphonylation of the NA-AchE conjugate through a unimolecular process (aging) [10, 11]. Due to the tetrahedral geometry of the phosphonate inhibitor in the initial AchE-S complex as well as in the covalent conjugate and the limited motion of the equatorial phosphorus substituents in the penta-coordinated transition state of the nucleophilic addition, the phosphonyl moiety in both the adducts may face similar regions of active center structural manifold. As a result, the same interactions which govern the stability of the initial complex may also be important in facilitating a characteristic chemical transformation of the covalent NA-AchE conjugate, the aging process [12, 13].

There are two distinct approaches available in theoretical studies to characterize the aforementioned biological events responsible for the observed toxic characteristics of the NAs. The first one is based on the structural characterization of the NA-bound cholinesterase complexes. Such structure-based approach generates an idea of the conformational requirement of the NAs during such interactions. Moreover, since many of these NAs are optically active, they exist in several enantiomeric forms (discussed later) depending to the number of chiral centers. The structure based approaches also lead to proper assignment of such enantiomers together with their conformational preference/s for the said biological property. The second approach is based on the direct reaction path analysis of these enzymatically controlled reactions to understand the complicacy involved in the aging process of the NA-AchE complexes.

10.2.1 Conformational Characteristics of NAs for Their Biological Activities

Several experimental [14–19] and molecular modeling [2–26] studies have been made on the enzyme-NA binding to understand the nature of their aging process, which has direct consequence on the toxicity of the NAs and their AchE inhibition properties. Studies on the rate processes on the aging of NA-inhibited human butyrylcholinesterase (hBchE) and cholinesterase (hAchE) [17] have shown that amongst the nerve gases DFP, cyclosarin and soman, the aging is fastest for the soman-inhibited complex. If the nerve agent is optically active, a particular enantiomer could be more active than the other. The mechanism of aging depends on the structure of the phosphonyl moiety of the NA. In the case of soman, sarin and DFP (O,O-diisopropyl phosphorofluoridate) inhibited AchE, this process occurs through C-O bond cleavage [14, 15]. Tabun (O-ethyl N,N-dimethyl-phosphoramidocyanide), on the other hand, has a –CN group attached to the P atom and the aging process occurs through P-N bond

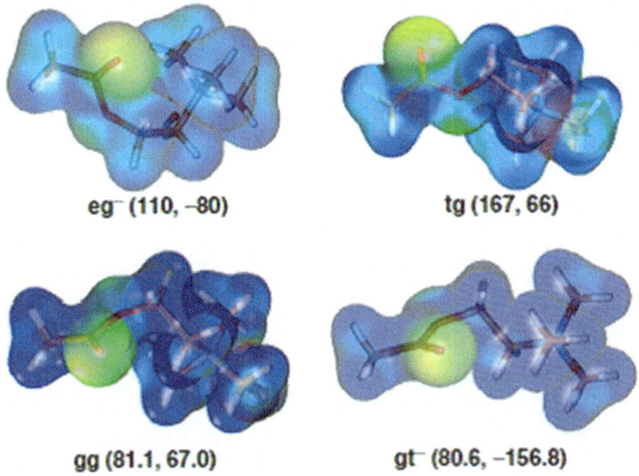

Fig. 10.2 Calculated MEP on the isodensity surfaces of Ach for its various low-energy conformers. The various colored regions on the surfaces are *deep blue* (highly positive, >0.1 a.u.), *light blue* (<0.1 a.u. and >0.05 a.u.), *green* (0.0 a.u.), *yellow* (<−0.1 a.u. and >−0.05 a.u.) and *red* (≥−0.1 a.u) and they represent isopotential MEP surfaces of different magnitudes (*as indicated*). The e^-g, tg, gt^-, and gg have maximum positive MEP values of 109.8, 111.4, 118.2, and 121.9 kcal mol^{-1} (*deep blue region*). The minimum negative MEP values of these conformers are within the range −10 to −20 kcal/mol (*greenish yellow region*) (Reproduced from Majumdar et al. [28], with kind permission from Taylor and Francis 2007)

cleavage [16]. Experiments on the rate of the aging processes of different nerve agents and the crystallographic studies on their aged structures [14] have shown that such reaction occurs at pH 7.5–8.0 and the presence of one water molecule is needed to facilitate such a process. The crystallographic studies on the aged crystals of soman, sarin and DFP [14] have further demonstrated that the oxygen atoms of NAs connected to the phosphorus atom are within hydrogen bonding distances of four potential donors in the catalytic sub-site of the enzyme. This suggests that the electrostatic forces significantly stabilize the aged enzyme.

A large number of theoretical investigations have so far been carried to figure out the conformational flexibility, their relation to inhibit the AchE inhibition and the probable pathways for such inhibition processes [2–26]. These were important to interpret the said inhibition properties of NAs, as they do not resemble Ach in many ways. Ach is a very flexible molecule with a possibility of very large number of conformers in gas-phase as well as in aqueous medium. Extensive conformation analysis at the density functional level of theories (DFT/B3LYP/6-31++g(d,p)) [27–30] has shown that the low energy conformers of Ach include eg^- (110, −80), gg (81.1. 67), gt^- (80.6, −156.8), and tg (167, 66) [26]. These low energy conformers of Ach dominate both in gas-phase as well as in aqueous medium. These conformers are shown in Fig. 10.2 along with their computed molecular electrostatic potential (MEP) surfaces. The orientation of these torsion angles is denoted as gauche (*g*), eclipsed (*e*) and trans (*t*) and these notations are used throughout the review to define various conformers. Several other theoretical analyses are also available on Ach. Vistoli and coworkers [31] have found the eg$^-$ (121.3, −68.4) to be the lowest energy conformer in the gas phase. The conformational analyses by Deakyne et al. [9] at the Hartree-Fock (HF/6-31G + (d)) followed by single point Möller-Plesset second order perturbation (MP2) theory [32] calculations have indicated e$^-$g (−113, 78) to be the lowest energy conformer in the gas phase. Their conformation studies on the hydrated Ach [9] further indicated that the minimum is shifted to tg (171, 73). The calculations at the DFT/B3LYP/6-31++G(d,p) [26] level indicate that although the eg$^-$ (110.2, −79.6)/ e$^-$g (−110.2, 79.6) are the lowest energy gas phase conformers with respect to the ΔH°gas (change of enthalpy) values, the calculated ΔG°_{298} (change of Gibbs free energy) values indicate the change of minima towards tg (167.1, 65.8)/t$^-$g$^-$ (−167.1, −65.8). These two are also the minimum energy conformers in aqueous medium, as predicted by Deakyne and coworkers [9]. Several experimental studies of Ach conformers (mostly X-ray crystallographic and NMR results) support the conformational flexibility of Ach [33–39] as observed in various theoretical studies.

The MEP surfaces of the various low energy Ach conformers (Fig. 10.2) reveal the weakly negative isopotential region around the acetoxy group, where one can expect weak hydrogen bonding with the NH part of the amino acids in the active centers [26]. The depth of this negative isopotential zone should increase when the oxyanion tetrahedral intermediate is formed, and the hydrogen bonding would be much stronger with the corresponding -NH part of the amino acid residues. The MEP surface actually reveals a strong positive isopotential region around the $-N^{+}(CH_3)_3$ group and the depth of the potential indicates that this site can go for strong cation-π interaction with the aromatic rich amino acid residues, since the active cavity contains tryptophan, tyrosine and aspartate as important members of the active cavity [4, 5].

The NAs sarin, soman, and tabun are optically active due to the presence of asymmetric P atom. Soman possesses a second asymmetric carbon center. Thus sarin, and tabun can have two enantiomers (R and S), while soman can have four (RR, SS, SR, RS) [40, 41]. Although this complicates the conformational properties of these molecules, the toxic properties are shown only by the S-enantiomers. The low energy conformers of sarin, soman, and tabun have some conformational similarity with Ach [26]. The computed MEP isopotential surfaces of the low energy conformers of sarin and soman for their different low energy conformers (Fig. 10.3) show the features which are similar to those of Ach [26]. The MEP surfaces of these two molecules are more or less similar in nature. The MEP surfaces around the hydrophobic part of the molecules show weakly positive isopotential zone, while the surfaces around O and F atoms are negative in character. The positive potential is maximized near the P atom (core region of the atom) and it is actually very high. But this high positive region quickly reduces and the value on the outer isodensity surface is quite low in comparison to Ach conformers (Fig. 10.3). Thus the hydrophobic part of these two molecules is not capable of binding with the aromatic residues of the cavity. The negative potential zone around the oxygen atoms is much stronger than fluorine and the situation is very favorable for hydrogen bonding. These similarities are important when the nerve agents

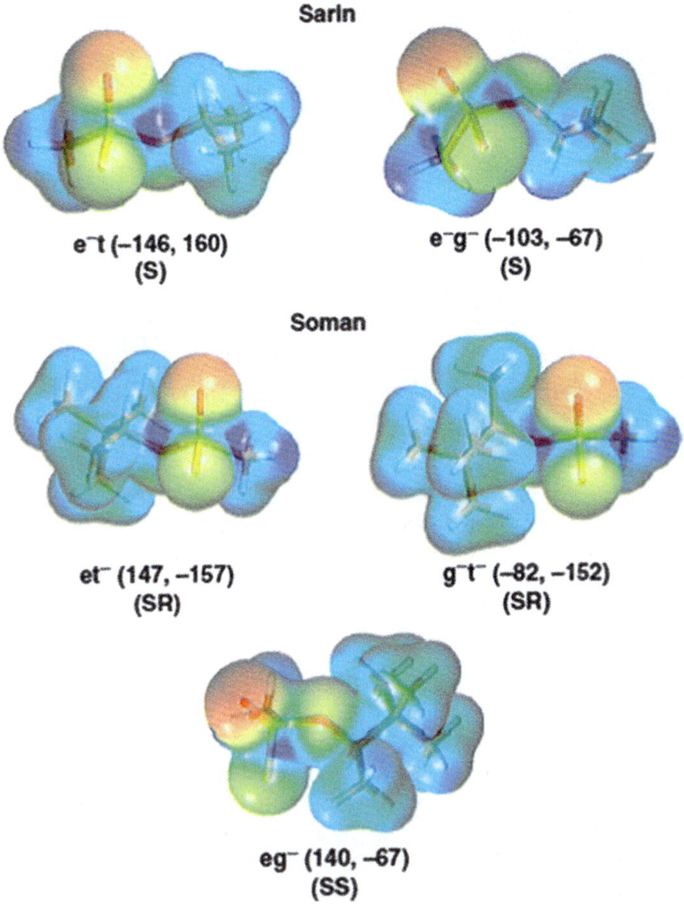

Fig. 10.3 Calculated MEP on the isodensity surfaces of sarin (S) and soman (SR and SS) for their various low-energy conformers. The definitions of colored regions are the same as in Fig. 10.2. The et^-(S) and e^-g^-(S) conformers of sarin (S) have the minimum MEP values of −47.5 and −51.5 kcal mol^{-1} (the *reddish yellow region*) respectively. The R enantiomers have similar values. The et^-(SR), g^-t^-(SR), and eg^-(SS) conformers of soman have the minimum MEP values of −47.6, −47.5, and −51.8 kcal/mol respectively (Reproduced from Majumdar et al. [28], with kind permission from Taylor and Francis 2007)

interact with the serine residue of the AchE active site. But before this reaction can occur, the nerve gases need to be trapped like Ach, and this binding process should also be competitive in nature. The structures of sarin and soman indicate they don't have any highly cationic center like Ach, and hence the contribution of cation-π type interaction is quite infeasible here.

The nature of interactions of Ach and sarin with AchE active cavity, as discussed above, is supported through theoretical interaction studies of these molecules with model AchE active cavity [42]. It was shown that Ach binds to the AchE cavity

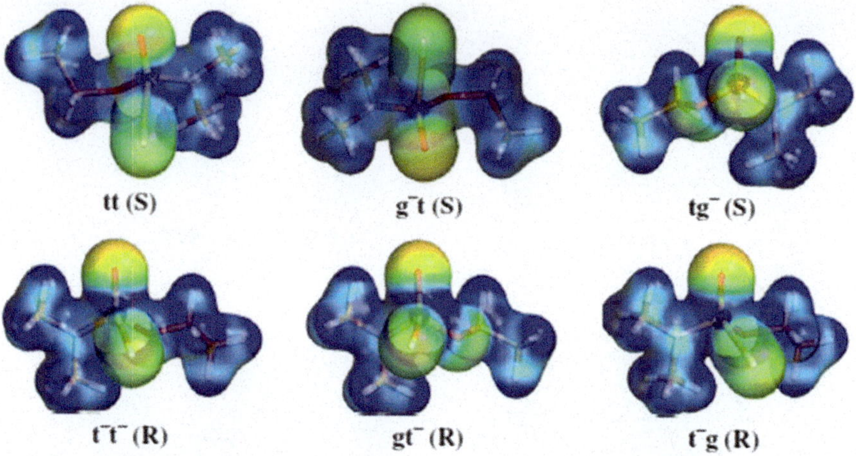

Fig. 10.4 Calculated MEP on the isodensity surface of tabun for its various low energy conformers. The definitions of the colored regions are the same as in Fig. 10.2. The minimum negative MEP values of these conformers are within the range −20 to −10 kcal/mol (*greenish yellow region*) (Reproduced from Paukku et al. [45], with the kind permission of Elsevier 2006)

through cation-π interactions. Sarin, on the other hand, binds to the active cavity through strong hydrogen bonds [42]. These observations are fully compatible with the results of MEP analyses on these molecules. Soman is not different from sarin and their MEP surfaces show similar characteristics. The earlier theoretical studies on the soman-inhibited AchE and chymotrypsin [25] showed that it's binding in the active cavity are through hydrogen bonds, which is compatible with the inferences drawn from MEP surface analyses of this molecule.

The results of calculations on AchE-Ach and AchE-sarin complexes on the formation of the pre-Michaelis complex formation [42] could be used to get an idea of the active conformers of Ach and sarin. These calculations revealed that the conformation of Ach in the AchE. . .Ach complex is eg^- with two possible geometries [τ_1: 117.5, τ_2: −74.0, τ_3: 174.7, τ_4: −166.4 ($\Delta E = 2.6$ kcal/mol) and τ_1: 109.8, τ_2: −77.3, τ_3: 172.8, τ_4: −166.7 ($\Delta E = 0.9$ kcal/mol)] [20, 40]. The S-enantiomer of sarin was found to form complex with AchE cavity and it adopted $g^- t^-$ conformation with two possible geometries [τ_1: −92.7, τ_2: −74.0 ($\Delta E = 2.6$ kcal/mol) and [τ_1: −106.1, τ_2: −178.7 ($\Delta E = 1.4$ kcal/mol)] [20, 42]. The determination of the exact geometric characteristics of the active conformers of substrate (Ach) or the inhibitors (sarin and soman) needs the knowledge of their structures in the non-aged enzyme-substrate/enzyme-inhibitor complexes. In recent years, several crystallographic studies are available on the non-aged structures of sarin, soman and tabun complexes with AchE [43, 44]. The observed non-aged structure of sarin compares well with its derived active structure from quantum-chemical studies.

The MEP isopotential surfaces of different low-energy tabun conformers (Fig. 10.4) [45] also reveal the similar situation like sarin and soman. The MEP surfaces are shown for the low-energy conformers of both R and S enantiomers. These MEP surfaces have some common features. The isopotential surface around the hydrophobic parts of these conformers shows weakly positive character. The surfaces around the O atoms and –CN group, on the other hand, are weakly negative. The crystallographic studies on the non-aged AchE-tabun complex [16, 44] have shown that the stereochemistry at the phosphorus center of tabun in this complex is R. It has been observed that during binding with the Ser-203 fragment of AchE, there is a stereochemical inversion at the P center of tabun and thus the active conformer of this NA is considered to be the S enantiomer. Earlier experimental data also supported this fact [16].

The MEP surfaces of the low energy conformers of NAs are useful to show configuration dependent toxic behavior of NAs. In the case of tabun, the situation is quite simple as the crystallographic structure of AchE-tabun complex is available [16]. A direct comparison of the negative and positive potential zones of the MEP surfaces of different conformers shows that tg^{-}(S) [Fig. 10.5b] correlates with the MEP surface of the tabun fragment in the enzyme-bound complex [Fig. 10.5a]. A comparison the geometry of the tg^{-}(S) conformer with the tabun part of the AchE-tabun structure in Fig. 10.5c reveals that the exact match of dihedral angle could be achieved through slight rotation of the torsion angles of the tg^{-}(S) conformer. The other low-energy conformers are not that close to this active conformer geometry. The resemblance of the tg^{-}(S) conformer is so close to the tabun fragment of the AchE-tabun non-aged structure, that this conformer could be considered as the active conformer of tabun [45].

10.2.2 Mechanistic Studies on the Reaction Paths of NA…AchE Interactions

The catalytic activity of AchE on neurotransmitter Ach has been shown to occur due to the termination of synaptic transmission at cholinergic synapses by catalyzing the breakdown of Ach. It has already been pointed out in the last section that the active site of AchE has a narrow and deep gorge with two separate ligand binding sites: the acylation site and the peripheral site. The acylation site has the catalytic triad composed of His, Ser and Glu amino acid residues. The inhibitors interfere in the catalytic activity of AchE by either through binding to the peripheral site to block the entrance of the active gorge [46–49] or through targeting directly on the active site to damage the enzyme activity [50, 51]. The inhibition activity is considered as the reason behind the toxic activity of NAs. Such activity of sarin is well characterized and it interacts with AchE by blocking the active center through the formation of a phosphonate ester bond to the serine residue of the active site. This is schematically shown in Fig. 10.6. During the phosphonylation reaction, the serine residue becomes more nucleophilic through transfer of its proton to the imidazole nitrogen of the histidine residue in the catalytic triad.

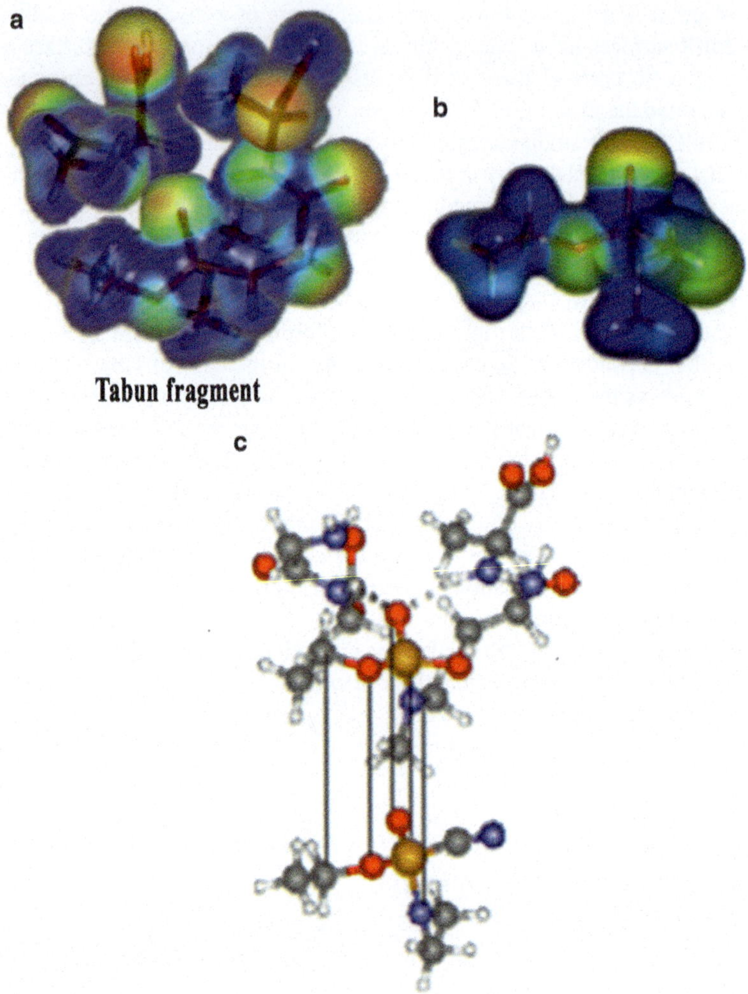

Fig. 10.5 (**a**) Calculated MEP on the isodensity surface of the selected part of AchE–tabun; (**b**) MEP surface tg$^-$(S) conformer of tabun; and (**c**) The geometric fit of the tabun fragment of AchE–tabun and tg$^-$(S) conformer from the MEP surface and torsion angle comparison. The definitions of the colored regions of the MEP surfaces are the same as in Fig. 10.2 (Reproduced from Paukku et al. [45], with the kind permission of the Elsevier 2006)

The resultant imidazolium ion is further stabilized by the polarizing effect of the carboxylate ion of the glutamic acid residue in the triad.

Several theoretical approaches have been made so far to understand the basic reactions of Ache-inhibition by NAs. The stereoselectivity of the phosphonylation reaction and the effects of the adduct configuration on the aging process were examined by four enantiomers of soman on human AchE (hAchE) and its selected active center mutants [17]. Rate-limiting P-O fission was reported in the

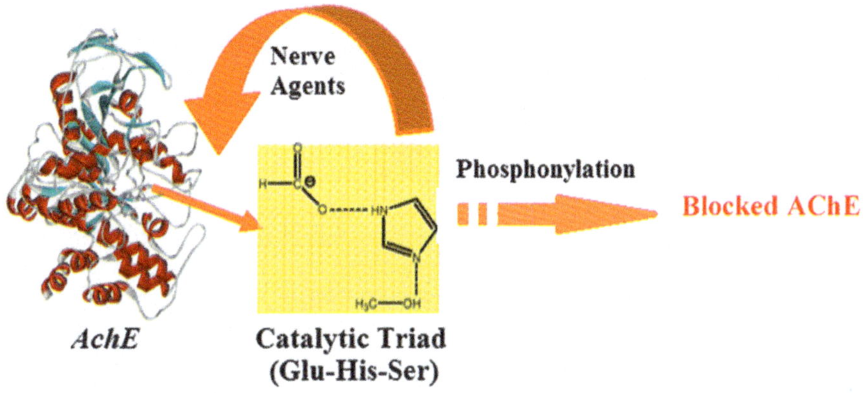

Fig. 10.6 Schematic representation of the phophonylation mechanism of sarin (NA in general) for AchE inhibition

self-stimulated inactivation of AchE by 4-nitrophenyl-2-propyl methylphosphonate [52]. The molecular dynamics (MD) simulations were used to assess the molecular origins of stereoselectivity of phosphonylation for the solution structures of the pentacoordinated and tetracoordinated P_SC_S and P_RC_S adducts of *Tc*AchE with soman [53]. The origins and diversity of the aging reaction in phophonate adducts of serine hydrolase enzymes including enzymes of AchE, trypsin, and chymotrypsin were studied with molecular mechanics and MD combined with semiempirical calculations [25]. Matrix-assisted laser desorption-ionization time-of-flight mass spectrometry analysis was performed to resolve aging reaction pathways for various nerve agent phosphyl adducts with the active site of AchE [54].

Model quantum-chemical calculation on the inhibition reactions of AchE by sarin has been studied in details in recent times [47–49, 51, 55]. These studies has enhanced the basic understanding of such reactions in microscopic details and one of the positive outcome of such mechanistic approach was the formulation of reaction paths for the NA-antidotes [56]. This quantum-chemical pathways for AchE inhibition of sarin would be discussed here in more detail as it has usually been found (see the previous section) that most of NAs follow similar reaction pathways in this respect.

The potential energy surfaces (PESs) for the phosphonylation by sarin of the active site serine of the catalytic triad of AchE have been studied [55] at the DFT level (B3LYP/6-311G(d,p)) of theory [27–30]. The serine residue was modeled by a methanol and a methoxide ion, respectively. Such models represent the related nucleophiles under two extreme reaction conditions [57]. Unlike the in-line displacement mechanisms, the results suggested that the phosphonylation process favors a major addition-elimination mechanism. More recently the further investigation has been carried out to show how the catalytic triad activates the serine residue in the phosphonylation reaction [55]. These investigations generated insights of the phosphonylation mechanisms between sarin and the catalytic triad of AchE. The more complex reaction models that included residues of histidine and glutamic acid, have been considered together with solvent effect using the polarizable continuum model (PCM) [58]. The model chemistry was studied by modeling all of the reaction ingredients in simpler molecular form. Sarin was modeled as O-methyl methylphosphonofluridate, serine residue as methanol (CH3OH) (labeled as Ser-m), histidine residue as an imidazole ring (labeled as His-m), and the glutamic acid residue as a formate anion (labeled as Glu-m) [55].

Three different reaction paths were investigated. In the first case, Ser-m and His-m are stabilized by hydrogen-bonding interactions and attack on the phosphorus of sarin takes place from the opposite side of fluoride. The H_γ proton from Ser-m is simultaneously transferred to the imidazole ring of histidine at the N_{e2} position. The protonated His-m then easily rotates to the direction of the fluoride ion of sarin to assist with the rupture of the P-F bond. The H_γ proton, which was transferred to histidine at the first nucleophilic step, remains bonded to the $N_{\varepsilon 2}$ of the imidazole ring until the phosphonylation reaction completes. This pathway reveals a critical addition-elimination reaction.

In the second pathway approach, the possible phosphonylation reaction takes place with no proton transfer from the imidazole ring of the His-m to the Glu-m of the glutamic acid residue. In the initial step, both the His-m and Ser-m retain their own protons ($H_{\delta 1}$ and H_γ). The Ser-m first transfers its H_γ proton to the histidine, while it carries out nucleophilic attack on sarin. After the deprotonation of His-m, the H_γ proton migrates and binds to O_1 of sarin, and the imidazole ring of histidine turns from the nucleophilic attacking side to the fluoride ion side. The histidine is then protonated again by H_γ. This H_γ proton later facilitates the departure of the fluoride ion from the serine…sarin adduct. The details of these two pathways are available in ref. [55] and are not further here discussed in details.

The most elaborate investigations were carried out in the third pathway. The entire catalytic triad Glu-m…His-m…Ser-m is taken into account. Six intermediates and five corresponding transition states are located. These are illustrated in Fig. 10.7 and the relative potential energy surfaces are shown in Fig. 10.8. In this pathway, the Ser-m is the reacting site, and the other two residues (Glu-m and His-m) facilitate the reaction. After serine is bonded to sarin through phosphorus, Glu-m and His-m alter their orientations as His-m does in the previous two pathways. The fluoride ion subsequently departs from the serine-sarin complex with the assistance of Glu-m and His-m. A simultaneous double proton transfer is favored among the triad throughout this pathway. Both the movements of the protons among the triad and the dynamic movement of histidine play important roles in this pathway.

The rationale of the mechanistic studies could be generalized in the following way.

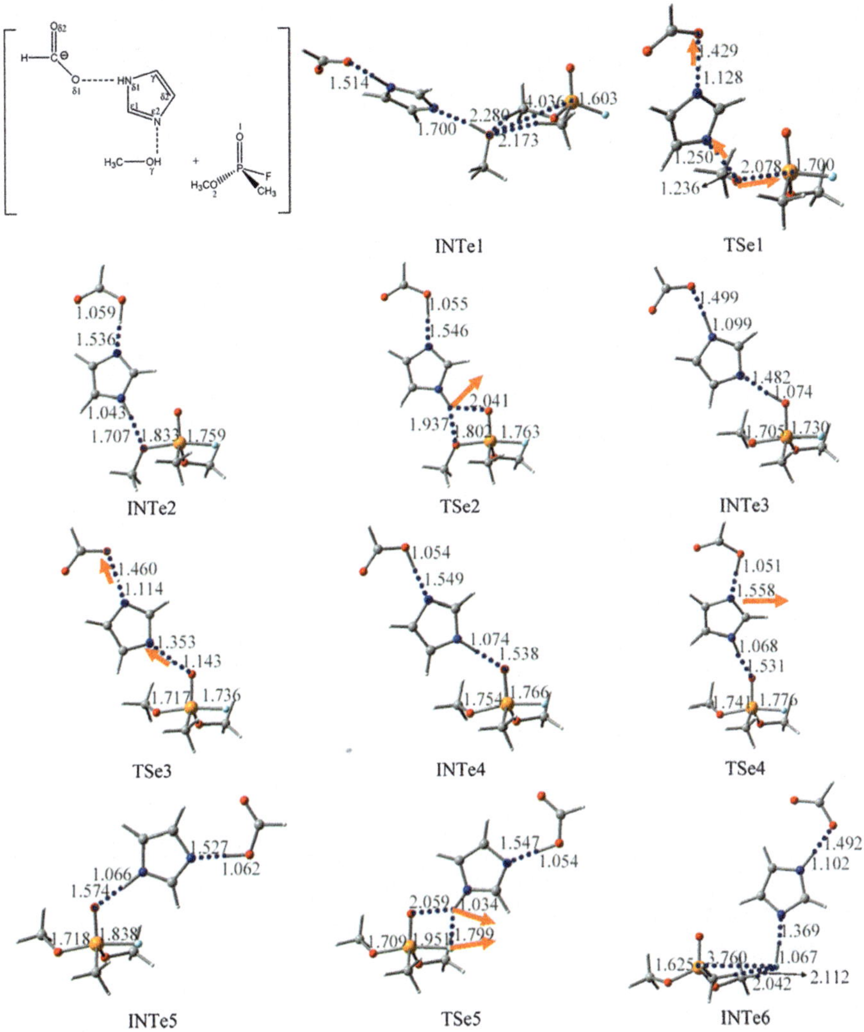

Fig. 10.7 Optimized structures of all intermediates and transition states for the reaction between the entire catalytic triad of AchE and sarin (DFT/B3LYP/6-311G(d,p) level) in the third pathway (as discussed in the text) (Reproduced from Wang et al. [55], with the kind permission of the American Chemical Society 2008)

1. In the addition-elimination reaction mechanism, the first step, which represents the nucleophilic addition of Ser-m to the phosphorus of sarin (addition process), is the rate-determining step.
2. The dynamic movement of His-m plays an important role in the reaction. The fluoride ion is much easier to be eliminated through the movement of histidine.

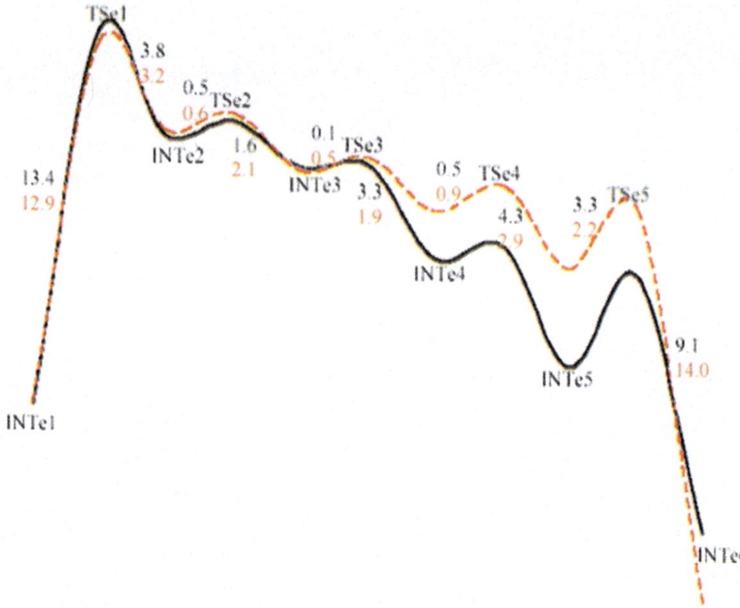

Fig. 10.8 Potential energy surfaces connecting different transition states and the intermediates in Fig. 10.7 (gas phase path is in *black* and PCM model results are in *orange*; energy units in kcal mol^{-1}) (Reproduced from Wang et al. [55], with the kind permission of the American Chemical Society 2008)

It should be noted that the histidine movement mechanism is still an issue. The present status is that the His motion is favored in the view point of energy.

3. A double proton transfer mechanism is proposed for the catalytic triad, and low-barrier hydrogen-bond is distinguished for two transition state structures (TSe1 and TSe3) in the third pathway (Figs. 10.7 and 10.8).
4. The effect of aqueous solvation generally lowers the activation energy and accelerates the phosphonylation reaction. Thus consideration of solvation effect is important in such reaction pathway studies.

10.3 The Adsorption/Binding of NAs on Various Surfaces

NAs have the property of getting adsorbed on various surfaces which include metal oxides, concrete, cement materials, clay materials and paints. There are two important consequences of such adsorption properties. Primarily the adsorption of NAs on man-made surfaces or clay materials (or soils) involves long-lasting hazardous effect due to environmental pollution. This poses an extra threat to the society, apart from the immediate mortal toxic effect of NAs on living organism. On the other hand, taking the advantage of the adsorption of NAs on various metal oxides, these materials could be destroyed through specific hydrolysis reactions. In the present

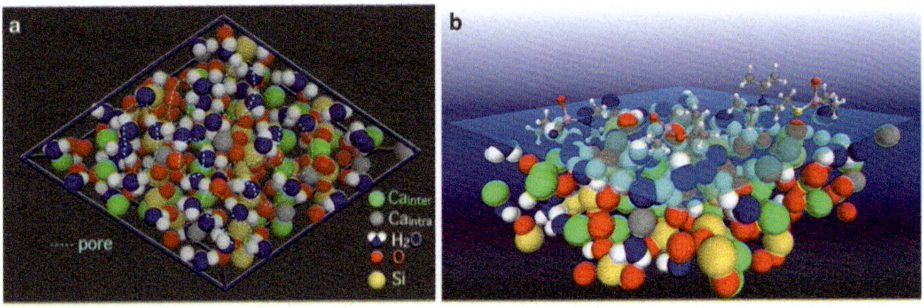

Fig. 10.9 Schematic representations of (**a**) pores in the structure of cement (the positions of pores (*Circled* by *green lines*) points to the possibility of multiple interaction sites) and (**b**) the interaction possibilities of NAs with the cement surface at the designates pore-positions

section we will discuss these two aspects of NA adsorption. The first part will cover the adsorption of NAs on the cement surfaces and paints using model systems. Such adsorption interactions will be compared against a NA-simulant and another chemical warfare agent mustard gas (HD). HD is a cytotoxic vesicant chemical warfare agent capable of forming large blisters on exposed skin and lungs. This gas is also capable of binding on surfaces causing long-lasting environmental problems and could have comparable adsorption properties with respect to the NAs. A discussion on the adsorption of sarin and soman with clay materials would also be discussed to complete the discussion on the environment pollution effect of NAs. In the second part of this section adsorption properties of NAs on metal oxides would be discussed with the probable use of these binding properties for their deactivation through hydrolysis reactions.

10.3.1 Adsorption of NAs on Cement Materials and Paints

10.3.1.1 Theoretical Investigation of the Interactions of Nerve Agents with Cement Components, and Concrete Surfaces

The structure of cement usually consists of pores where NAs or NA-simulants could interact and the possibility is there that they might also diffuse through the pores, if the size permits. A typical schematic model of such pores and the nature of interactions are shown in Fig. 10.9. The cement components include tri-calcium silicate (Ca_3SiO_5), di-calcium silicate (Ca_2SiO_4), tricalcium aluminate ($Ca_3Al_2O_6$), tricalcium aluminoferrite ($Ca_4Al_2Fe_2O_{10}$), and gypsum ($CaSO_4$, $2H_2O$). There are not many quantitative studies available on the adsorption of NAs and related species on such materials. The present discussion is based on the theoretical findings through quantum chemical model studies. The initial theoretical studies would discuss interaction analyses of NA and NA-simulants with the calcium silicate components (in cluster form) to pin point proper binding sites. The other part of such interaction studies would involve inclusion of surfaces of cements

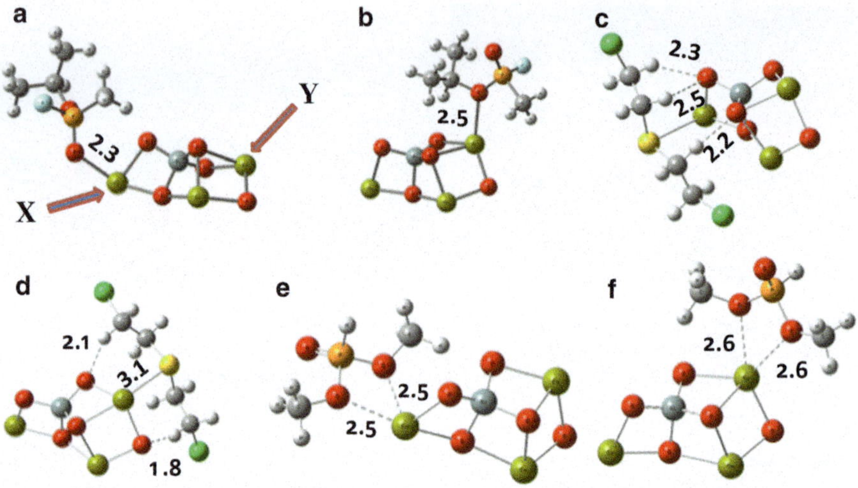

Fig. 10.10 Optimized structures of Ca_3SiO_5...Sarin (**a**, **b**), Ca_3SiO_5...HD (**c**, **d**), and Ca_3SiO_5...DMMP (**e**, **f**) at the DFT/B3LYP/6-31G(d) level of theory. The probable interaction sites of Ca_3SiO_5 are shown by *arrows* in figure (**a**). The computed interaction distances in these complexes are indicated in the figures (in Å)

containing the pores for binding (assumed to be responsible for the diffusion of such materials). The interactions were mostly investigated at the DFT [27] and semiempirical PM6 level of theories [59].

There are two exposed calcium atoms available for binding with the sarin, HD and DMMP. This is shown in Fig. 10.10 with arrows (sites X and Y). While sarin and DMMP uses the oxygen atom attached to the phosphorus to bind with the exposed calcium atoms, HD binds through the sulfur atom. The structures of such complexes are shown in Fig. 10.10 with the magnitude of binding energies (ΔE_B) in Table 10.1. The interactions are always stronger when the NAs/NA simulants bind to the site X. The binding with site Y is relatively weaker. These are evident from the interaction distances of various sites (Fig. 10.10). Although the strengths of interaction are comparable between sarin and DMMP, there is a basic difference in their nature. While sarin interacts thorough it's $P = O$ oxygen, DMMP used both the oxygen atoms attached to phosphorus, for such interactions (Fig. 10.10a–d). Interactions of HD with the exposed Ca atoms are relatively weaker because of primary Ca....S interactions. But there are also some weak hydrogen bonding interactions (due to the Cl...H interactions, Figs. 10.10e, f), and such interactions makes the total ΔE_B more or less equivalent to those of the sarin/DMMP cases.

Ca_2SiO_4, unlike Ca_3SiO_5 offers multiple binding possibilities due to planar and pyramidal nature of the silicon site (Figs. 10.11 and 10.12). Four possible interactions are possible for both sarin and HD (Fig. 10.11), while six structural possibilities are there for Ca_2SiO_4...DMMP interactions (Fig. 10.12). The ΔE_B for the most stable interactions in all the three cases (Table 10.1) are more or less similar, although the nature of bindings are a bit different due the presence of multiple weak

Table 10.1 Binding energies (ΔE_B, kcal/mol)[a] of sarin, mustard gas (HD), and DMMP with Ca_3SiO_5, Ca_2SiO_4 and interaction energies (ΔE_{int})[b] of these gases with model cement-pore

System	Structure reference	Structure	$\Delta E_B/\Delta E_{int}$
Ca_3SiO_5...Sarin	Fig. 10.10	A	−39.90
		B	−18.70
Ca_3SiO_5...HD	Fig. 10.10	C	−27.20
		D	−24.60
Ca_3SiO_5...DMMP	Fig. 10.10	E	−37.60
		F	−26.60
Ca_2SiO_4...Sarin	Fig. 10.11	A	−39.80
		B	−46.20
		C	−45.26
		D	−44.83
Ca_2SiO_4...HD	Fig. 10.11	E	−25.88
		F	−20.57
		G	−36.44
		H	−21.96
Ca_2SiO_4...DMMP	Fig. 10.12	A	−43.28
		B	−38.25
		C	−43.40
		D	−36.24
		E	42.85
		F	−47.12
Cement-pore...Sarin	Fig. 10.14	A	−72.40
Cement-pore...HD	Fig. 10.14	B	−42.25
Cement-pore...DMMP	Fig. 10.14	C	−279.85

[a]Except the cases of Fig. 10.14, all the energies are counterpoise corrected
[b]The interaction energies are not counterpoise corrected as the results are of semiempirical calculations (PM6) origin

hydrogen bonding interactions (Figs. 10.11 and 10.12). These weak interactions also control the relative stability of a particular complex in such interactions. The magnitudes of these interaction distances are available in both Figs. 10.11 and 10.12. The Ca...O (Fig. 10.11) and Ca...S (Fig. 10.12) are always the most strongly interacting sites. This is followed by weak hydrogen bonding interactions. The Ca_2SiO_4...HD interactions are much weaker than those of sarin or DMMP cases since the Ca...S interactions are much weaker than Ca...O. The weak hydrogen bonding interactions do not help to make ΔE_B stronger (Table 10.1). If the hydrogen bonding interactions are on the stronger side, it can provide some extra relative stability over the other complexes. For example both sarin and DMMP in the complexes B (Fig. 10.11b) and F (Fig. 10.12f) interact almost equally strongly with Ca_2SiO_4, but the interactions in DMMP case is slightly stronger because of the P-H...O hydrogen bonding effect. The sarin complex of Fig. 10.11b has much weaker O-H...O hydrogen bonding due the vertical orientation of O-H oxygen with respect to the second oxygen (responsible for the hydrogen bond).

The nature of interactions for all the complexes depends on the charge redistribution due to the approaching NAs/NA-simulants towards Ca_3SiO_5/Ca_2SiO_4. This

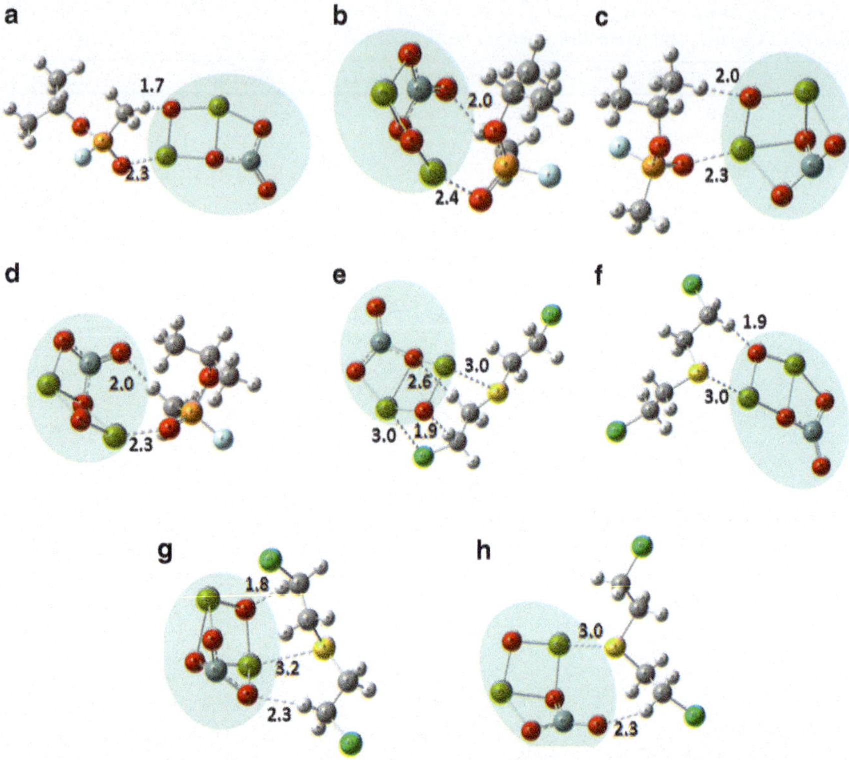

Fig. 10.11 Optimized structures of Ca_2SiO_4...Sarin (**a–d**) and Ca_2SiO_4...HD (**e–h**) complexes at the DFT/B3LYP/6-31G(d) level of theory. The Ca_2SiO_4 orientation in each of the complexes is shown within the *shaded region*. The computed interaction distances in these complexes are indicated in the figures (in Å)

could be understood from an analysis of the molecular electrostatic potential (MEP) surfaces of the respective complexes. Figure 10.13 presents the MEP surfaces of the most strongly bound complexes with Ca_3SiO_5 and Ca_2SiO_4 as discussed above (Figs. 10.10, 10.11, 10.12 and Table 10.1). The P = O oxygen atom of sarin/DMMP binds to the more positively charged Ca atom of the Ca_3SiO_5 and Ca_2SiO_4 molecules. This interaction is followed by negative charge migration to the neighboring oxygen centers (yellow shades around oxygen atoms – indicating negative MEP) (Fig. 10.13a, c, d, e). In the case of HD similar effect is originated through the S...Ca interactions (Fig. 10.13b, e). Such charge migrations are responsible for the observed hydrogen bond formation in the complexes. The other important feature observed from the MEP surfaces is that the basic interaction patterns in such complexes are similar and this generates more or less equivalent ΔE_B trend in such complexes.

There is need to figure out whether these simple interaction pictures could be transferred to the real surface interactions. Since the NA/NA-simulants are

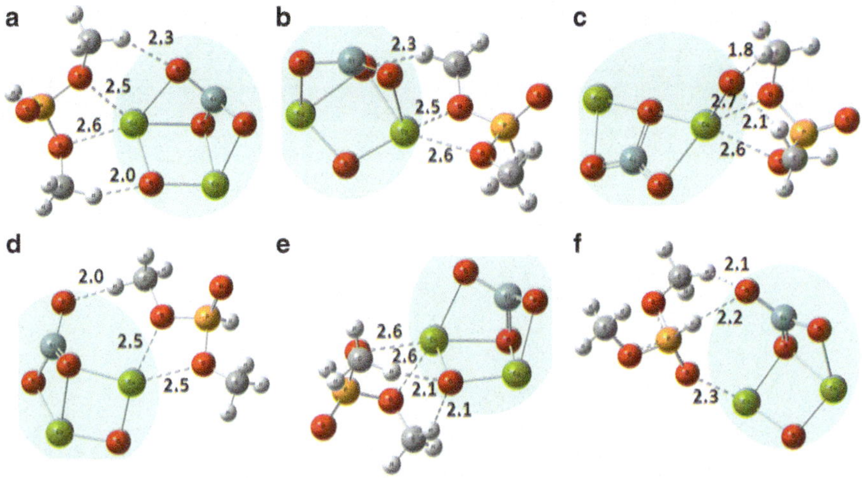

Fig. 10.12 Optimized structures of Ca_2SiO_4...DMMP (**a–f**) complexes at the DFT/B3LYP/6-31G (d) level of theory. The Ca_2SiO_4 orientation in each of the complexes is shown within the *shaded region*. The computed interaction distances in these complexes are indicated in the figures (in Å)

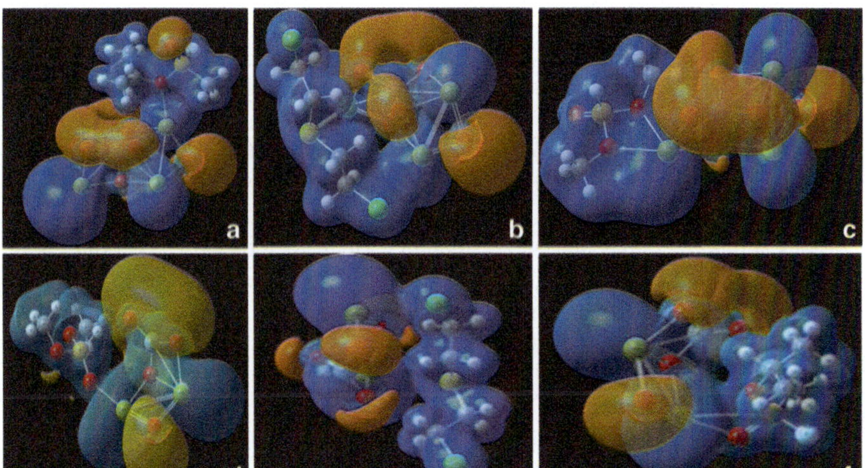

Fig. 10.13 MEP iso-surfaces of the most strongly interacting complexes of (**a**) Ca_3SiO_5...sarin (**b**) Ca_3SiO_5...HD, (**c**) Ca_3SiO_5...DMMP, (**d**) Ca_2SiO_4...Sarin, (**e**) Ca_2SiO_4...HD and (**f**) Ca_2SiO_4...DMMP at the DFT/B3LYP/6-31G(d) level. *Blue shades* are positive potential zones while *yellow shades* indicate negative ones. The *green shades* indicate boundary between negative and positive potential zones

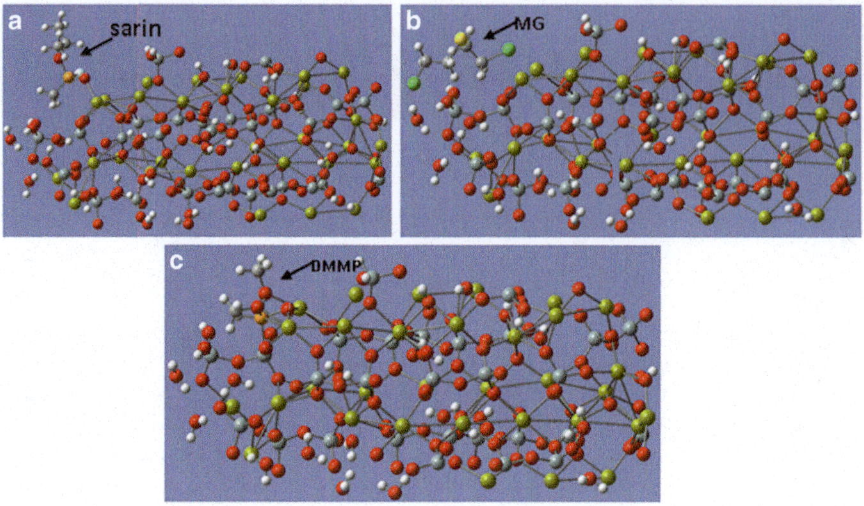

Fig. 10.14 Cement pore adducts of sarin, HD, and DMMP at the semi-empirical PM6 level of theory

supposed interact with the cement-pores, model structure from the available structure of cement [60] have been generated to monitor the nature of interactions of such sites with sarin, HD and DMMP. Only one pore with biggest radius was chosen for such investigation and the interaction energy calculations were carried out at the semi-empirical PM6 [59] level keeping the pore structure frozen at its original geometry. The open edge positions were terminated by hydrogen atoms to minimize the edge effect. The results of computation are summarized in Fig. 10.14 and interaction energy (ΔE_{int}) values are available in Table 10.1. Since these interaction calculations were carried out on partially frozen structures, exact ΔE_B values in such cases cannot be reported. The results (ΔE_{int}) reveal some interesting features of binding of the NAs and NA-simulant to the cement pore site. All of these molecules are binding to the multiple sites (due to their availability on the surface). The P = O oxygen of sarin is binding to the two calcium sites, and considering that they are almost equivalent, the individual interactions amounts to ~36 kcal/mol. This interaction energy is similar to the binding with the calcium silicates. The nature of HD…cement-pore interactions are also similar to the binding with calcium silicates (Table 10.1). One of the Cl atoms of HD binds to the two calcium atoms (like sarin), while the other develops hydrogen bond with the nearest available water molecule. The interaction energy is still much less than the case of sarin. The important feature in both the cases is that they do not fit with the pore size and thus cannot enter within the pore. The case of DMMP is very different. It fits to the pore size and thus interacts very strongly with multiple binding sites. As a result the interactions in this case are very strong and not comparable with sarin or HD. Thus, while DMMP, HD and sarin show similarity of interactions during binding with a single molecule (like calcium silicates), DMMP is not a very efficient NA-simulant when the nature of interactions with cement pores need to be investigated for comparative studies. This is a very preliminary result in this area and computational studies are not available with higher level of theories. More advanced computations are needed to find the validity of such observations.

10.3.1.2 Theoretical Investigation of the Interactions of Nerve Agents with Components of Paints

Paint materials also can act as host to bind NAs and depending on the strength of binding such occlusion of NAs is capable of imposing long lasting environmental hazard. A wide variety of raw materials are used in manufacturing of paints and they can be grouped according to their function. Paints usually have two primary components viz., polymeric binders and pigments. Polymeric binders are resinous materials like acrylics, polyurethanes, polyesters etc. They could be both synthetic as well as natural resins (such as alkyds, acrylics, vinyl-acrylics, vinyl acetate/ ethylene (VAE), polyurethanes, polyesters, melamine resins, epoxy, or oils). We will discuss here the binding of NAs with the most commonly used polyurethane and epoxy pre-polymer binders.

Polyurethane (PU) is a polymer composed of a chain of organic units joined by carbamate (urethane) links and are formed by combining two or higher functional monomers [61]. Epoxy paints are usually used in industrial and automotive applications since they are more heat resistant than latex-based and alkyd-based paints [62]. The interactions of NAs with polymeric materials have not been studied theoretically in details. In the resent review several recent investigations on the interactions of nerve agents with the most commonly used polyurethane and epoxy pre-polymer binders in the paints would be discussed. It focuses on the theoretical binding analysis of sarin and HD on the dimer, trimer and tetramer units of urethane ($n = 2$, 3 and 4) and the epoxy pre-polymers starting (from $n = 1$ to $n = 4$). All of these discussions on binding properties are based on the results of computational studies using DFT/M06-2X [27, 63] level of theories.

The binding of sarin and HD with epoxy polymeric units (designated as Epoxy-1, Epoxy-2, Epoxy-3, and Epoxy-4 for $n = 1$–4) are basically weak in nature (Table 10.2). The optimized structures of these complexes are shown in Fig. 10.15. The complexes Epoxy-2...sarin and Epoxy-4...sarin seem to have similar type of spiral arrangements and such an orientation of the epoxy systems around sarin is responsible for the relatively stronger interactions in these complexes. The major stabilizing forces could be the multiple hydrogen bonding (like O-H···O, C-H···O and C-H···F). The interactions in Epoxy-1...sarin and Epoxy-3...sarin complexes are much weaker (−4.4 and −4.0 kcal/mol, Table 10.2) as sarin is sitting over the epoxy moiety through very weak hydrogen bonding interactions and the other stabilizing factors available in the cases of Epoxy-2...sarin and Epoxy-4...sarin interactions are absent in these complexes. HD also shows similar weak interactions with the epoxy polymers (Fig. 10.15, Table 10.2). They are less strongly bound to the different epoxy polymeric units than the corresponding sarin complexes. The Epoxy-4...HD

Table 10.2 Binding energies (ΔE_B, kcal/mol)[a] of sarin, and mustard gas (HD) with Epoxy (n = 1–4) and polyurethane (PU; n = 2–4) monomer/polymeric units

System[b]	ΔE_B	System[b]	ΔE_B
Epoxy-1. . .Sarin	−4.40	Epoxy-1. . .HD	−3.90
Epoxy-2. . .Sarin	−13.60	Epoxy-2. . .HD	−7.00
Epoxy-3. . .Sarin	−4.00	Epoxy-3. . .HD	−10.60
Epoxy-4. . .Sarin	−18.70	Epoxy-4. . .HD	−18.70
PU-2. . .Sarin	−8.90	PU-2. . .HD	−14.50
PU-3. . .Sarin	−25.30	PU3. . .HD	−19.10
PU-4. . .Sarin	−26.90	–	–

[a]All the interaction energies are counterpoise corrected
[b]Refer to Figs. 10.15 and 10.16 for the definitions of the respective structures

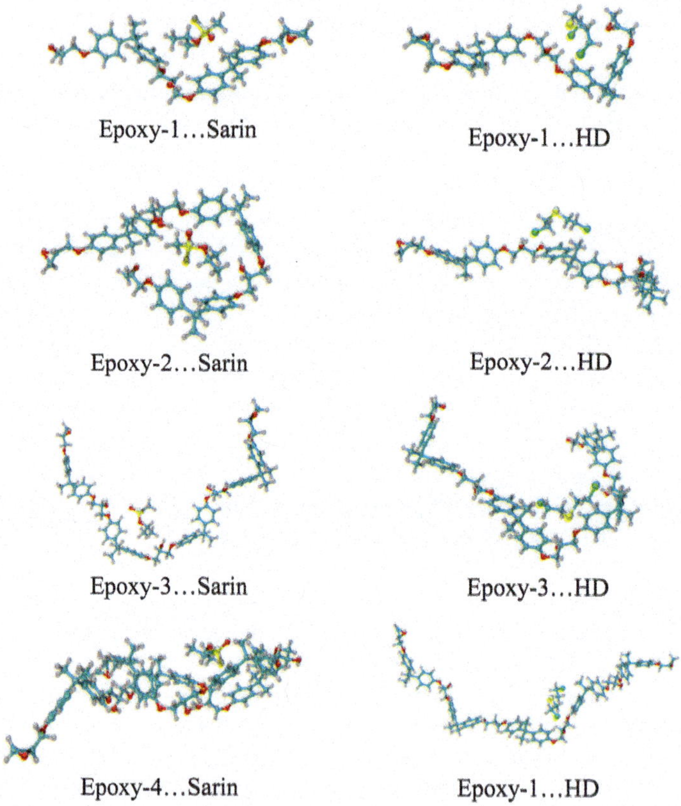

Fig. 10.15 M06-2X/6-31G(d) level optimized structures of the complexes of epoxy pre-polymers (Epoxy-1, Epoxy-2, Epoxy-3 and Epoxy-4) with sarin and HD

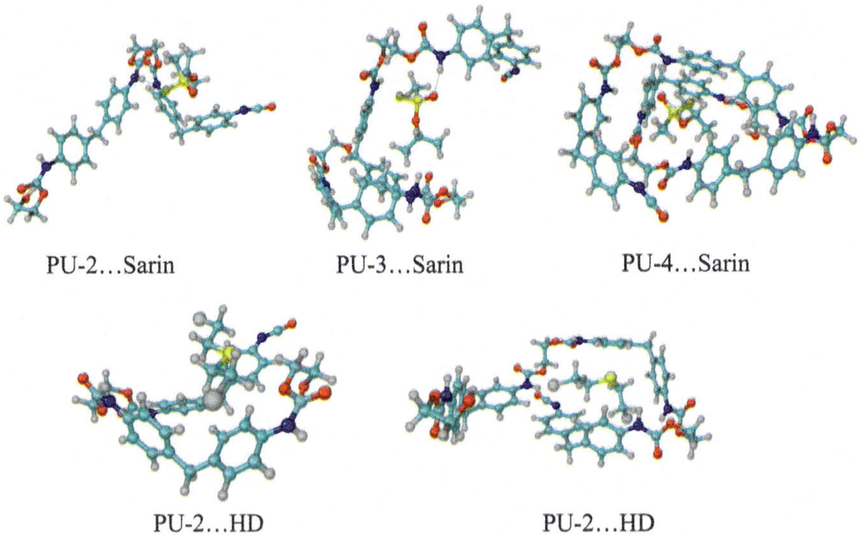

Fig. 10.16 M06-2X/6-31G(d) level optimized structures of the complexes di-, tri-, and tetra-urethane (PU-2, PU-3 and PU-4)...sarin and PU-2 and PU-3...HD

complex is less strongly bound than the Epoxy-3...HD and Epoxy-2...D complexes. Multiple H-bonds due to structural deformation (like sarin complexes) are actually the stabilizing forces for such complexes.

The optimized structures of PU (n = 2–4) are more puckered when n is increased. As a consequence of such structural effect, when sarin and HD interacts with higher polymeric units of PU, they are more and more surrounded by the polymeric chains and the interactions become stronger. The ΔE_B values in both the cases increase regularly with the increase of the size of the polymeric units (Table 10.2). The interactions of HD with PU-2 and PU-3 are much weaker than the corresponding sarin complexes. The PU-4...sarin interaction is the strongest here. Sarin also forms much stronger complexes with di-, tri- and tetra-urethane than with the corresponding epoxy polymeric units. Such stronger interaction involving PU is due to multiple inter- and intra-molecular H-bonding and C–H···O interactions. These are evident from the comparison of the optimized structures of epoxy...sarin (Fig. 10.15) and PU-sarin (Fig. 10.16) complexes. The interactions in PU-2...HD complex are stronger than PU-2...sarin, while it is weaker in PU-3-HD with respect to the corresponding PU-3...sarin (Table 10.2). The optimized structures of PU...HD complexes have similar structural features like the PU—sarin complexes (Fig. 10.16) and the interactions in such complexes are also stronger than the corresponding epoxy...HD complexes.

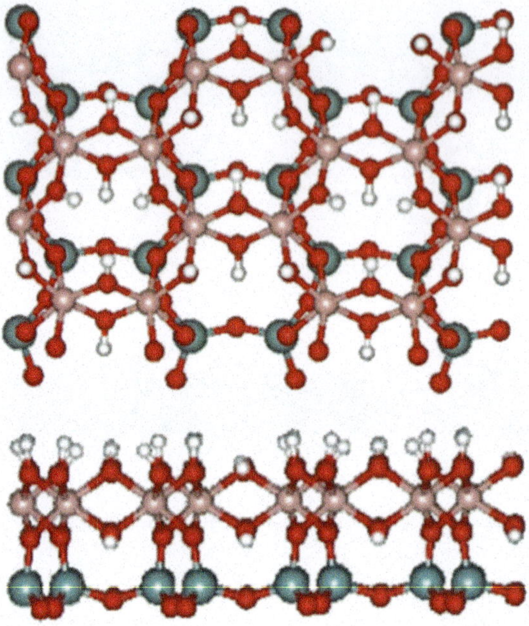

Fig. 10.17 Optimized structure of the isolated 1:1 kaolinite layer: top (*upper*) and side (*lower*) views (Reproduced from Michalkova et al. [65], with the kind permission of the American Chemical Society 2004)

10.3.1.3 Adsorption of NAs on Clay Materials

Clay minerals are naturally occurring adsorbents of NAs [64]. They primarily exist as very small particles (~2 μm in diameter) with a high specific surface area. Their structural diversities provide the functionality that facilitates adsorption of ions and molecules (Fig. 10.17) [66]. Adsorption studies of sarin on dickite (a 1:1 dioctahedral clay mineral) on the octahedral and tetrahedral sites indicated that the polarization of the organic molecule is more significant in the case of adsorption on the octahedral side [65, 67]. The computed adsorption energies of sarin (GB) and soman (GD) on the octahedral surface of dickite were found to be about −16.0 and −15.0 kcals/mol respectively. A larger model of the dickite was used for such data collection and the computations were carried out using ONIOM (B3LYP/6-31 g(d,p):PM3) level of theory [68]. Similar calculations on the adsorption of GB and GD on the tetrahedral surface showed the respective interactions to be much weaker (−7.0 and −9.0 kcal/mol respectively). These adsorption energies are quite weak and could be comparable to the adsorption energies of NAs on cement and paint surfaces (discussed in the previous section).

In the case of adsorption on the octahedral surface, significant contributions of the electrostatic and delocalization were found in the total interactions for both the NAs [65]. In the case of the tetrahedral surface, the contribution of the Hartree-Fock term to the total interaction energy is repulsive. It originates from repulsive interactions between the P = O oxygen atom of sarin (and soman) and the basal oxygen atoms of the tetrahedral surface of dickite. The correlation forces play important role in the total interactions of sarin and soman on the tetrahedral surface. An additional local minimum of adsorbed sarin, on the tetrahedral and octahedral surface of dickite, was also found. They were occurring due to the involvement of the fluorine atom (of NAs) in intermolecular interactions. The hydrogen-bonds are weaker with respect to those occurring due to the participation of the P = O oxygen atom and hence such minima are relatively less stable.

10.3.2 Interactions of NAs with Metal Oxides

Metal oxides offer a much stronger adsorption surface (with respect to the previously discussed cases) for the NAs in both pure and hydroxylated conditions [2, 69–78] due to availability of the larger surface area and considerable basic character. The NAs mostly bind to the oxides through the P = O bonds and quantum-chemical studied on the adsorption of NAs on metal oxide surfaces have shown that there could be two different ways through which such interactions can take place. If the NAs bind to the oxide surfaces through the edges, the binding is quite strong. This was observed when sarin was allowed to bind to a single unit of model magnesium oxide surface [79]. Both physisorption and chemisorption (destructive adsorption) were observed in such computational studies [79]. The physisorption was interpreted due to the formation of hydrogen bonds and ion-dipole and dipole-dipole interactions between adsorbed sarin and the surface. The chemisorption, on the other hand, occurs due to the formation of covalent bonds between the sarin and the surface. The adsorption energy of the most stable chemisorbed system was found to be −50.0 kcals/mol [79] (MP2/6-31 g(d) level of theory [32]). It was also found that in the case of hydroxylated MgO, the adsorption of sarin is much weaker (−39.3 kcals/mol at the MP2/6-31 g(d) level) [79].

Four different type of clusters were studied to monitor the edge-effect on the interactions of tabun (GA) with CaO (modeled as Ca_4O_4) (Fig. 10.18) and hydroxylated CaO (modeled as $Ca_4O_4(OH)_2H_2$) and partially hydroxylated CaO (modeled as $Ca_4O_4(OH)H$) (Fig. 10.19) [80]. The computations of the adsorption of tabun on these model non-hydroxylated and hydroxylated CaO indicated that GA is strongly bonded (chemisorbed) to the non-hydroxylated CaO (A1-GA) through the P = O bond (ΔE_{int} = −87 kcal/mol). Although the interactions in the second model of this complex (A2-GA) is weaker (ΔE_{int} = −35 kcal/mol) and is not energetically favorable. In the hydroxylated complexes (B-GA, ΔE_{int} = −26.7 kcal/mol; and C-GA, ΔE_{int} = −27.4 kcal/mol), the interactions are competitive and weak in nature (physisorption). Adsorption of NA-simulants dimethylmethylphosphonate (DMMP) and trimethylphosphate (TMP) on similar non-hydroxylated and hydroxylated CaO clusters were further investigated to confirm the effect of edges of CaO clusters on the binding nature of NAs studied so far [81]. The results indicated that

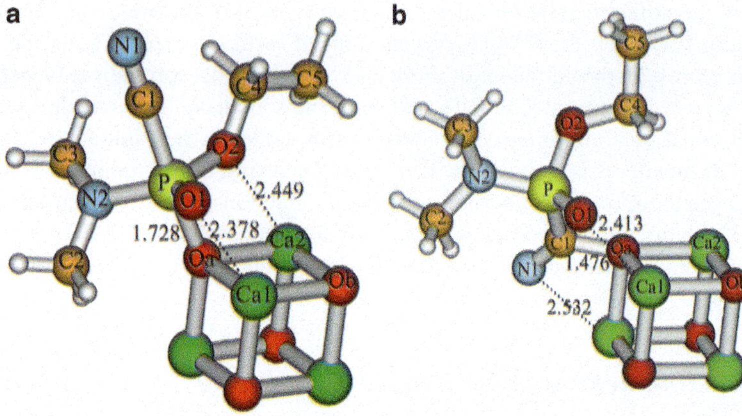

Fig. 10.18 The optimized geometry of tabun (DFT/B3LYP/6-31G(d)) adsorbed on the non-hydroxylated CaO cluster in (**a**) A1-GA and (**b**) A2-GA complexes (Reproduced from Michalkova et al. [80], with the kind permission of Elsevier 2007)

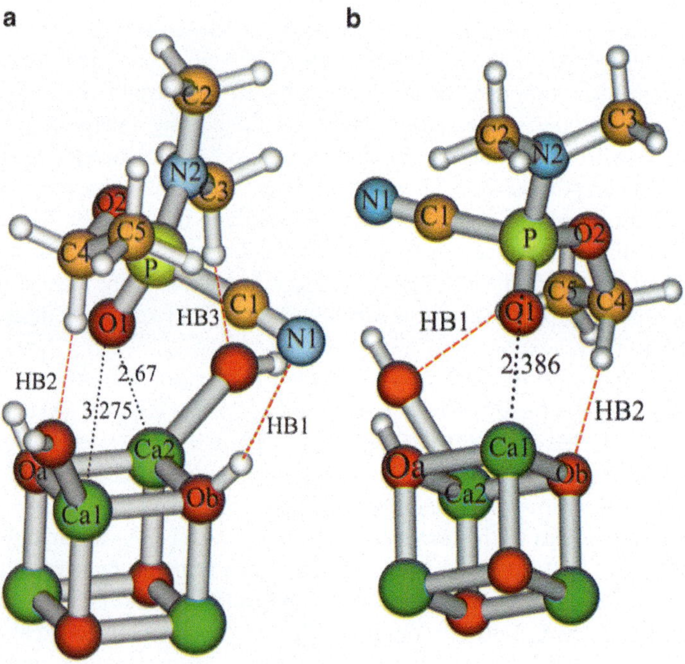

Fig. 10.19 The optimized geometry of tabun (DFT/B3LYP/6-31G(d)) adsorbed on the hydroxylated CaO cluster in (**a**) B-GA and (**b**) C-GA complexes (Reproduced from Michalkova et al. [80], with the kind permission of Elsevier 2007)

the most energetically favorable adsorption of DMMP and TMP occurs at the configuration where the oxygen atoms of the P = O and methoxy groups point towards the Ca atom of the cluster-edge and the P atom points towards the O atom forming a P-O bond. Interestingly, this optimal configuration was revealed for both the non-hydroxylated and partially hydroxylated CaO-DMMP and CaO-TMP complexes [81]. The study concluded that the number of surface hydroxyl groups on the CaO models governs the adsorption of studied compounds. The calculations provided details of the interactions and revealed that DMMP and TMP form much less stable complexes when interacted to completely hydroxylated CaO surface. In such a case the adsorption occurs only via weak electrostatic interactions and H-bonding to the surface oxygen atoms and hydroxyl groups. This was also observed in the case CaO...Tabun complexes [80].

There could be a second kind of adsorption of NAs if the edge positions of the metals oxides could be avoided. In such case the interaction energies of the complexes could be much weaker. The adsorptions of the NA diisopropylfluorophosphate (DFP) was investigated on larger metal oxide $M_{16}O_{16}$ (M = Mg, Ca) using the DFT and MP2 techniques [82]. The cluster structures was chosen in such a way that they resembled the (001) surface. The geometries of the fully optimized $M_{16}O_{16}$...DFP (M = Mg, Ca) complexes showed that they could serve as a model for studying the adsorption DFP on the respective oxide surfaces. The computed structures have shown that primary adsorption of DFP on these MgO/CaO clusters occur through oxygen atom of the P = O bond (like the previous small structure cases), but the structure did not have the multiple bonding characteristic of the edge-effect. The calculated interaction energies are much lower ($Mg_{16}O_{16}$...DFP, $\Delta E_{int} = -19.5$ kcal/mol; $Ca_{16}O_{16}$...DFP, $\Delta E_{int} = -10.0$ kcal/mol - DFT/B3LYP/LANL2DZ level) [82]. DFP is actually physisorbed on these surfaces and adsorption on the model MgO surface is stronger.

There is another important interaction studies of GA and NA-simulants (DMMP and TMP) on the $(ZnO)_n$ (n = 4, 18, 60) surfaces [83]. The sizes of the surfaces were found to affect the strength of interactions to a large extent. GA is bound more tightly to the ZnO surface than DMMP. The predicted adsorption energies of ZnO-GA and ZnO-GA are almost 50 % lower than it was calculated for the CaO-GA and CaO-DMMP systems due to the avoidance of the edge-effect ($\Delta E_{int} = -54.1$ kcal/mol, DFT/B3LYP/LANL2DZ). If the edge-effect is fully reduced, as was the case of $Zn_{60}O_{60}$...GA complex, the ΔE_{int} was found to be further reduced (−46.3 kcal/mol) [83].

10.4 Deactivation of NAs

The previous discussions on the NA-adsorption properties on cement and paint materials surmises that they do not bind strongly on such materials and it is also not very probable that NAs would penetrate inside the cement-pores due to size factors. Such properties of NAs could be used to destroy those using moderately stronger external binders and subsequent chemical processing (viz. hydrolysis) of the bound materials. Metal oxides have long been advocated as a reasonable material for such purposes because of their basic properties. The investigation of the binding characteristics of NAs with metal oxides and hydroxylated metal oxides showed that they could be used as effective binders for the NAs adsorbed on various surfaces [69–83]. Before going to the detailed review of hydrolysis processes of NA. . .metal oxide bound complexes, we first summarize several other available NA-destruction techniques. This primary discussion would show the pros and cons of the various methods used for NA-deactivation and would point to the probable superiority and easiness of NA-destruction techniques using chemical reactions on metal oxide surfaces.

There are several aspects of NA-deactivation. The most important factor is the time for deactivation. If the stockpile could be shifted to the destruction facility, sufficient time could be provided for such deactivation. In case of the on spot deactivation, the reactions should be faster with less side products (warranting further treatment). These basic criteria have generated several deactivation procedures (involving a wide spectrum of materials (chemicals) to carry out such reactions) as listed below.

10.4.1 Incineration

The NAs are destroyed thermally by treating with excess oxygen [84]. The end products depend on the nature of the NAs. The G-type NAs (sarin, soman etc.) produce P_2O_5 together with CO_2, H_2O, and HF. The V-type NAs (e.g., VX) produce SO_2 and NO in addition to P_2O_5, CO_2, and H_2O. The pollutant end products (e.g., P_2O_5, SO_2 etc.) need further treatment in such procedures.

10.4.2 Hydrolysis by Alkalis

The hydrolysis by alkalis (e.g., NaOH) usually has a mixed effect. Sarin, e.g., is fully neutralized [84]. VX on the other hand produces both toxic and non-toxic end products [85]. These end products need further treatment for complete deactivation. There are also further problems of dumping the materials and to tackle the corrosion of the reactor containers (due to use of supercritical water at high temperature and pressure) [86].

10.4.3 Decomposition Reactions

These reactions actually form the basis of more advanced deactivation procedures of NAs. The NAs are known to undergo hydrolysis reactions in aqueous medium under acidic/basic conditions [87–89]. These reactions are quite slow (half-life ~60 h) [90] but could be enhanced under several conditions. These include the use of peroxide (known as perhydrolysis of NAs) or bleach (e.g., mixture of Ca (OCl)Cl, $Ca(OCl)_2/CaO/MgO$ together with other active ingredients like detergent water, sodium citrate and NaH_2PO_4) [69] as catalyst and various metal-based approaches (where the metal catalyzes the reaction) [91] to hydrolyze the NAs. Maintenance of pH and temperature is essential in such reactions. The half-life of the reactions is substantially reduced (half-life ~15 min) [92], although some end toxic products need further treatment to generate non-toxic products.

10.4.4 Biotechnological Degradation and Related Processes

A series of techniques have been developed depending on the degradative binding nature of NAs with various kinds of cells (including human, animal, and bacterium), isolated enzymes, and antibodies [2]. The degradation of phosphates and phosphinates by cyclodextrine [93] and halogen oxide (e.g., BrO^-) [94] assisted reactions are also explored for the possible NA-destruction. These techniques are still under developing stage.

10.4.5 Surface Chemistry

A variety of species were treated with emphasis on oxidation, degradation, and hydrolysis to develop potential materials for NA-destruction [95]. Several NA-simulant (e.g., dimethylmethylphosphonate, DMMP) were studied for degradation on bare metals and solid nanoparticles [2]. The hydrolysis of various NAs on metal oxide nanoparticle (e.g., MgO, CaO, Al_2O_3) surfaces has also been investigated [77, 96]. Several metal nanoparticles (mostly Ag), deposited over zeolites were also used for such studies in recent times [97]. The hydrolysis on oxide surfaces are usually considered as effective. We are going to elaborate the various aspects of NA…metal oxide interactions and NA-deactivation in details as it leads to one of the most promising way to destroy NAs without any toxic environmental side effects.

10.4.6 Detailed Analysis of the Hydrolysis Reactions of Metal Oxide Bound NAs for Their Deactivation

The previous discussions on the results of model calculations of sarin on MgO and tabun on CaO clusters (both pristine and hydroxylated) have shown that they bind less strongly with the hydroxylated clusters [79, 80]. Relatively weak binding in the

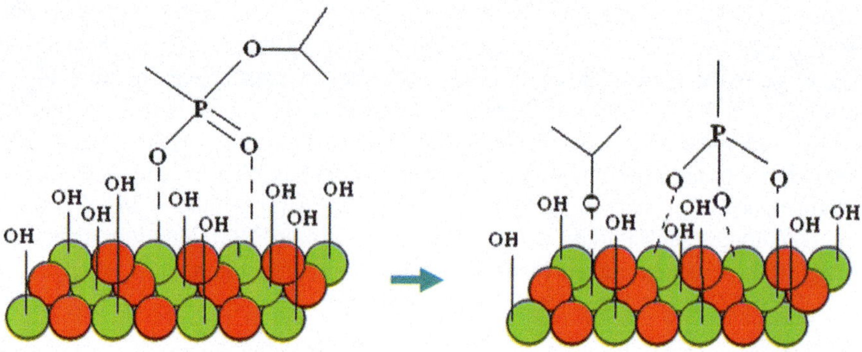

Fig. 10.20 Schematic representation of the degradation of sarin on hydroxylated metal oxide surface. The scheme is designed for the divalent metal (Mg, Ca, Zn etc.). In the case of trivalent metals, the scheme would me more or less similar with some modifications

case of hydroxylated clusters usually facilitates the hydrolysis reaction. The general hydrolysis steps on oxide surfaces are schematically shown in Fig. 10.20. These steps are hypothetical, but in some cases (e.g., hydrolysis on Al_2O_3) could be confirmed from the available NMR data [96].

With the above background of degradative adsorption of NAs on hydroxylated metal oxide surface, we are going to discuss the specific cases of sarin, on the hydroxylated MgO, surfaces. Both complete and partial hydroxylation of these nanoparticles is possible through exposure to humid air [77, 96]. The partially hydroxylated nanoparticles would be considered as it leads to convenient hydrolysis. With the properly designed model chemistry, the binding studies on the partially hydroxylated MgO would enable us to follow the course of the degradation reactions. The reaction path analysis would not only confirm the validity of the reaction paths with the probable location of the transition states, but they would also be useful to predict the efficiency of these materials as NA-degradation agents. Moreover the complete study of such reactions would enable one to determine the factors (e.g., defect position on the surfaces, or doping of other foreign metals) that would enhance these reactions on such surfaces (which are the ultimate goal of such material designing).

Recently sarin degradation was investigated on the brucite [$Mg(OH)_2$] surface using DFT techniques [98]. We are presenting below the similar degradation reaction of sarin on partially hydroxylated MgO clusters of various sizes. Sarin, in these studies, is treated as a model system. The $-CH(CH_3)_2$ group (Fig. 10.1) is replaced by $-CH_3$. Such modeling is logical, as it does not affect the course of the reaction (see Fig. 10.20). Four models of MgO clusters were constructed as $Mg(OH)_2$, $Mg_4O_4(OH)_2$, $Mg_6O_6(OH)_2$, and $Mg_8O_8(OH)_2$. Sarin was allowed to react with these four hydroxylated MgO clusters. Although these reactions are known to occur [77, 78, 96], there are not many theoretical studies available in this directions except on model brucite surface [98]. These preliminary mechanistic studies are reported here at the semi-empirical PM3 [99, 100] level of theory. Full geometry

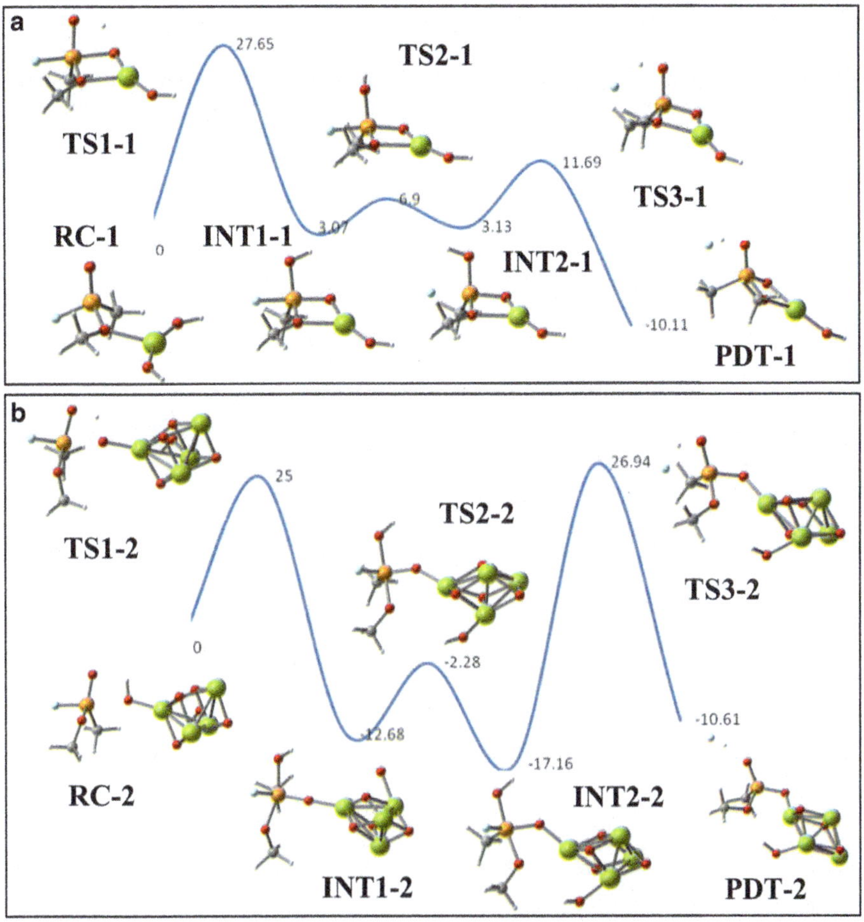

Fig. 10.21 Computed reaction paths of the hydrolysis of sarin by (**a**) $Mg(OH)_2$ and (**b**) $Mg_4O_4(OH)_2$ at the semi-empirical PM3 level of theory. The reaction complex (RC), different intermediates (INT), transition states (TS) and the end product (PDT) are marked in the figures with their respective relative energies (with respect to RC)

optimizations of the model systems, harmonic frequency analyses and intrinsic reaction coordinate (IRC) analysis were carried out to ascertain both the stationary points and the transition states. All of these reactions report the mechanism of the primary step for the elimination F from sarin as HF (as shown in Fig. 10.20). The reaction paths for the HF elimination of sarin using $Mg(OH)_2$ and $Mg_4O_4(OH)_2$ are mechanistically similar (Figs. 10.21a, b). The initial reactant complex in both the cases (RC-1 and RC-2) is formed through P. . .O interactions and they are of similar type (2.11 Å in both RC-1 and RC-2). The dissimilarities lie in the rate determining step. In the case of $Mg(OH)_2$. . .sarin reaction (Fig. 10.21a), the primary addition step in the first stage is the rate-determining step of the entire pathway. The energy barrier of this step is estimated to be ~28.0 kcal/mol. The next step is the elimination of fluorine from the sarin and the final products are HF and the HO-Mg-O...CH_3O-P(O)-CH_3 complex. The energy barrier for this elimination is less than 9.0 kcal/mol (Fig. 10.21a). There is another intermediate (INT2-1) prior to this HF elimination step. It is connected to the primary adduct through a low-barrier transition state (TS2-1). In the case of sarin...$Mg_4O_4(OH)_2$ the addition step in the first stage has an energy barrier of 25 kcal/mol. But this is not the rate determining step. The final elimination step of HF is the rate determining step with an energy barrier of ~44 kcal/mol (Fig. 10.21b). The final products in this reaction are HF and Mg_4O_4(OH)O...CH_3O-P(O)-CH_3 complex. The other features of this reaction are similar to that of sarin...$Mg(OH)_2$ reaction.

The reaction steps of sarin with higher clusters $Mg_6O_6(OH)_2$ and $Mg_8O_8(OH)_2$ are bit different than those of the previous two cases. Although the initial RC in $Mg_6O_6(OH)_2$...sarin interactions (RC-3, Fig. 10.22a) is like the previous two cases, the P...O distance is longer (2.17 Å). The nature of interactions in initial RC of the $Mg_8O_8(OH)_2$...sarin complex (RC-4, Fig. 10.22b) is a bit different. The main stabilizing interaction is due to the O...H (of CH_3; distance 2.58 Å). The primary adduct in both of these two cases undergo direct HF elimination through a transition state (Fig. 10.22). In the case of $Mg_6O_6(OH)_2$...sarin interaction, this is the rate determining step (barrier height is ~38 kcal/mol) (Fig. 10.22a). The transition state for the primary adduct (TS1-4) is the rate determining step in the $Mg_8O_8(OH)_2$...sarin reaction (barrier height is ~41 kcal/mol) (Fig. 10.22b). The end complexes in both cases (after HF elimination) are Mg_6O_6(OH)O... CH_3O-P(O)-CH_3 and Mg_8O_8(OH)O ... CH_3O-P(O)-CH_3 respectively (Fig. 10.22a, b). All of these end products as represented in Figs. 10.21 and 10.22 are in conformity with the HF elimination product in Fig. 10.20 and sarin...brucite reaction case [97]. It is to be mentioned in this connection that in the reaction paths of Figs. 10.21a, b and 10.22a, the initial RC has higher energy than the final end product. These reactions could be interpreted as exothermic in nature. The reaction paths in Fig. 10.22b, on the other hand shows that the initial RC is slightly more stable (0.06 kcal/mol) than the final end product. The general interpretation is that that the overall reaction could be slightly endothermic, although considering the level of theory used (PM3) such small energy differences between the reactive ingredients cannot be considered as fully reliable.

It is to be noted from the above results that different models of the MgO clusters play an important role in the energy profile of the reaction pathways for the hydrolysis of sarin. For further understanding the intrinsic mechanisms of the hydrolysis of sarin by partially hydroxylated MgO clusters, more models of MgO clusters are needed to be examined at higher levels of theory. This is needed, as the semi-empirical techniques usually overestimate the reaction barrier by almost one and half times than the higher correlation methods. In that respect, although the estimated barriers are higher, they would be much reduced at higher level of theories and predict the feasibility of such reactions. These reactions are in fact known to occur [2] and the present theoretical results generate the reaction steps, which are not very well studied previously.

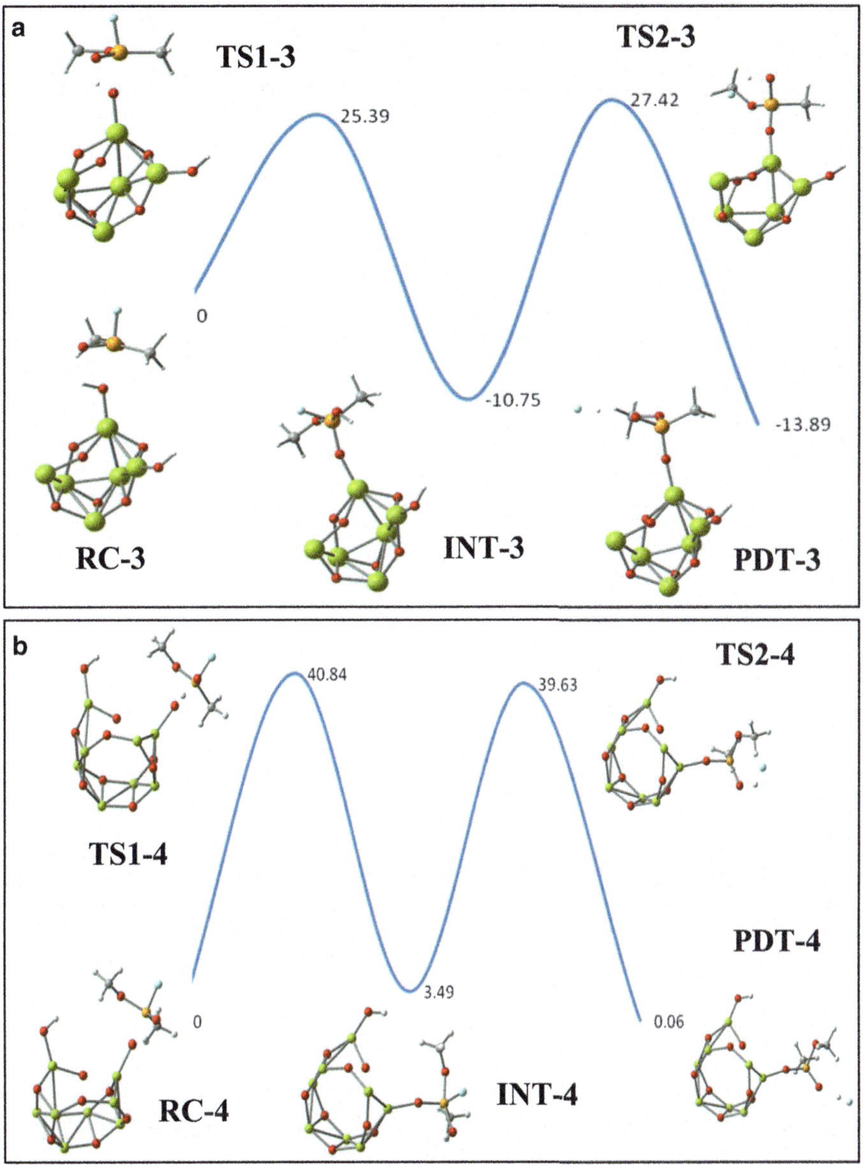

Fig. 10.22 Computed reaction paths of the hydrolysis of sarin by (**a**) $Mg_6O_6(OH)_2$ and (**b**) $Mg_8O_8(OH)_2$ at the semi-empirical PM3 level of theory. The reaction complex (RC), different intermediates (INT), transition states (TS) and the end product (PDT) are marked in the figures with their respective relative energies (with respect to RC)

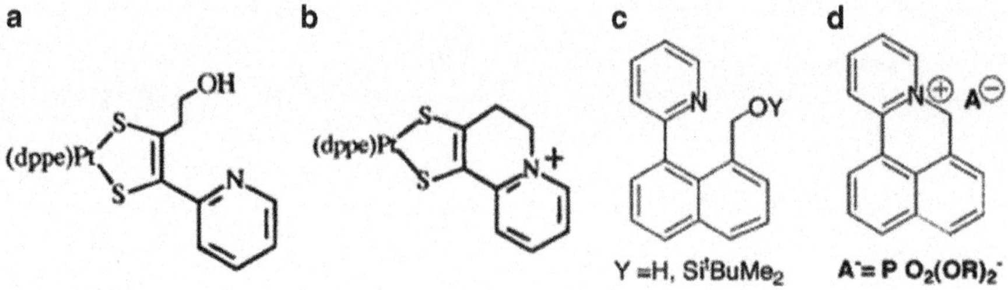

Fig. 10.23 Structures of (**a**) the open-ring platinum 1,2 –endothiolate complex, (**b**) corresponding ring-closed complex, (**c**) open-ring naphthalene-derivative (chemosensor) and (**d**) the corresponding closed-ring system used in the detection of NAs. The *closed-ring structures* are due the reactions of the open-ring systems with NAs with the enhanced luminescence and phosphorescence properties

10.4.7 Detection of NAs

There are various techniques and materials to detect NAs. A detailed discussion is available in a recent review [2]. We are going to discuss here about most commonly used techniques and materials available with the emphasis on the theoretically studied properties of materials used in the NA-detection through Resonance-Raman technique. In most of the cases NAs are needed to be detected on the spot. The early instrumental techniques (e.g., gas chromatography, mass spectrometry, and NMR spectroscopy) are more or less reliable for such detection but are expensive and non-portable to the site [2]. In last 15 years several developments have been achieved to detect NAs. They, in general, meet the basic criteria of a detector and we present below several important developments in this respect. These detection procedures are based on the chemical behavior of NAs (organophosphates in general) towards certain compounds (chemophores) and the spectroscopic behavior of the end-products compared to the reactants.

The initial approach in this direction used luminescent detection of phosphates esters in solution and in the gas-phase using platinum 1,2-endothiolate complex with an appended alcohol (Fig. 10.23a) (with very weak emission at 650 nm). The phosphate esters used were NA-simulants and 23A on exposure to such esters (at room temperature) is converted to a closed ring luminophore (Fig. 10.23b) with an emission (fluorescence) at 650 nm (excitation energy 470 nm). This emission is very strong compared to 23A (almost 25 times stronger) and is a specific characteristic of organophosphates (including NAs) [101]. Catalytic hydration of fluorophosphonate with a porous silicon interferometer is another way of detecting NAs (through the reflectivity spectra) [102]. DFP was actually used as test material. A blue shift and decrease in intensity of the Fabry-Pérot interference fringes due to the NA (DFP) hydrolysis is used as a tool for detection [23]. The most important development in this direction was available when a highly sensitive chemosensor (Fig. 10.23c, λ_{max}: 291 nm, emission: 385 nm) was synthesized. It undergoes ring closure reaction due to the exposure of NA (DFP was used in the actual experiment) (molecule 23D, λ_{max}: 275, 340 nm, emission: 465 nm) with a high red-shift (~80 nm) in the fluorescence (with increased intensity) with respect to 23C [103].

Resonance Raman spectroscopy (RRS) and its variant surface-enhanced Raman scattering (SERS) has evolved into an important spectroscopic tool to detect molecules. In contrast to normal Raman scattering, these techniques allow the detection of molecules with high sensitivity. Raman scattering of a particular mode in a molecule could be enhanced up to 10^6 times in RRS when excitation into an absorption band of the molecule under study is applied [104, 105]. Several in-silico investigations were carried out to find the characteristic vibrational mode/s of adsorbed NAs on different oxide and oxide-supported gold clusters [106]. As discussed earlier (Sect. 10.3.2) the test NA (DFP) is physisorbed on small MgO ($Mg_{16}O_{16}$) and CaO ($Ca_{16}O_{16}$) clusters through the oxygen of the P = O bond. It has been observed that the *Raman intensity of the P = O stretching mode* ($RI_{P=O}$) shows enhancement with respect to the isolated DFP [82]. The intensity enhancement is interpreted due to chemical effect, as oxides are not the materials to cause plasmonic resonance. It has been computed through quantum-chemical calculations that Au and metal oxide ($Mg_{16}O_{16}$, $Ca_{16}O_{16}$ and their single oxygen defect clusters) supported Au clusters can trap DFP and the $RI_{P=O}$ is higher than those of the isolated metal oxide clusters. This enhancement of intensity due to the chemical effect is the key factor to select materials to generate effective $RI_{P=O}$ in RRS. When this molecule is adsorbed on such clusters, the Raman signal regarding intensity of the P = O stretching mode could be really enhanced many times with respect to the isolated DFP molecule. This is usually done by generating Raman spectra using the excitation wave-length around the specific λ_{max} of gold clusters (or the oxide supported gold clusters) and such frequency dependent Raman signal (usually called RRS) is a typical characteristic of these NAs and could be used for their detection [106]. The RRS spectra have the further advantage of not only enhancing this $RI_{P=O}$, but it also cuts down the vibrational frequencies of other modes within a large window. The in-silico analysis on the RRS spectra (Fig. 10.24) of DFP on Au_{20}-cluster showed this remarkable feature. This property would be quite unique for the NAs under certain experimental condition and could generate a promising site-specific technique for their detection in the gas-phase.

In present days diode lasers are available in various frequency ranges (e.g., 380 nm, 485 nm, 532 nm, 633 nm, 670 nm, 780 nm and 960 nm to name a few of them) and since the absorption spectra of the gold nanoclusters [106] fall within this range, it is possible to set up of such an experimental device to measure the RRS of DFP (for P = O intensity) at a particular excitation. Since most of the nerve agents known so far has Raman active P = O bond, this selective enhancement of $RI_{P=O}$ could be considered as a unique feature of this class of molecules. Furthermore, since the gold/silver clusters as well as the oxide-supported gold/silver clusters are all known in solid state, experimental measurements are possible in solid state to detect the NAs through SERS or RRS using this unique property of P = O stretching mode. The normal vibrational frequencies also in such cases are capable of generating highly intense P = O stretching mode. However, the advantage of Raman spectroscopy is that the vibrational frequencies of the Raman active modes

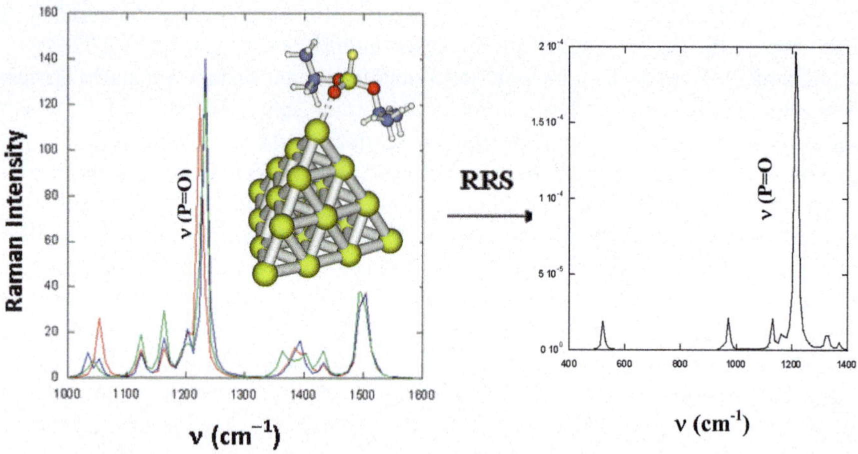

Fig. 10.24 Schematic representation of the Raman spectra and RRS of DFP adsorbed on Au-cluster. The RRS spectra shows how the $\nu(P = O)$ of DFP could be assigned as a characteristic of the NA (DFP) (The detailed studies are available in Ref. [106])

are only visible and it cuts off most of the other frequencies in a broad range and in SERS/RRS this P = O stretching mode could be made more selective.

10.5 Concluding Remarks

The present review is primarily motivated by a necessity to introduce the recent advancements in the research of NAs covering their conformational, biological and surface binding properties. The review is based on the results of state of the art quantum-chemical methods and our goal is to corroborate efficacy of such methods to the cases which are challenging to investigate experimentally. There are several reviews covering the experiments on these subjects, and it seems appropriate to highlight the usefulness of *in silico* techniques, since the use of NAs are getting restricted in direct experiments. The review, of course, uses experimental data (wherever available) to compare with the theoretical results. After a brief review of the various NA-properties, the conformational and biological reactivity of NAs are discussed in details with a focus on the enantiomeric preference (together with their conformations). The basic objective is to show how NAs compete with Ach to inhibit their reactions with AchE. The second part of the review concerns the adsorption of NAs on various surfaces including cement, paints and clay materials. These processes are usually not strong enough to keep NAs on such surfaces for a long time, but they are a potential source of environmental hazard due to the after-exposure effects of NAs. The basic materials which could behave as a stronger binder to these NAs with respect to their primary occlusion on the cement/paint surfaces could be used to remove and deactivate them. The third part of the review discusses the properties of metal oxides for such purposes along with other

NA-deactivation techniques. The adsorption of NAs on different surfaces has also the characteristic of enhancing their Raman-active P = O stretching mode. This feature could be used to detect NAs using SERS/RRS spectroscopy. The final section of the review discusses the possibility of NA-detection using such techniques along with other available potential methods.

Abbreviations

Ach	Acetylcholine
AchE	Acetylcholinesterase
CW	Chemical warfare
DFP	O,O-diisopropylfluorophosphate
DFT	Density Functional Theory
DMMP	Dimethyl phosphonate
GA	Tabun
GB	Sarin
GD	Soman
Glu	Glutamic acid
hAchE	Human AchE
hBchE	Human butylcholinesterase
HD	Mustard gas
HF	Hartree-Fock
His	Histidine
IR	Infrared
MEP	Molecular Electrostatic Potential
MP2	Möller-Plesset second-order perturbation
NA	Nerve-agent
NMR	Nuclear Magnetic Resonance
ONIOM	Our-own N-layered Integrated Molecular Orbital and Molecular Mechanics
OP	Organophosphorus
PCM	Polarizable Continuum Model
PU	Polyurethane
RRS	Resonance Raman Spectroscopy
Ser	Serine
SERS	Surface Enhanced Raman Scattering
TcAcHE	Torpedo Californica AchE
TMP	Trimethyl phosphate

Acknowledgments The authors acknowledge the support of NSF CREST (No.: HRD-0833178), EPSCOR (Award No. 362492-190200-01\NSFEPS-0903787) and SERRI (Award No. 400001122261) grants. One of the authors (S.R.) acknowledges the financial support by a statutory activity subsidy from Polish Ministry of Science and Technology of Higher Education for the Faculty of Chemistry of Wroclaw University of Technology.

References

1. Toy ADF, Walsh EN (1987) Phosphorus chemistry in everyday living, 2nd edn. ACS, Washington, DC
2. Kim K, Tasy OG, Atwood DA, Churchill DG (2011) Chem Rev 111:5345–5403
3. Malany S, Sawai M, Sikorski SR, Seravalli J, Quinn DM, Radić Z, Taylor P, Kronman C, Velan B, Shafferman A (2000) J Am Chem Soc 122:2981–2987
4. Sussman JL, Harel M, Frowlow F, Oefner C, Goodman A, Toker L, Silman I (1991) Science 253:872
5. Harel M, Quinn DM, Nair HK, Silman I, Sussman JL (1996) J Am Chem Soc 118:2340
6. Ariel N, Ordentlich A, Barak D, Bino T, Velan B, Shafferman A (1998) Biochem J 335:95
7. Gentry MK, Doctor BP (1991) In: Mossoulie J, Bacou F, Bernard E, Chattonet A, Doctor BP, Quinn DM (eds) Cholinesterases: structure, function, mechanism, genetics, and cell biology. American Chemical Society, Washington, DC, p 394
8. Taylor P, Lappi S (1975) Biochemistry 14:1989
9. Deakyne CA, Mout-Ner (Moutnet) M (1999) J Am Chem Soc 121:1546
10. Taylor P (2001) In: Hardman JG, Limbird LE (eds) Goodman and Gilman's the pharmaceutical basis of therapeutics, 10th edn. McGraw-Hill, New York, p 175
11. Benchop HP, Keijer JH (1966) Biochim Biophys Acta 128:586
12. Ordenlich A, Kronman C, Barak D, Stein D, Ariel N, Mercus D, Valen B, Shafferman A (1993) FEBS Lett 334:215
13. Hosea WA, Radić Z, Tsigelny I, Berman HA, Quinn DM, Taylor P (1996) Biochemistry 35:10995
14. Millard CB, Kryger G, Ordentlich A, Greenblatt HM, Harel M, Raves ML, Segall Y, Barak D, Shafferman A, Silman I, Sussman JL (1999) Biochemistry 38:7032
15. Nachon F, Asojo OA, Brogstahl GEO, Masson P, Lockridge O (2005) Biochemistry 44:1154
16. Ekström F, Akfur C, Tunemalm A-K, Lundberg S (2006) Biochemistry 45:74
17. Ordentlich A, Barak D, Kronmann C, Benschop HP, De Jong LPA, Ariel N, Barak R, Segall Y, Velan B, Shafferman A (1999) Biochemistry 38:3055
18. Jennings LL, Malecki M, Komives EA, Taylor P (2003) Biochemistry 42:11083
19. Hill CG, Li W-S, Thoden JB, Holden HM, Raushel FM (2003) J Am Chem Soc 125:8990
20. Michel HO, Hackley BE Jr, Berkowitz L, List G, Hackley EB, Gillian W, Pankau M (1967) Arch Biochem Biophys 121:29
21. Ordentlich A, Barak D, Kronmann C, Ariel N, Segall Y, Velan B, Shafferman A (1998) J Biol Chem 273:19509
22. Masson P, Fortier P-L, Albarte C, Froment M-T, Bartels CF, Lockridge O (1997) Biochem J 327:601
23. Ordentlich A, Barak D, Kronmann C, Ariel N, Segall Y, Velan B, Shafferman A (1996) J Biol Chem 271:11953
24. Segall Y, Waysbort D, Barak D, Ariel N, Doctor BP, Grunwald J, Ashani Y (1993) Biochemistry 32:13441
25. Bencsura A, Enyedy I, Kovach IM (1995) Biochemistry 34:8989
26. Majumdar D, Roszak S, Leszczynski J (2007) J Mol Phys 105:2551
27. Parr RG, Yang W (1989) Density functional theory of atoms and molecules. Oxford University Press, New York
28. Becke AD (1993) J Chem Phys 98:5648
29. Vosko SH, Wilk L, Nausair M (1980) Can J Phys 58:1200
30. Lee C, Wang W, Parr RG (1988) Phys Rev B 37:785
31. Vistoli G, Pedretti A, Testa L, Testa B (2002) J Am Chem Soc 124:7472
32. Møller C, Plesset M (1934) Phys Rev 46:618
33. Cothia C, Pauling P (1969) Nature (London) 223:919
34. Cothia C, Pauling P (1970) Proc Natl Acad Sci U S A 65:477
35. Svinning T, Sorum H (1975) Acta Crystallogr 31:1581

36. Jensen B (1975) Acta Chem Scand B 29:531
37. Jensen B (1982) Acta Crystallogr B 38:1982
38. Culvenor CCJ, Ham NS (1966) Chem Commun 15:537
39. Herdkoltz JK, Sass RL (1970) BioChem Biophys Res Commun 40:583
40. Walker ARH, Suenram RD, Samuels A, Jensen J, Ellzy MW, Lochner JM, Zeorka D (2001) J Mol Spectrosc 207:77
41. Suenram RD, DaBeu RS, Walker ARH, Lavrich RJ, Plusquellic DF, Ellzy MW, Lochner JM, Cash L, Jensen JO, Samules AC (2004) J Mol Spectrosc 224:176
42. Majumdar D, Roszak S, Leszczynski J (2006) J Phys Chem B 110:13597
43. Fleming CD, Edwards CC, Kirby SD, Maxwell DM, Potter PM, Cerasoli DM, Redinbo MR (2007) Biochemistry 46:5063
44. Sanson B, Nachon F, Colletier J-P, Froment M-T, Toker L, Greenblatt HM, Sussman JL, Ashani Y, Masson P, Silman I, Wiek M (2009) J Med Chem 52:7593
45. Paukku Y, Michalkova A, Majumdar D, Leszczynski J (2006) Chem Phys Lett 422:317
46. Haviv H, Wong DM, Greenblatt HM, Carlier PR, Pang Y-P, Silman I, Sussman JLJ (2005) Am Chem Soc 127:11029–11036
47. Wang J, Gu J, Leszczynski J (2005) J Phys Chem B 109:13761
48. Wang J, Gu J, Leszczynski J (2006) Chem Phys Lett 431:149
49. Wang J, Gu J, Leszczynski J (2006) Biomol Struct Dyn 24:139
50. Xu Y, Shen J, Luo X, Silman I, Sussman JL, Chen K, Jiang H (2003) J Am Chem Soc 125:11340
51. Wang J, Roszak S, Gu J, Leszczynski J (2005) J Phys Chem B 109:1006
52. Bennet AJ, Kovach IM, Schowen RL (1988) J Am Chem Soc 110:7892
53. Bencsura A, Enyedy IY, Kovach IM (1996) J Am Chem Soc 118:8531
54. Elhanany E, Ordentlich A, Dgany O, Kaplan D, Segall Y, Barak R, Velan B, Shafferman A (2001) Chem Res Toxicol 14:912
55. Wang J, Gu J, Leszczynski J (2008) J Phys Chem B 112:3485
56. Wang J, Gu J, Leszczynski J, Feliks M, Sokalski WA (2007) J Phys Chem B 111:2404
57. Wang J, Gu J, Leszczynski J (2006) J Phys Chem B 110:7567
58. Cossi M, Barone V, Cammi R, Tomasi J (1996) Chem Phys Lett 255:327
59. Stewart JJP (2007) J Mol Model 13:1173
60. Pellenq RJ-M, Kushimac A, Shahsavarib R, Van Vlietd KJ, Buehlerb MJ, Yi S, Ulmb F-J (2009) Proc Natl Acad Sci U S A 106:16102
61. Seymour RB, Kauffman GB (1992) J Chem Educ 69:909
62. Clayton AM (1987) Epoxy resins: chemistry and technology, 2nd edn. Marcel Dekker, New York
63. Zhao Y, Truhlar DG (2008) Theor Chem Acc 120:215
64. Michalkova A, Leszczynski J (2009) In: Leszczynski J, Shukla M (eds) Practical aspects of computational chemistry: methods, concepts and applications. Springer, New York, p 277
65. Michalkova A, Gorb L, Ilchenko M, Zhikol OA, Shishkin OV, Leszczynski J (2004) J Phys Chem B 108:1918
66. Benco L, Tunega D, Hafner J, Lischka H (2001) Am Mineral 86:1057
67. Michalkova A, Martinez J, Zhikol OA, Gorb L, Shishkin OV, Leszczynska D, Leszczynski J (2006) J Phys Chem B 110:21175
68. Svensson M, Humbel S, Froese RDJ, Matsubara T, Sieber S, Morokuma K (1996) J Phys Chem 100:19357
69. Yang Y-C, Baker JA, Ward JR (1992) Chem Rev 92:1729
70. Yang Y-C (1999) Acc Chem Res 32:109
71. Yang Y-C, Szafraniec LL, Beaudry WTJ (1993) Org Chem 58:6964
72. Koper O, Lucas E, Klabunde KJ (1999) J Appl Toxicol 19:S59
73. Klabunde KJ, Stark J, Koper O, Mohs C, Park DG, Decker S, Jiang Y, Lagadic I, Zhang DJ (1996) Phys Chem 100:12142
74. Stark JV, Park DG, Lagadic I, Klabunde KJ (1996) Chem Mater 8:1904

75. Lin S-T, Klabunde KJ (1985) Langmuir 1:600
76. Li Y-X, Koper OB, Atteya M, Klabunde KJ (1992) Chem Mater 4:323
77. Wagner GW, Koper OB, Lucas E, Decker S, Klabunde KJ (2000) J Phys Chem B 104:5118
78. Wagner GW, Bartram PW, Koper OB, Klabunde KJ (1999) J Phys Chem B 103:3225
79. Michalkova A, Ilchenko M, Gorb L, Leszczynski J (2004) J Phys Chem B 108:5294
80. Michalkova A, Paukku Y, Majumdar D, Leszczynski J (2007) Chem Phys Lett 438:72
81. Paukku Y, Michalkova A, Leszczynski J (2008) Struct Chem 19:307
82. Kolodziejczyk W, Majumdar D, Roszak S, Leszczynski J (2007) Chem Phys Lett 450:138
83. Paukku Y, Michalkova A, Leszczynski J (2009) J Phys Chem C 113:1474
84. NRC recommendations for the disposal of chemical agents and munitions (1994) National Academy Press, Washington, DC
85. Groenewold GS (2010) Main Group Chem 9:221
86. U.S. Army's Chemical Materials Agency, Richmond. http://www.cma.army.mil/bluegrass.aspx
87. Farquharson S, Inscore FE, Christesen S (2006) Top Appl Phys 103:447
88. Gustafson RL, Martell AE (1962) J Am Chem Soc 84:2309
89. Ward JR, Yang YC, Wilson RB Jr, Burrows WD, Ackerman LL (1988) Bioorg Chem 16:12
90. Munro NB, Talmage SS, Griffin GD, Waters LC, Watson AP, King JF, Hauschild V (1999) Environ Health Perspect 107:933
91. Smith BM (2008) Chem Soc Rev 37:470
92. Holm FW (1998) NATO Sci Ser 1 22:159
93. Van Hooidonk C, Breebaart-Hansen JC, Recl AE (1970) Trav Chim Pays-Bas 89:289
94. Morales-Rojas H, Moss RA (2002) Chem Rev 102:2497
95. Ekerdt JG, Klabunde KJ, Shapley JR, White JM, Yates JT Jr (1988) J Phys Chem 92:6182
96. Wagner GW, Porcell LR, O'Connor RJ, Munavalli S, Carnes CL, Kapoor PN, Klabunde KJ (2001) J Am Chem Soc 123:1636
97. Gordon WO, Tissue BM, Morris JR (2007) J Phys Chem C 111:3233
98. Vaiss VS, Borges I Jr, Leitão AA (2011) J Phys Chem C 115:24937
99. Stewart JJP (1989) J Comput Chem 10:209
100. Stewart JJP (1989) J Comput Chem 10:221
101. Van Houten KA, Heath DC, Pilato RS (1998) J Am Chem Soc 120:12359
102. Sohn H, Létant S, Sailor MJ, Trogler WC (2000) J Am Chem Soc 122:5399
103. Zhang S-W, Swager TM (2003) J Am Chem Soc 125:3420
104. Mayers AB (1996) Chem Rev 96:911
105. Rousseau DL, Friedman JM, Williams PF (1979) In: Weber A (ed) Raman spectroscopy of gases and liquids, vol 11, Topics in current physics. Springer, Berlin, p 203
106. Majumdar D, Roszak S, Leszczynski J (2010) J Phys Chem A 114:4340

Chapter 11
Valence Anions of DNA-Related Systems in the Gas Phase: Computational and Anion Photoelectron Spectroscopy Studies

Piotr Storoniak, Haopeng Wang, Yeon Jae Ko, Xiang Li, Sarah T. Stokes, Soren Eustis, Kit H. Bowen, and Janusz Rak

Abstract Formation of stable radical anion is one of the most apparent event resulting from the interaction of a biomolecule with an electron. Although, the dipole-bound (DB) anions of nucleobases (NBs) prevail in the gas phase as indicated by negative ion photoelectron spectroscopy (PES), even relatively weak interactions such as those present in the uracil-water complexes are sufficient to render the valence bound (VB) anions adiabatically stable. Moreover, since the electron clouds of dipole bound anionic states are much more diffuse than those of valence bound anions, they are strongly destabilized with respect to the latter in condensed phase. This is why VB anions rather than DB anions of nucleobases are more relevant for biological systems, i.e., in particular for DNA. In this review article, we discuss molecular factors governing the stability of valence anions of nucleobases. On the basis of PES measurements and quantum chemical

P. Storoniak • J. Rak (✉)
Department of Chemistry, University of Gdańsk, Wita Stwosza 63, Gdańsk 80-308, Poland
e-mail: piotr.storoniak@ug.edu.pl; janusz.rak@ug.edu.pl

H. Wang
Department of Chemistry, Johns Hopkins University, Baltimore, MD 21218, USA

Department of Chemistry, Georgetown University,
37th & O St NW, Washington, DC 20057, USA

Y.J. Ko
Department of Chemistry, Johns Hopkins University, Baltimore, MD 21218, USA

The Rowland Institute at Harvard, 100 Edwin H. Land Boulevard,
Cambridge, MA 02142, USA

X. Li
Department of Chemistry, Johns Hopkins University, Baltimore, MD 21218, USA

Center for Research and Exploration in Space Science & Technology, University of Maryland Baltimore County, 1000 Hilltop Circle, Baltimore, MD 21250, USA

S.T. Stokes • S. Eustis • K.H. Bowen
Department of Chemistry, Johns Hopkins University, Baltimore, MD 21218, USA

J. Leszczynski and M.K. Shukla (eds.), *Practical Aspects of Computational Chemistry III*, DOI 10.1007/978-1-4899-7445-7_11,

calculations, we demonstrate that tautomerisation leading to the very rare tautomers of NBs renders the valence anions of nucleobases adiabatically stable. Moreover, we present how the stability of VB anions increases on the transition from NBs to nucleotides. On the other hand, studying anionic complexes of nucleobases with inorganic and organic proton donors, other nucleobases and nucleosides, we emphasize the importance of interactions within double stranded DNA as well as with species present in the environment in which DNA is always immersed under biological conditions. We show that in the complexes of NBs with sufficiently acidic proton donors, electron attachment frequently induces barrier-free proton transfer (BFPT) leading to a significant stabilization of anionic states in nucleobases. Our discussion is closing with a summary, including open questions on the influence of interactions between DNA and proteins on the stability and fate of anionic species induced by electrons in DNA.

11.1 Introduction

Radiotherapy is the most common modality employed in the treatment of cancer. Indeed, about 50 % of cancer patients are irradiated with ionizing radiation (IR) in some point of their treatment [1]. As far as cell death by ionizing radiation is concerned, the DNA molecule is the most important target. Although IR deposits its energy in a random fashion, damaging all molecules within the cell, there are multiple copies of most molecules and many of them undergo a continuous and rapid turnover that limits the damaging effects of IR. On the other hand, since there are only two copies of DNA per cell, its turnover is very limited, and the molecule itself is crucial for all cellular functions. Indeed with eukaryotic cells that contain their DNA in the nucleus, little lethal damage is observed as long as IR damage is limited to the membrane or cytoplasm. However, there is a dramatic increase in cellular death in instances where the ionizing radiation reaches the nucleus [2].

IR interacts with DNA either directly causing ionizations/excitations or indirectly via the products of radiolysis of the molecules present in the environment. In the cell, which contains 70–80 % of water, the indirect action of IR far exceeds the direct effects. Indeed, water radiolysis results in hydroxyl and hydrogen radicals as well as in hydrated electrons. Previous studies with scavenger molecules indicate that almost all indirect DNA damage is due to attack by the highly reactive hydroxyl radicals ($OH^{\bullet}$). In addition, the reducing counterparts of $OH^{\bullet}$, i.e. hydrated electrons, which are generated by IR in the similar amounts to hydroxyl radicals [3], are relatively ineffective at inducing lethal DNA strand breaks [4]. Nevertheless, the seminal paper by Sanche and co-workers [5], published in 2000, shed new light on the possible role of electrons in DNA damage. These researches demonstrated that the bombardment of plasmid DNA deposited on a metal surface in ultra-high vacuum with low-energy electrons (below 20 eV), having specific energies, leads to the formation of significant amounts of single- and double strand breaks. This discovery resulted in flourish of papers on the interactions between electrons and DNA components as well as between electrons and DNA itself (for reviews on the experimental and theoretical studies see e.g. [6, 7] and [8, 9], respectively).

Although solvated electrons seem to be relatively inefficient as far as the damage of genomic DNA is concerned, it is possible to sensitize a biopolymer by labeling it with modified nucleosides that have larger electron affinity than the original nucleobases [10, 11]. For instance 5-bromouracil (BrU), which is effectively incorporated into cellular DNA during cell division increases cell killing by 2–3 fold [12] in comparison with native cells. The presence of BrU in DNA enables the solvated electron to interact with the modified base, triggering dissociative electron attachment, which itself is followed by secondary chemical reactions that ultimately lead to DNA damage [13–16]. It is also worth mentioning that some of electrons released by IR bring an important contribution to the so-called non-scavengable damage that originates from the direct ionization of water present in a hydration layer tightly bound to the DNA molecule [17].

Hence, the above mentioned facts show why studies of interactions between electrons and DNA related systems are important. Certainly, in order to comprehend and accordingly control IR-induced DNA damage at the cellular level, one has to learn first the details of electron attachment to DNA components, notably to nucleobases, as well as the fate of negative ions formed in the electron attachment process. Even though long range electron transfer in double stranded helix seems to be well proved [18], electron-induced DNA damage is a local phenomenon concerning a particular base, sugar or phosphate unit. Therefore, the electrophilic properties of individual nucleotides are the determining factor for DNA's response to electrons. Although particular nucleotides interact with others in the DNA molecule and these interactions modify, to some extent, their characteristics, the intrinsic electrophilic properties of isolated nucleobases, nucleosides and nucleotides, as observed in the gas phase, are the main determinants of DNA behavior towards excess electrons.

As indicated by negative ion photoelectron spectroscopy (PES), all nucleobases except guanine (due to the difficulty of obtaining sufficiently high pressure of this species without isomerization or decomposition) form DB anions in the gas phase. In order to observe the other type of nucleobase anions, i.e. the VB anions, an additional component besides the nucleobase is necessary: a water molecule, noble gas atom or proton donor, interaction with which makes the valence bound anion more stable than the dipole bound one. Indeed, a number of computational studies confirmed that the valence bound anions of isolated nucleobases are somewhat less stable than their dipole bound counterparts [19]. However, since the DB anionic states are much more diffuse than the VB anions, they are strongly destabilized with respect to the latter in condensed phase [20], especially in liquid water in which short- and long-range order is due to numerous hydrogen bonds between the water molecules. This is why the DB anions of nucleobases are not directly relevant for biological systems.

In this chapter, we review the results of a successful combination of the PES experiments with quantum chemical calculations. For the reasons mentioned above we limited the following discussion to only valence anions of systems involving nucleobases. First, we briefly characterize negative ion photoelectron spectroscopy and show its main advantages/disadvantages. Then, the usefulness and limitations of various quantum chemical approaches employed for studying the electron bonding by the discussed systems will be briefly reviewed. The methodological section will be followed by a discussion of the results of PES measurements combined with quantum chemical calculations on the anions of nucleobases, dimeric and trimeric complexes of nucleobases with inorganic and organic proton donors, nucleobase pairs, nucleosides, nucleotides, and finally on the homodimers of pyrimidine nucleosides. Ultimately, the discussion will be concluded with a summary, including open questions about the influence of interactions between DNA and proteins on the stability and fate of anionic species induced by electrons in DNA.

11.2 Methods

11.2.1 Negative Ion Photoelectron Spectroscopy

Anion photoelectron spectroscopy (PES) is conducted by crossing a mass-selected, negative ion beam with a fixed-energy photon beam and analyzing the energies of the resultant photodetached electrons. This technique is governed by the well-known energy-conserving relationship, $h\nu = \text{EBE} + \text{EKE}$, where $h\nu$, EBE, and EKE are the photon energy, electron binding energy (photodetachment transition energy), and the electron kinetic energy, respectively. Typically, anion photoelectron spectra provide the adiabatic electron affinity (AEA), which is the energy difference between the ground states of the anion and its neutral counterpart, the vertical detachment energy (VDE), which is the energy between the ground state of the anion and its corresponding neutral at the geometry of the anion, the electronic state spacings of the neutrals' excited electronic states, and in some cases through a Franck-Condon analysis, structural information about the anion.

The anion photoelectron spectrometer used in most of the work described herein consists of a laser-based anion source, a linear time-of-flight mass spectrometer, a mass gate, a momentum decelerator, a neodymium-doped yttrium aluminum garnet (Nd:YAG) laser operated at third harmonic (355 nm) for photodetachment, and a magnetic bottle, electron energy analyzer with a resolution of 35 meV at EKE = 1 eV. A schematic is shown in Fig. 11.1. Most photoelectron spectra described here were calibrated against the well-known photoelectron spectrum of Cu^- [21]. Several anion sources are available for bringing biological molecules into the gas phase as intact (parent) anions. The most valuable one, which we developed, is the photodesorption/photoemission anion source. In it, a weak infrared laser beam first desorbs delicate, involatile molecules (most DNA component molecules are in this class) from an infrared absorber into the gas phase. Very nearby, a second higher energy, typically visible, laser interacts with a low work function metal to produce low energy photoelectrons. A split second after these two components co-mingle in the gas phase producing anions by electron attachment, a burst of very cold helium cools and

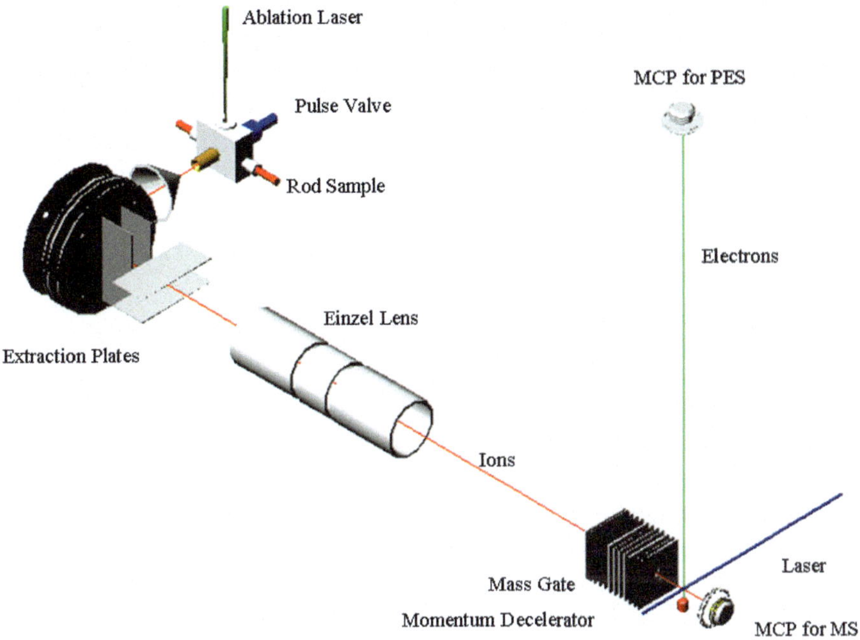

Fig. 11.1 Schematic of a typical anion photoelectron spectrometer

entrains them, directing them into subsequent chambers where they are extracted in to the time of flight mass spectrometer stage of the anion photoelectron spectrometer.

PES provides a unique tool for studying the intrinsic electrophilic properties of DNA components, free of interactions with their surroundings. Apart from investigations related to individual molecules, external effects, such as solvation, may be scrutinized gradually using mixtures of different solvents with compounds of interest. In the early days of this technique's application to the study of DNA component anions, electron binding was explored only to small systems, e.g., uracil with water or noble gases. In the course of time the anionic systems were expanded and finally such complex systems as dimers of nucleosides are investigated. An advantage of gas phase experimental studies of biological molecules is that the results can be directly compared with theoretical predictions, as will be presented below.

11.2.2 Computational Methods

The AEA and VDE depicted in Fig. 11.2 are accessible experimentally with the help of the negative anion PES technique. These characteristics can also be calculated using a number of quantum chemical (QM) methods.

The vertical detachment energy is evaluated as a difference between the electronic energy of the neutral (X) and anionic species (X^-) at the geometry of the fully

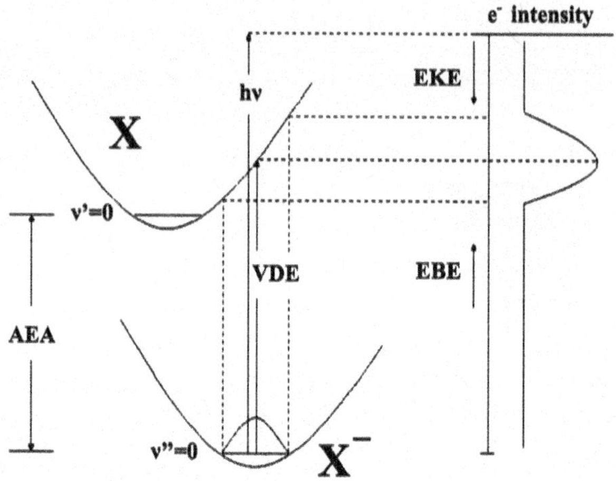

Fig. 11.2 Energetic relationships between VDE, and AEA

relaxed anion. The difference in the electronic energies of the neutral and the anion at their corresponding fully relaxed structures is a measure of AEA_E. In order to obtain electron affinity values in the free energy scale (AEA_G), the electronic energies of X and X^- have to be corrected for zero-point vibration terms (ZPE), thermal contributions to energy, pV term, and entropy term. These corrections usually do not change the AEA_E value by more than 2–3 kcal/mol. For DNA fragments, ZPE corrections often increase their AEA by 0.1 eV or even more. Therefore ZPE, thermal contribution to energy, and entropy terms may affect the sign of AEA for weakly bound nucleobase anions and as a consequence change a qualitative picture [22, 23] concerning the stability/instability of the species under consideration. These terms are usually calculated in the rigid rotor-harmonic oscillator approximation at $T = 298$ K and $p = 1$ atm [9]. Yet another parameter related to energy difference between the neutral and anionic potential energy surfaces reflects the propensity of the relaxed neutral to bind the excess electron and is dubbed a vertical electron affinity (EA_v/VEA). It can be measured using the electron transmission spectroscopy technique, and it shows the ability of a neutral to attach an electron. VEA is calculated as a difference between the electronic energy of the neutral (X) and anionic species (X^-) at the relaxed geometry of the neutral.

A number of QM methods have been used so far to evaluate the above defined energy differences between potential energy surfaces of X and X^- [24]. Regardless of the employed basis set the Hartree-Fock model always predicts significantly negative AEA for uracil, a pyrimidine base which seems to possess the strongest ability to bind an electron among all nucleobases. An inclusion of electron correlation energy at the most economic MP2 (Møller-Plesset perturbation theory truncated at the second order) level diminishes the adiabatic instability of uracil valence anion. Depending on basis set, the MP2 AEAs of uracil span the range of −0.5 eV (6-31+G(d)//6-31G(d) basis set) [25] to −0.1 eV (a value extrapolated to the basis set limit) [23]. At the level of one of the most accurate correlation approaches employed in practice, CCSD(T) (coupled cluster with single, double and iterative triple contributions), the ZPE corrected AEA is slightly positive suggesting thus that the valence anion of uracil is adiabatically bound – a conclusion that agrees with the RET (Rydberg Electron Transfer) measurements [26] and the AEA value originating from the extrapolation of experimental AEAs for water clusters of uracil, $U(H_2O)_n$, to n equal to zero [27]. Also composite schemes that estimate the electronic energy at the basis sets limit, as Gn, predict positive values of adiabatic electron affinity for uracil. The G4 estimate of AEA amounts to 0.11 eV [28] while that of G4MP2 to as much as 0.16 eV [29]. Similarly, the CASPT2 calculations give the AEA of uracil valence anion as large as 0.1 eV, the value being within experimental limit [30, 31].

As discussed above, highly correlated wave-function methods are able to predict the electrophilic properties of uracil within experimental accuracy. Unfortunately, the cost of highly accurate calculations is impractical for all but the smallest systems. Indeed, a formal scaling of CCSD(T), CISDT or CCSDTQ is M^7, M^8, and M^{10}, respectively, where M stands for the number of atomic orbitals [32]. Therefore, it seems that density functional models (DFT), that formally scales as M^4 and include electron correlation energy, are the methods of choice for studies related to electron attachment to DNA subunits. Taking again the AEA of uracil as an example, it is worth noticing that the most popular B3LYP model (the correlation functional of Lee, Yang, and Parr [33] in conjunction with Becke's three-parameter hybrid exchange functional [34]) predicts the AEA of valence anion of uracil to be 0.14 eV with a TZVP basis set [35], 0.22 eV when the 6-311++G(d,p) basis set is applied [35] and to 0.18 eV with the 6-311+G(2df,p) [36]. Even with the relatively small basis set, 6-31+G(d), the B3LYP method predicts a similar AEA for uracil (0.18 eV) [37]. Apparently, as oppose to the conventional QM methods, basis set dependence of the DFT energies is fairly small and the double-ζ with polarization basis set augmented with diffuse functions is sufficient to obtain an acceptable accuracy. While the newly developed M05-2X [38] and M06-2X [39] functionals provide AEAs for the five nucleic acid bases that best fit the G4 predictions, the B3LYP/DZP++ approach overestimates these AEAs by ca. 0.1 eV systematically. This is also worth to emphasize that all three functionals provide AEA estimations consistent with the experimental values [26, 27, 40].

The PES technique, which is the main source of experimental data reviewed in the current article, usually tends to create the most stable form of a given anion [19]. Therefore, in order to interpret PES spectra using a computational approach, one has to scrutinize the conformational space accessible to a studied system in order to identify the most stable configurations. If the system of interest is a molecule having several tautomeric forms [22, 41–48] or a conformationally flexible complex [9], consisting of two or more monomers, the configurational space that has to be probed within calculations may be enormous [49], which makes the task of identifying the most stable form(s) non-trivial. This is also an argument that support the usage of cost-effective DFT methods rather than accurate but very expensive coupled-cluster approach for such type of studies. In the following we will demonstrate how computational search for the most stable anion(s) results in a convincing interpretation of a PES spectrum.

11.3 Anion Radicals of Individual Nucleobases

The earliest ab initio molecular orbital calculations at the ROHF/3-21G level of theory followed by single point calculations (in the 6-31G(d) and 6-31+G(d) basis sets) predicted negative values for electron affinities for the valence anions of the canonical DNA nucleobases implying that such anions are unstable in the gas phase and the least negative AEA was found for thymine (−0.99 eV) [50]. Later calculations revealed that both thymine and uracil are characterized by significantly negative AEAs among nucleobases (close to −1 eV at the ROHF/6-31+G(d)//6-31G(d) level and −0.5 eV at the ROMP2/6-31+G(d)//6-31G(d) level) [25]. Similarly, using several large basis sets and the Hartree-Fock method, Bachorz and co-workers [23] calculated negative AEA values for uracil that span −0.67 to −0.91 eV. Also, high level ab initio calculations at the coupled-cluster level of theory with single, double, and perturbative triple excitations with the augmented correlation-consistent polarized valence double-zeta basis set (CCSD(T)/aug-cc-pVDZ) indicate that the valence anion of canonical uracil is adiabatically unbound by 0.051 eV with respect to the neutral [41], in agreement with the former prediction of Dolgounitcheva et al. [51] who applied a many body perturbation approach, coupled-cluster method, and electron-propagator theory to calculate the stability of the valence bound anion of uracil. The CCSD(T)/aug-cc-pVDZ approach was also applied for the canonical thymine [22] and predicted its valence anion to be unstable in respect to neutral by 0.11 eV. Only explicitly correlated calculation at the MP2 level (MP2-R12) in conjunction with conventional coupled-cluster theory with single, double, and perturbative triple excitations and inclusion of a zero-point vibrational energy correction, determined in the harmonic approximation at the RI-MP2 level of theory led to the conclusion that the valence anion of the canonical uracil is adiabatically stable, although by as small as 0.04 eV [23]. The MP2 calculations for cytosine indicate its valence bound anions to be adiabatically unstable. Namely, the MP2/6-311++G(2df,2p) estimates clearly show that isolated canonical cytosine and its conventional imino-oxo and amino-hydroxy tautomers do not support adiabatically stable anions [52]. At this level of theory, the AEA of −0.382 eV was reported for the canonical amino-oxo cytosine. Moreover, the values of −0.450 and −0.522 eV were found for two rotamers of the imino-oxo form and −0.853 and −0.549 eV for two rotamers of the amino-hydroxy tautomer. At the same time, the valence anions of the amino-oxo and two imino-oxo tautomers were found to be vertically bound with respect to the neutral with the VDEs of 0.271, 0.204, and 0.142 eV, respectively. One more estimate concerning the valence anion of the canonical cytosine was obtained at the MP4/6-31++G(d,p)//MP2/6-31++G(d,p) level where the VDE of 0.102 eV and AEA of −0.562 eV was calculated. At the same level of theory the respective values for the amino-hydroxy tautomer was equal to −0.495 and −1.062 eV. Purine nucleobases seem to be even less prone to electron attachment than pyrimidine ones. Indeed, VDE and AEA determined at the relatively accurate CCSD(T)/AVDZ//MP2/AVDZ level for the canonical isomer of guanine are 0.585 and −0.459 eV, respectively [47]. At this level of theory stable valence anion of canonical adenine was not found [44].

In general, the AEAs resulting from the wave-function methods do not support the occurrence of the conventional valence anions of nucleobases in the gas phase almost independently on the method and basis set used [23, 25]. In contrast to wave-function methods, various exchange-correlation functionals predict mostly positive values of AEA for the valence anion of canonical uracil and thymine and typically small but negative values for cytosine. For both purines, its DFT calculated AEA was found negative, with that for adenine being the lowest among all five nucleobases [35, 36, 53]. Comparing experimental data to the DFT results obtained using the B3LYP functional and various basis sets, Li et al. concluded that the best estimates of the AEA values of the canonical pyrimidines are: 0.2 eV (U), 0.22 eV (T) and −0.05 eV (C), whereas for the purines: −0.35 eV (A) and −0.75 eV (G) [54]. Similarly, adiabatic electron affinities (with zero-point correction) of valence anions of canonical nucleobases at the B3LYP/6-311++G(d,p) level reported by Russo et al. amount to: 0.22 eV (U), 0.18 eV (T), 0.006 (C), −0.004 eV (G) and −0.26 eV (A) [35]. Finally, the respective AEAs resulting from the B3LYP/DZP++ calculations were calculated to be 0.24, 0.20, 0.03, −0.12 and −0.28 eV for U, T, C, G and A, respectively, [53]. The above mentioned order of the AEA values for canonical nucleobases was also confirmed by the other density functionals and various basis sets [8].

Since all nucleobases possess dipole moment larger than 2.5 D [19] they may form the so-called dipole bound (DB) anions in addition to conventional valence bound anionic states (VBs), in which the excess electron is attached predominantly to the positive end of the molecular dipole in the former case. In the mid-1990s, the calculations of Oyler and Adamowicz predicted stable dipole-bound anions for isolated uracil and thymine, with adiabatic electron affinities of 0.086 and 0.088 eV, respectively [55, 56]. The most accurate AEA for the formation of the DB state of uracil and thymine was reported relatively recently by Bachorz et al. [41], 0.071 eV, and by Mazurkiewicz et al. [22], 0.053 eV, respectively. The Adamowicz group also estimated the dipole-bound AEA of the amino-oxo cytosine and two rotamers of the amino-hydroxy tautomer to be 0.058, 0.022, and 0.006 eV [57]. The above mentioned theoretical predictions are in good accordance with the experimental findings. Indeed, shortly after the Adamowicz reports, Hendricks et al. [58] discovered, using the negative ion photoelectron spectroscopy, dipole bound anions of uracil and thymine. The experimental AEA values were found to be 0.093 ± 0.007 for uracil and 0.069 ± 0.007 eV for thymine. Similarly, the RET spectroscopy measurements carried out by Desfrançois et al. [59] confirmed the existence of the dipole-bound states of isolated uracil and thymine (AEAs 0.054 ± 0.035 and 0.068 ± 0.020 eV, respectively) as well as that of adenine (AEA equal to 0.012 ± 0.005 eV). The RET experiments also suggested a simultaneous formation of covalently bound anions of U and T [26, 59]. Using an extrapolation of AEA values with the number of water molecules in a given nucleobases-$(H_2O)_n$ complex, Schiedt et al.[27] extrapolated the electron affinity of the valence anion of bare uracil, thymine and cytosine to be 0.15 ± 0.12, 0.12 ± 0.12, and 0.13 ± 0.12 eV, respectively, which suggested an adiabatic stability of the VB anionic states of gaseous pyrimidines. These measurements were later found to significantly over-estimate AEA values due to inadequate Franck-Condon overlap in the spectra. The RET studies on the gas-phase

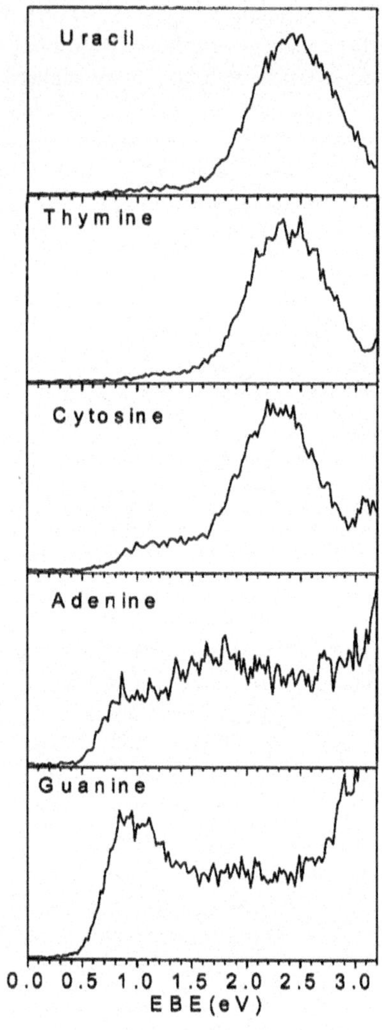

Fig. 11.3 The photoelectron spectra of all five nucleobases (Reprinted with permission from Ref. [48]. Copyright 2007, American Institute of Physics)

uracil-argon clusters were in line with uracil having a slightly positive AEA, since they found the adiabatic electron affinity for the valence uracil anion to be 0.062 ± 0.032 eV [26]. The latter value agrees well with the most accurate theoretical estimate of the AEA of uracil calculated by Bachorz et al. [41]. Here, it is worth noticing that the AEA of the valence bound anion of uracil is smaller than that for its dipole-bound state [41] which may explain why only dipole bound anions of nucleobases are observed in the gas phase [26, 58, 59].

Despite the difficulties in preparing VB anions of nucleobases, because of their very low or even negative AEA values, in 2007 Bowen's group reported the observation and study of VB anions for all nucleobases. However, with electron binding energies above 2 eV (see Fig. 11.3), based on quantum chemical calculations, it was clear that these anions did not originate from the canonical forms of nucleobases. These were very rare tautomers of nucleobase anions.

11.3.1 Anions of Very Rare Tautomers

11.3.1.1 Uracil

The photoelectron spectrum of uracil [48] anions features a broad peak with a maximum at an electron binding energy (EBE) of ~2.5 eV. (see PES for uracil in Fig. 11.3). The shape and position of the spectrum is an indicative of a valence-bound anion/anions. Indeed, a photoelectron spectrum of a dipole-bound state usually represents a narrow peak with the maximum at low EBEs – well below 0.5 eV. Two years earlier, Bachorz et al. [41] conducted computational studies on the stability of various tautomers of anionic uracil. They scrutinized, at the B3LYP/6-31++G(d,p) level, 75 tautomers/rotamers of uracil. On the basis of these DFT prescreening, the most stable valence anionic structures were selected for the further optimization at the MP2/AVDZ level. The final energies, obtained at the CCSD(T)/AVDZ level, were corrected for the MP2/AVDZ zero-point vibrations for the canonical and imino-oxo valence anions in which nitrogen site was deprotonated and the C5 or C6 atom protonated [41].

The most stable valence anion, aU_{n1c5}, which has a proton transferred from N1H to C5, was characterized by the VDE of 1.27 eV (see Table 11.1 and Fig. 11.4). This structure is unrelated to any of the most stable tautomers of the neutral uracil. The anion aU_{n1c5} was found to be adiabatically stable with respect to the canonical neutral, nU_{can}, by 3.9 kcal $\cdot$ mol^{-1} in terms of electronic energy (this corresponds to $AEA_E = 0.17$ eV). It is also more stable by 2.4 and 5.2 kcal $\cdot$ mol^{-1} than canonical forms, the dipole-bound ($aU_{can}{}^{dbs}$) and valence anion ($aU_{can}{}^{val}$), respectively [41].

The VDE values for $aU_{can}{}^{dbs}$ and $aU_{can}{}^{val}$ are 0.073 and 0.506 eV, while their AEAs calculated with respect to the canonical neutral are 0.071 and −0.051 eV, respectively (see Table 11.1). However, the photoelectron spectrum of uracil does not support existence of aU_{n1c5}, $aU_{can}{}^{dbs}$ and $aU_{can}{}^{val}$ because the intensity of the signal is negligible at the EBE values corresponding to VDEs predicted for these structures.

The maximum of the main feature of the photoelectron spectrum of uracil at ~2.5 eV is reproduced by the VDE obtained for the tautomer aU_{n3c5}, shown in Fig. 11.4, which is only 0.21 kcal $\cdot$ mol^{-1} less stable than $aU_{can}{}^{val}$ and which exhibits an adiabatic electron affinity (−0.06 eV) similar to $aU_{can}{}^{val}$. Moreover, the adiabatic electron affinity for aU_{n3c5}, 1.2 eV, calculated with respect to corresponding proton-transferred neutral is the value which most closely fits the AEA read from the spectrum. AEAs of the remaining tautomers calculated with respect to their neutral counterparts in the range from −0.3 to 0.7 eV are definitely too small to be comparable with the experimental picture (Table 11.1 presents the AEA values defined with respect to the canonical neutral). However, aU_{n3c5} is unstable by as much as 5.3 kcal $\cdot$ mol^{-1} relative to the most stable, aU_{n1c5} which ultimately calls into question whether the aU_{n3c5} tautomer is viable.

Table 11.1 Theoretical adiabatic electron affinities (AEA) and electron vertical detachment energies (VDE) for the tautomers of uracil and thymine. The AEA values are defined with respect to the canonical neutral. The values of AEA_G mean the relative stability of the anion with respect to the canonical neutral in terms of Gibbs free energy. AEA and VDE are given in eV

	VDE					
	aU_{can}^{val}	aU_{can}^{dbs}	aU_{n1c5}	aU_{n1c6}	aU_{n3c5}	aU_{n3c6}
Ref. [41][a]	0.506	0.073	1.267	2.822	2.499	
Ref. [42][b]	0.579		1.379	3.089	2.578	3.924
	aT_{can}^{val}	aT_{can}^{dbs}	aT_{n1c5}	aT_{n1c6}	aT_{n3c5}	aT_{n3c6}
Ref. [22][a]	0.457	0.055	1.251	2.816	2.458	3.449
	AEA (AEA_G)					
	aU_{can}^{val}	aU_{can}^{dbs}	aU_{n1c5}	aU_{n1c6}	aU_{n3c5}	aU_{n3c6}
Ref. [41][a]	−0.051 (−0.034)	0.071 (0.069)	0.171 (0.183)	−0.275 (−0.278)	−0.060 (−0.043)	
Ref. [42][b]	0.021		0.135	−0.311	−0.101	−0.370
	aT_{can}^{val}	aT_{can}^{dbs}	aT_{n1c5}	aT_{n1c6}	aT_{n3c5}	aT_{n3c6}
Ref. [22][a]	−0.087 (−0.037)	0.053	0.115 (0.146)	−0.234 (−0.172)	−0.121 (−0.087)	−0.248 (−0.193)

[a]Calculated at the CCSD(T)/aug-cc-pVDZ//MP2/aug-cc-pVDZ and CCSD(T)/aug-cc-pVDZ + diff//MP2/aug-cc-pVDZ + diff level for the valence anions and dipole bound anion, respectively. A "diff" stands for additional diffuse functions centered on the H(C6) atom of the canonical uracil or thymine. Zero point energy, thermal corrections and entropy term obtained at the MP2/aug-cc-pVDZ level for T = 298 K and p = 1 atm

[b]Calculated as the sum of incremental contributions from various level of theory, where HF and correlation energies (up to CCSD(T)) are extrapolated to the basis set limit; zero point energy calculated at the MP2/aug-cc-pVTZ level

Subsequently, very accurate ab initio calculations [42] did not change the relative stability order of uracil anionic tautomers presented in Ref. [41] and therefore have not answered the question why the most preferable thermodynamically valence anion is not visible in the photoelectron spectrum at EBE ~1.4 eV. The refined results confirmed that the most stable valence anion results from a proton transfer from N1H to C5 (aU_{n1c5}). This tautomer is adiabatically bound with respect to the canonical neutral by only 0.135 eV (AEA), and its VDE is as large as 1.4 eV (see Table 11.1). The second most stable valence anion is that of the canonical tautomer (AEA = 0.021 eV and VDE = 0.58 eV), and the third most stable valence anion is that of the tautomer resulting from a proton transfer from N3H to C5 (AEA = −0.101 eV and VDE = 2.6 eV). The valence anions of the five tautomers analyzed differ in stability by less than 0.6 eV, and the smallest VDE amounts to 0.58 while the largest to 3.9 eV [42].

11.3.1.2 Thymine

The photoelectron spectrum of $T^{\bullet-}$ obtained with 3.493 eV photons is very similar to that of $U^{\bullet-}$ but shifted by ~0.1 eV towards smaller electron binding energies (see Fig. 11.3).[48] It constitutes a broad band that starts at 1.0 eV, and ends at

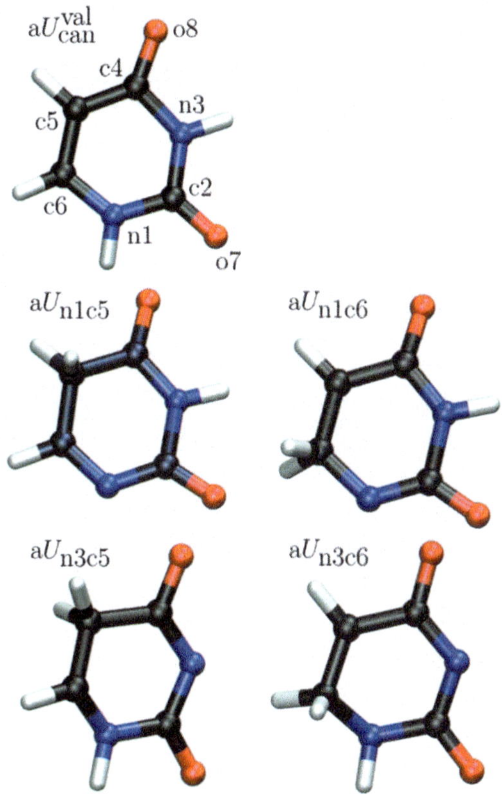

Fig. 11.4 The tautomers of uracil anion considered in Ref. [42] (Reprinted with permission. Copyright 2008, American Institute of Physics)

3.1 eV and has a maximum at ~2.4 eV. The shift of the spectrum of $T^{\bullet-}$ when compared to the spectrum of $U^{\bullet-}$ results from the electron donating effect of the methyl group.

Employing the computational methodology of Ref. [41], Mazurkiewicz et al. [22] demonstrated that the similarity between the $T^{\bullet-}$ and $U^{\bullet-}$ spectra may be connected with the appearance of the analogous tautomeric forms of those nucleobases under the experimental conditions. Thus, on the basis of the prescreening at the B3LYP/6-31++G(d,p) level, 5 out of 77 possible tautomers were selected for the MP2 optimizations. Their final single point energies were obtained at the CCSD(T)/aug-cc-pVDZ level using the optimal MP2 geometries.

Indeed, in close resemblance to the results for $U^{\bullet-}$, the most stable valence anion appeared to be aT_{n1c5} (N1 to C5 intramolecular proton transfer anion tautomer; see Fig. 11.5), which is adiabatically stable with respect to the canonical neutral nT_{can} by 0.115 eV (2.65 kcal/mol) and characterized by a VDE of 1.25 eV. The anion aT_{n1c5} is more stable by 1.4 and 4.7 kcal · mol^{-1} than the canonical anions, the dipole-bound ($aT_{can}{}^{dbs}$) and valence anion ($aT_{can}{}^{val}$), respectively. In analogy to the results for $U^{\bullet-}$, the stability of the considered anionic tautomers of thymine decreases in the following order: $aT_{can}{}^{dbs}$, $aT_{can}{}^{val}$, aT_{n3c5}, aT_{n1c6} and aT_{n3c6}

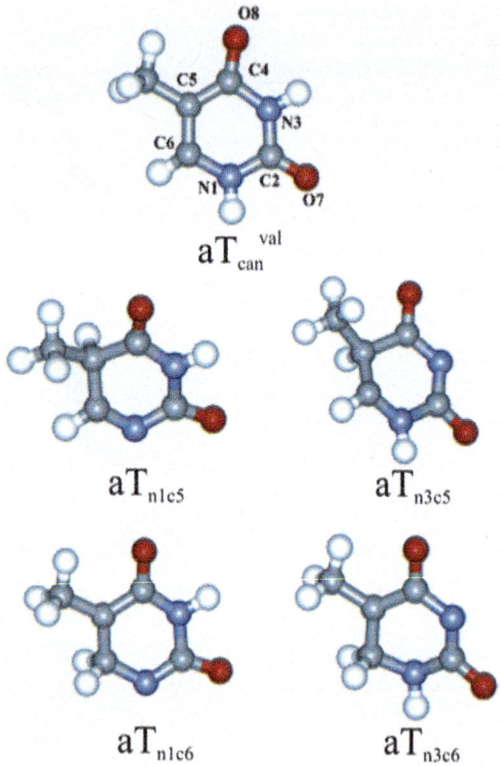

Fig. 11.5 The tautomers of thymine anion considered in Ref. [22] (Reprinted with permission. Copyright 2006 American Chemical Society)

(see Fig. 11.5). For the canonical valence anion, a VDE of 0.46 eV was predicted while for the remaining valence tautomers the VDE values span the range from 2.5 to 3.5 eV (see Table 11.1). In turn, dipole bound-type anionic state of thymine was characterized by VDE equal to 0.055 eV.

The photoelectron spectrum does not support the existence of the most stable aT_{n1c5} nor the canonical anions ($aT_{can}{}^{dbs}$, $aT_{can}{}^{val}$) as indicated by the lack of a feature in the ~0–1.5 eV range. Instead, the third most stable among valence anions, aT_{n3c5} with a VDE of 2.46 eV, can be associated with the main feature of the PES spectrum [48].

The above described studies demonstrate that uracil and thymine may exist as stable valence anions in the gas phase. These anions occur as tautomeric forms resulting from N-to-C proton transfer. Such neutral tautomers are very unstable and they can exist only due to favorable interactions with the excess electron. The detection of these tautomers in the photoelectron experiment was possible due to the use of a specifically designed laser vaporization ion source, which inherently made more energy available than some other sources during the anion formation process.

A unique feature of tautomers with carbon sites protonated is the high stability of valence anions accompanied by their instability after removing of the excess electron. Indeed, the six-member ring of aT_{n3c5} and aT_{n3c6} transforms after electron detachment to a linear or a bicyclo structure, respectively. Such barrier-free decomposition involving anion radicals of nucleobases may be of interest in the context of investigation of mechanisms of electron-induced DNA damage.

11.3.1.3 Cytosine

The PES of $C^{\bullet-}$ represents a broad band at high EBE values with a maximum at 2.3 eV [48]. Clearly, there is a resemblance with the spectra of the other anionic pyrimidine bases, $U^{\bullet-}$ and $T^{\bullet-}$(see Fig. 11.3). The spectrum of $C^{\bullet-}$ differs, however, from these of $U^{\bullet-}$ and $T^{\bullet-}$ with a low intensity shoulder extending from 0.6 to 1.7 eV. Additionally, the relative intensities of the higher EBE band and the lower EBE shoulder change with source conditions which indicates that the two bands correspond to two different tautomers of $C^{\bullet-}$ being in equilibrium [48].

The computational interpretation of the photoelectron spectrum was performed by Li et al. [48] following a procedure of Ref. [41]. That is, in the first step, several low-energy anionic structures, likely to be responsible for the PES shape, were selected on the basis of the B3LYP/6-31++G(d,p) calculations from a large number of conceivable tautomers. The final structures and harmonic frequencies of selected geometries were determined at the second order Møller-Plesset level of theory. The electronic energies were calculated at the coupled cluster level of theory with single, double, and perturbative triple excitations. The augmented, polarized, correlation-consistent basis set of double-zeta quality has been used at this stage.

The results of above described calculations revealed that the most stable valence anion (labeled in the original work as $aC_{n1n3/n4c5}$) has two protons shifted within its structure, namely from N1 to N3 and from N4 to C5. Less stable, by 1.8 kcal $\cdot$ mol^{-1}, is the tautomer aC_{n4c5} with intramolecular proton transfer from N4 to C5. The valence canonical anion, aC_{can}, is separated by 4.0 kcal $\cdot$ mol^{-1} from the most stable $aC_{n1n3/n4c5}$. For the $aC_{n1n3/n4c5}$, aC_{n4c5} and aC_{can} tautomers the calculated VDEs are 1.10, 2.34 and 0.39 eV, respectively. Thus, the aC_{n4c5} tautomer may be assigned to the main feature in the photoelectron spectrum and the most stable $aC_{n1n3/n4c5}$ tautomer, to the lower intensity hump. The existence of the canonical anion radical is not confirmed by the spectrum, which remains in agreement with the negative AEA values predicted earlier for this form at various levels of theory [36, 52–54, 57]. The negative AEAs were also calculated for the conventional imino-oxo and amino-hydroxy tautomers with proton transferred to N3 or O2, respectively and these tautomers do not show up in the spectrum as well [52, 57]. The appearance of the above-mentioned three types of anions under the experimental conditions would be manifested by the feature in the range of the EBE from 0.1 to 0.3 eV in the photoelectron spectrum.

The finding that a tautomeric form of isolated cytosine, with a proton transferred from N4 to the C5 atom, supports that the adiabatically stable anion radical is consistent with the earlier studies by Harańczyk et al. [43] devoted to the stabilization effect of electron attachment to the very rare tautomers of 1-methylcytosine

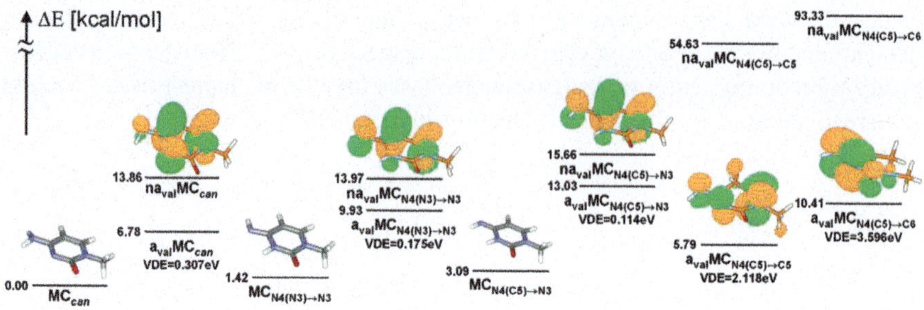

Fig. 11.6 The most relevant tautomers of anionic 1-methylcytosine (Reprinted with permission from Ref. [43]. Copyright 2005 American Chemical Society)

(see Fig. 11.6). When methylated in the N1 position, cytosine is expected to exhibit slightly smaller electron affinity than its non-substituted form. At the same level of theory (CCSD(T)/AVDZ//MP2/AVDZ) it was confirmed that the most stable valence anion of methylated cytosine is an imino-oxo tautomer, $a_{val}MC_{N4(C5)\rightarrow C5}$, (see Fig. 11.6) which is the methylated equivalent of aC_{n4c5}. The $a_{val}MC_{N4(C5)\rightarrow C5}$ anion characterized by a VDE of 2.12 eV is more stable by 1.0 kcal/mol than the canonical anion $a_{val}MC_{can}$ (VDE = 0.31 eV). Another low-lying anionic tautomer with a carbon atom protonated, $a_{val}MC_{N4(C5)\rightarrow C6}$, is 3.6 kcal/mol less stable than $a_{val}MC_{can}$ and its VDE amounts to 3.60 eV.

The two conventional imino-oxo tautomers with N3 protonated: $MC_{N4(N3)\rightarrow N3}$ and $MC_{N4(C5)\rightarrow N3}$ also support vertically bound valence anionic states with VDEs of 0.18 and 0.11 eV, respectively (see Fig. 11.6). They are, however, less stable than the anion of the canonical tautomer by 3.2 and 6.3 kcal/mol, respectively, which is indicative that such species should not be populated in the gas phase.

It should be noted that at the B3LYP/6-31++G(d,p) level of theory $a_{val}MC_{N4(C5)\rightarrow C5}$ is characterized by VDE = 2.29 eV, which perfectly reproduces the maximum at the photoelectron spectrum, being 2.5 kcal/mol less stable than canonical anion. In turn, the B3LYP/6-31++G(d,p)(5d) value of the VDE predicted for $a_{val}MC_{can}$ amounts to 0.14 eV [43].

All the five anions considered are adiabatically unbound with respect to their corresponding neutrals with the CCSD(T) values of AEA of −0.192, −0.261, and −0.323 eV for MC_{can}, $MC_{N4(N3)\rightarrow N3}$, and $MC_{N4(C5)\rightarrow N3}$, respectively. The neutrals $MC_{N4(C5)\rightarrow C5}$ and $MC_{N4(C5)\rightarrow C6}$ do not support minima for six-member ring structures because of the barrier-free ring transformation to a linear or bicyclo structure. The values of AEA reported for these tautomers of −0.149 and −0.358 eV, respectively, were calculated with respect to the neutral MC_{can}.

The highest stability of the anionic tautomers with a carbon atom protonated implies that the excess electrons might affect the structure of DNA and RNA by favoring these very rare tautomers of nucleic acid bases. It may be anticipated that the electron detachment from these anions would introduce a strand distortion and contribute to DNA/RNA damage.

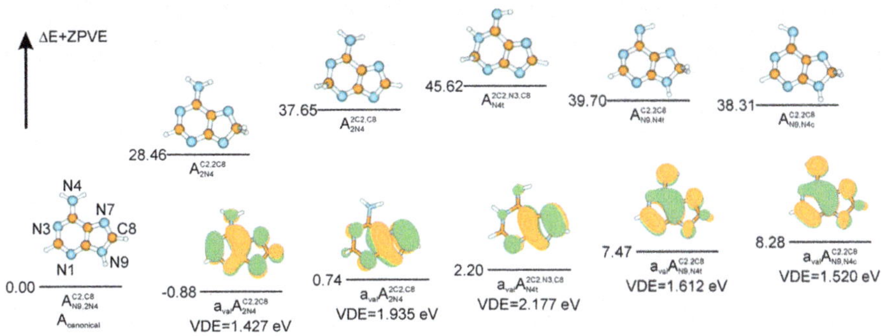

Fig. 11.7 The most stable tautomers of adenine and their relative electronic energies (ΔE_{ZPVE} in kcal/mol) and electron vertical detachment energies (VDE in eV) (Reproduced from Ref. [44]. Copyright 2007 National Academy of Sciences, USA)

As far as the mechanism of the formation of the N4 → C5 and N4 → C6 tautomers of MC is concerned, the barriers for the unimolecular tautomerizations are too high to be probable at standard conditions [43]. Formation of the "very rare" anionic tautomers might proceed through intermolecular proton transfers. Another possibility is a two-step process, with the dissociative electron attachment to MC_{can} followed by a barrier-free attachment of a hydrogen atom to a carbon atom of the deprotonated MC_{can}.

11.3.1.4 Adenine

The photoelectron spectrum of the adenine anion was measured with 3.493 eV photons, and the resulting spectrum is presented in Fig. 11.3 [48]. A broad band has its onset at ~0.5 eV and displays a plateau extending from ca. 0.7 to 3.0 eV. The spectrum of $A^{\bullet-}$ is clearly due to the presence of several valence-bound anions.

The calculations performed at the CCSD(T)/AVDZ//MP2/AVDZ level revealed that the most stable tautomer, $a_{val}A_{2N4}{}^{C2,2C8}$, that features intramolecular proton transfer from N9 to C8 (see Fig. 11.7) is characterized by a VDE of 1.43 eV [44]. This valence anion radical is adiabatically bound with respect to the canonical neutral $A_{N9,2N4}{}^{C2,C8}$ by 0.88 kcal/mol (AEA = 0.038 eV). The second low-energy anion, $a_{val}A_{2N4}{}^{2C2,C8}$, is formed in the course of proton transfer from N9 to C2, whereas the third one, $a_{val}A_{N4t}{}^{2C2,N3,C8}$ results from N9 → C2 and N4 → N3 proton transfers. The two latter tautomers were found adiabatically unbound by 0.74 and 2.2 kcal/mol (AEAs equal to −0.095 and −0.32 eV), but vertically bound (VDEs equal to 1.94 and 2.18 eV, respectively). These three anion radicals resulting from enamine-imine transformation of the canonical adenine should dominate the gaseous mixture of $A^{\bullet-}$. The B3LYP/6-31++G(d,p) VDEs for tautomers $a_{val}A_{2N4}{}^{C2,2C8}$, $a_{val}A_{2N4}{}^{2C2,C8}$ and $a_{val}A_{N4t}{}^{2C2,N3,C8}$ were calculated to be 1.8, 2.20 and 2.60 eV, respectively, while their DFT AEAs were 1.82, −0.44 and −1.97 eV, respectively [44].

The finding that there are tautomeric forms of adenine supporting valence anions with electron detachment energies as large as 2 eV is surprising in the view of previous studies always reporting negative electron affinities for the valence bound state of adenine [25, 35, 36, 53, 54].

Similarly to the very rare tautomers of other nucleobases, the very rare anionic tautomers of adenine might be formed via dissociative electron attachment followed by a hydrogen atom attachment to a carbon atom of the deprotonated adenine.

11.3.1.5 Guanine

Guanine together with adenine are believed to have much smaller electron affinity than the pyrimidine bases, as indicated by the experimental [60] and computational studies [36, 46, 53, 54]. As a consequence, pyrimidines rather than the purine bases have been considered so far as the sites that trap excess electrons in DNA [61–63].

The N-to-C tautomers of anionic guanine, having the AEA values larger than those of the pyrimidine bases, were proposed for the first time within a computational study by Haranczyk et al. [45]. Later on, the photoelectron spectrum of guanine anions, presented in Fig. 11.3, was measured with 3.493 eV photons [48]. The spectrum consists of a broad band with the onset at ~0.5 eV and the maximum at 0.8–1.1 eV. Its shape and position suggest that electrons are detached from the valence-bound states and the broadness of the PES signal suggests the presence of several adiabatically stable anions of guanine.

In order to identify the tautomeric species of $G^{\bullet -}$ responsible for the PES feature, the geometries of 499 tautomers have been optimized using the B3LYP/6-31++G(d,p) method [47]. The DFT calculations enabled 14 adiabatically bound anionic tautomers to be identified. Subsequent calculations at the CCSD(T)/AVDZ//MP2/AVDZ level demonstrated that 13 tautomers support adiabatically bound anions and 7 of them, with AEA values ranging from 0.17 to 0.37 eV, are adiabatically more strongly bound than any of pyrimidine bases studied so far. The selected tautomers of $G^{\bullet -}$, important for the interpretation of photoelectron spectrum, are presented in Fig. 11.8 and their AEAs and VDEs are given in Table 11.2.

The canonical tautomer (G) as well as the most stable neutral tautomer (GN), see Fig. 11.8, have also been considered at the CCSD(T)/AVDZ//MP2/AVDZ level. The valence anions based on the most stable neutral tautomers G and GN are adiabatically unbound by ~0.5 eV and their VDE is equal to 0.59 and 0.21 eV, respectively (see Table 11.2) [47].

The VDE values for adiabatically bound tautomeric anions of guanine span a range of 1.1–2.4 eV, which overlaps the range of EBEs in which the PES spectrum has a significant intensity (Fig. 11.3). Dissociative electron attachment (DEA) was suggested as the most probable way of the formation of the anionic tautomers of guanine. The population of neutral guanine in the gas phase should be dominated by two tautomers: GN (70 %) and G (30 %). The DEA pathway most likely favors formation of these anionic tautomers that are separated from G or GN by one DEA

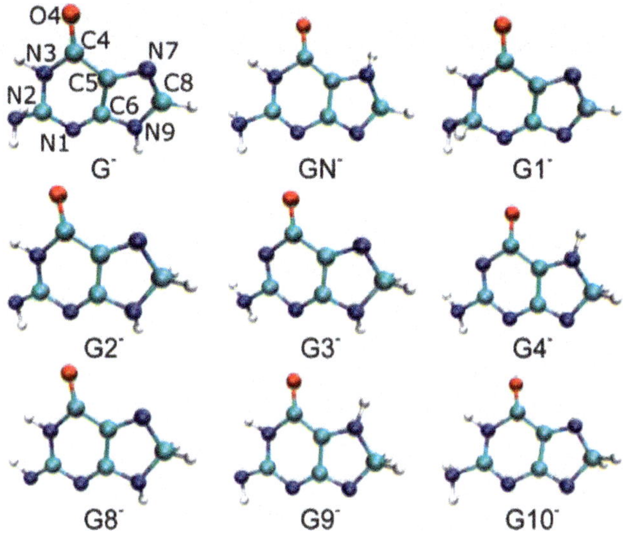

Fig. 11.8 The most relevant tautomers of anionic guanine (Reprinted with permission from Ref. [47]. Copyright 2007 American Chemical Society)

Table 11.2 AEAs and VDEs (in eV) for selected anion radical tautomers of guanine in the gas phase calculated at the CCSD(T)/AVDZ//MP2/AVDZ level [47]

Tautomer	AEA	VDE
G	−0.459	0.585
GN	−0.503	0.212
G1	0.369	2.426
G2	0.365	1.604
G3	0.304	1.699
G4	0.278	2.205
G8	0.165	1.617
G9	0.116	2.427
G10	0.104	1.137

elementary step. For the dominating neutral tautomer, GN, four adiabatically bound anions can be formed by a single DEA step followed by a H atom attachment, namely $G1^{\bullet -}$, $G4^{\bullet -}$, $G9^{\bullet -}$, and $G10^{\bullet -}$. From the second most populated neutral tautomer, G, five adiabatically bound anions might be formed by a single DEA step: $G1^{\bullet -}$, $G3^{\bullet -}$, $G8^{\bullet -}$, and $G10^{\bullet -}$. The $G10^{\bullet -}$ and $G1^{\bullet -}$ structures can be formed from both G and GN. The PES spectrum seems to be in accord with the theoretical findings. Indeed, there is a local maximum in the PES spectrum at ~1.0 eV and the calculated VDE of $G10^{\bullet -}$ is 1.14 eV. A fingerprint of $G1^{\bullet -}$ is expected at 2.43 eV, but it would be masked by the unidentified feature that develops for EBEs larger than 2.7 eV. The only anionic tautomers that have VDEs in the 1.2–2.3 eV range and can be associated with a single DEA step are $G2^{\bullet -}$, $G3^{\bullet -}$, and $G8^{\bullet -}$. They all can be formed from the less populated G but not from the GN tautomer. This might explain why the intensity of the PES spectrum is smaller in the 1.2–2.3 eV range.

Several guanine tautomers are adiabatically more strongly bound than any pyrimidine base studied so far, which suggests that guanine might be the strongest excess electron acceptor within a DNA strand [47]. This might explain why the probability of capturing of an excess electron increases with the number of guanines in short single and double strands of DNA [64].

11.4 Anion Radical Complexes of Nucleobases with Various Inorganic and Organic Proton Donors

Intra- and intermolecular tautomerizations involving nucleic acid bases have long been suggested as critical steps in mutations of DNA [65]. Intramolecular proton transfer (PT) reactions have been studied for both isolated and hydrated nucleic acid bases [66]. Proton transfer induced by electron attachment may also take place in DNA. To elucidate the fate of primary anionic states generated in DNA through irradiation with low-energy electrons, it is necessary to determine factors governing the occurrence of proton transfer between anionic nucleobases and potential proton donors in the surrounding. The successful deployment of the PES technique for the observation of the anionic nucleobases complexed with various proton donors enabled the exploration of PT processes occurring in such anion radical systems.

In the current section, we review photoelectron spectra measured for anionic complexes of uracil with weak inorganic (H_2O, HCN and H_2S) and organic proton donors (alcohols, aminoacids and formic acid). In addition to that, we describe the photoelectron spectra of thymine-amino acid and thymine-formic acid dimers. Finally, we discuss the photoelectron spectra of cytosine and adenine complexed with formic acid. The computational characteristics concerning dimers of nucleobases and various model inorganic/organic proton donors were consistently obtained at the DFT level with B3LYP and MPW1K functionals and 6-31++G(d,p) basis set. Some geometries were also recalculated at the MP2/6-31++G(d,p) level. A comparison between the CCSD(T)/AVDZ result for the vertical detachment energy (VDE) of the valence π^* anionic state of an isolated uracil with that originating from the B3LYP, MPW1K and MP2 (all with the 6-31++G(d,p) basis set) levels leads to the conclusion that the calculated VDE is overestimated by 0.2 and 0.3 eV at the B3LYP and MPW1K level, respectively, and underestimated by 0.1 eV at the MP2 level [67, 68]. Therefore, B3LYP, MPW1K, and MP2 VDEs discussed in this sections are incremented by −0.2, −0.3, and +0.1 eV, respectively.

11.4.1 Complexes of Uracil with Inorganic Acids

The photoelectron spectrum of $(U\cdots HCN)^{\bullet-}$ [68] is presented in Fig. 11.9 and compared to the spectra of $(U\cdots H_2O)^{\bullet-}$ (from Ref. [69]) and $(U\cdots H_2S)^{\bullet-}$ (from

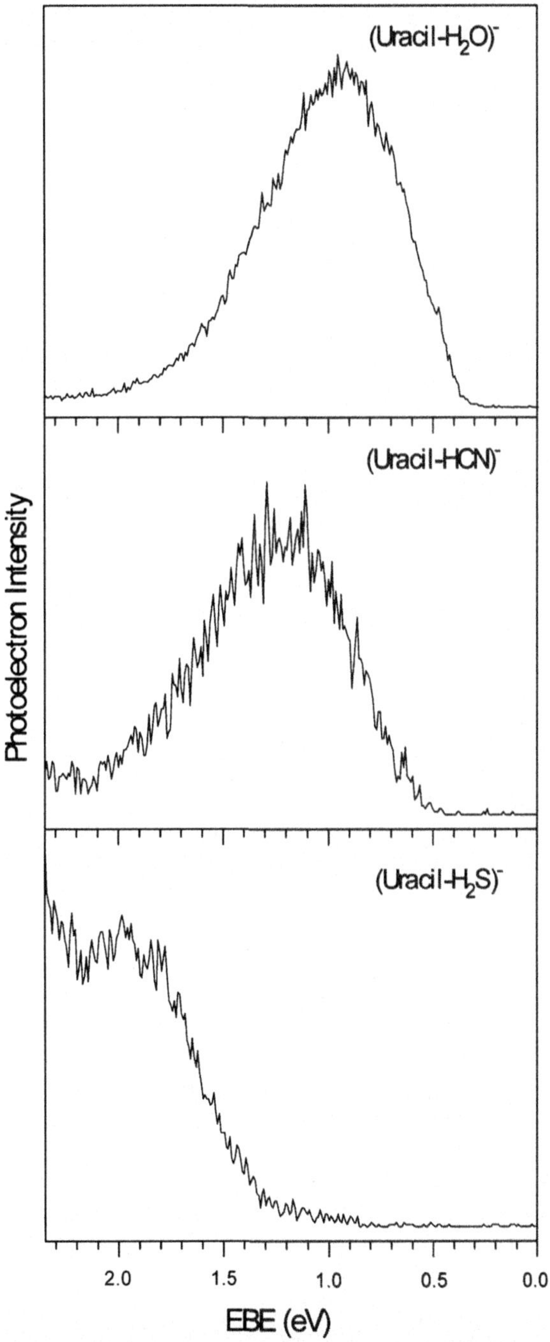

Fig. 11.9 Photoelectron spectra of uracil···H_2O dimer anion (top), uracil-HCN dimer anion (*middle*), and uracil···H_2S dimer anion (*bottom*) recorded with 2.540 eV/photon (Reprinted with permission from Ref. [68])

Ref. [70]). These spectra were recorded with 2.54 eV photons. The spectrum of $(U\cdots HCN)^{\bullet-}$ has a maximum at about 1.1–1.2 eV. The spectrum of $(U\cdots H_2O)^{\bullet-}$ is similar to that of $(U\cdots HCN)^{\bullet-}$, but its maximum is shifted by about 0.2 eV to smaller electron binding energy (maximum at ca. 0.9 eV). The spectrum of $(U\cdots H_2S)^{\bullet-}$ differs from the two and shows a broad feature with a maximum between 1.7 and 2.1 eV.

According to Refs. [42] and [41], the valence and dipole-bound anionic states of uracil are characterized by a calculated VDE value of 0.58 and 0.073 eV, respectively. Thus it may be concluded that only the valence anionic states are observed in the experiment, since the measured values of VDE for $(U\cdots \text{inorganic acid})^{\bullet-}$ are far too large for the dipole-bound anionic state of uracil solvated by the inorganic acid. The shift in photoelectron spectra of $(U\cdots HCN)^{\bullet-}$ and $(U\cdots H_2O)^{\bullet-}$ may be explained by a stronger electrostatic stabilization of an excess electron localized on uracil in the former complex, as the dipole moment of HCN is larger than that of H_2O. In case of both dimers, $(U\cdots HCN)^{\bullet-}$ and $(U\cdots H_2O)^{\bullet-}$, the spectra suggest a uracil anion radical solvated by the neutral H_2O or HCN molecule.

In contrast, the high experimental VDE values of $(U\cdots H_2S)^{\bullet-}$ clearly indicate that the signal is not due to the complex of $U^{\bullet-}$ solvated by H_2S. An extra stabilization in the anionic complex of uracil with H_2S is achieved by a barrier-free intermolecular proton transfer. In view of similar deprotonation enthalpies for HCN and H_2S, the difference is attributed to differences in hydrogen bonds formed by HCN and H_2S with uracil. Hydrogen bonding in $(U\cdots HCN)^{\bullet-}$ apparently fails to provide as much stabilization as in $(U\cdots H_2S)^{\bullet-}$. A common feature of anionic wave functions of the discussed complexes is that the excess electron is localized on a π^* orbital of uracil, in close resemblance to the valence anionic state of isolated uracil.

The computations indicated that the global minimum corresponds to the complex labeled as $aUHCN_{O8.C5}$ in which the O8 atom of uracil is involved in a hydrogen bond with HCN approaching it from the C5 side of the O8 atom (the most important structures are presented in Fig. 11.10). The VDE values calculated for this complex, 1.1 and 1.2 eV, at the MP2 and B3LYP level, respectively, are consistent with the maximum on the photoelectron spectrum. Interestingly, the DFT model predicts stable local minima in which proton is transferred on the O8 atom of uracil. The VDE values for such minima are predicted to be 1.9 and 2.0 eV, hence larger by 0.8 eV than the most stable $(U\cdots HCN)^{\bullet-}$. These PT minima are less stable than the global minimum by 1.8–4.2 kcal/mol. The existence of such $(UH\cdots CN)^{\bullet-}$ structures is not confirmed by the photoelectron spectrum. Moreover, at the MP2 level of theory only non-PT minima have been found.

The most stable anionic complexes of U and H_2S are characterized by a BFPT (barrier-free proton transfer) from the acid to the O8 atom of uracil; see Fig. 11.11 [70]. A driving force for the proton transfer is to stabilize the excess negative charge, which is primarily localized in the O8-C4-C5-C6 region. In consequence of the extra stabilization of the excess electron provided by the transferred proton, the values of VDE for the $aUH_2S_{O8.C5}$ and $aUH_2S_{O8.N3}$ structures are larger by ~1.5 eV than those for the valence anion of an isolated uracil. The B3LYP/6-31++G(d,p) values of VDE for these structures are 2.17 and 2.08 eV, respectively. After

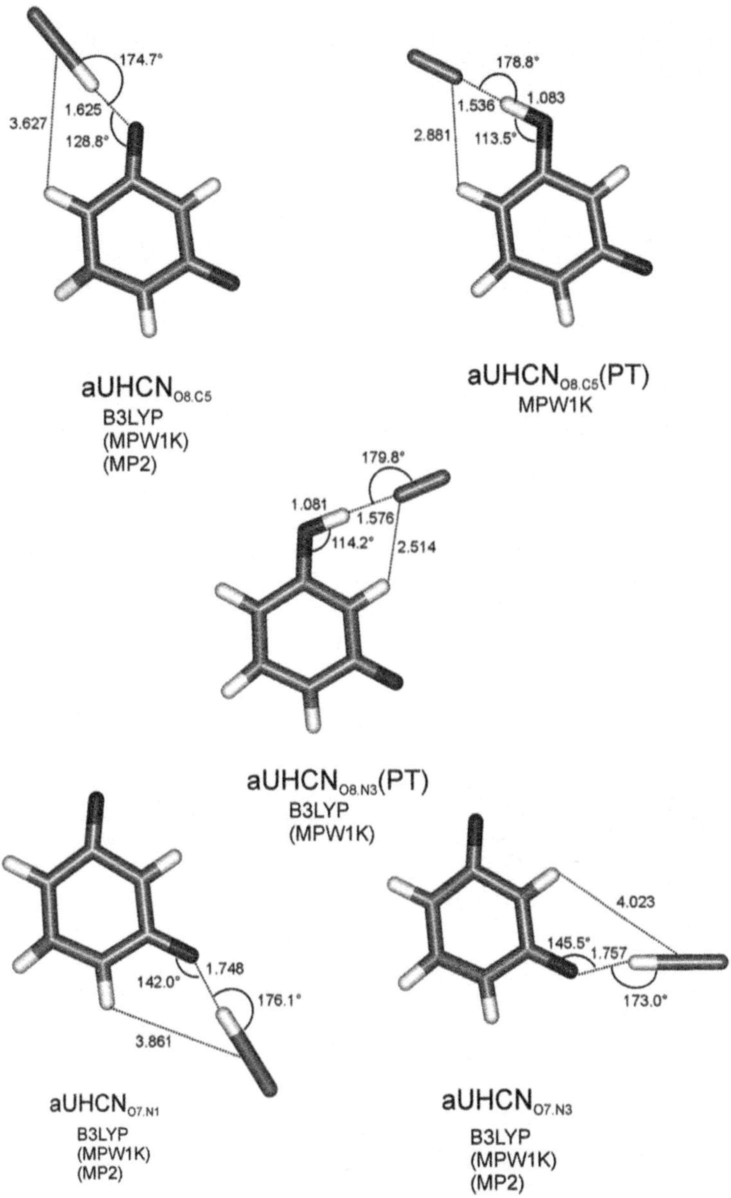

Fig. 11.10 The optimized structures of anionic UHCN complexes with the most important geometrical features. The aUHCN$_{O8.N3}$ was not found (Reprinted with permission from Ref. [68])

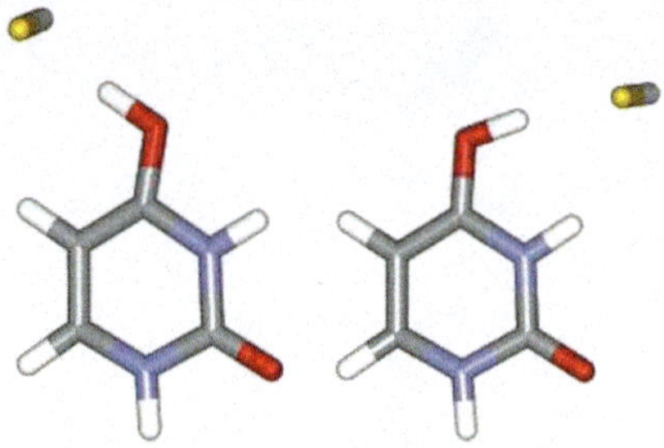

Fig. 11.11 Optimized structures of anionic complexes H_2S that underwent intermolecular barrier-free proton transfer (Reprinted with permission from Ref. [70]. Copyright 2003 American Chemical Society)

correcting the VDEs for aUH$_2$S downward by 0.2 eV, the resulting values, 1.97 and 1.88 eV, coincide with a maximum of the photoelectron spectrum peak at ~1.9 eV.

Consistent with the experimental picture, no BFPT was found for the $(U\cdots H_2O)^{\bullet-}$ complexes from H_2O to either O8 or O7 of uracil. The calculated values of VDE for the $(U\cdots H_2O)^{\bullet-}$ structures span the range of 0.96–1.19 eV (0.76–0.99 eV after correcting downward), and they are in good agreement with a maximum of the photoelectron peak at 0.9 eV. These predictions are in good agreement with the previous computational results from the Ortiz group [71].

11.4.2 Complexes of Uracil and Thymine with Amino Acids

Studies on the attachment of an excess electron to the complexes of uracil/thymine with amino acids addressed the possibility of electron-induced mutations in DNA-peptide complexes. The anionic complexes of uracil or thymine with glycine and uracil with alanine were investigated by means of photoelectron spectroscopy combined with computational approach [67, 72, 73]. The photoelectron spectra for the uracil···alanine (UA), uracil···glycine (UG), and thymine···glycine (TG) anionic complexes has been recorded with 2.54 eV photons (see Fig. 11.12).

The spectra are similar to each other and exhibit a broad, structure less feature. For $(UA)^{\bullet-}$, the maximum occurs between 1.6 and 2.1 eV, while for $(UG)^{\bullet-}$ between 1.7 and 2.0 eV. In the spectrum of $(TG)^{\bullet-}$, the maximum is observed at EBE = 1.6–1.9 eV. When compared to the spectrum of $(UG)^{\bullet-}$, the photoelectron spectrum of $(UA)^{\bullet-}$ has additional intensity in the EBE region between ~0.5 and 1.4 eV. These experimental VDE values are too large for the dipole-bound anionic states. Thus, the spectra cannot be attributed to intact $U^{\bullet-}$ or $T^{\bullet-}$ solvated by an amino acid. A dipole-bound anionic amino acid solvated by a nucleobase is also improbable, because the most stable conformer of canonical alanine and glycine does not bind an electron as the measured electron affinities of alanine and glycine are ~ −1.8 and ~ −1.9 eV, respectively.[74]

Lastly, the widths of the main spectral features for $(UA)^{\bullet-}$ and $(UG)^{\bullet-}$ are much greater than of the photoelectron features for the valence anionic state of uracil

Fig. 11.12 Photoelectron spectra of (**a**) uracil–alanine dimer anion and (**b**) uracil–glycine dimer anion and (**c**) thymine-glycine. All spectra recorded with 2.540 eV/photon (Figure 11.12a reprinted with permission from Ref. [72]. Copyright 2004, American Institute of Physics. Figure 11.12b, c from Ref. [73] -reproduced by permission of the PCCP Owner Societies)

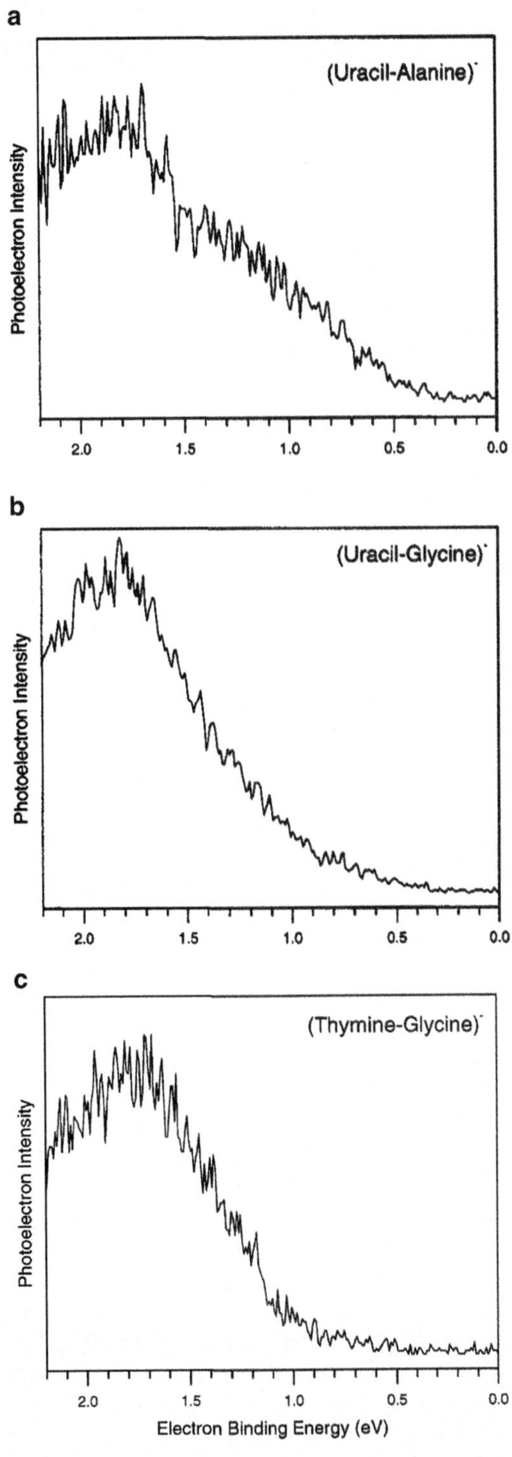

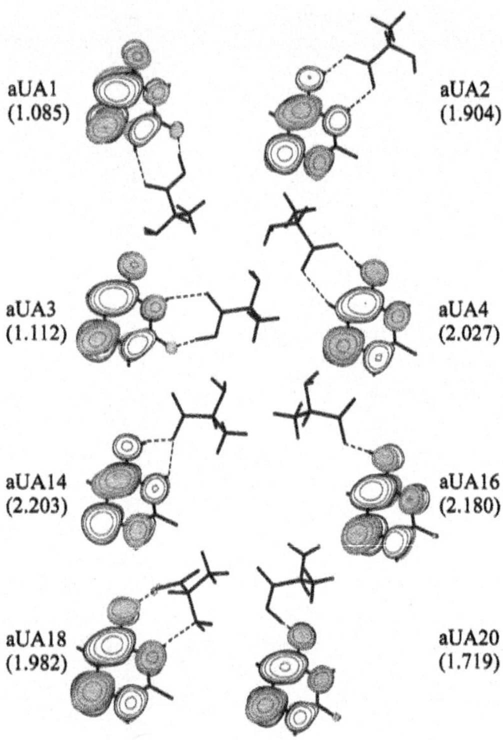

Fig. 11.13 The structure and excess electron charge distribution in the uracil···alanine complexes. The orbitals were plotted with a contour line spacing of 0.03 bohr$^{-3/2}$. The B3LYP/6-31++G(d,p) values of electron vertical detachment energies are in parentheses (in eV) (Reprinted with permission from Ref. [72]. Copyright 2004, American Institute of Physics)

solvated by either a single water molecule or a xenon atom [69]. Representative structures of anions of hydrogen-bonded uracil···alanine complexes originating from the B3LYP/6-31++G(d,p) calculations are displayed in Fig. 11.13 [72].

The most stable structure of the anionic complex is aUA2 which features double hydrogen bonds involving proton acceptor O8 and proton donor N3H sites of uracil and the carboxylic group of alanine. This structure is characterized by a BFPT from the carboxylic group to the O8 atom of uracil (see Fig. 11.13). These complexes can be viewed as a neutral radical of hydrogenated uracil solvated by a deprotonated alanine. The VDE value for aUA2 is predicted to be 1.9 eV (1.7 eV after correcting downward). The AEA calculated with respect to its neutral amounts to 0.7 eV and perfectly reproduces AEA read from the spectrum. Five out of eight most relevant anionic complexes shown in Fig. 11.13 are proton transferred structures. These are aUAn ($n = 2, 4, 14, 16, 18$). Proton transfer from the carboxylic group of alanine to the O8 atom of uracil provides extra stabilization of the excess charge as reflected by their calculated VDEs spanning a range of 1.9–2.2 eV. After correcting downward by 0.2 eV, the resulting range of 1.7–2.0 eV coincides well with the broad peak in the photoelectron spectrum of $UA^{\bullet-}$, see Fig. 11.12. Indeed, for the structures with BFPT, i.e., aUAn ($n = 2, 4, 14, 16, 18$) E_{stab} varies from −1.21 to −0.92 eV, whereas for the structures without BFPT, i.e., aUAn ($n = 1, 3, 20$), the values of E_{stab} are smaller: $-0.94\ eV < E_{stab} < -0.80\ eV$. The second, third, and fourth most stable anionic structures are UA4, UA14, and UA16, respectively. They are separated from the most stable anionic structure aUA2 by only 0.1 eV. Due to BFPT they exhibit significant stability and their theoretical VDEs are as large as 1.8–2.0 eV (after correcting downward), see Fig. 11.12. The adiabatic electron affinities, calculated with respect to the parent neutral structures were found to be 0.7–0.9 eV. The barrier-free nature of the proton transfer process has been confirmed using the MPW1K functional as well as the MP2 method. Preference to transfer a proton to the O8 site is much larger than to the O7 atom. In fact, no BFPT occurring to the O7 atom of uracil was identified. This may be related to the fact that an excess electron on the π^* orbital is not localized in the neighborhood of the O7 atom [67]. Complexes with no BFPT, exemplified by aUA1 and aUA3, are characterized by the calculated values of VDE of only 0.9 eV after correcting downward. The calculated vertical detachment energies for structures of this type are in a range of 0.9–1.7 eV, and some of these structures may contribute to the broad photoelectron peak.

In view of the similarity between main features in the photoelectron spectra of the anionic uracil···alanine and uracil···glycine dimers, the similarity between theoretical results for both complexes is not surprising. The representative structures of $(UG)^{\bullet-}$ complexes obtained at the B3LYP/6-31++G(d,p) level are shown in Fig. 11.14.

The lowest energy anionic structure identified for uracil···glycine, labeled as aUG2, is analogous to the lowest structure found for uracil-alanine complexes (aUA2). Again, the most stable anionic complex aUG2 is the BFPT structure, where the electron induces intermolecular proton transfer from COOH to the O8 atom. The VDE value for the optimal anionic structure of aUG2 is 1.7 eV (after the correction) and AEA amounts to 0.8 eV. The remaining most stable anionic dimers (aUG4, aUG14, and aUG16) are also characterized by a BFPT from the carboxylic group of glycine to the O8 atom of uracil. The calculated values of VDE for these structures span a range of 1.7–2.0 eV (after shifting down by 0.2 eV) which coincides well with the broad peak in the photoelectron spectrum, see Fig. 11.12 [67].

Similarly to computational results for UA, some uracil···glycine complexes do not undergo a BFPT upon attachment of an excess electron. These are mainly geometries with glycine coordinated to the O7 atom of uracil (their VDEs are in a range of 0.9–1.5 eV). A preference to transfer a proton to the O8 site is larger than to the O7 site, though several structures have been identified with the O7 site protonated. The protonation of this site requires, however, overcoming a barrier on the anionic potential energy surface and the structure with the O7 site protonated is less stable than the corresponding structure with the carboxylic group intact. The protonation of the O8 or O7 site increases the value of VDE and the effect is more pronounced for the former site. Aforementioned results suggest that alanine acts in a similar way as glycine does in its anionic dimeric complexes with uracil [67, 72]. Thus, the BFPT process does not seem to be sensitive to the nature of the amino acid's hydrophobic residual group.

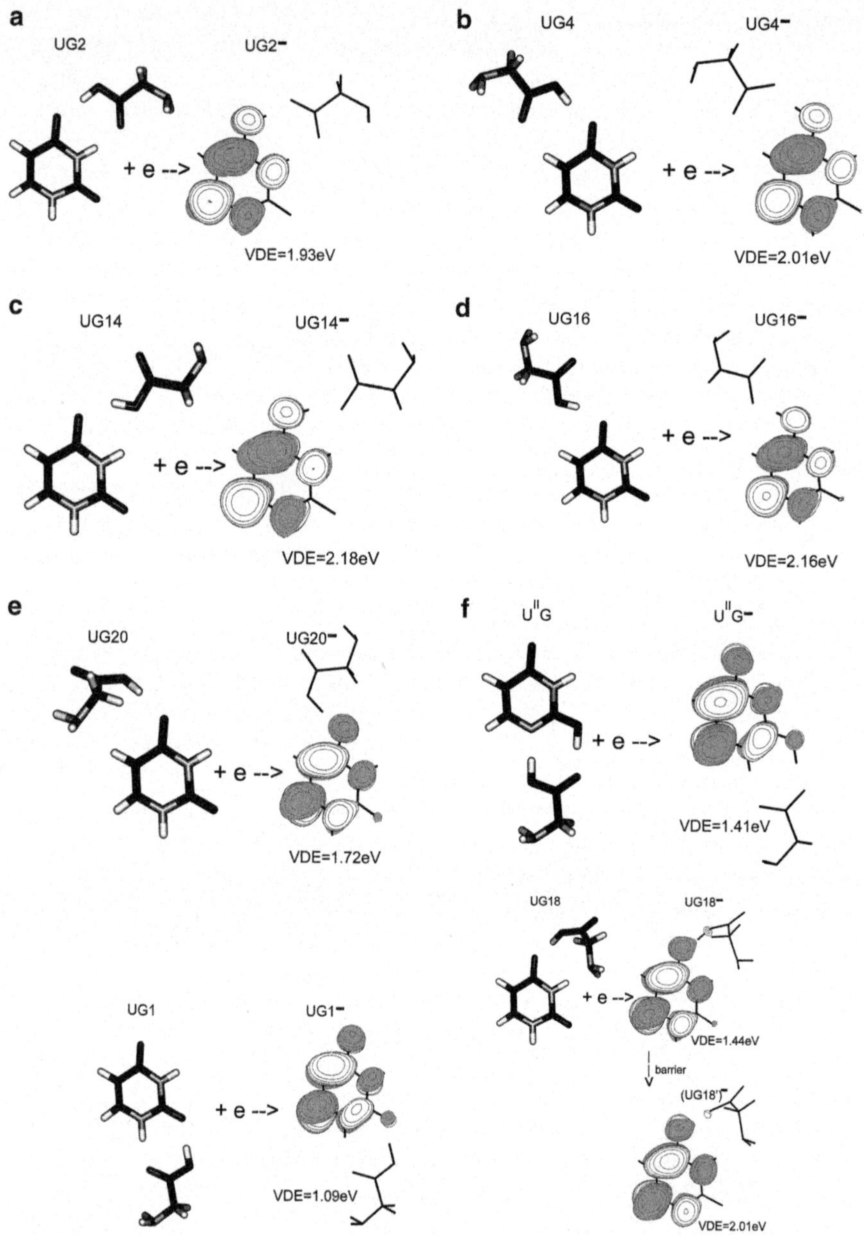

Fig. 11.14 The structure and excess electron charge distribution in the uracil···glycine complexes. The orbitals were plotted with a contour line spacing of 0.03 bohr$^{-3/2}$. The B3LYP/6-31++G(d,p) values of electron vertical detachment energies are in parentheses (in eV) (Reproduced from Ref. [67] with kind permission from Springer Science and Business Media)

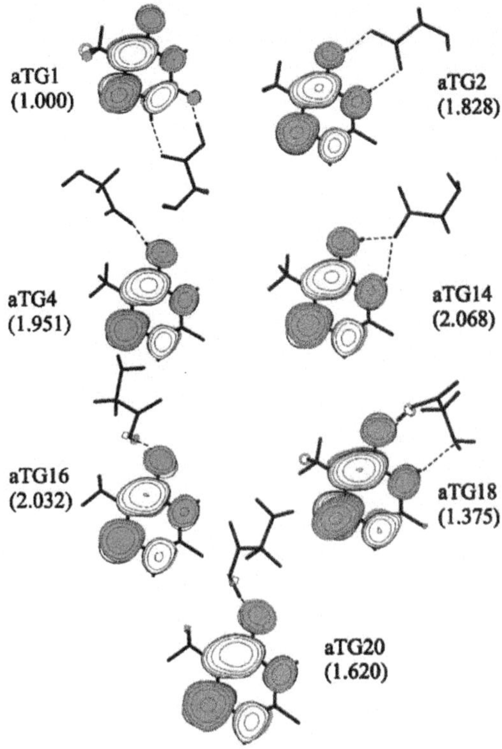

Fig. 11.15 The structure and excess electron charge distribution in the anionic thymine···glycine complexes. The orbitals were plotted with a contour line spacing of 0.03 bohr$^{-3/2}$. The B3LYP/6-31++G(d,p) values of electron vertical detachment energies are in parentheses (in eV) (Reproduced from Ref. [73] by permission of the PCCP Owner Societies)

In turn, experimental and computational investigations of anionic complexes of thymine···glycine $(TG)^{\bullet-}$ explored the effect of methylation of the C5 site of uracil. The photoelectron spectrum registered for $(TG)^{\bullet-}$ closely resembles those for $(UG)^{\bullet-}$ suggesting that thymine in anionic complexes with glycine behaves very much like uracil in anionic complexes with glycine (Fig. 11.12). Indeed, computational results revealed that BFPT in $(TG)^{\bullet-}$ is also operative in analogy to the anionic uracil···glycine complexes. However, there is also a difference in the photoelectron spectrum of $(TG)^{\bullet-}$ relative to that of $(UG)^{\bullet-}$. When compared to the spectrum of $(UG)^{\bullet-}$, the photoelectron spectrum of $(TG)^{\bullet-}$ appears to be shifted by about 0.1 eV towards smaller values of electron binding energy. This can be attributed to intrinsic electrophilic properties of thymine and uracil since similar difference in the calculated VDEs for isolated anions of thymine and uracil, i.e. 0.46 and 0.58 eV was reported in the past [22, 42]. Representative structures resulting from B3LYP/6-31++G(d,p) calculations for conceivable thymine···glycine arrangements are shown in Fig. 11.15.

The most stable anionic $(TG)^{\bullet-}$ structure, which is likely to be responsible for the main feature in the photoelectron spectrum at EBE = 1.6–1.9 eV is labeled as aTG2 [73]. Within this dimer a BFPT occurs from the carboxylic group of glycine to the O8 atom of thymine in analogy to that reported for the most stable $(UG)^{\bullet-}$ dimer. The

B3LYP value of the VDE for aTG2 is 1.6 eV (after correcting downward by 0.2 eV). For comparison, the calculated value of the VDE for the aTG1 complex, which does not undergo intermolecular proton transfer, is 0.8 eV (after the 0.2 correction). For aTG1 and aTG2 structures the calculated values of adiabatic electron affinity are 0.44 and 0.68 eV. It was found that, in addition to aTG2, to the main feature in the photoelectron spectra might contribute other anionic structures susceptible to BFPT, aTG*n* ($n = 4$, 14, 16), characterized by the VDE in a 1.8–1.9 eV and AEAs in 0.5–0.6 eV range. A difference by about 0.1 eV in the maximum of the main feature differences in photoelectron spectra for $(TG)^{\bullet-}$ and $(UG)^{\bullet-}$ was confirmed by the B3LYP/6-31++G(d,p) results. Experimental and computational results consistently demonstrate that not only uracil but also thymine is susceptible to barrier-free intermolecular proton transfer [73].

11.4.3 Complexes of Uracil with Alcohols

The occurrence of intermolecular proton transfer in anion radical complexes of nucleobases with different proton donors can be viewed as the effect of an interplay between the deprotonation energy of proton donor (HA), protonation energy of nucleobase (NB), and the intermolecular stabilization energy. A small variation in any of these parameters can alter the $NB^{\bullet-}\cdots HA \leftrightarrow NBH^{\bullet}\cdots A^{-}$ equilibrium. For the proton transfer to occur, first of all the stabilizing interaction in the $NBH^{\bullet}\cdots A^{-}$ system needs to compensate the barriers of the PT process leading to non-interacting products. In addition, stabilization of $NBH^{\bullet}\cdots A^{-}$ must be at least as much as that in the untransformed $NB^{\bullet-}\cdots HA$ complex.

The relationships between the deprotonation energy of proton donor and complex stability as well as its VDE were characterized in the work devoted to complexes between uracil and a series of 18 alcohols with deprotonation enthalpy (HDP) varied in a systematic manner [75]. It was found out that a HDP smaller than 14.3 eV is required for BFPT with the product being $UH^{\bullet}\cdots OR^{-}$. Two minima coexist on the anionic energy surface for 14.8 eV < HDP < 14.3 eV. These minima correspond to the $UH^{\bullet}\cdots OR^{-}$ and $U^{\bullet-}\cdots HOR$ structures. For ROH's with deprotonation enthalpies above 14.8 eV only the non-PT $U^{\bullet-}\cdots HOR$ minimum exists on the potential energy surface.

The calculated values of VDE as well as the stabilization energy of anionic uracil···alcohol complexes systematically increase as the gas-phase acidity of the alcohol increases [75]. The alcohol is preferably bound to the C5 side of O8 for complexes without intermolecular proton transfer and to the N3 side of O8 for the complexes with intermolecular proton transfer. The same behavior of proton donor solvating uracil was observed in $U\cdots H_2O$ and $U\cdots H_2S$ dimers [70]. The anionic complexes without intermolecular proton transfer have only one strong hydrogen bond involving the O8 of U. Thus, the C5 side of O8 is a preferable binding direction for the proton donor, because the gas-phase basicity of $U^{\bullet-}$ is larger on the C5 side than the N3 side of O8. The anionic complexes with intermolecular proton transfer have an additional hydrogen bond, which involves a deprotonated weak acid (i.e., an anion) and the N3H or the C5H of U. This hydrogen bond is stronger for the former than the latter and provides an additional stabilization reflected in the reported values of E_{stab}. Theoretical results were supported by the photoelectron spectra (measured with 2.54 eV photons) of anionic complexes of uracil with 3 of these alcohols. The differences between spectra of uracil with ethanol, 2,2,3,3,3-pentafluoropropanol and 1,1,1,3,3,3-hexafluoro-2-propanol illustrate the change in the structure of anionic dimers as a function of the gas-phase acidity of ROH. The photoelectron spectra of aforementioned complexes denoted as aA1U, aA6U and aA10U, respectively, are shown in Fig. 11.16.

The main features in the photoelectron spectra of aA1U and aA6U have maxima at about 1.0 and 1.3 eV, respectively. The spectrum of aA10U is structureless with two broad features: a larger intensity peak at 2.0–2.2 eV and a lower intensity peak at 1.3–1.6 eV. The B3LYP values of VDE for the aA1U complex amount to 1.13 and 1.18 eV for the N3 and C5 sides of O8, respectively When corrected downward by 0.2 eV they approximately overlap with the broad maximum of the PES spectrum of this complex at 1.0 eV (see Fig. 11.16). Similarly, the B3LYP values of VDE for the aA6U complex are 1.45 and 1.47 eV for the C5 and N3 sides of O8, respectively. After the −0.2 eV correction, they approximately overlap with the broad maximum of the PES spectrum of this complex at 1.3 eV. Based on the qualitative agreement between computational and experimental results, the non-PT arrangement ($U^{\bullet-}$···HOR) was assigned to the complexes aA1U and aA6U. In turn, the spectrum of $(U\cdots1,1,1,3,3,3\text{-hexafluoro-2-propanol})^{\bullet-}$ was interpreted in terms of complexes of U with the four isomers of this alcohol. It was found that BFPT develops in one of these isomers. Because of the small differences in DFT stabilities of the four complexes, additional MP2 calculations were conducted for aA10U. The final VDE values for the three most stable structures span the range of 1.9–2.0 eV (MP2 values with +0.13 eV correction) and correlate with the higher intensity feature in the PES spectrum. The VDE value for the least stable structure of 1.4 eV coincides with the lower intensity feature in the PES spectrum [75].

11.4.4 Complexes of Nucleobases with Formic Acid

In this section we review four papers in which formic acid was employed to investigate the effects exerted by particular proton donor on the stability of the valence anions of nucleobases. The anion photoelectron spectroscopy (PES) measurements and quantum chemical calculations are reported for anionic complexes of formic acid (FA) with uracil, thymine [76], 1-methylcytosine [77], adenine and 9-methyladenine [78, 79]. Formic acid can be considered as the simplest model of a weak organic acid. For the anionic complexes of formic acid with uracil and thymine, very different photoelectron spectra were recorded featuring broad signals with maxima at ~1.7 and ~1.1 eV, respectively (see Fig. 11.17). The maximum of

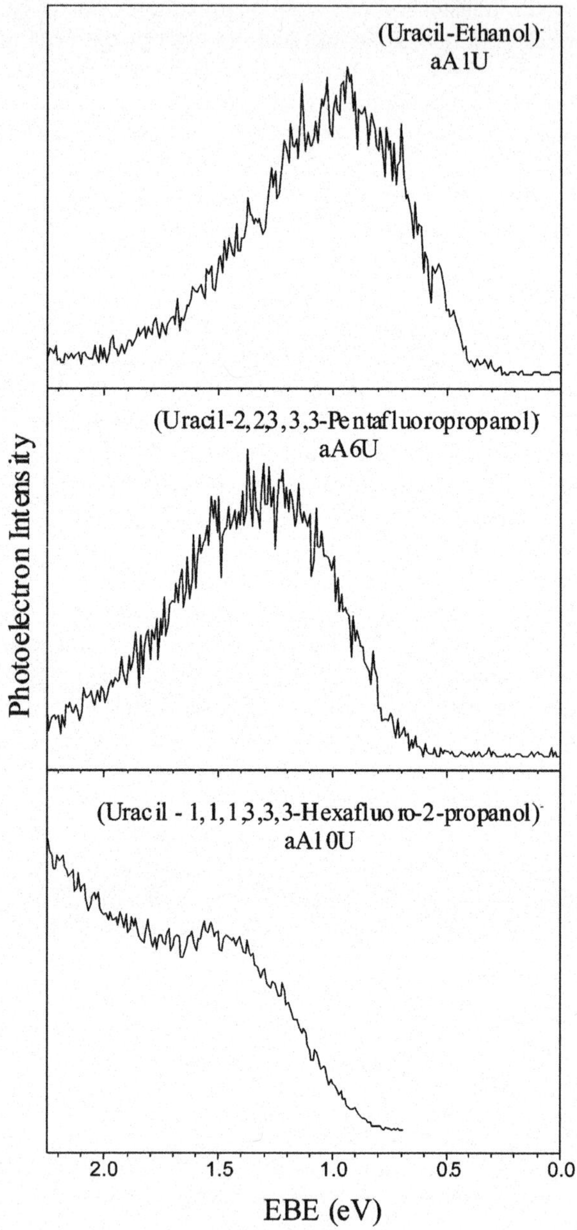

Fig. 11.16 Photoelectron spectra of uracil···alcohol complexes recorded with 2.54 eV photons (Reprinted with permission from Ref. [75]. Copyright 2005 American Chemical Society)

(T···FA)$^{\bullet-}$ is substantially broader than for (U···FA)$^{\bullet-}$ and spreads from 0.5 to 2 eV [76].

The results of the B3LYP/6-31++G(d,p) calculations suggest that the four most Stable U···FA anionic structures undergo BFPT from the carboxylic group of FA to the O8 atom of U$^{\bullet-}$ [76]. The B3LYP values of VDE for these U···FA structures are

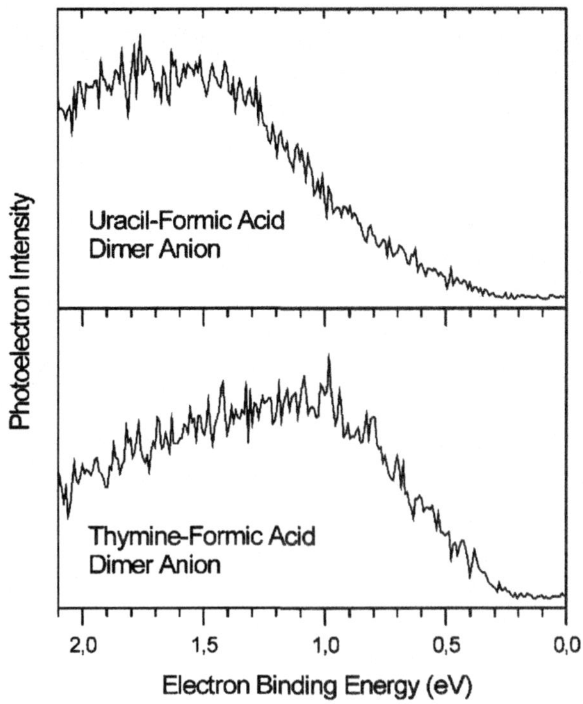

Fig. 11.17 Photoelectron spectra of uracil···formic acid and thymine···formic acid complexes recorded with 2.54 eV photons (Reprinted with permission from Ref. [76]. Copyright 2004 American Chemical Society)

as large as 1.7–1.9 eV (after correcting downward by 0.2 eV) which coincides with the maximum of the broad PES peak. No BFPT was found for dimers in which FA is coordinated to O7 of $U^{\bullet-}$, and their values of VDE are smaller by 0.7 eV than for the structures with BFPT. Due to the small values of Gibbs stabilization energies, structures with no BFPT should not be populated in the anionic beam. Identical arrangements were identified for the most stable $(T{\cdots}FA)^{\bullet-}$ complexes with VDEs in the range of 1.6–1.8 eV (after correction). The difference in spectra registered for $(U{\cdots}FA)^{\bullet-}$ and $(T{\cdots}FA)^{\bullet-}$ was attributed to the appearance of thymine···formic acid dimer, which exhibits resistance to BFPT upon electron attachment. This structure, labeled as aTF16, depicted in Fig. 11.18, does not undergo proton transfer probably because of steric protection provided by the methyl group of thymine. The aTF16 structure may be responsible for the lower energy part of the broad signal of the $(T{\cdots}FA)^{\bullet-}$ spectrum as indicated by a calculated VDE value equal to 1.3 eV. Aforementioned studies confirm that both nucleic acid bases undergo barrier-free proton transfer in anionic complexes with formic acid. The difference in experimental spectra of $(U{\cdots}FA)^{\bullet-}$ and $(T{\cdots}FA)^{\bullet-}$ indicates that the methyl group of thymine interferes with the intermolecular proton transfer [76].

In order to analyze how the stability of the nucleobase anion increases with the number of solvent molecules, complexes of 1-methylcytosine (mC) with FA were studied in 1:1 and 1:2 stoichiometries [77]. The methylation of N1 in cytosine served the purpose to mimic the nucleobase attached to the sugar in nucleic acid.

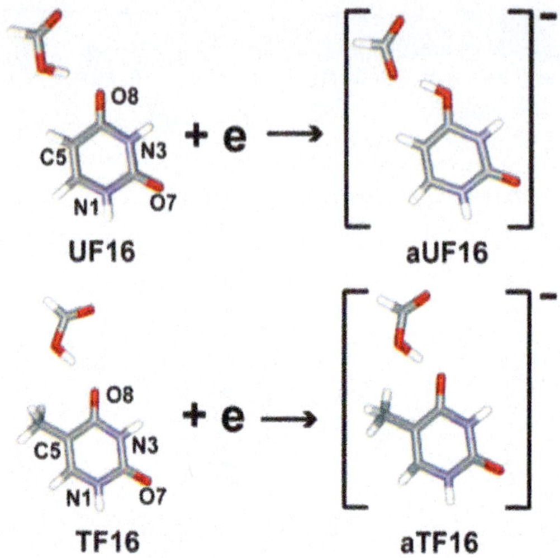

Fig. 11.18 At the B3LYP and MP2 levels, barrier-free proton transfer upon an excess electron attachment occurs for the UF16 structure (*top*) but it does not for the TF16 structure (*bottom*) (Reprinted with permission from Ref. [76]. Copyright 2004 American Chemical Society)

Photoelectron spectra of $[(mC)\cdots(FA)]^{\bullet-}$ and $[(mC)\cdots(FA)_2]^{\bullet-}$ are shown in Fig. 11.19. Both consist of a broad band that has an onset at about 1.1 and 1.6 eV, respectively. The maxima of the PES signals, which correspond to the experimental VDEs, are at about 1.85 and 2.1 eV, respectively. Both spectra having the maximum intensities above 1.5 eV are indicative of high stabilization of the excess electron within complexes; most likely by proton transfer. Moreover, the shift towards higher EBE values of the onset and maximum of the PES signal registered for $[(mC)\cdots(FA)_2]^{\bullet-}$ reveals that the trimeric complexes are better electron scavengers than the respective dimers.

The attachment of an electron to the binary complexes leads to the formation of adiabatically stable valence anions (see Fig. 11.20) [77].

Similarly to anionic complexes of uracil with various proton donors, in all mC···FA complexes, the excess electron attaches to the π^* orbital localized on the cytosine moiety. Moreover, when formic acid interacts with the N3 atom of the canonical mC or the N8 atom of the imino tautomer, the excess electron induces a barrier-free proton transfer that leads to additional stabilization. This leads to a higher stability of the PT complexes when compared to their non-PT counterparts. The results of the MP2 and MPW1K optimizations confirm the BFPT predicted at the B3LYP level [77].

The two most stable anionic proton-transferred structures, $a[mC_c]_{N3}{}^{N8}FA\_pt$ and $a[mC_tc5]_{N8}FA\_pt$, should contribute to the shape and position of the experimental spectrum. For the first anion ($AEA_G = 0.9$ eV), $HCOO^-$ lies in the plane of the mC radical, while in the second, it is located out of the mC plane and interacts via its delocalized bond system ($AEA_G = 1.1$ eV). The B3LYP/6-31++G(d,p) VDE values of the two low-energy anions amount to 1.61 and 1.91 eV, respectively. For better accuracy, additional calculations for these geometries were conducted at the CCSD(T)/aug-cc-pVDZ level and VDEs diminished to 1.53 and 1.76 eV, respectively.

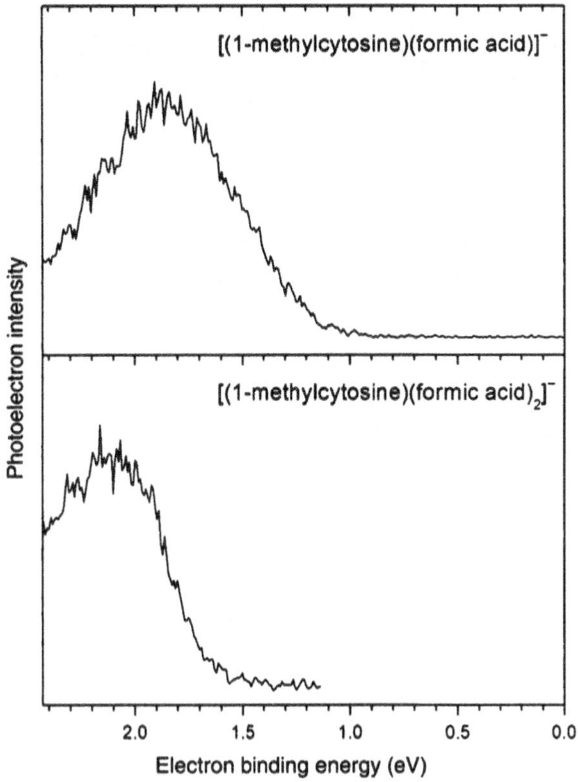

Fig. 11.19 Photoelectron spectra of [(1-methylcytosine)···(formic acid)]$^{\bullet-}$ and [(1-methylcytosine)···(formic acid)2]$^{\bullet-}$ recorded with 2.54 eV photons (Reproduced from Ref. [77] by permission of the publisher [Taylor & Francis Ltd, http://www.tandf.co.uk/journals])

The B3LYP calculations carried out for the anionic complexes of 1:2 stoichiometry identified five valence-type anions shown in Fig. 11.21. The attachment of an excess electron induces BFPT in all studied trimers and their stabilities suggest that only two structures, labeled as a$[mC_c]_{N3O7}{}^{N8CH3}$2FA_pt and a$[mC_tc5]_{O7N8}{}^{CH3}$2FA_pt, should be present under experimental conditions. These trimers are characterized at the B3LYP/6-31++G(d,p) level by VDEs equal to 1.98 and 2.2 eV, respectively [77].

Despite that it is well established the gaseous canonical adenine does not support stable valence anions (the AEA value for adenine is substantially negative), interactions with solvent molecules might stabilize its valence anions [27, 60]. Indeed, it was observed in RET experiment that two molecules of water stabilize the conventional anion of adenine in the gas phase [60]. Later Bowen's group measured photoelectron spectra for the hydrated adenine cluster anions $A^{\bullet-}(H_2O)_{n=1-7}$ [40]. The measured VDEs of the studied clusters increase with the number of water coordinated to adenine from 0.78 to 1.69 eV.

The stabilization of valence anions of purines in the course of intermolecular proton transfer has been demonstrated for the first time by Mazurkiewicz et al. [78] in the report devoted to the effect of electron attachment to the dimers of adenine (A) and 9-methyladenine (MA) with formic acid (FA). Indeed, photoelectron

Fig. 11.20 Optimized structures of [(1-methylcytosine)···(formic acid)]$^{\bullet-}$ complexes and their singly occupied molecular orbitals plotted with a contour value of 0.03 bohr$^{-3/2}$ (Reproduced from Ref. [77] by permission of the publisher [Taylor & Francis Ltd, http://www.tandf.co.uk/journals])

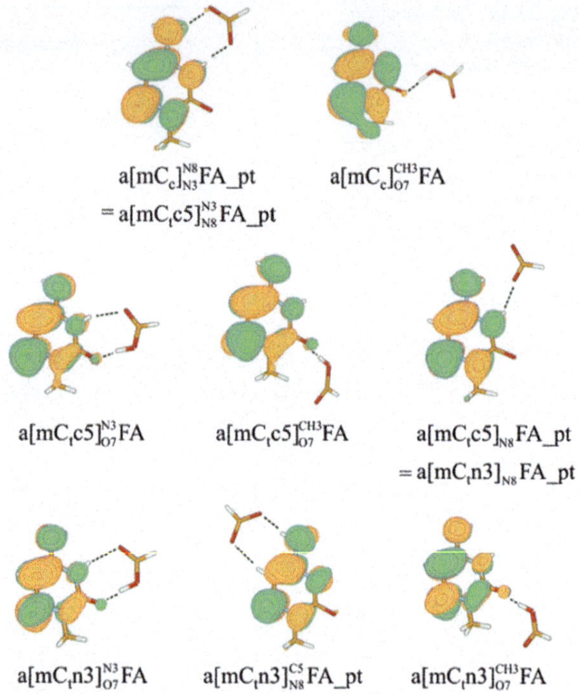

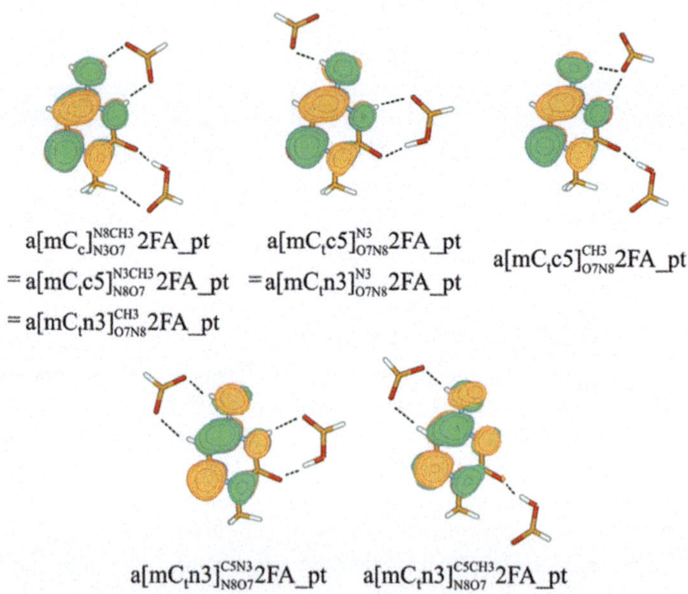

Fig. 11.21 Optimized structures of [(1-methylcytosine)···(formic acid)$_2$]$^{\bullet-}$ complexes and their singly occupied molecular orbitals plotted with a contour value of 0.03 bohr$^{-3/2}$ (Reproduced from Ref. [77] by permission of the publisher [Taylor & Francis Ltd, http://www.tandf.co.uk/journals])

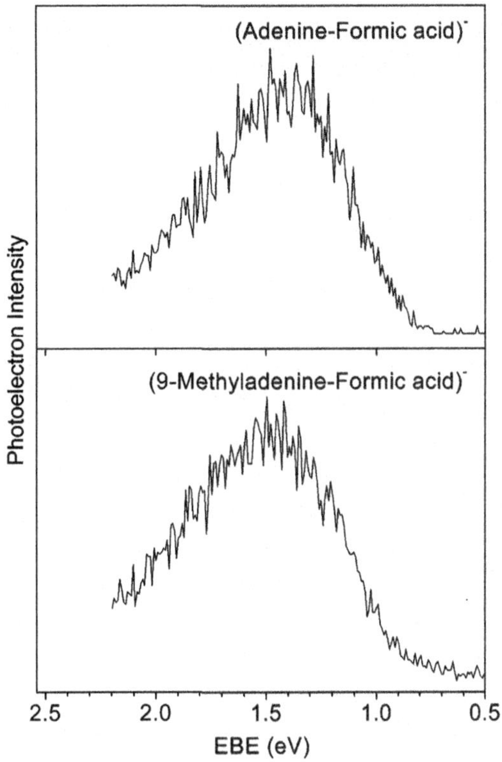

Fig. 11.22 Photoelectron spectrum of adenine ···formic acid and 9-methyl-adenine···formic acid anions measured with 2.54 eV electrons (Reprinted with permission from Ref. [78]. Copyright 2007 American Chemical Society)

spectra depicted in Fig. 11.22 indicate that A···FA and MA···FA form stable valence-type anions rather than dipole-bound states. The spectra cover broad ranges of binding energies, 1.0–2.0 eV, with a maximum at 1.45 and 1.50 eV for the $(A\cdots FA)^{\bullet-}$ and $(MA\cdots FA)^{\bullet-}$ complexes, respectively [78].

A common feature of the anionic wave function is that the excess electron is localized on a π* orbital of adenine and distributed evenly over its proton-acceptor centers (see Fig. 11.23). The attachment of an excess electron leads to a spontaneous proton transfer from the OH group of FA to a proton-acceptor site of A (MA) [78].

All the identified anions are cyclic double hydrogen-bonded complexes and are adiabatically stable by 0.3–0.7 eV with respect to the neutral complexes (in terms of AEA_G) [78]. The most stable dimers, involving non-methylated and methylated adenine, denoted as $aA_{N3}{}^{N9}FA$ and $aMA_{N7}{}^{N10}FA$, are characterized by VDEs of 1.21 and 1.67 eV, respectively. Applying an increment of −0.2 eV, one obtains 1.01 and 1.47 eV, respectively. The latter value matches perfectly the maximum of the signal in the spectrum of $(MA\cdots FA)^{\bullet-}$ at 1.5 eV, whereas the former is inconsistent with the experimental VDE value of ~1.4 eV for $(A\cdots FA)^{\bullet-}$. The problem of calculating reliable VDEs for the AFA complexes at the B3LYP level of theory was traced back to the deficiency of the DFT method to predict reliable geometries

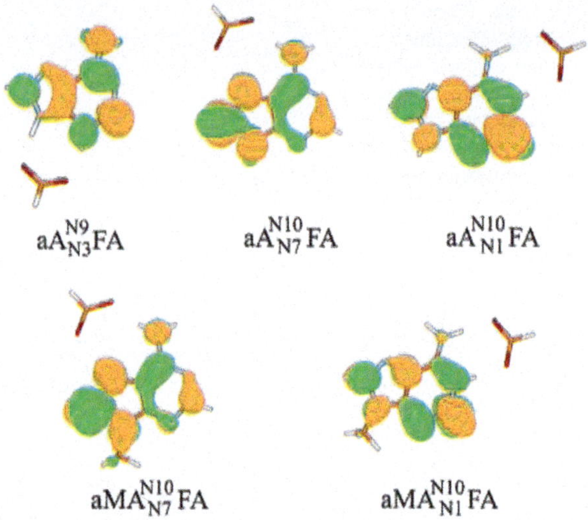

Fig. 11.23 Anionic AFA and MAFA complexes and their singly occupied molecular orbital plotted with a spacing 0.03 bohr$^{-3/2}$ (Reprinted with permission from Ref. [78]. Copyright 2007 American Chemical Society)

for certain anionic configurations. The VDE values, 1.38 and 1.36 eV calculated at the MP2/aug-cc-pVDZ level for $aA_{N3}{}^{N9}FA$ and $aMA_{N7}{}^{N10}FA$, respectively, reproduce better the maximum of the dominant feature observed in the respective spectra.

The studies concerning the anionic dimers (1:1) of adenine/9-methyladenine with formic acid have been extended in other report by Mazurkiewicz et al. [79] to systems with more complex stoichiometries. Thus, the effect of electron attachment to the complexes of A(or MA) with FA in 1:2 and 1:3 proportions was discussed. The photoelectron spectra registered for 1:2 complexes $(A{\cdots}FA_2)^{\bullet-}$ and $(MA{\cdots}FA_2)^{\bullet-}$ unquestionably indicate the presence of valence-type anions under the experimental conditions. However, the authors were not able to measure spectra for anionic tetramers which were explored purely by means of computational methods. Spectra for trimers cover a broad range of binding energies, 1.4–2.3 eV, with a plateau between 1.9 and 2.3 eV (see Fig. 11.24). Because the maximum intensity at EBE larger than 1.5 eV in photoelectron spectra was typically observed for the complexes of nucleobases with various proton donors stabilized by the barrierless proton transfer, the same phenomenon was expected here [79].

Indeed, as calculations demonstrated, in each anionic 1:2 and 1:3 complex, two FA molecules coordinated to adenine spontaneously transfer protons to the base (see Fig. 11.25) [79]. It turned out that the double-BFPT reactions are predicted to be barrier-free also at the MPW1K and MP2 levels for $(A{\cdots}FA_2)^{\bullet-}$ and $(MA{\cdots}FA_2)^{\bullet-}$. For the most stable tetrameric complex $aA(FA)_3{}^{IPT/IIPT/III}$ a small barrier separating single- and double-PT was found in terms of electronic energy, which disappears in terms of Gibbs free energy. For the 1:3 anion radical structure with three protons transferred does not exist. Explanation of this finding is that after a second PT, the adenine acquires positive charge and subsequent accepting of the third proton by this purine is thermodynamically unfavorable.

The AEA_G values, in a range of 0.8–1.23 eV, calculated for the 1:2 anionic complexes are approximately two times larger than the ones predicted for the respective pair of 1:1 anions (the AEA_G values previously reported for hydrogen-

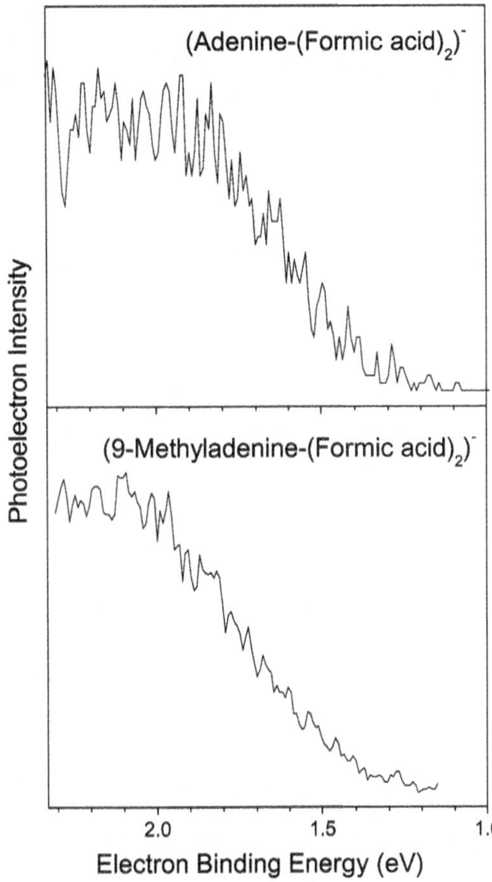

Fig. 11.24 Photoelectron spectra of [adenine···(formic acid)$_2$]$^{\bullet-}$ and [9-methyl-adenine···(formic acid)$_2$]$^{\bullet-}$ measured with 2.54 eV photons (Reprinted from Ref. [79]. Copyright 2007, with permission from Elsevier)

bonded dimers of A(MA) do not exceed 0.67 eV) [78]. In turn, the theoretical AEA_G values for the 1:3 complexes are even higher (~1.2 eV).

Based on the relative stabilities and the VDEs estimated on various levels of theory, it was found that the main contribution to the signal displayed in the spectrum of $(A{\cdots}FA_2)^{\bullet-}$ comes from the less stable of the two isomers present in the gas phase, $aA(FA)_2^{IPT/IIIPT}$, that has a VDE value of ~2.0 eV resulting from the different theoretical models. The second anion, $aA(FA)_2^{IPT/IIPT}$, more stable by 0.8 kcal/mol in terms of G, with a VDE of ~2.7 eV, is out of the spectrum range. The signal visible in the $(MA{\cdots}FA_2)^{\bullet-}$ is assigned to $aMA(FA)_2^{IIPT/IIIPT}$ having the VDE of ~2.2 eV.

Calculations suggesting that it is impossible to experimentally observe the tetrameric anions is well explained by the VDE of $aA(FA)_3^{IPT/IIPT/III}$ being 2.86 eV, since is larger than the energy of the photon (2.54 eV) used in the PES experiment.

The above described combined experimental/computational studies demonstrated that the simultaneous involvement of several molecules capable of forming hydrogen

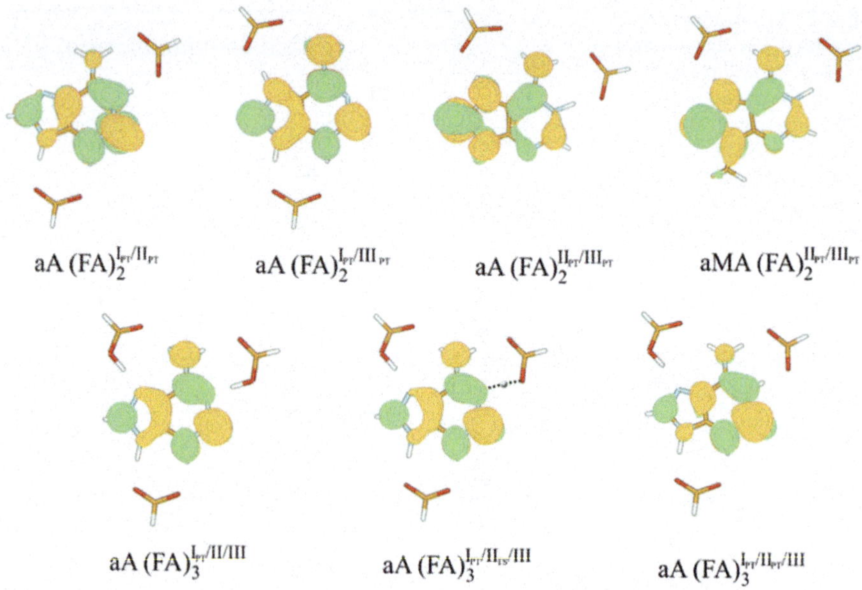

Fig. 11.25 Structures of [adenine···(formic acid)$_2$]$^{\bullet-}$, [9-methyladenine···(formic acid)$_2$]$^{\bullet-}$ and [adenine···(formic acid)$_3$]$^{\bullet-}$ complexes and their singly occupied molecular orbital plotted with a spacing 0.03 bohr$^{-3/2}$ (Reprinted from Ref. [79]. Copyright 2007, with permission from Elsevier)

bonds with adenine increases its readiness to bind an excess electron when compared to 1:1 complexes. Secondly, it was shown that the more species are involved in hydrogen bonding with a nucleobase, the more stable is its valence anion. Therefore, it may be anticipated that the interaction between purines bound in DNA and various species from the physiological environment (water, histones, repair or replication enzymes, etc.) might counterbalance the larger electron affinities of free pyrimidines and as a consequence, both types of nucleic bases (purines and pyrimidines) could play a crucial role in DNA damage induced by low energy electrons.

11.5 Anion Radical Complexes of Complementary Base Pairs

The Watson-Crick AT [80–83] and GC [49, 50, 82, 84–87] base pairs were also shown to form stable valence anions in the gas phase, in which the excess charge is located entirely at the pyrimidine moiety. The hydrogen bonding between complementary bases increases the AEA values of pyrimidine nucleobase. For example, at the B3LYP/DZP++ level the AEA of the Watson-Crick AT pair is 0.36 eV (zero-point vibrationally corrected). This AEA is about 0.16 eV larger than the corresponding electron affinity obtained at the same level for thymine [80]. On the

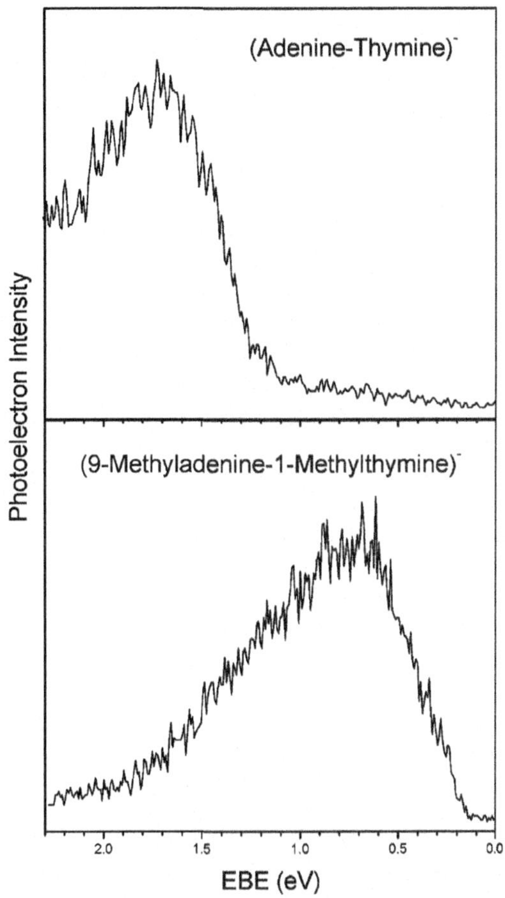

Fig. 11.26 Photoelectron spectra of (adenine···thymine)$^{\bullet-}$ and (9-methyladenine···1-methylthymine)$^{\bullet-}$ recorded with 2.54 eV photons (Reprinted with permission from Ref. [83]. Copyright 2005 American Chemical Society)

other hand, the AEA of the Watson-Crick GC pair, calculated at the B3LYP/DZP++ level by Richardson et al. [86] is equal to 0.6 eV, which is substantially higher when compared to the AEA value of –0.09 eV predicted for cytosine at the same level of theory [53].

11.5.1 Adenine···Thymine and 9-methyladenine···1-methylthymine

The photoelectron spectra of adenine···thymine, (AT$^{\bullet-}$) and 9-methyladenine···1-methylthymine (MAMT$^{\bullet-}$) are very different as can be seen in Fig. 11.26 [83]. Indeed, the photoelectron spectrum of AT$^{\bullet-}$ consists of a broad peak with maximum at ~1.7 eV, while the spectrum of MAMT$^{\bullet-}$, consists of a broad peak with a maximum at ~0.7 eV. Recalling the results of previous reports on the anion

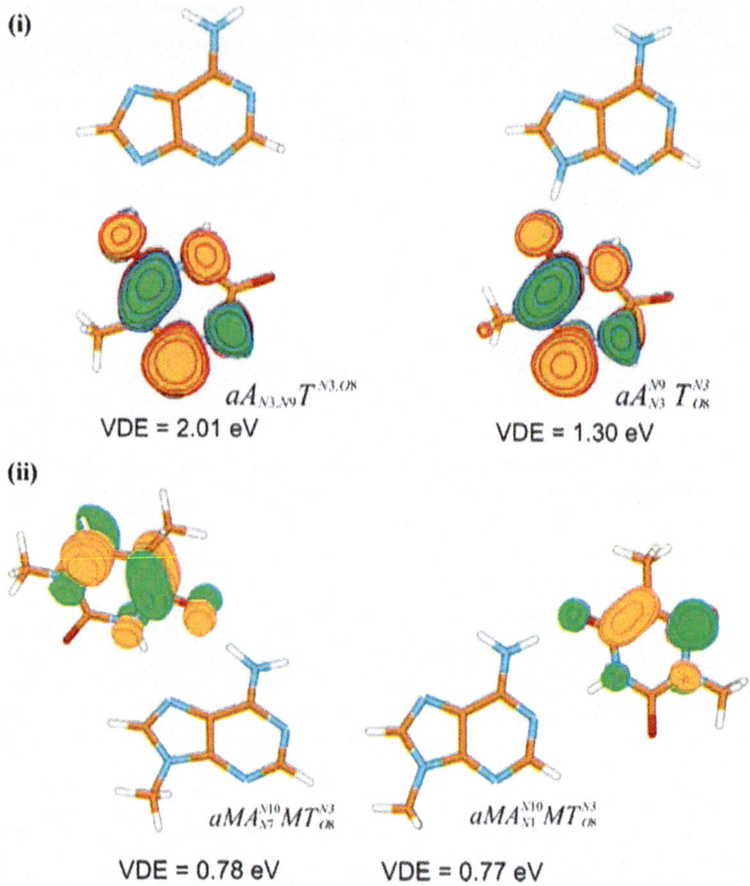

Fig. 11.27 Unpaired electron orbital plotted with a contour line spacing of 0.03 bohr$^{-3/2}$ for the two most stable anions of (*i*) AT pair and (*ii*) MAMT pair (Reprinted with permission from Ref. [83]. Copyright 2005 American Chemical Society)

radical complexes involving nucleobases, one could conclude that the shape of the spectrum of AT$^{\bullet-}$ is typical for structures stabilized by intermolecular proton transfer, for example (T···FA)$^{\bullet-}$ [76]. Instead, the MAMT$^{\bullet-}$ spectral feature resembles those of thymine (VDE of 0.8 eV; [27]) or adenine (VDE of 0.78 eV; [40]) solvated by one molecule of water. The spectra of Fig. 11.26 also provide the estimations of the AEA values for AT$^{\bullet-}$and MAMT$^{\bullet-}$ which are equal to ~1.1 eV and 0.2–0.4 eV, respectively.

According to the B3LYP/6-31+G(d,p) calculations for AT$^{\bullet-}$, the global minimum on the anionic potential energy surface results from proton transfer from the N9H of adenine to O8 of thymine [83]. This structure, labeled $aA_{N3N9}T^{N3,O8}$, is shown in Fig. 11.27 together with other relevant geometries. Corresponding VDEs predicted the for $aA_{N3N9}T^{N3,O8}$ structure and the second most stable $aA_{N3}{}^{N9}T_{O8}{}^{N3}$ (without PT) are 2.0 and 1.3 eV, respectively (uncorrected values). The difference in the thermodynamic stability of both complexes is small so that both structures are believed to contribute to the maximum of the spectrum at 1.7 eV. The AEA values estimated for both dimers are found similar (~0.6 eV) [83].

The most stable anionic complexes for methylated bases correspond to the Hoogsteen $aMA_{N7}{}^{N10}MT_{O8}{}^{N3}$eV) and Watson-Crick $aMA_{N1}{}^{N10}MT_{O8}{}^{N3}$ structures, with the former being more stable by 2.0 and 1.2 kcal/mol in terms of E and G, respectively; see Fig. 11.27 [83]. Both structures are characterized by AEAs 0.33–0.35 eV and an identical VDE of 0.8 eV at the B3LYP/6-31+G(d,p) level. The results are consistent with the maximum of the photoelectron spectrum peak for $MAMT^{\bullet-}$ at 0.7 eV (see Fig. 11.26).

A common feature of anionic wave functions for the AT and MAMT complexes is that the excess electron is localized on a π^* orbital of thymine, mainly in the O8-C4-C5-C6 region of the nucleobase, in close resemblance to the valence anionic state of the isolated thymine [73]. The charge distribution in anionic complexes determined the most favorable proton acceptor and donor pair of T, O8 and N3H. This is reflected in the hydrogen bonding pattern in the most stable anions. The most favorable binding sites of adenine are the same in the neutral and anionic complexes: N3 and N9H.

The most important finding of this joint experimental/computational study is that the excess electron attaching to MAMT does not induce intermolecular proton transfer [83]. The lack of PT is apparently connected with the methylation of the N9 position of adenine. Since in DNA the N9 is also blocked because it is involved in glycosidic bond between adenine and sugar, the MAMT may be considered as the more realistic model of the complementary base pair in DNA than the non-methylated dimer. Thus, electron attachment to AT within DNA may trigger protonation at the C6 position of thymine, which in turn can abstract a hydrogen from its own or neighboring sugar and ultimately lead to a single-strand break. Indeed, a high yield of 5,6-dihydrothymine was observed experimentally in the reaction of electrons with thymine bound in DNA [88].

11.5.2 Guanine···Cytosine and 9-methylguanine···1-methylcytosine

While the $MAMT^{\bullet-}$ and $AT^{\bullet-}$ complexes differ in their propensity to undergo intermolecular proton transfer, both the $MGMC^{\bullet-}$ and $GC^{\bullet-}$ valence anions exist as proton-transferred structures in the gas phase [49, 87]. The photoelectron spectrum of $MGMC^{\bullet-}$ occurs as a single broad band with maximum at approximately 2.0 eV, while the $GC^{\bullet-}$ feature consists of two bands: one of lower intensity with maximum around 2.0 eV and another with higher intensity and maximum at approximately 2.6 eV. Both spectra are shown in Fig. 11.28.

When it comes to non-methylated guanine···cytosine complexes, the relative thermodynamic stabilities obtained at the DFT level suggest that 3 anionic structures are likely to show up in the photoelectron experiment [49]. Their geometries

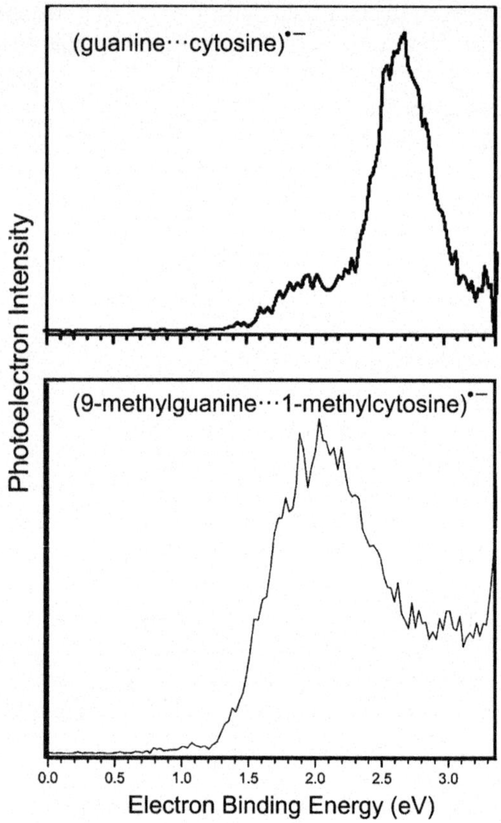

Fig. 11.28 Photoelectron spectra of (guanine···cytosine)$^{\bullet-}$ and (9-methylguanine···1-methylcytosine)$^{\bullet-}$ recorded with 3.49 eV photons (Reprinted with permission from Ref. [49] [Copyright 2010 John Wiley and Sons] and from Ref. [87] [Copyright 2009 American Chemical Society])

are presented in Fig. 11.29. The stabilization energies and Gibbs free energies calculated at the DFT level for these structures are within a very narrow range and their stabilities decrease in the following order Gf_{O10}-Ca_{N3} > Gc_{N1}-Ca_{N3} > Ga_{N1}-Ca_{N3}. The most stable complexes are degenerate in terms of G and the third one (which is the canonical proton-transferred WC structure) is separated from them by barely 1.3 kcal/mol. The relative stability order was confirmed at the MP2 level. According to the MP2 calculations the second and third anionic dimer are 0.5 and 2.7 kcal/mol less stable than Gf_{O10}-Ca_{N3}. Hence, the two maxima observed in the PES spectrum can be assigned to the two isomers of $GC^{\bullet-}$, namely, Gf_{O10}-Ca_{N3} and Gc_{N1}-Ca_{N3}. The former is responsible for the feature at ~2.6 eV and the latter for the ~2 eV one. Despite significant difference in VDEs predicted for both complexes, their adiabatic affinities are almost identical (~0.7 eV) [49].

The corresponding B3LYP VDEs of 2.42 and 2.07 eV turned out to be slightly underestimated. The problems with obtaining realistic VDEs for certain systems at the DFT level were already reported previously [78]. The discrepancies between calculated and experimental VDEs were rationalized by unsatisfactory geometries predicted at the DFT level. In such cases the correct prediction of the VDE values

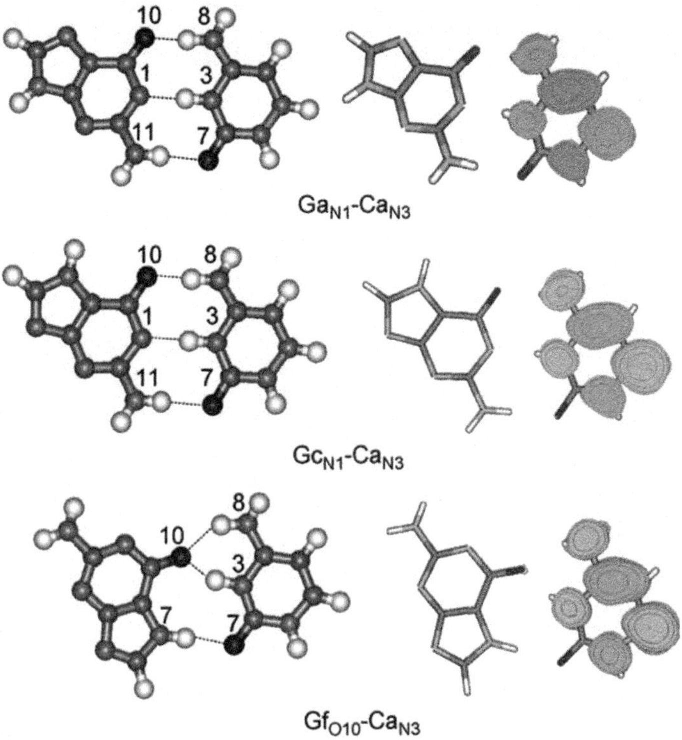

Fig. 11.29 B3LYP/6-31++G(d,p) geometries for the three most stable (guanine···cytosine)$^{\bullet-}$ base pairs along with their singly occupied molecular orbitals (Reprinted with permission from Ref. [49]. Copyright 2010 John Wiley and Sons)

can be achieved employing MP2 optimized geometries. Indeed, the recalculated B3LYP VDEs for the most stable GC$^{\bullet-}$ in their MP2 geometries are 2.63 and 2.31 eV for Gf_{O10}-Ca_{N3} and Gc_{N1}-Ca_{N3}, respectively. Finally, it can be noted that as in the AT$^{\bullet-}$ base pair, [83] the most populated GC$^{\bullet-}$ anion, Gf_{O10}-Ca_{N3}, is not biologically relevant.

As far as the MGMC$^{\bullet-}$ anionic dimer is concerned, a systematic search over its conformational space converged to five anionic structures depicted in Fig. 11.30 [87]. Two of them, a1MGMC and a1MGMC_SPT2, are more stable than the other anions by at least 14 kcal/mol. Both geometries originate from electron attachment to the neutral WC complex. The difference between these structures lies in proton transfer between N1(G) and N3(C) which is developed in a1MGMC_SPT2. This proton-transferred structure is more stable than the regular WC form by barely 2.9 and 1.3 kcal/mol in terms of energy and free energy. From the relative stabilities it may be assumed that only these two anions should be present in detectable amounts in the equilibrated gaseous mixture. These two geometries are separated by a negligible kinetic barrier. The VDE value of 1.93 eV predicted at the B3LYP level of theory for a1MGMC_SPT2 matches well the maximum at the photoelectron spectrum. To refine this value, the most stable complex was reoptimized at the

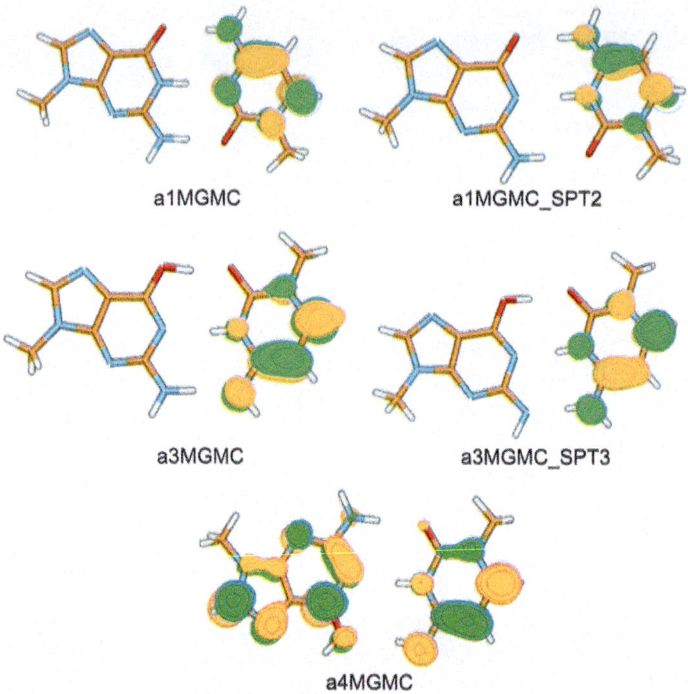

Fig. 11.30 Unpaired electron orbitals plotted with a contour line spacing of 0.03 bohr$^{-3/2}$ for the four most stable complexes of (9-methylguanine···1-methylcytosine)$^{\bullet-}$ (Reprinted with permission from Ref. [87]. Copyright 2009 American Chemical Society)

MP2 level. The VDE value calculated for the MP2 optimized geometry appeared to be 2.12 eV [87].

The above-mentioned data suggest that the attachment of excess electron to GC incorporated in double-stranded DNA should lead to the most stable proton-transferred configuration. This is in contrast to the behavior of the AT base pair in DNA, where the formation of reactive anion of thymine may promote hydrogen abstraction from a sugar and ultimately a strand damage. In case of GC, the strand damage could be suppressed by the extra stabilization resulting from the proton transfer from G to C.

11.5.3 9-Methyladenine···1-methylthymine Base Pair Solvated by Formic Acid

In addition to the of 9-methyladenine···1-methylthymine base pair (MAMT), Bowen's group recorded the photoelectron spectrum of MAMT interacting with the third component, formic acid [89]. The spectrum of (MAMT···FA)$^{\bullet-}$ (see Fig. 11.31) displays a broad band having an onset at an electron binding energy of ~0.7 eV and a maximum at ~1.9 eV. The position of the maximum of the signal at

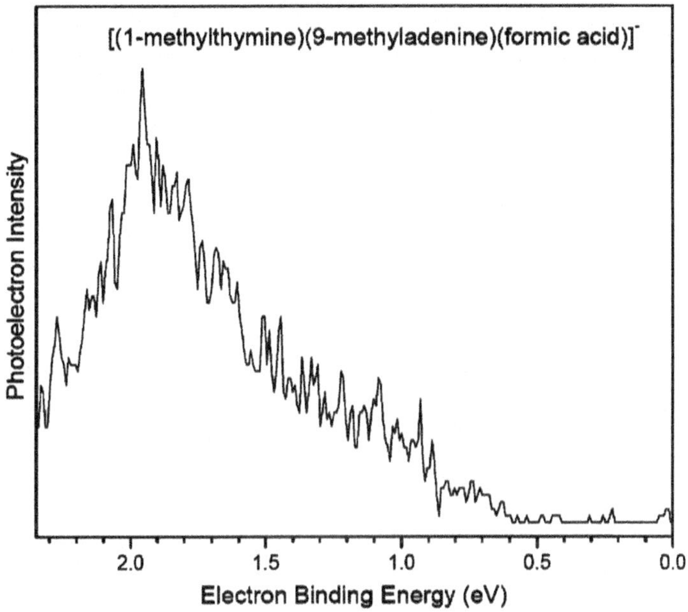

Fig. 11.31 Photoelectron spectrum of [(1-methylthymine)···(9-methyladenine)···(formic acid)]$^{\bullet-}$ recorded with 2.54 eV photons (Reprinted with permission from Ref. [89]. Copyright 2010 American Chemical Society)

the high EBE values, as well as the width of the main feature is typical for proton-transferred systems. The spectrum of (MAMT···FA)$^{\bullet-}$ closely resembles that of AT$^{\bullet-}$ with the maximum at ~1.9 eV [83]. Let us recall that the AT$^{\bullet-}$ spectrum was interpreted as due to the structure resulting from low energy barrier proton transfer from the N9–H site of the neutral adenine to the O8 atom of anionic thymine [83]. From the fact that experimental VDEs for AT$^{\bullet-}$ and for (MAMT···FA)$^{\bullet-}$ differ by only 0.2 eV it could be deduced that proton transfer induced by electron attachment also takes place in the anionic (MAMT···FA)$^{\bullet-}$ complex.

Interactions between formic acid and the MAMT dimers provide strong stabilization of the complexes as reflected by the B3LYP/6-31++G(d,p) level adiabatic electron affinities [89]. The AEA_G values of 47 considered trimeric configurations span a range of 0.37–0.97 eV surpassing AEA_Gs calculated for the MAMT configurations (0.26–0.36 eV) at a similar level of theory. Based on the relative thermodynamic stabilities, seven anion radicals with VDEs of 1.3–2.06 eV have been identified to be responsible for the shape of the photoelectron spectrum (see Fig. 11.32). The most stable anions were predicted for those trimers, where the hydroxyl group of formic acid interacts with the O8 atom of thymine. Due to the relative electron affinities of canonical forms of particular nucleobases, an extra electron has the tendency to localize on thymine and then its O8 atom becomes one of the thymine atomic centers characterized by the highest density of the excess

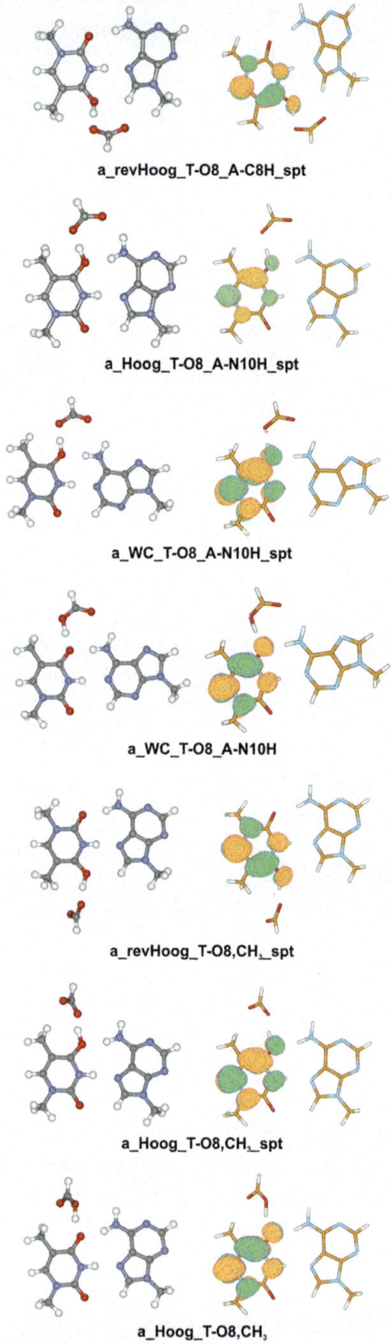

Fig. 11.32 Optimized structures of the seven low energy (MAMT···FA)$^{\bullet-}$ complexes and their singly occupied molecular orbitals plotted with a contour value of 0.03 bohr$^{-3/2}$ (Reprinted with permission from Ref. [89]. Copyright 2010 American Chemical Society)

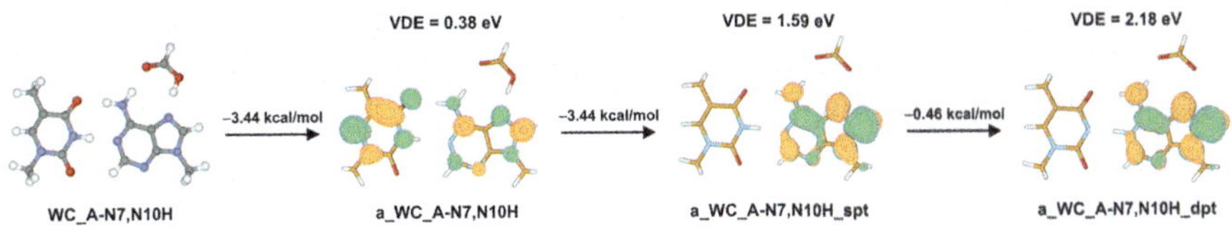

Fig. 11.33 Possible proton transfer reactions triggered by electron attachment to the WC_A-N7, N10H complex. The numbers above the *arrows* indicate the difference in the electronic energy between product and substrate (Reprinted with permission from Ref. [89]. Copyright 2010 American Chemical Society)

electron. This explains the observed increased stability of the configurations involving FA···O8(MT) interactions [89].

The most stable anions are biologically relevant reversed-Hoogsteen and Hoogsteen structures. These are a_revHoog_T-O8_A-C8H_spt is (VDE of 2.06 eV) and a_Hoog_T-O8_A-N10H_spt (VDE of 1.96 eV) [89]. The latter is less stable than the a_revHoog_T-O8_AC8H_spt anion by only 0.72 kcal/mol in terms of the free energy. The VDEs of the above-mentioned anions reproduce very well the observed maximum in the PES spectrum, 1.9 eV. In both complexes proton transfer between formic acid and the anionic base pair takes place and appears as an important factor deciding on the anionic trimer stability. Indeed, the a_revHoog_T-O8_A-C8H_spt and a_Hoog_T-O8_A-N10H_spt anions of the largest adiabatic electron affinities are also the most stable in terms of the electronic and free energy.

Electron attachment to the mentioned above neutral complexes triggers barrier free proton transfer; thus, only PT structures are minima on the potential energy surface, while their non-PT counterparts, a_revHoog_T-O8_A-C8H and a_Hoog_T-O8_A-N10H, do not exist. For the other complexes, both the non-PT and PT geometries frequently support adiabatically stable anions. This, for instance, holds for the complexes where formic acid interacts with adenine. There are also two non-PT structures in the pool of seven low energy anions shown in Fig. 11.32, a_WC_T-O8_AN10H and a_Hoog_T-O8,CH3, being vertically stable by 1.3 and 1.35 eV. The above VDEs values could be assigned to the low-intensity but broad feature in the range of 1.0–1.5 eV (see Fig. 11.31).

The important finding of the investigations concerning 9-methyladenine···1-methylthymine complexes with formic acid is that proton transfer resulting from the electron attachment takes place in the most stable anionic geometries [89]. Moreover, in those anionic trimers where adenine interacts with FA, proton transfer leads to a partial or almost complete electron shift from the initial localization on thymine to adenine and gaining of the negative charge by adenine promotes a second proton transfer from thymine to adenine. The role of the second PT would be additional stabilization of an electron residing on the adenine moiety. Indeed, the process of electron attachment triggering two consecutive proton transfers is thermodynamically feasible. As illustrated in Fig. 11.33 attachment of an electron leads to the a_WC_A-N7,N10H anion adiabatically stable by 3.4 kcal/mol [89]. The first proton transfer, from FA, is favored again by 3.4 kcal/mol and leads to the anion in which the electron is almost completely localized on the adenine molecule. Finally, the second proton transfer from the N3H of thymine to the N1 atom of adenine provides additional stabilization by 0.5 kcal/mol and leads to a complete localization of an extra electron on adenine. Thus, if adenine in DNA interacts with a sufficiently acidic external proton donor, then intermolecular proton transfers may occur and an unpaired electron becomes completely localized on adenine. A similar conclusion has also been drawn from other studies [78, 90]. This finding questions the commonly accepted paradigm that in DNA electrons localize on pyrimidines rather than on purines when the cell is exposed to high energy radiation.

11.6 Nucleosides

Gas phase, parent intact valence anions of nucleosides, were also shown to exist as stable anions characterized by higher AEAs and VDEs than the corresponding isolated nucleobases. Bowen's group measured the photoelectron spectra of three deoxyribonucleoside anions, i.e., 2′-deoxythymidine$^{\bullet-}$ (VDE = 0.89 eV), 2′-deoxycytidine$^{\bullet-}$ (VDE = 0.87 eV), and 2′-deoxyadenosine$^{\bullet-}$ (VDE = 1.14 eV) plus four ribonucleoside anions, i.e., uridine$^{\bullet-}$ (VDE = 1.39 eV), cytidine$^{\bullet-}$(VDE = 1.08 eV), adenosine$^{\bullet-}$(VDE = 1.40 eV), and guanosine$^{\bullet-}$ (VDE = ~1.6 eV) [91]. Figure 11.34 depicts the photoelectron spectra for the above mentioned nucleosides. There is a close resemblance between the photoelectron spectra of deoxynucleoside and ribonucleoside which share the same nucleobase moieties. In other words, the spectra of cytidine$^{\bullet-}$ and deoxycytidine$^{\bullet-}$ look very similar although the maximum on the EBE scale of deoxycytidine$^{\bullet-}$ is shifted by ~0.2 eV towards the smaller values. Similarly, the maximum on the spectrum of 2′-deoxyadenosine$^{\bullet-}$ is shifted by ~0.3 eV with respect to the position of maximum on the spectrum of adenosine$^{\bullet-}$ (Fig. 11.34).

Experimental results were compared with the earlier theoretical predictions at the B3LYP level employing the DZP++[93] and 6–311G(d) basis sets [94]. The calculated AEAs and VDEs for 2′-deoxythymidine$^{\bullet-}$and 2′-deoxycytidine$^{\bullet-}$were in excellent agreement with those extracted from spectra. The largest discrepancy between theory and experiment was found for 2′-deoxyadenosine$^{\bullet-}$, namely calculated AEAs of dA$^{\bullet-}$ were found to be close to zero (0.06 eV and −0.04) whereas the experimental value is ~0.7 eV. In turn, theoretical VDE, 0.91 eV, predicted for dA$^{\bullet-}$ reasonably matches the experimental value of 1.14 eV. The computational and photoelectron results for 2′-deoxyguanosine$^{\bullet-}$ were not compared to each other due to technical problems with measuring the PES spectrum of 2′-deoxyguanosine$^{\bullet-}$. From the above theoretical and experimental characteristics it could be concluded that gaseous nucleosides observed by Stokes et al. upon electron attachment become canonical valence-bound anion radicals. Moreover the VDE values around 1 eV for deoxyribonucleosides and in the range from 1.1 to 1.6 eV for ribonucleosides are too small to suggest PT in these anions [91].

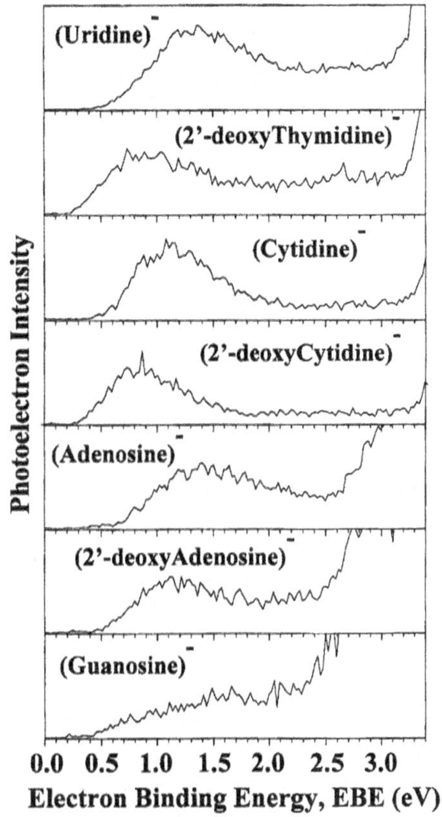

Fig. 11.34 The photoelectron spectra of the seven nucleoside parent anions. These spectra were recorded with 3.49 eV photons (Reprinted with permission from Ref. [91]. Copyright 2007, American Institute of Physics)

11.7 Anion Radicals of Nucleotides (Adenosine-5′-Monophosphate and 2′-Deoxyadenosine-5′-Monophosphate)

As the extension of studies on electrophilic properties of gas-phase DNA-related systems comprising adenine, Bowen et al. measured, using 3.49 eV photons, the parent, valence anions of two nucleotides: adenosine-5′-monophosphate (AMPH), and 2′-deoxyadenosine-5′-monophosphate (dAMPH) [95]. The photoelectron spectra of both anionic nucleotides, which are displayed in Fig. 11.35, are typical for valence anions. The photoelectron spectra of $(AMPH)^{\bullet-}$ and $(dAMPH)^{\bullet-}$ are similar but their maxima are shifted relative to one another by ~0.1 eV. The VDE and AEA values read from the spectrum of $(dAMPH)^{\bullet-}$ are 1.81 and 1.2 eV, respectively, whereas the corresponding characteristics of $(AMPH)^{\bullet-}$ are 1.9 and 1.4 eV [95].

The interpretation of the experimental picture of $(dAMPH)^{\bullet-}$ was provided by Kobyłecka et al. who employed the DFT model at the B3LYP/6-31++G(d,p) level [92]. The authors demonstrated that the four representative neutral conformations

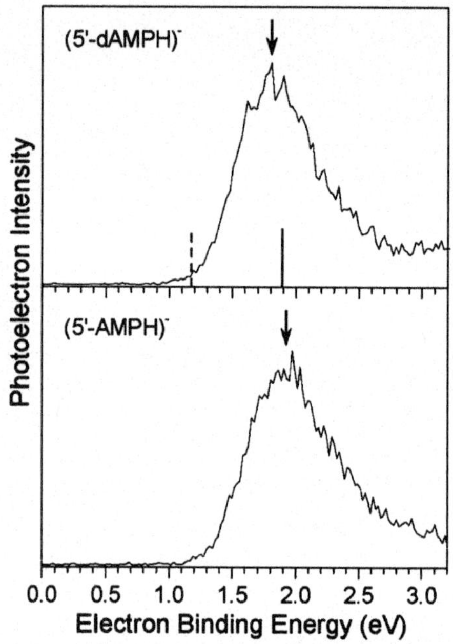

Fig. 11.35 The photoelectron spectra of $(AMPH)^{\bullet-}$ and $(dAMPH)^{\bullet-}$, both recorded with 3.49 eV photons. The *arrows* mark the experimentally determined VDE values, while for the $(dAMPH)^{\bullet-}$ spectrum, the solid vertical line denotes the theoretically determined VDE value and the dashed vertical line denotes the theoretically determined AEA value (Reprinted with permission from Ref. [95]. Copyright 2008, American Institute of Physics)

of dAMPH are able to bind electron forming adiabatically stable radical anions with AEAs within a range of 0.13–0.73 eV and of 0.2–0.9 eV in terms of AEA_E and AEA_G, respectively. Out of four anionic geometries, the most stable anion, characterized by VDE and AEA_G equal to 1.62 and 0.9 eV, respectively, revealed intramolecular barrier-free proton transfer induced by electron attachment to the neutral *south-syn* conformer. The anionic geometries and their singly occupied molecular orbital plots are presented in Fig. 11.36 [92]. It can be noted that the type of anionic state (the valence, dipole bound or mixed) depends on the internal hydrogen bonds arrangement. The *anti* geometries (stabilized by the intramolecular ((base)C8H···O5′(sugar) hydrogen bond) support dipole bound state while *syn* conformations ((base)N3···HO(phosphate) bond) form valence or mixed anions. Because the theoretical VDE for $dAMPH^{\bullet-}$ equal to ~1.62 eV clearly underestimated the experimental VDE of 1.81 eV, the authors performed additional conformational search for the most stable *south-syn* conformer. As a result, the feature observed in the photoelectron spectrum was attributed to the structure denoted as *a_south-syn_VI* (see Fig. 11.37) [92]. The *a_south-syn_VI* nucleotide is stabilized by two intramolecular hydrogen bonds, that is, (base)N3···HO(phosphate) and (sugar)3′OH···O(phosphate) bonds, the former featuring proton transfer from phosphate hydroxyl group to the N3 site of adenine. The calculated at the B3LYP/6-31++G(d,p) level VDE and AEA_G for this conformer, 1.89 and 1.19 eV, respectively, reproduce well the feature on the photoelectron spectrum of $(dAMPH)^{\bullet-}$.

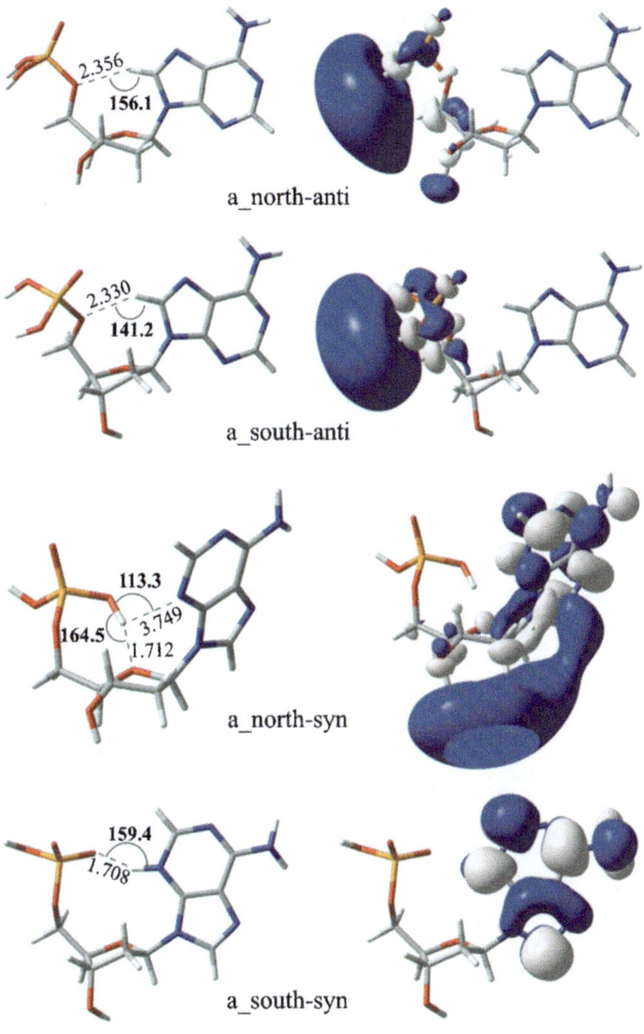

Fig. 11.36 The anionic geometries obtained in the course of geometry optimization that started from the neutral configurations of dAMPH together with their singly occupied molecular orbital. The X···H distance and X···HY angle are given in Å and degrees, respectively (Reprinted with permission from Ref. [92]. Copyright 2008, American Institute of Physics)

Stokes et al. [95] and Kobyłecka et al. [92] were the first who identified intramolecular BFPT phenomenon in an anionic nucleotide. Those experimental-computational studies demonstrated that in contrast to the isolated canonical adenine, this nucleobase forms a stable anion when incorporated into the nucleotide. Therefore, damage of nucleic acids induced by low-energy electrons might begin with anionic states localized not only on pyrimidines but also on purines. In other words, purines in a DNA strand may exhibit electron scavenging properties similar

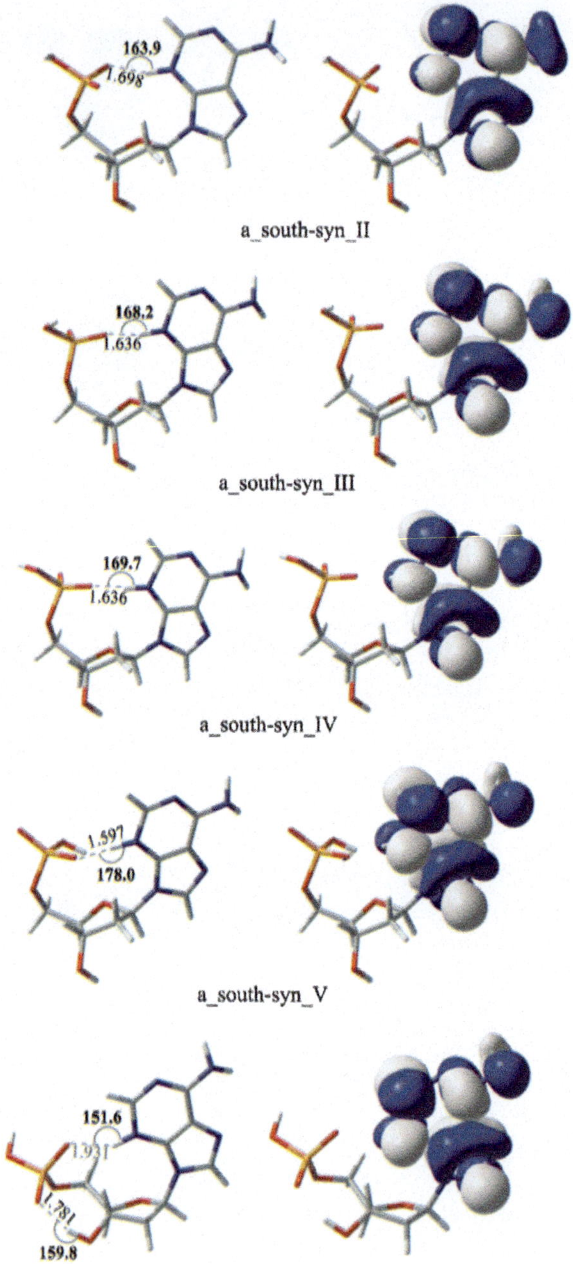

Fig. 11.37 The selected anionic geometries belonging to the south-syn family of conformations and their singly occupied molecular orbital. The X···H distance and X···HY angle are given in Å and degrees, respectively (Reprinted with permission from Ref. [92]. Copyright 2008, American Institute of Physics)

to pyrimidines. It should be noted that the *south-syn* anionic conformer of the dAMPH may occur in a single-stranded form of DNA under the physiological conditions.

11.8 Homodimeric Complexes of Nucleobases

Development of a source for generating parent (intact) anions of various nucleosides, nucleotides and other biomolecules has allowed Bowen's group to investigate homogeneous nucleobase dimer anions, i.e., $U_2^{\bullet-}$, $T_2^{\bullet-}$, $C_2^{\bullet-}$, $A_2^{\bullet-}$, and $G_2^{\bullet-}$ [96]. The photoelectron spectra of the five homogeneous nucleobase dimer anions generated with an infrared desorption/photoemission source and measured on the pulsed apparatus are presented in Fig. 11.38. The photoelectron spectra of the three pyrimidine dimer anions exhibit a dominant broad peak, before their intensities gradually increase toward the high electron binding energy (EBE) ends of their spectra.

Generally, their intensity thresholds occur at EBE ~0.7 eV. In case of $T_2^{\bullet-}$, the maximum is at EBE ~1.6 eV and a weaker intensity, broad peak is also discernible at EBE ~2.3 eV. In the case of $U_2^{\bullet-}$, the maximum intensity is at ~2.2 eV and additionally there is a shoulder on the low EBE side of the main peak. The spectrum of $C_2^{\bullet-}$ displays low-intensity but distinct peak at ~1.3 eV and a broad, structureless signal at the higher EBE range [96].

The photoelectron spectra of the two purine dimer anions [see Fig. 11.38d, e] have similar spectral patterns, i.e. each exhibit two peaks. In both spectra the peaks are separated by ~0.9 eV, and exhibit about the same intensity ratios. They differ, however, in that the spectrum of $A_2^{\bullet-}$ is shifted to higher EBE, by ~0.5 eV, relative to the spectrum of $G_2^{\bullet-}$. The peaks in the spectrum of $A^{\bullet-}$ occur at ~1.3 eV (lower intensity) and ~2.7 eV (higher intensity). Based on the B3LYP/6-31++G(d,p) level calculations, the peaks at EBE ~2.2 eV in the photoelectron spectra of dimeric thymine and uracil, were assigned to proton-transferred complexes [96]. Both complexes have analogous hydrogen bonding pattern N1H···O8/O7···HN3, where the proton is shifted between bases from N1 to O8. On the other hand, the peak centered at ~1.6 eV in the spectrum of $T_2^{\bullet-}$ was assigned to the structure similar to the above configurations but without proton transfer.

11.9 Anion Radicals of Homodimers of Pyrimidine Nucleosides

As was demonstrated in the previous paragraphs of this review, the electron induced proton transfer is a common phenomenon in the gaseous complexes of nucleobases hydrogen bonded to various proton donors. Moreover, it was confirmed by

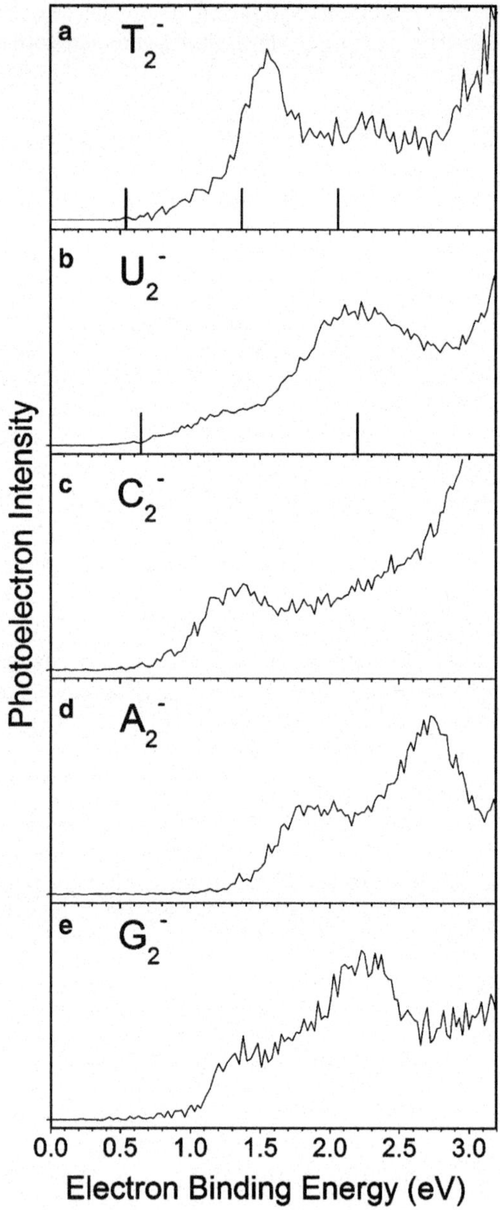

Fig. 11.38 Photoelectron spectra of homogeneous nucleobase dimer anions (**a**)–(**e**) recorded with 355 nm (3.49 eV) photons. VDE calculations of $T_2^{\bullet -}$ and $U_2^{\bullet -}$ are shown as stick spectra in (**a**) and (**b**), respectively (Reproduced from Ref. [96] by permission of the PCCP Owner Societies)

experimental and computational methods, that in the gaseous clusters composed of canonical adenine/9-methyladenine and two or three formic acid molecules, attachment of electron triggers double-BFPT [79]. The possibility of two consecutive proton transfers involving 9-methyladenine was also predicted in anionic trimers of 9-methyladenine···1-methylthymine···FA [89]. It is generally assumed that a driving force for proton transfer is the stabilization of the excess electron localized on a π* orbital of a nucleobase. The anionic complexes stabilized by intermolecular PT are characterized by VDE values above 1.5 eV, typically around 2 eV. Increase in VDEs coincides with the number of intermolecular proton transfers within complexes [79]. In addition to intermolecular proton transfer between nucleobases and solvating species, electron induced intramolecular proton transfer was disclosed within adenosine-5′-monophosphate. This single PT from the phosphate moiety to the nucleobase in $(dAMPH)^{\bullet-}$ was shown to provide a vertical stability of ~1.9 eV [92, 95]. In the light of the above, it is surprising that the non-PT structures rather than PT ones are responsible for the PES signals at the EBEs close and above 2 eV in nucleoside homodimers: uridine, 2′-deoxythymidine and 2′-deoxycytidine.

The photoelectron spectra of intact, $(rU)_2^{\bullet-}$, $(dT)_2^{\bullet-}$ and $(dC)_2^{\bullet-}$ homodimer radical anions were measured using infrared desorption/photoemission anion generation and pulsed laser photodetachment with 3.49 eV photons [97–99]. The shapes of these spectra indicate that all three complexes exist as stable valence-bound anions, and the broad spectral features suggest the presence of more than one anion radical structure in the gas phase.

The spectrum of uridine homodimer radical anions exhibits a broad signal with a threshold at ~1.2 eV (as mentioned earlier, threshold value is an approximation of the AEA_G) and a maximum intensity at 2.0–2.5 eV (see Fig. 11.39) [97]. The shape of photoelectron spectrum of monomeric uridine anions $(rU^{\bullet-})$ registered previously resembles that of dimers $(rU)_2^{\bullet-}$. However, dimerization process shifts the $(rU)_2^{\bullet-}$ PES maximum by around 0.8 eV to the higher EBEs when compared to $rU^{\bullet-}$. On the other hand, the photoelectron spectra of $(uracil)_2^{\bullet-}$ and uridine homodimer $(rU)_2^{\bullet-}$ are characterized by the maximum intensity lying within the same EBE range.

The photoelectron spectrum of the thymidine homodimer anion, $(dT)_2^{\bullet-}$, presented in Fig. 11.40, has the onset of the spectrum at EBE ~1.1 eV [98]. The spectrum features a broad peak over the range of 1.9–2.2 eV, with a maximum at ~2 eV. Finally, the photoelectron spectrum of 2′-deoxycytidine homodimer anion, $(dC)_2^{\bullet-}$, reveals a noticeable narrower peak (with maximum at EBE between 1.6 and 1.9 eV) than two foregoing spectra, with a threshold at ~1.2 eV (see Fig. 11.41) [99]. In order to identify the most stable anionic complexes which should dominate in the gas phase and to be responsible for the shape of the PES spectra, computational DFT studies were performed at the B3LYP/6-31++G(d,p) level.

When it comes to the computational investigation of uridine and thymidine homodimers, due to the structural similarity of monomers, both systems behave similarly [97, 98]. On the basis of the relative thermodynamic stabilities of conceivable complexes and their propensity to attach and detach the excess electron, it was found that in the most stable structures of $(rU)_2^{\bullet-}$ and $(dT)_2^{\bullet-}$ one monomer (hosting an excess electron on the base moiety) is hydrogen-bonded via O8 to the sugar hydroxyl groups of the other nucleoside (the arrangement denoted as nucleobase···sugar). Surprisingly, the hydrogen-bonded radical anionic complexes utilizing the typical uracil/thymine proton-donor N3H center and the proton-acceptor O8, O7 sites appeared to be highly unstable in comparison with the structures featuring uracil/thymine···ribose interactions.

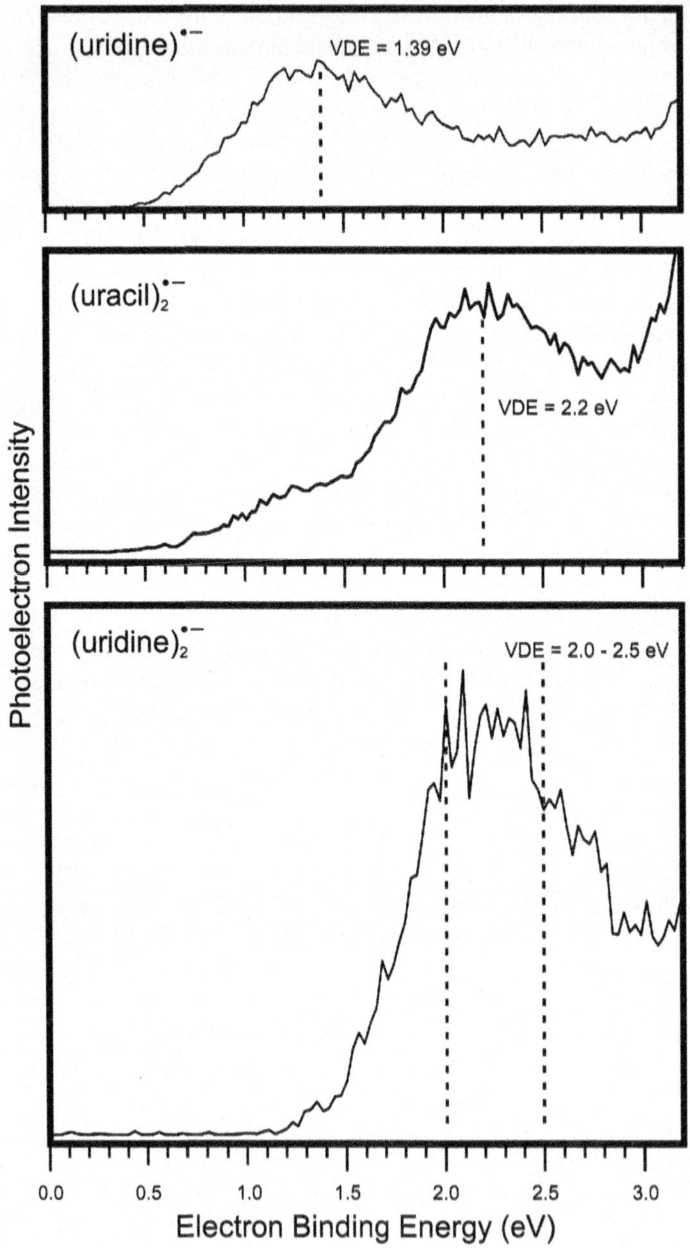

Fig. 11.39 Comparison of the anion photoelectron spectra of uridine$^{\bullet-}$ from Ref. [91], (uracil)$_2^{\bullet-}$ from Ref. [96], and (uridine)$_2^{\bullet-}$ from Ref. [97], recorded with 3.49 eV photons (Reprinted with permissions)

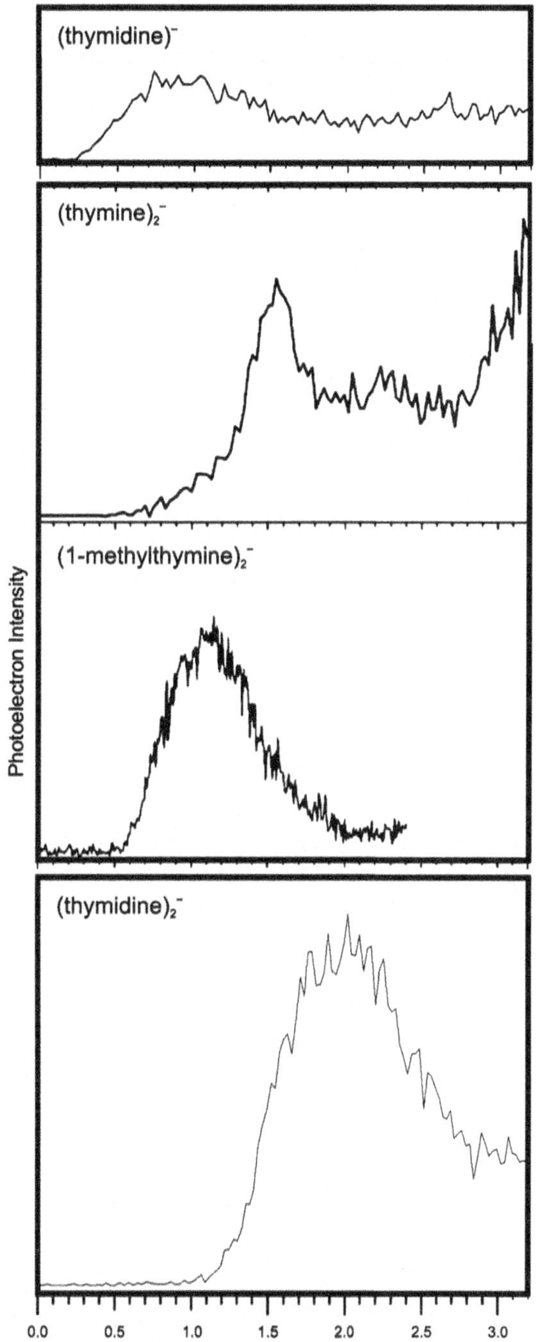

Fig. 11.40 Comparison of the anion photoelectron spectra of 2′-deoxythymidine$^{\bullet-}$ from Ref. [91], (thymine)$_2^{\bullet-}$ and (1-methylthymine)$_2^{\bullet-}$ from Ref. [96] and (2′-deoxythymidine)$_2^{\bullet-}$ from Ref. [98], recorded with 3.49 eV photons (Reprinted with permissions)

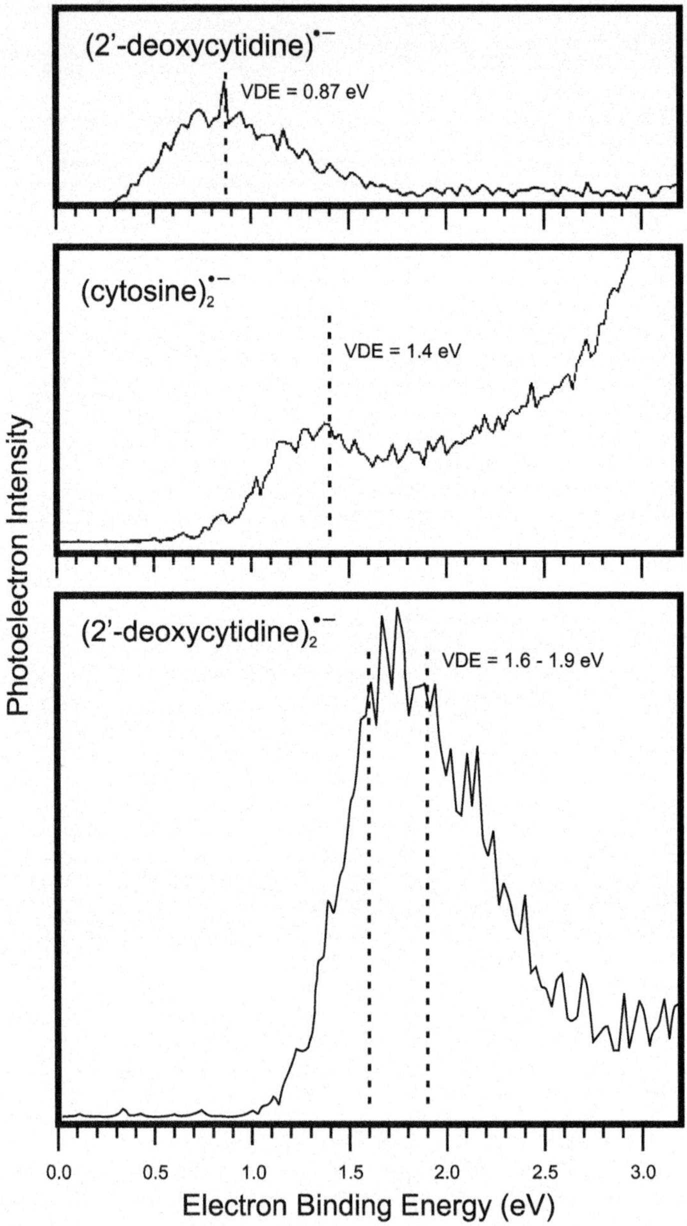

Fig. 11.41 Comparison of the anion photoelectron spectra of 2′-deoxycytidine$^{\bullet-}$ from Ref. [91], $(\text{cytosine})_2^{\bullet-}$ from Ref. [96] and $(2'\text{-deoxycytidine})_2^{\bullet-}$ from Ref. [99], recorded with 3.49 eV photons (Reproduced from Ref. [74]. Reprinted with permissions)

The minimum structures resulting from the B3LYP/6-31++G(d,p) calculations for $(rU)_2^{\bullet-}$ and $(dT)_2^{\bullet-}$ are shown in Fig. 11.42 [97] and Fig. 11.43 [98], respectively.

A common feature of the investigated anionic structures is uneven charge distribution, which can be perceived as the solvation of the negatively charged nucleoside, with the electron density localized entirely within the base moiety, by the neutral monomer. For the seven most stable uracil···ribose structures, labeled as ax_O8, x = 1–7, the calculated adiabatic stabilities are in the range from 1.41 to 1.46 eV, while VDEs assume values from 2.26 to 2.46 eV (values corrected by −0.15 eV).[97] In the most stable structures, a1_O8 and a2_O8, contributing to the photoelectron feature, the nucleoside hosting the excess charge on the base moiety forms a bifurcated hydrogen bond via its O8 atom with two hydroxyl groups of the other neutral nucleoside's ribose. Calculated VDEs of 2.46 and 2.3 eV fit perfectly the experimental VDE. Also, predicted for a1_O8 and a2_O8 adiabatic electron affinities, a value of 1.46 eV match reasonably the threshold of the photoelectron signal.

The least stable member of uracil···ribose family is separated from the next structure (which is a8_O8-O8-pt, see Fig. 11.42) by ~2 and 3 kcal/mol in terms of E and G, respectively [97]. The anionic a8_O8-O8-pt and a9_O8-O7-pt structures are stabilized by electron-induced intermolecular proton transfer (PT) occurring between the complementary uracil moieties. Despite favorable stabilization free energies and positive AEA_Gs of ~1 eV, such PT structures are not the lowest energy configurations and, therefore, should not be populated in the spectrum of $(rU)_2^{\bullet-}$. Similarly, the relative stabilities of non-PT uracil···uracil complexes (a10 to a14 in Fig. 11.42) indicate that these complexes do not show up in the spectrum.

In analogy to the uridine homodimers, calculations for $(dT)_2^{\bullet-}$ revealed that the most stable anions were two base···deoxyribose structures, a1_O8 and a2_O8, characterized by almost identical AEA_Gs of ~1 eV and identical VDEs of ~ 2 eV [98]. The absence of the hydroxyl group at the 3′ position of the sugar does not make a possible formation of the bifurcated hydrogen bond between the base and the sugar as in the most stable complexes of $(rU)_2^{\bullet-}$. Therefore in the most relevant configurations of $(dT)_2^{\bullet-}$, thymidine monomers are bonded through a single O8···HO-C3′ hydrogen bond (see Fig. 11.43). The VDE values for the above-mentioned thymine···deoxyribose dimers are consistent with the position of the maximum of the PES signal at ~2 eV. The onset of the PES signal is around 1.1 eV and is in very good agreement with the theoretical AEA_G obtained for the most stable a1_O8 and a2_O8 geometries. The difference in absolute energies and free energies between the most stable thymine···deoxyribose and thymine···thymine geometries is approximately 9 and 7 kcal/mol, respectively.

Anionic complexes where PT takes place are characterized by similar adiabatic electron affinities (AEA_G ~0.6 eV) and similar vertical stabilities (VDE ~2.3 eV). The non-PT structures have weaker electron acceptor sites than their corresponding PT counterparts as reflected by substantially smaller VDEs (~1.6 eV). The calculated VDE values of the PT complexes are comparable to the thymine···deoxyribose dimers, suggesting a similar magnitude of excess charge stabilization in the complexes. The calculated AEA_G values, however, confirmed that the anionic structures

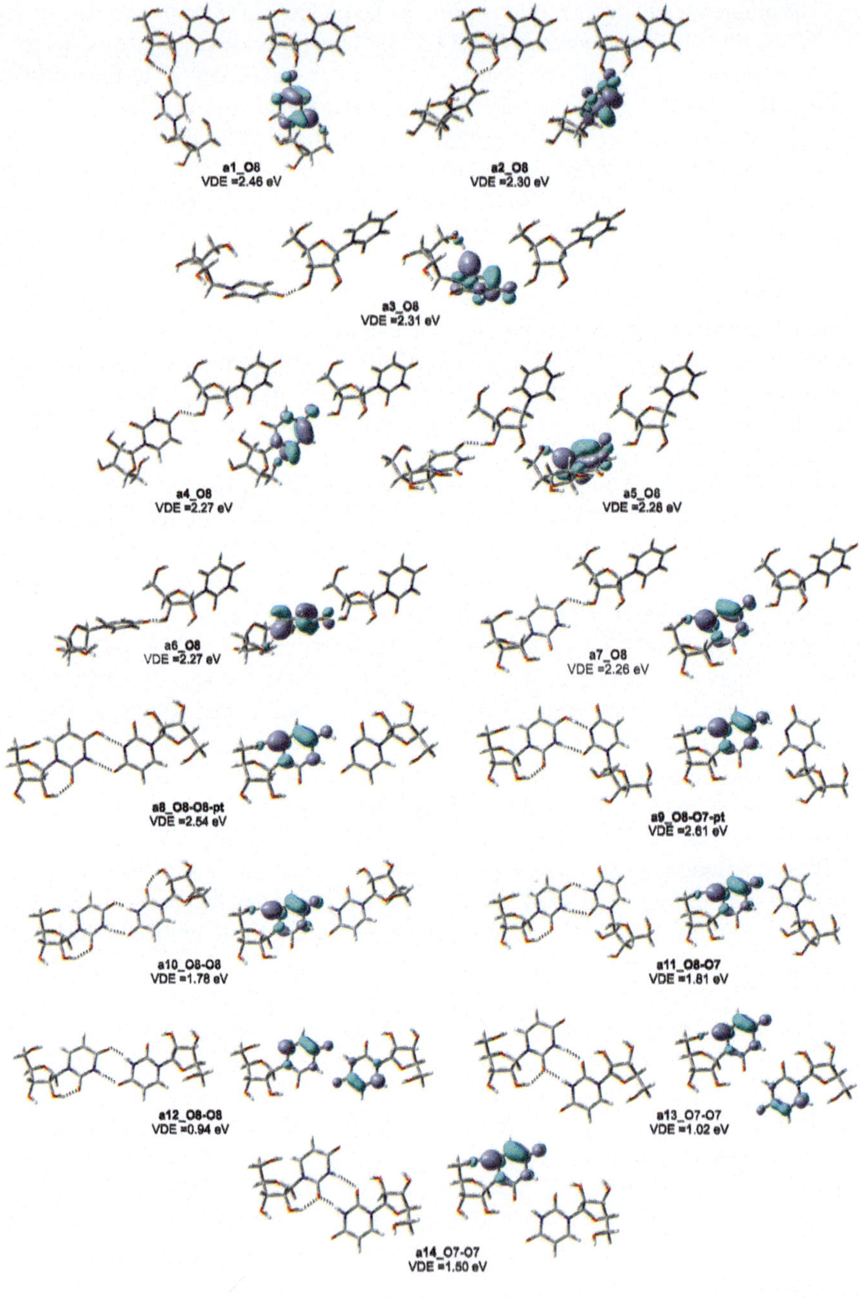

Fig. 11.42 Structures of anion radical uridine dimers optimized at the B3LYP/6-31++G(d,p) level with corresponding VDE values and their singly occupied molecular orbitals plotted with a contour value of 0.05 bohr$^{-3/2}$ (Reprinted with permission from Ref. [97]. Copyright 2012, American Institute of Physics)

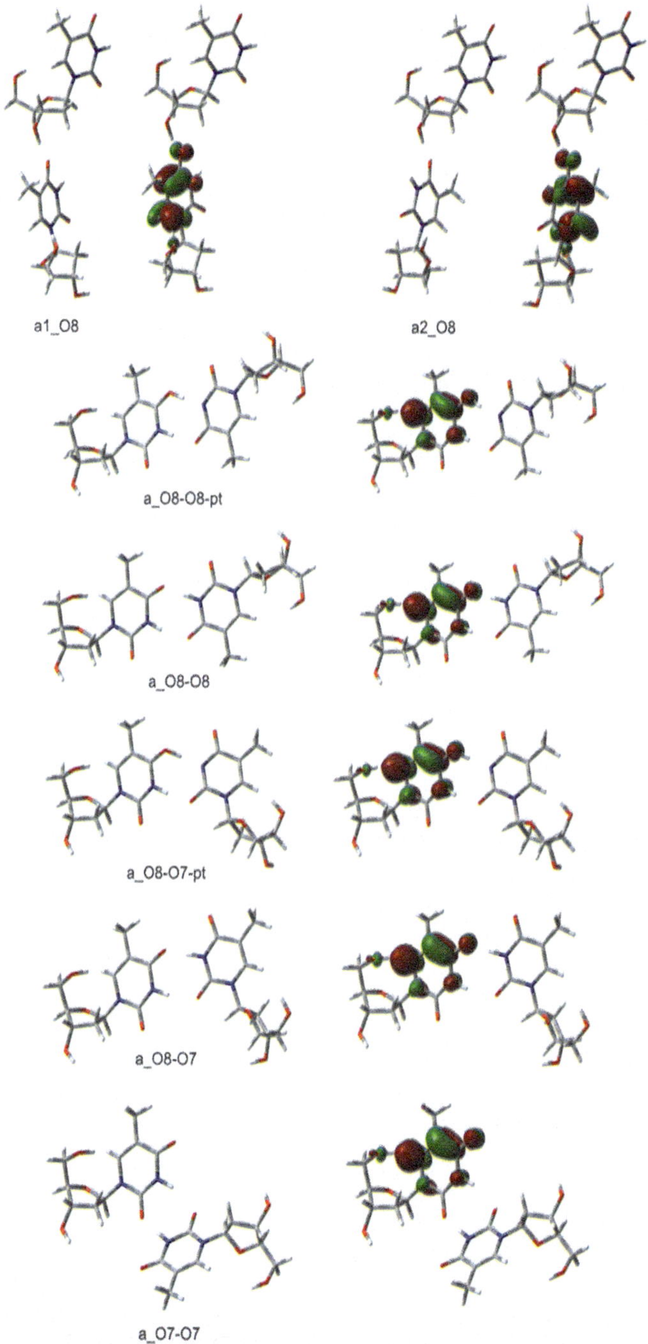

Fig. 11.43 Structures of anion radical 2′-deoxythymidine homodimers optimized at the B3LYP/6-31++G(d,p) level and their singly occupied molecular orbitals plotted with a contour value of 0.05 bohr$^{-3/2}$ (Reprinted with permission from Ref. [98]. Copyright 2012 American Chemical Society)

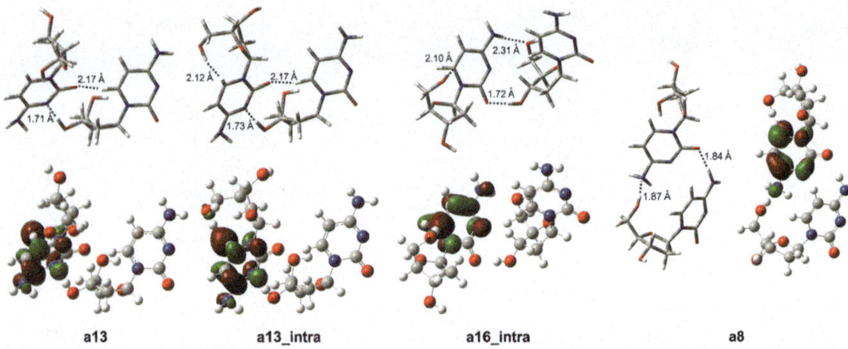

Fig. 11.44 Structures of the most stable anion radical 2′-deoxycytidine homodimers optimized at the B3LYP/6-31++G(d,p) level with corresponding VDE values and their singly occupied molecular orbitals plotted with a contour value of 0.05 bohr$^{-3/2}$ (Reproduced from Ref. [99])

belonging to the thymine···thymine family should not be competitive in the gas phase [98].

As far as $(dC)_2^{\bullet-}$ is concerned, the most stable anionic complexes found at the B3LYP/6-31++G(d,p) level, a13, a13_intra, a16_intra and a8, are displayed in Fig. 11.44 [99]. Their adiabatic stabilities (AEA_G) span the range of ~0.8–1 eV, and for the a13 structure, which is expected to dominate in the gas phase, AEA_G equals to 0.97 eV. The theoretical VDEs calculated for the structures of Fig. 11.44 correlate perfectly with the maximum EBE of 1.6–1.9 eV (see Fig. 11.41). The highest VDE values among the most stable structures, 1.95 and 1.79 eV, are attributed to the a13_intra and a16_intra geometries, where one of the monomers features an intramolecular H-bond. It was found that the formation of the intramolecular O5′–H···C6 bond in 2′-deoxycytidine monomer stabilizes the excess charge and shifts the VDE to higher values by 0.39 eV with respect to the VDE value of nucleoside without an internal H-bond [99].

Homodimers of $(dC)_2^{\bullet-}$ depicted in Fig. 11.44 are stabilized by two hydrogen bonds involving both sugars' and cytosines' sites. The H-bond pattern is identical in the two most stable anionic structures (a13 and a13_intra) [99]. The only difference lies in the conformation of one of nucleoside. Thus, a13 and a13_intra are held together by the sugar···base, O3′–H···N3, and base···base, C6-H···O7, interactions. In the third most stable complex, a16_intra, one of nucleosides utilizes two of the sugar's hydroxyl groups (its O3′-H serves as a proton donor and O5′ as a proton acceptor) to bind to the nucleobase of the second nucleoside. The least stable structure, a8, employs deoxyribose's 5′-end hydroxyl group as a proton donor to form a H-bond with the N8 of cytosine in the second nucleoside. The finding that the strongest stabilization of the anionic nucleoside dimer structures arises from sugar···base interaction agrees with the results for the uridine and thymidine homodimer anions [97, 98]. However, the most relevant $(dC)_2^{\bullet-}$ complexes are more diversified geometrically when compared to the most relevant ax_O8 structures of $(rU)_2^{\bullet-}$ and $(dT)_2^{\bullet-}$.

In general, the vertical stability of homodimer anion radicals, $(rU)_2^{\bullet-}$, $(dT)_2^{\bullet-}$ and $(dC)_2^{\bullet-}$, originate from intermolecular interactions between the negatively charged monomer and the neutral monomer and at the same time from the presence of the sugar moiety in the anionic nucleoside. Comparison of photoelectron spectra registered for homodimeric anion radicals of pyrimidine nucleosides with previously measured spectra for systems involving pyrimidines enables us to discuss the stabilizing effects resulting from inter- and intramolecular interactions occurring in the systems of interest.

The maxima of photoelectron signal for $(dT)_2^{\bullet-}$ and $(dC)_2^{\bullet-}$ are significantly shifted toward higher EBEs with respect to those of single nucleosides, 2′-deoxythymidine$^{\bullet-}$ and 2′-deoxycytidine$^{\bullet-}$, respectively, as well as with respect to anionic dimers of nucleobases, $(\text{thymine})_2^{\bullet-}$ and $(\text{cytosine})_2^{\bullet-}$ (see Figs. 11.40 and 11.41). Similarly, the photoelectron spectrum of $(rU)_2^{\bullet-}$ is shifted to higher EBEs with respect to the radical anionic uridine monomer ($rU^{\bullet-}$) but it closely resembles the spectrum of $(\text{uracil})_2^{\bullet-}$. A similarity between the two spectra in the EBEs range from 2 to 2.5 eV may be interpreted as resulting from BFPT structures present in gaseous mixtures of ($rU^{\bullet-}$) and $(\text{uracil})_2^{\bullet-}$.

By comparing the anion photoelectron spectra of 2′-deoxythymidine$^{\bullet-}$ from Ref. [91] with $(\text{1-methylthymine}_2)^{\bullet-}$ from Ref. [96], it is possible to estimate the stabilization effects related to the formation of hydrogen bonds and presence of the sugar moiety. Thus, comparing the VDE of 0.457 eV for isolated valence bound thymine$^{\bullet-}$(only theoretical VDE is available here) [22] with the VDE of 1.15 eV registered for $(\text{1-methylthymine})_2^{\bullet-}$, one can conclude that hydrogen bonding shifts the VDE values by ~0.69 eV toward higher EBEs. Comparing, in turn, the experimental VDE for thymine$^{\bullet-}$(0.457 eV) to that of 2′-deoxythymidine$^{\bullet-}$ (0.89 eV) it can be seen that the presence of sugar is reflected by the VDE shift of ~0.43 eV. Assuming that both effects are additive, one could estimate that the maximum of the PES signal for $(2'\text{-deoxythymidine})_2^{\bullet-}$ should be observed at ~1.6 eV. Although this value corresponds well to the calculated VDE for the a_O8–O8 complex (1.55 eV), the experimental PES spectrum speaks against the above prediction. Indeed, the most stable anionic dimers are stabilized by interactions between charged thymine of one deoxynucleoside and the deoxyribose moiety of the other, and are characterized by a VDE of ca. 2 eV. Similarly, by comparing the vertical stability of cytosine homodimer, $(\text{cytosine})_2^{\bullet-}$, (VDE = 1.4 eV) [96] and that of the isolated cytosine anion radical, cytosine$^{\bullet-}$ (theoretical VDE = 0.39 eV), [48] one can estimate that intermolecular interaction shifts VDE by ~1 eV. The comparison between the vertical stability of 2′-deoxycytidine$^{\bullet-}$ (VDE = 0.87 eV) [91] and cytosine$^{\bullet-}$ predicts the influence of the sugar moiety attached to N1 of cytosine on the photoelectron maximum to be ~0.48 eV. On the premise that both effects are additive, one could estimate that the maximum of the photoelectron signal for $(dC)_2^{\bullet-}$ should be observed at an EBE of ~1.87 eV. Finally, repeating the above procedure for uracil, one obtains a VDE shift of 0.62 eV for the dimerization process and 0.81 eV resulting from the presence of sugar. Thus, the VDE estimated for dimeric nucleoside, $(rU)_2^{\bullet-}$, in which particular nucleosides interacts via nucleobases should be approximately 2 eV.

11.10 Summary

The anions of nucleobase-related systems are of paramount importance, since they play crucial roles in DNA damage induced by radiotherapy, which is the most common modality used in anticancer therapy. Indeed, many solvated electrons are generated in water interacting with ionizing radiation, and these electrons can interact with the DNA double helix, attaching mainly to nucleobases as they possess the highest electron affinity among the DNA components, this ultimately leading to DNA damage. In fact, Sanche and coworkers demonstrated, already over a decade ago, that single- and double-strand breaks are formed in plasmid DNA irradiated with low energy electrons. Information concerning characteristics of anionic states of nucleobases, nucleosides, and nucleotides is, therefore, a prerequisite to understanding the mechanisms that are triggered by electron attachment to DNA.

Electron transmission spectroscopy experiments in the gas phase demonstrate that all nucleobases form so-called resonance anions, i.e. unstable anionic states which capture electrons temporarily and as a consequence, have extremely short lifetimes of the order of 10^{-14} s. On the other hand, photoelectron spectroscopy measurements show that nucleobases support dipole-bound anions of relatively low stability. These facts suggest that conventional (valence) anions of nucleobases do not exist in the gas phase. However, numerous quantum chemical calculations support that marginally stable valence anions may exist at least for pyrimidine bases, although their stability is lower than that of DB anions. This explains why photoelectron spectroscopy, which usually tends to create the most stable form of a given anion, only detects DB states. Here, it is worth noticing that with the advent of new, more powerful ion sources, PES is capable of measuring spectra of valence anions for all nucleobases. It was however demonstrated that those anions do not originate from the canonical bases; but are supported by so-called very rare tautomers resulting from an internal proton transfer between N and C sites. Despite the fact that the neutral forms of these tautomers are extremely unstable, they support adiabatically stable valence anions which are significantly more stable than the valence anions of canonical or rare tautomers, and therefore have a chance to occur in DNA. If such anionic tautomers were actually formed in DNA, a subsequent electron detachment would lead to strand damage as the neutral species may immediately open their heterocyclic ring or form bicyclo derivatives.

Although the valence bound anions of nucleobases are weakly bound or unbound in the gas phase, even relatively weak interactions as those present in the uracil···water complex are sufficient to render the uracil valence anion adiabatically stable. This effect is especially visible in complexes consisting of nucleobases and species having proton-donor properties where electron attachment frequently induces BFPT/PT, leading to the most stable anionic configurations. Indeed, both the PES experiments and the results of quantum chemical calculations strongly suggest that electron attachment to complexes of nucleobases with proton donors of sufficient acidity leads to a barrier-free proton transfer (BFPT) to a nucleobase with the products being a neutral radical of hydrogenated nucleobase and an anionic counterpart. The propensity of the anionic complexes of nucleobases to BFPT, resulting in the formation of neutral radicals of hydrogenated bases may be relevant to DNA and RNA damage by low-energy electrons. For instance, the neutral radical UH• with the O8 atom protonated, cannot form a hydrogen bond with adenine, as dictated by the Watson–Crick pairing scheme. An anion of UH• radical might also react with an adjacent deoxyribose molecule triggering strand breaks in DNA. In a cellular environment, nucleobases interact not only with complementary bases, being proton-donors, but also with water, low molecular weight species, and a variety of proteins such as histones, DNA replication/repair enzymes. All these species stabilize the valence anionic states formed in DNA either via electrostatic interactions or through BFPT processes.

The influence of intra- and intermolecular interactions on the stability of VB nucleobase anions is illustrated by the PES experiment and DFT calculations on nucleosides, nucleotides and nucleoside dimers. Although VB anions of pyrimidine nucleobases are weakly bound in the gas phase, all nucleosides support stable anions with VDE that spans the 1.0–1.6 eV range. Thus, the presence of a sugar moiety in close vicinity to a nucleobase is sufficient to make the VB anions of nucleobases stable. Even larger stabilization is observed in nucleotides, where electron attachment induces PT as was demonstrated for adenosine nucleotide. Finally, intermolecular interactions between the sugar moiety of one nucleoside and the anionic nucleobase of the other nucleoside in an anionic homodimer lead to unexpectedly strong stabilization, comparable to that observed in BFPT-stabilized anionic complexes.

A large body of experimental and computational data described in the current review demonstrate the paramount importance of valence bound anions for understanding and controlling electron induced DNA damage. Although much is already known about the formation of such species in DNA and their fate ultimately leading to DNA damage, partially due to the synergetic combination of photoelectron spectroscopy studies and quantum chemical modeling, a still deeper understanding can be reached. As was emphasized above, interactions between nucleobase anions and surrounding molecules strongly influence the final outcome of the electron attachment process. It is worth noting that in the cellular environment, DNA continuously interacts with proteins and these interactions may influence both the primary electron attachment process and the secondary reactions leading to DNA damage. Hydrogen bonding between nucleobases and amino acids side chains is believed to be one of the most important types of interactions responsible for the formation of DNA–protein complexes [100–102]. This situation calls for systematic PES/quantum chemical studies on the excess electron attachment to small model systems mimicking the interactions between particular bases/base pairs and models of amino acid side chains. Such investigations are under way in our laboratories.

Acknowledgements The authors thank Lidia Chomicz and Justyna Wiczk for critical proofreading of the manuscript. Most of the experimental work described in this review was supported by the (US) National Science Foundation under grant no. CHE-1111693 (KHB). This work was also supported by the Polish Ministry of Science and Higher Education (MNiSW), Grant No. DS/530-8221-D186-13 (J.R.).

References

1. Joiner M, Van der Kogel A (eds) (2009) Basic clinical radiobiology, 4th edn. Hodder Arnold, London.
2. Warters RL, Hofer KG, Harris CR, Smith JM (1977) Curr Top Radiat Res Q 12:389
3. von Sonntag C (1987) The chemical basis of radiation biology. Taylor & Francis, London.
4. Michael BD, O'Neill P (2000) Science 287:1603
5. Boudaiffa B, Cloutier P, Hunting D, Huels MA, Sanche L (2000) Science 287:1658
6. Sanche L (2008) In: Shukla M, Leszczynski J (eds) Radiation induced molecular phenomena in nucleic acid: a comprehensive theoretical and experimental analysis. Challenges and advances in computational chemistry and physics. Springer, Dordrecht, p 531
7. Sanche L (2009) In: Greenberg MM (ed) Radical and radical ion reactivity in nucleic acid chemistry. Wiley series of reactive intermediates in chemistry and biology. Wiley, Hoboken, p 239
8. Gu J, Leszczynski J, Schaefer HF (2012) Chem Rev 112:5603
9. Rak J, Mazurkiewicz K, Kobyłecka M, Storoniak P, Haranczyk M, Dąbkowska I, Bachorz RA, Gutowski M, Radisic D, Stokes ST, Eustis SN, Wang D, Li X, Ko YJ, Bowen KH (2008) In: Shukla M, Leszczynski J (eds) Radiation induced molecular phenomena in nucleic acid: a comprehensive theoretical and experimental analysis. Challenges and advances in computational chemistry and physics. Springer, Dordrecht, p 619
10. Djordjevic B, Szybalski W (1960) J Exp Med 112:509
11. Erickson RL, Szybalski W (1963) Radiat Res 20:252
12. Brust D, Feden J, Farnsworth J, Amir C, Broaddus WC, Valerie K (2000) Cancer Gene Ther 7:778
13. Cecchini S, Girouard S, Huels MA, Sanche L, Hunting DJ (2005) Biochemistry 44:1932
14. Chomicz L, Rak J, Storoniak P (2012) J Phys Chem B 116:5612
15. Park Y, Polska K, Rak J, Wagner JR, Sanche L (2012) J Phys Chem B 116:9676
16. Polska K, Rak J, Bass AD, Cloutier P, Sanche L (2012) J Chem Phys 136:075101
17. Lehnert S (2008) Biomolecular action of ionizing radiation. Taylor & Francis, New York, p 123
18. Voityuk AA (2006) In: Sponer J, Lankas F (eds) Computational modeling of charge transfer in DNA in computational studies of RNA and DNA. Challenges and advances in computational chemistry and physics, vol 2. Springer, Dordrecht, p 485
19. Svozil D, Jungwirth P, Havlas Z (2004) Collect Czech Chem Commun 69:1395
20. Sevilla MD, Besler B, Colson AO (1994) J Phys Chem 98:2215
21. Neumark DM, Lykke KR, Andersen T, Lineberger WC (1985) Phys Rev A: Gen Phys 32:1890
22. Mazurkiewicz K, Bachorz RA, Gutowski M, Rak J (2006) J Phys Chem B 110:24696
23. Bachorz RA, Klopper W, Gutowski M (2007) J Chem Phys 126:085101
24. Rienstra-Kiracofe JC, Tschumper GS, Schaefer HF, Nandi S, Ellison GB (2002) Chem Rev 102:231
25. Sevilla MD, Besler B, Colson A (1995) J Phys Chem 99:1060
26. Desfrançois C, Periquet V, Bouteiller Y, Schermann JP (1998) J Phys Chem A 102:1274
27. Schiedt J, Weinkauf R, Neumark DM, Schlag EW (1998) Chem Phys 239:511
28. Curtiss LA, Redfern PC, Raghavachari KJ (2007) Chem Phys 126:084108
29. Curtiss LA, Redfern PC, Raghavachari KJ (2007) Chem Phys 127:124105
30. Andersson K, Malmqvist P-A, Roos BO (1992) J Chem Phys 96:1218
31. Roca-Sanjuan D, Merchan M, Serrano-Andres L, Rubio MJ (2008) Chem Phys 129:095104
32. Jensen F (2007) Introduction to computational chemistry second edition. Wiley, Chichester p 184.
33. Lee C, Yang W, Parr RG (1998) Phys Rev B 37:785; Miehlich B, Savin A, Stoll H, Preuss H (1989) Chem Phys Lett 157:200
34. Becke AD (1993) J Chem Phys 98:5648

35. Russo N, Toscano M, Grand A (2000) J Comput Chem 21:1243
36. Wetmore SD, Boyd RJ, Eriksson LA (2000) Chem Phys Lett 322:129
37. Saettel NJ, Wiest OJ (2001) J Am Chem Soc 123:2693
38. Zhao Y, Schultz NE, Truhlar DG (2006) J Chem Theory Comput 2:364
39. Zhao Y, Truhlar DG (2008) Theor Chem Acc 120:215
40. Eustis S, Wang D, Lyapustina S, Bowen KH (2007) J Chem Phys 127:224309
41. Bachorz RA, Rak J, Gutowski M (2005) Phys Chem Chem Phys 7:2116
42. Bachorz RA, Klopper W, Gutowski M, Li X, Bowen KH (2008) J Chem Phys 129:054309
43. Harańczyk M, Rak J, Gutowski M (2005) J Phys Chem A 109:11495
44. Haranczyk M, Gutowski M, Li X, Bowen KH (2007) Proc Natl Acad Sci U S A 104:4804
45. Haranczyk M, Gutowski M (2005) Angew Chem Int Ed 44:6585
46. Haranczyk M, Gutowski M (2005) J Am Chem Soc 127:699
47. Haranczyk M, Gutowski M, Li X, Bowen KH (2007) J Phys Chem B 111:14073
48. Li X, Bowen KH, Haranczyk M, Bachorz RA, Mazurkiewicz K, Rak J, Gutowski M (2007) J Chem Phys 127:174309
49. Szyperska A, Rak J, Leszczynski J, Li X, Ko YJ, Wang H, Bowen KH (2010) Chem Phys Chem 11:880
50. Colson A, Besler B, Close DM, Sevilla MD (1992) J Phys Chem 96:661
51. Dolgounitcheva O, Zakrzewski VG, Ortiz JV (1999) Chem Phys Lett 307:220
52. Dolgounitcheva O, Zakrzewski VG, Ortiz JV (2001) J Phys Chem A 105:8782
53. Wesolowski SS, Leininger ML, Pentchev PN, Schaefer HF (2001) J Am Chem Soc 123:4023
54. Li X, Cai Z, Sevilla MD (2002) J Phys Chem A 106:1596
55. Oyler NA, Adamowicz L (1993) J Phys Chem 97:11122
56. Oyler NA, Adamowicz L (1994) Chem Phys Lett 219:223
57. Smith DMA, Jalbout AF, Smets J, Adamowicz L (2000) Chem Phys 260:45
58. Hendricks JH, Lyapustina SA, de Clercq HL, Snodgrass JT, Bowen KH (1996) J Chem Phys 104:7788
59. Desfrançois C, Abdoul-Carime H, Schermann JP (1996) J Chem Phys 104:7792
60. Periquet V, Moreau A, Carles S, Schermann J, Desfrançois C (2000) J Electron Spectrosc Relat Phenom 106:141
61. Simons J (2006) Acc Chem Res 39:772
62. Dabkowska I, Rak J, Gutowski M (2005) Eur Phys J D 35:429
63. Bao X, Wang J, Gu J, Leszczynski J (2006) Proc Natl Acad Sci U S A 103:5658
64. Ray SG, Daube SS, Naaman R (2005) Proc Natl Acad Sci U S A 102:15
65. Löwdin PO (1963) Rev Mod Phys 35:724
66. Desfrançois C, Carles S, Schermann JP (2000) Chem Rev 100:3943
67. Gutowski M, Dąbkowska I, Rak J, Xu S, Nilles JM, Radisic D, Bowen KH (2002) Eur Phys J D 20:431
68. Harańczyk M, Rak J, Gutowski M, Radisic D, Stokes ST, Nilles JM, Bowen KH (2004) Isr J Chem 44:157
69. Hendricks JH, Lyapustina SA, de Clercq HL, Bowen KH (1998) J Chem Phys 108:8
70. Harańczyk M, Bachorz R, Rak J, Gutowski M, Radisic D, Stokes S, Nilles JM, Bowen KH (2003) J Phys Chem B 107:7889
71. Dolgounitcheva O, Zakrzewski VG, Ortiz JV (1999) J Phys Chem A 103:7912
72. Dąbkowska I, Rak J, Gutowski M, Nilles JM, Radisic D, Bowen KH (2004) J Chem Phys 120:6064
73. Dąbkowska I, Rak J, Gutowski M, Radisic D, Stokes ST, Nilles JM, Bowen KH (2004) Phys Chem Chem Phys 6:4351
74. Aflatooni K, Hitt B, Gallup GA, Burrow PD (2001) J Chem Phys 115:6489
75. Haranczyk M, Rak J, Gutowski M, Radisic D, Stokes ST, Bowen KH (2005) J Phys Chem B 109:13383
76. Harańczyk M, Dąbkowska I, Rak J, Gutowski M, Nilles JM, Stokes ST, Radisic D, Bowen KH (2004) J Phys Chem B 108:6919

77. Ko YJ, Wang H, Radisic D, Stokes ST, Eustis SN, Bowen KH, Mazurkiewicz K, Storoniak P, Kowalczyk A, Haranczyk M, Gutowski M, Rak J (2010) Mol Phys 108:2621
78. Mazurkiewicz K, Haranczyk M, Gutowski M, Rak J, Radisic D, Eustis SN, Wang D, Bowen KH (2007) J Am Chem Soc 129:1216
79. Mazurkiewicz K, Haranczyk M, Storoniak P, Gutowski M, Rak J, Radisic D, Eustis SN, Wang D, Bowen KH (2007) Chem Phys 342:215
80. Richardson NA, Wesolowski SS, Schaefer HF (2003) J Phys Chem B 107:848
81. Al-Jihad I, Smets J, Adamowicz L (2000) J Phys Chem A 104:2994
82. Kumar A, Knapp-Mohammady M, Mishra PC, Suhai S (2004) J Comput Chem 25:1047
83. Radisic D, Bowen KH, Dąbkowska I, Storoniak P, Rak J, Gutowski M (2005) J Am Chem Soc 127:6443
84. Smets J, Jalbout AF, Adamowicz L (2001) Chem Phys Lett 342:342
85. Li X, Cai Z, Sevilla MD (2001) J Phys Chem B 105:10115
86. Richardson NA, Wesolowski SS, Schaefer HF (2002) J Am Chem Soc 124:10163
87. Szyperska A, Rak J, Leszczynski J, Li X, Ko YJ, Wang H, Bowen KH (2009) J Am Chem Soc 131:2663
88. Falcone JM, Becker D, Sevilla MD, Swarts SG (2005) Radiat Phys Chem 72:257
89. Storoniak P, Mazurkiewicz K, Haranczyk M, Gutowski M, Rak J, Eustis SN, Ko YJ, Wang H, Bowen KH (2010) J Phys Chem B 114:11353
90. Mazurkiewicz K, Harańczyk M, Gutowski M, Rak J (2007) Int J Quantum Chem 107:2224
91. Stokes ST, Li X, Grubisic A, Ko YJ, Bowen KH (2007) J Chem Phys 127:084321
92. Kobyłecka M, Gu J, Rak J, Leszczynski J (2008) J Chem Phys 128:044315
93. Richardson NA, Gu J, Wang S, Xie Y, Schaefer HF (2004) J Am Chem Soc 126:4404
94. Li X, Sanche L, Sevilla MD (2006) Rad Res 165:721
95. Stokes ST, Grubisic A, Li X, Ko YJ, Bowen KH (2008) J Chem Phys 128:044314
96. Ko YJ, Wang H, Cao R, Radisic D, Eustis SN, Stokes ST, Lyapustina S, Tian SX, Bowen KH (2010) Phys Chem Chem Phys 12:3535
97. Ko YJ, Storoniak P, Wang H, Bowen KH, Rak J (2012) J Chem Phys 137:205101
98. Storoniak P, Rak J, Ko YJ, Wang H, Bowen KH (2012) J Phys Chem B 116:13975
99. Storoniak P, Rak J, Ko YJ, Wang H, Bowen KH (2013) J Chem Phys 139:075101
100. Seeman NC, Rosenberg JM, Rich A (1976) Proc Natl Acad Sci U S A 73:804
101. Cheng AC, Chen WW, Fuhrmann CN, Frankel AD (2003) J Mol Biol 327:781
102. Luscombe NM, Laskowski RA, Thornton JM (2001) Nucl Acids Res 29:2860

Chapter 12
Hints from Computational Chemistry: Mechanisms of Transformations of Simple Species into Purine and Adenine by Feasible Abiotic Processes

Jing Wang, Jiande Gu, and Jerzy Leszczynski

Abstract The chemical evolution of biomolecules such as nucleobases and their analogues from simple, one carbon containing molecules under abiotic conditions is a puzzle closely connected to the origin of life. Theoretical elucidation of the abiotic reaction routes leading from basic molecules cyanide acid (HCN) and formamide (H_2NCHO) to the formation of purine and adenine is reviewed here. The mechanism of three pathways: from formamide dimer via pyrimidine to purine, from AICN (4-aminoimidazole-5-carboxamidine) to adenine, and from formamide to purine and adenine, are discussed. Based on the comparison of step-by-step mechanism of the reaction pathways, in the addition reaction formamide is suggested to be more reactive than HCN. Beside its simplicity, the formamide self-catalyzed mechanism is energetically more viable than either water-catalyzed mechanism or non-catalyzed process. Moreover, this self-catalyzed mechanism is able to explain the ratio of purine to adenine observed in experiments. The formamide self-catalyzed mechanism for the route leading from formamide to purine and/or adenine is most likely for the formation of adenine (and purine) in the formamide solutions in the early stage of the earth.

J. Wang • J. Leszczynski (✉)
Department of Chemistry and Biochemistry, Interdisciplinary Center for Nanotoxicity, Jackson State University, Jackson, MS 39217, USA
e-mail: jerzy@icnanotox.org

J. Gu (✉)
Drug Design & Discovery Center, State Key Laboratory of Drug Research, Shanghai Institute of Materia Medica, Chinese Academy of Sciences, Shanghai 201203, China
e-mail: jiande@icnanotox.org

J. Leszczynski and M.K. Shukla (eds.), *Practical Aspects of Computational Chemistry III*, DOI 10.1007/978-1-4899-7445-7_12,

12.1 Introduction

The 1950s witness the extraordinary discoveries related to the origin of life on the earth. The discovery of the structure of DNA by Watson and Crick in 1953 indicated that the secret of life is hidden in the nucleobases [1, 2]. In the same year, Miller's experiment demonstrated that amino acids can be derived from the simple mixtures of water, methane, ammonia, and hydrogen under the harsh conditions that mimic the early stage of earth [3]. Although no DNA or RNA bases had been identified in the Miller's experiment, careful analyses revealed many of the chemical reactions that occurred in these experiments, especially, the gases in the "atmosphere" reacted to form a pool of simple organic molecular building blocks including hydrogen cyanide (HCN) [4]. Late, in 1961, by mixing hydrogen cyanide and ammonia in an aqueous solution Juan Oró found that in addition to amino acids, adenine, one of the five nitrogen-containing bases present in RNA and DNA, could also be produced from these chemicals [5, 6]. On the other hand, one important molecule that can be chemically assembled from the Miller's experiments is NH_2COH (formamide). The major pathway of the formation of this molecule includes the reaction between water and hydrogen cyanide [7, 8]. As a ubiquitous molecule, formamide has been identified as one of the most common molecules of the Universe. It has been detected in comets, satellites, and interstellar medium. Formamide has been found to be a potential abiotic source of nucleobases and their analogues [9–12]. Purine has been reported to be generated in high yield simply by heating formamide to near its boiling point in laboratories [13, 14]. During the last 15 years, all five nucleobases and their analogues have been synthesized from formamide in the presence of various catalysts [8, 15, 16]. One-pot synthesis of cytosine starting from formamide was reported in 2001 [17, 18]. Synthesis of thymine and adenine from formamide in the presence of titanium dioxide was performed in 2003 [19]. Uracil and hypoxanthine, along with adenine and cytosine, were generated from formamide in the presence of montmorillonites in 2004 [20]. Guanine was derived from formamide with UV light irradiation process in the presence of catalysts in 2006 and 2010 [21, 22]. Thus, these experiments indicate that under plausible prebiotic conditions, important constituents of proteins and nucleic acids could have been synthesized on the early earth.

Different mechanistic pathways from formamide to purine and adenine have been proposed based on the experimental data [8, 18, 22–29]. One of the suggested mechanistic routes progresses through a pyrimidine intermediate to purine (Scheme 12.1) [8, 25–27]. Pathways from formamide to adenine have been theorized to proceed though pyrimidine [8, 18]. Other considered reaction route to purine suggested that a relatively stable complex diaminomaleonitrile (DAMN) intermediate is important in this process [22]. DAMN provides the ring junction scaffold leading to an imidazole, a five-member ring compound, that subsequently forming a bicyclic purine. Also, a slightly different route for the abiotic syntheses of both purine and adenine from formamide has been proposed based on recent experimental investigations [23]. In this pathway (Scheme 12.2), purine (or adenine) is formed through a five-membered ring intermediate, namely

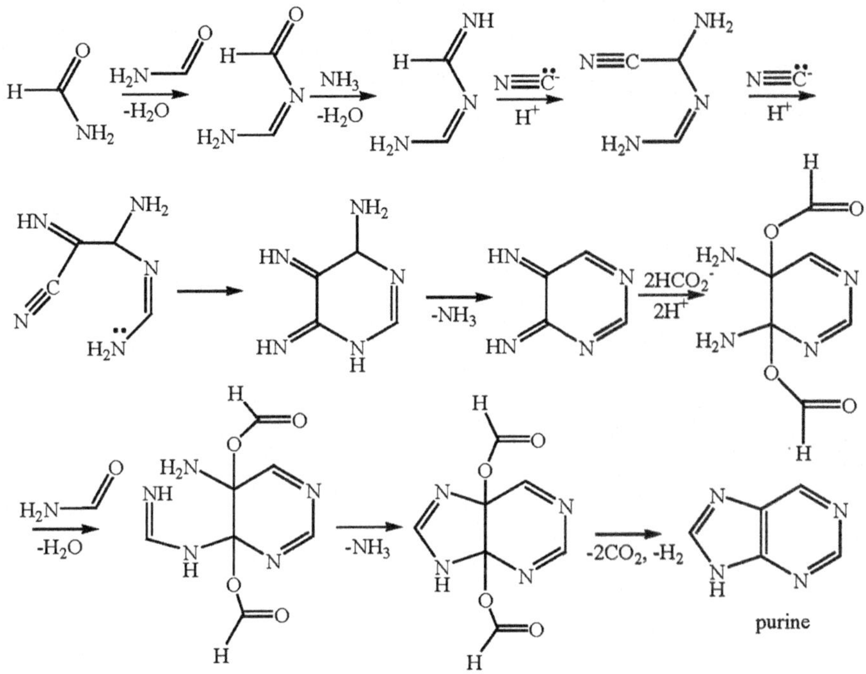

Scheme 12.1 Proposed route that progresses from formamide through a pyrimidine intermediate to purine (Redraw based on Ref. [25–27])

5-aminoimidazole (for purine) or 4-aminoimidazole-5-carboxamidine (AICN, for adenine) [23]. Analogous reaction route from HCN to adenine through DAMN and AICN intermediates has been outlined and analyzed based on various experimental investigations (Scheme 12.3) [5, 24, 30–36].

Complementary theoretical studies of reaction mechanisms are able to provide details for all species involved in the studied reaction including intermediates and transition states formed along the reaction pathways. Information revealed in quantum chemistry computational studies allows screening and assorting the most feasible route among various possible reaction pathways. However, finding a viable, thermodynamically realistic, step-by-step mechanism that can account for the formation of nucleobases is still a challenge for computational studies. On the other hand, the efforts engaged in accomplishing this challenge are paid off. The details revealed in the reaction pathway obtained from the quantum chemistry computational studies cast a new light on understanding the chemical evolution of biomolecules at their primitive stage.

We shall focus this review on a potential physico-chemical frame in which the formation of the nucleobases is based on the application of reactants as simple as possible. This consideration is derived from the hypotheses that life is a robust physico-chemical process. Therefore, the start and the formation of the basic key materials of life such as nucleobases and their analogues have to be based on both easily available raw materials and reasonable energy demanding chemical

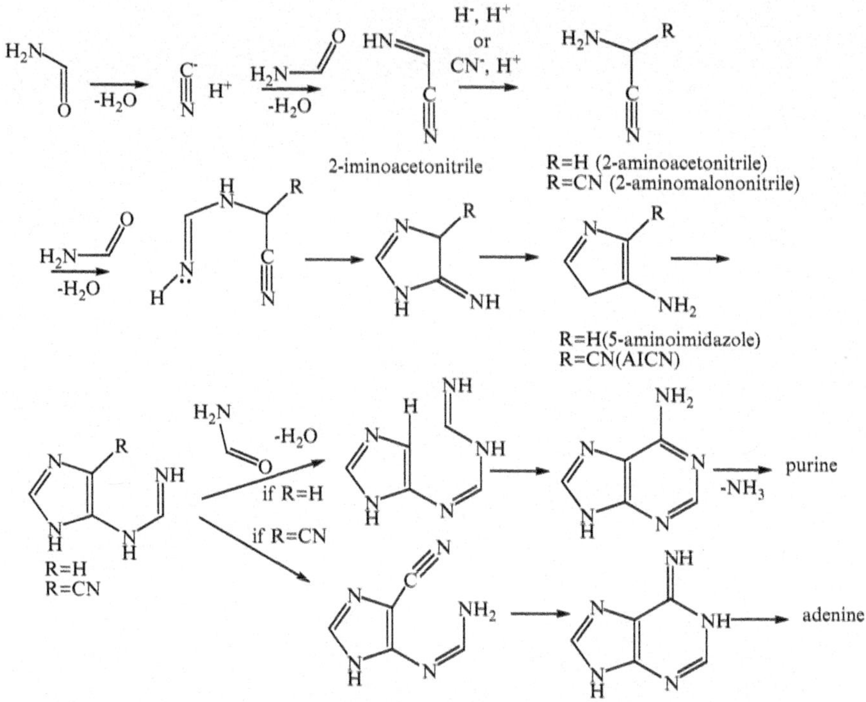

Scheme 12.2 Proposed pathways that transfer formamide through a five-member ring intermediate aminoimidazole into purine and adenine (Redraw based on Ref. [23])

processes. In this sense, both HCN and formamide should be good starting raw materials.

Among various quantum chemistry methods the B3LYP approach [37–39] of the density functional theory (DFT) has been widely accepted in reaction mechanism pathway investigations. For the studies reviewed here, the activation energy barriers predicted for different reaction pathways using the B3LYP method are close to those calculated using the high level couple-cluster theory CCSD(T) [40–42] method (with differences less than 3 kcal/mol) [43–45].

Below we discuss the details of computationally revealed reaction pathways that use simple reactants and yield nucleobases.

12.2 From Formamide Dimer to Purine: Route Through a Pyrimidine Intermediate

Based on the studies conducted by Yamada et al. [24, 26, 27] the route leading to pyrimidines starts from the addition of HCN to a formamide dimer (**1_1**, Fig. 12.1). Therefore, instead of formamide, a formamide dimer is adopted as a raw material in the computational studies conducted by Sponer's group [29].

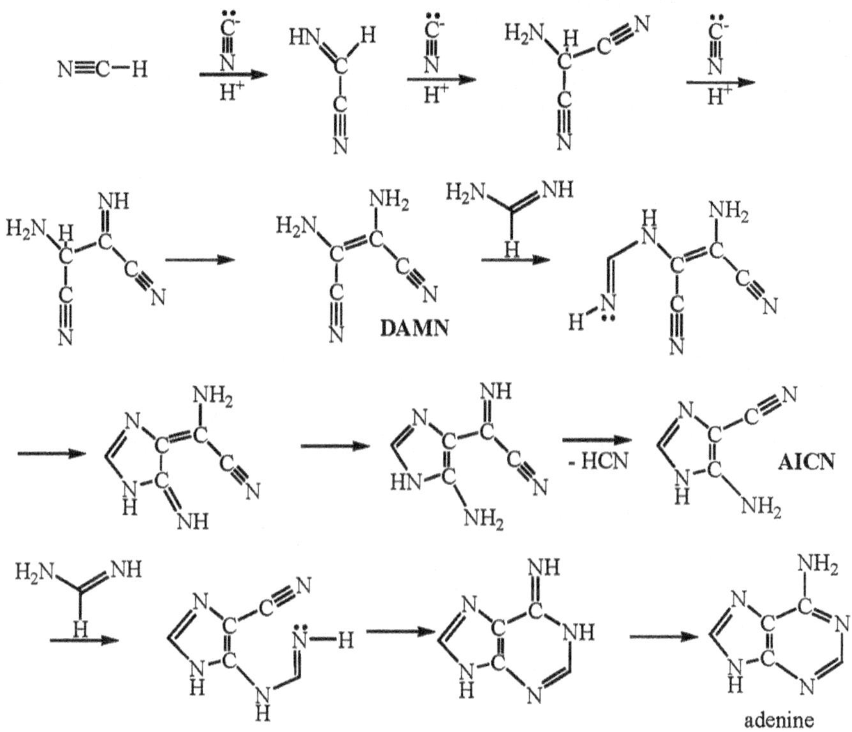

Scheme 12.3 Proposed pathway that transfers HCN through a five-member ring intermediate aminoimidazole into adenine (Redraw based on Ref. [30–36])

12.2.1 From Formamide Dimer to Pyrimidine

Four major steps are included in this reaction (see Fig. 12.2): (a) Reaction of formamide dimer with HCN; (b) CN anion attack on 3-Hydorxy-2,4,6-triaza-hex5-en-1-one; (c) Formation of the six-membered ring; and (d) Dehydration leading to double-bond formation.

12.2.1.1 Reaction of Formamide Dimer with HCN

Without catalysts, HCN attacks the -NH_2 group of the formamide dimer (**1_1**) to generate 3-Hydorxy-2,4,6-triaza-hex5-en-1-one (**1_2**). This requires a substantial activation energy. The energy barrier corresponding to this step is predicted to be 61.9 kcal/mol in the gas phase by using the B3LYP functional with 6-311++G (2d,2p) basis set [46]. Inclusion of bulk solvent effects (mimicked by the C-PCM approximation [47, 48]) slightly reduces this activation barrier (60.3 kcal/mol – formamide solution) and (57.9 kcal/mol – aqueous solution). Catalyzed by a water

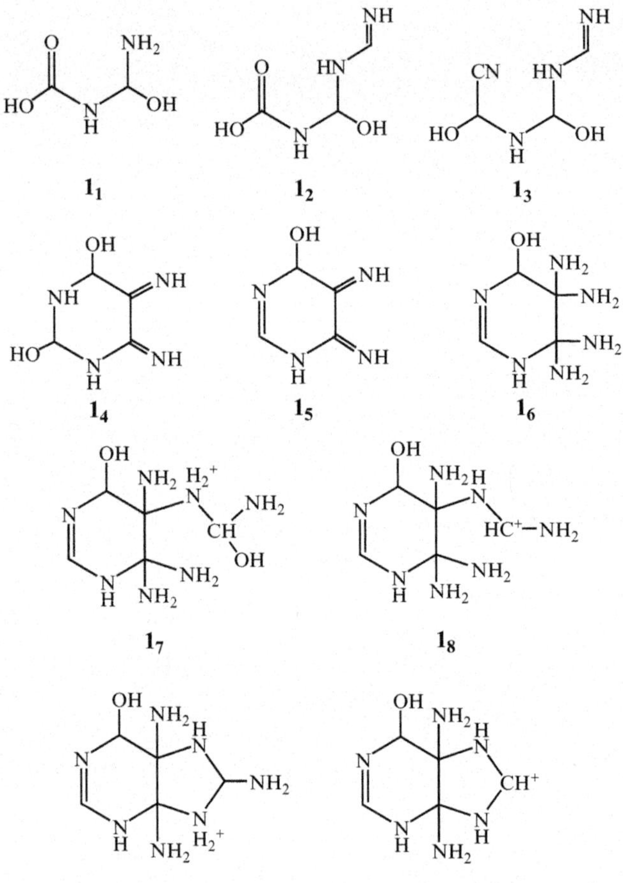

Fig. 12.1 Intermediate compounds formed along the reaction route leading from formamide dimer to purine. (redraw based on Ref. [29])

molecule (or an ammonia molecule), the speed of this reaction step can be increased significantly. In the presence of the catalytic water, the corresponding activation energy barrier is predicted to be 43.8, 37.7, and 35.6 kcal/mol in the gas phase, formamide solution, and aqueous solution, respectively. The catalytic effects of ammonia are found to be less efficient as compared to water.

12.2.1.2 CN Anion Attack on 3-Hydorxy-2,4,6-triaza-hex5-en-1-one

Due to the fact that they were not able to locate a plausible transition state, Sponer's group suggested an anionic mechanism, in which CN anion attacks on the C1 position of the intermediate $\mathbf{1_2}$ (3-Hydorxy-2,4,6- triaza-hex5-en-1-one), generating

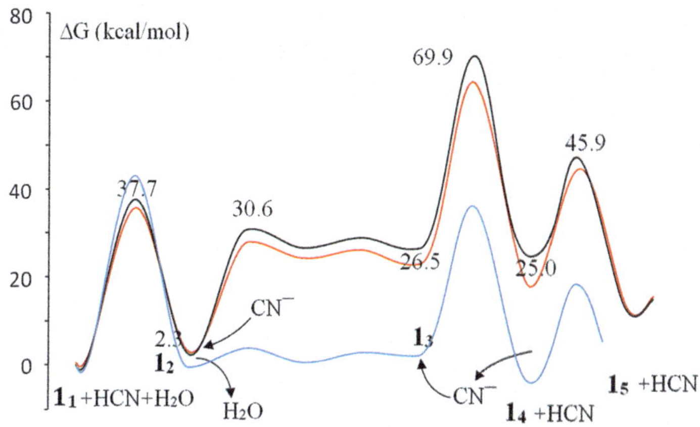

Fig. 12.2 Free energy profiles along the pathway leading to the formation of pyrimidine from formamide dimer [29]. Results based on the B3LYP/6-311++G(2d,2p) approach. C-PCM model was adopted to mimic solvents. *Black*: in formamide solutions; *Red*: in aqueous solutions; *Blue*: in gas phase

O(5) deprotonated 3,5-dihydroxy-5-cyano-2,4-diaza-pentan-1-imine (**1**$_3$). This reaction seems easy in the gas phase and the corresponding activation energy is estimated to be 4.3 kcal/mol [29]. However, the solvent effects affect the reaction rate significantly. The activation energy is 28.3 kcal/mol in bulk formamide, and 24.1 kcal/mol in bulk water. One possible reason for this great difference is that a negative charged system is sensitive to a polarizable medium.

12.2.1.3 Formation of the Six-Membered Ring

A CN anion is suggested to catalyze the formation of the six-membered ring from 3,5-dihydroxy-5-cyano-2,4-diaza-pentan-1-imine. In this process, the role of the CN anion is to receive a proton from the imino carbon C1 and, subsequently, the nucleophilic attack of the cyano carbon (at C5 position of **1**$_3$) on the imino carbon leads to the formation of the six-membered heterocyclic ring compound 2,6-dihydroxy-4,5-diimino-hexahydropyrimidine (**1**$_4$ in Fig. 12.1). The activation energy corresponding to this ring-closing step is calculated to be 33.7 kcal/mol in the gas phase. Inclusion of the solvent effects of the polarizable medium increases the energy barrier to 43.7 kcal/mol (in formamide solution), and 41.5 kcal/mol (in aqueous solution).

12.2.1.4 Dehydration Leading to Double-Bond Formation

Dehydration of **1**$_4$ at the C2– position is catalyzed by HCN. This process consists of two elementary steps as suggested by Sponer's group [29]. A HCN molecule protonates the hydroxyl group at C2 in the first step, which leads to the elimination of a water molecule from the reaction complex, resulting in a carbocationic center at C2. Subsequently, this carbocation is stabilized by spontaneous proton loss from the neighboring N1, initiated by the CN anion formed in the previous reaction step, and leading the formation of $\mathbf{1_5}$ (Fig. 12.1). This dehydration step requires activation energy of 21.6, 20.9, and 19.2 kcal/mol in gas phase, bulk formamide, and water, respectively.

The free energy profiles along the pathways corresponding to the formation of pyrimidine from formamide dimer are summarized and depicted in Fig. 12.2. The rate controlling step in the gas phase is the first step: HCN attacking the -NH_2 group of the formamide dimer. However, in formamide solutions it is the step corresponding to the formation of the six-membered ring from 3,5-dihydroxy-5-cyano-2,4-diaza-pentan-1-imine (the third step).

12.2.2 From Pyrimidine to Purine

This section of reaction comprises (a) addition of ammonia to the ketimino groups at C4 and C5 of 6-hydroxy-4,5-diimino-3,4,5,6-tetrahydro-pyrimidine, (b) formiminylation (formylation and dehydration), (c) ring-closing, and (d) deamination.

12.2.2.1 Addition of Ammonia to the Ketimino Groups at C4 and C5 of $\mathbf{1_5}$

Although there is no detail of the reaction mechanism of this step given, by comparison to the reaction of ammonia with carbonyl model compounds Sponer et al. concluded that this reaction proceeds with a low activation energy with the assistance of a catalytic water molecule. Addition of two ammonia molecules to $\mathbf{1_5}$ generates an important intermediate 6-hydroxy-4,4,5,5-tetramino-3,4,5,6-tetrahydropyrimidine ($\mathbf{1_6}$, Fig. 12.1).

12.2.2.2 Formiminylation of 6-Hydroxy-4,4,5,5-tetramino-3,4,5,6-tetrahydropyrimidine

Formylation of $\mathbf{1_6}$ in the assistance of catalytic proton requires relatively low activation energy (12.7 kcal/mol in gas phase, 14.1 kcal/mol in bulk formamide, and 12.7 kcal/mol in bulk water). However, subsequent dehydration of the formylated intermediate $\mathbf{1_7}$ (5-protonated 5,6,6-triamino-5-(amino(hydroxy) methylamino)-1,4,5,6-tetrahydro-pyrimidin-4-ol) has been found to be a kinetically retardant step, yielding the intermediate $\mathbf{1_8}$ and a water molecule. The corresponding activation energy is predicted to be 32.0, 29.7, and 29.9 kcal/mol in gas phase, bulk formamide, and bulk water, respectively.

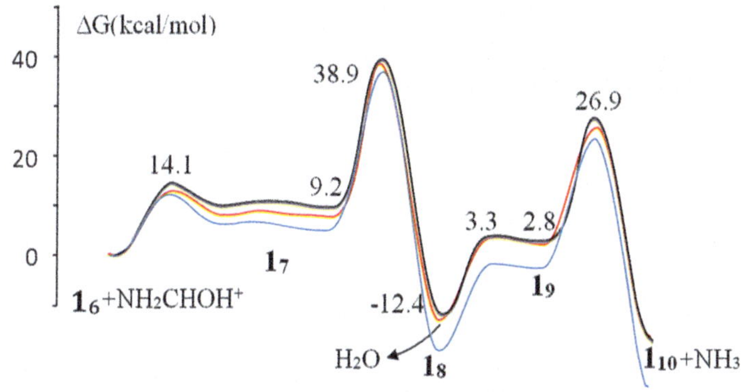

Fig. 12.3 Free energy profiles along the pathway leading to the formation of purine from pyrimidine derivative [29]. Results based on the B3LYP/6-311++G(2d,2p) approach. C-PCM model was adopted to mimic solvents. *Black*: in formamide solutions; *Red*: in aqueous solutions; *Blue*: in gas phase

12.2.2.3 Ring-Closing

Intramolecular attack of the positively charged carbon of $\mathbf{1_8}$ at the adjacent amino group proceeds the dehydration of $\mathbf{1_7}$, generating the bicyclic cationic intermediate 4,5,8-triamino-4,5,6,7,8,9,9′-heptahydro-3H-purin-6-ol $\mathbf{1_9}$ (see Fig. 12.1). The activation energy for the ring-closing step amounts to 16.8 kcal/mol in gas phase, 15.6 kcal/mol in formamide solutions, and 15.3 kcal/mol in aqueous solutions.

12.2.2.4 Deamination of $\mathbf{1_9}$

Deamination occurs at the C8 position of the intermediate $\mathbf{1_9}$, yielding the N9-protonated 4,5-diamino-4,5,6,7-tetrahydro-3*H*-purin-6-ol ($\mathbf{1_{10}}$). The corresponding activation energy barrier is estimated to be 25.3 kcal/mol in gas phase (24.1 and 23.8 kcal/mol in bulk formamide and water, respectively). The cationic intermediate ($\mathbf{1_{10}}$) has been considered as the precursor of purine bases as concluded by Sponer et al. [29].

The free energy profiles along the pathways from pyrimidine to purine are summarized and depicted in Fig. 12.3.

Although the mechanism of the route from formamide to purine via a pyrimidine intermediate is far from sophisticated, two pieces of important information can be drawn from here:

1. The existence of catalytic water molecule is crucial for the formation of the pyrimidine intermediate, which is consistent with the conclusion of earlier study of reaction mechanism of the formation of adenine from HCN based route conducted by Schleyer's group [43].

2. The solvent effects of a polarizable medium have little influence on the activation energy barriers in the studied reactions, albeit the solvent effects may change the total energy of the charged complexes.

12.3 From AICN to Adenine: Route Through AICN Intermediate

The reaction of the HCN tetramer, AICN (Fig. 12.4), with a fifth HCN molecule is the last crucial stage to give adenine, as suggested by Schleyer's group [43]. It is believed to be the key, rate-limiting step in the general adenine formation pathway starting from HCN [44]. Major elementary processes in this route are: (a) addition of a HCN to AICN; (b) *syn-anti* hydrogen transfer at the imino group of the HCN-AICN adduct; (c) six-membered ring closure; and (d) intra-molecular proton transfer.

12.3.1 Addition of a HCN to AICN

Addition of a HCN to the NH_2 group of AICN is found to be not able to proceed directly in the absence of a catalyst. The energy barrier of this reaction is predicted to be as high as 60.4 kcal/mol at the B3LYP/6-311+G(d,p) level of theory. Thus, an uncatalyzed mechanism for the generation of adenine from HCN along this route is unlikely. The catalytic participation of a water molecule, in the addition of HCN to the NH_2 group of AICN, is able to reduce the activation energy down to

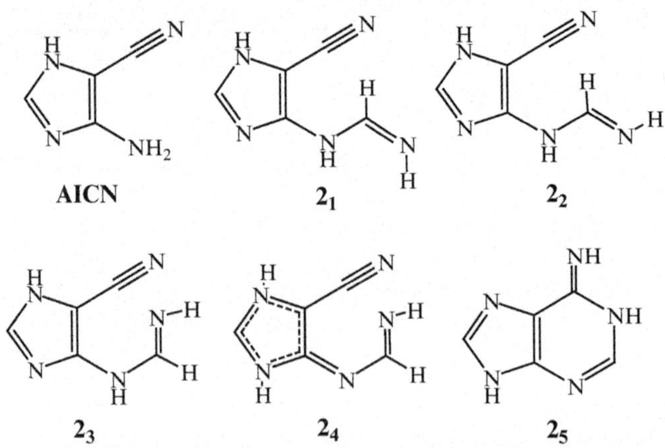

Fig. 12.4 Intermediate compounds formed along the HCN based reaction route from AICN to adenine

38.0 kcal/mol. The catalytic water molecule in the transition state transfers its proton to the HCN nitrogen, forming the new bond between the AICN amine nitrogen and the carbon of HCN. Inclusion of second water molecule does not improve the reaction rate of this step. Also, replacement of water with ammonia in the catalytic reaction has no significant effect. However, inclusion of the polarizable medium (aqueous solutions) effects in the reduction of calculated activation energy barrier (PCM model [49, 50]) by about 4 kcal/mol.

12.3.2 syn-anti *Hydrogen Transfer*

The H atoms at the C=N group of the HCN-AICN adduct, *N*-(5-cyano-1*H*-imidazol-4-yl)formamidine (**2_1**), are in the *anti*-form. Steric effects prohibit the formation of the six-membered ring directly from the *anti*-form **2_1**. Therefore, *syn-anti* hydrogen transfer at the imino group of the HCN-AICN has to proceed. The energy barrier of the *syn-anti* hydrogen transfer step is evaluated to be 28.1 kcal/mol in aqueous solutions, with the assistance of a catalytic water molecule. The subsequent structure rearrangement results in the *syn*-form **2_3**, which is ready to precede the process of six-membered ring closure (see Fig. 12.4).

12.3.3 Six-Membered Ring Closure

Two water molecules are needed to assist the six-membered ring closure. One water molecule participates in the proton transfer (from the secondary amino group of formamidine to N3 of the imidazole group) directly as a catalyst, while the second H_2O assists the N side of the imino moiety of formamidine in attacking the C end of cyano group of imidazole (resulting 1H-purin-6(9H)-imine, **2_5**) by forming hydrogen bonding. With the assistance of two catalytic water molecules, the energy barrier of the six-membered ring closure reaction is estimated to be 22.4 kcal/mol (at the B3LYP/6-311+G(d,p) level of theory) [43].

12.3.4 Intra-molecular Proton Transfer

Proton transfer from N1 to the imine of **2_5** generates the final product – adenine. With the assistance of catalytic water molecule, this intra-molecular proton transfer is easy, and the corresponding energy barrier amounts to only 9.9 kcal/mol.

The water molecule catalyzed reaction profiles for the formation of adenine from AICN are depicted in Fig. 12.5. The rate-controlling step in this mechanism is the first step: addition of a HCN to the NH_2 group of AICN. The study conducted by Schleyer's group suggested that ammonia could also serve as a catalyst along this reaction mechanism [43]. Moreover, their study demonstrated that the energy

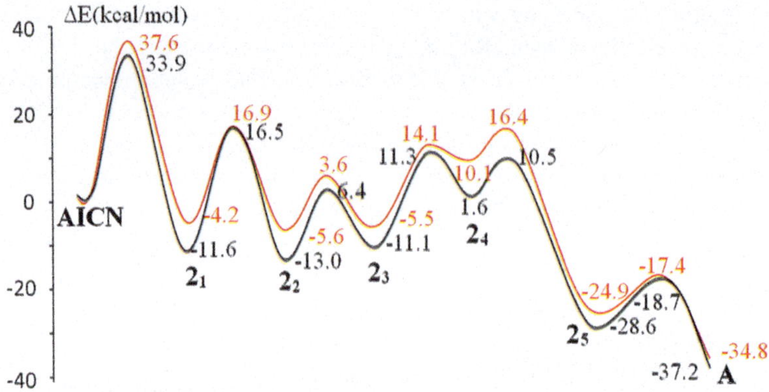

Fig. 12.5 Energy profiles along the pathway leading to the formation of adenine from AICN [43]. Results based on the B3LYP/6-311+G(d,p) approach. PCM model was adopted to mimic aqueous solutions. *Black*: in aqueous solutions; *Red*: in gas phase

barriers estimated in the gas phase might be 3 ~ 4 kcal/mol higher, as compared to the PCM calculations. The corresponding energy barrier indicates that this reaction route is viable from the viewpoint of energy. However, it should be noted that the mechanism of the formation of AICN from the basic materials such as HCN or formamide is still unsolved.

12.4 From Formamide to Purine: Route Through Aminoimidazole Intermediate

An entire mechanistic pathway that transforms formamide to purine through a five-membered ring intermediate has been explored by density functional theory calculations [44, 45]. Fascinatingly, all necessary reactants, including catalysts, are generated from a single initial compound – formamide. The catalytic effect of formamide itself has been found to be much more significant than that of water. Several important reaction steps are involved in this mechanistic route: dehydration of formamide, formylation-dehydration, Leuckart reduction, five- and six-membered ring closure, and deamination.

12.4.1 From Formamide to HCN: Dehydration of Formamide

The mechanistic route to the dehydration of formamide includes three major steps: (1) a proton transfer from N to O to form the imidic acid tautomer of formamide; (2) a proton exchange between one imidic acid and one amide form of formamide,

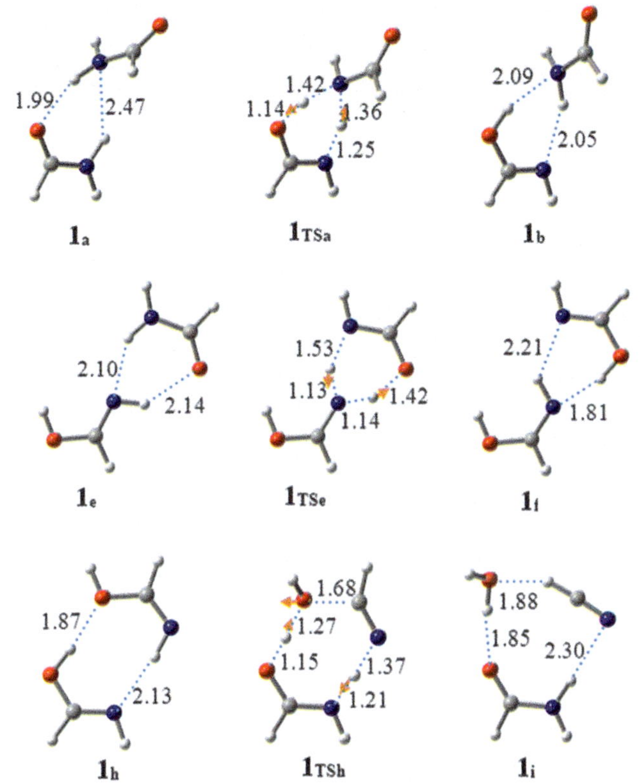

Fig. 12.6 The optimized structures of the complexes classified as local minima and the transition states on the potential energy surface in the self-catalyzed dehydration of formamide at B3LYP/6-311G(d,p) level. Atomic distance in Å. *Orange arrows* represent the vibrational mode corresponding to the single imaginary frequency in the transition states. Color representations are: *red* for oxygen, *blue* for nitrogen, *grey* for carbon, and *white* for hydrogen

resulting in two imidic acid formamides; and (3) an interaction between two imidic acids yielding a formamide, a water molecule, and a HCN (Fig. 12.6). These three main steps are connected through molecular rotation and re-orientation steps. The activation energy barriers in the self-catalytic mechanistic route are significantly lower than those in other pathways. The zero-point energy corrected activation energy barriers corresponding to the three major steps are evaluated to be 26.1, 24.2, and 18.7 kcal/mol, respectively (Fig. 12.7) [44, 45]. In comparison, other studies based on the tautomerization-intra-molecular proton transfer mechanism have calculated an activation energy barrier of 44.7 kcal/mol for the tautomerization of formamide to the imidic acid, and 58.3 ~ 59.1 kcal/mol (at different levels of theory) for the rate-controlling dehydration of formamide [44, 51, 52]. Although the water-assisted mechanism for the dehydration of formamide greatly reduces the activation energy barrier to 33.5 kcal/mol [44, 51, 52],

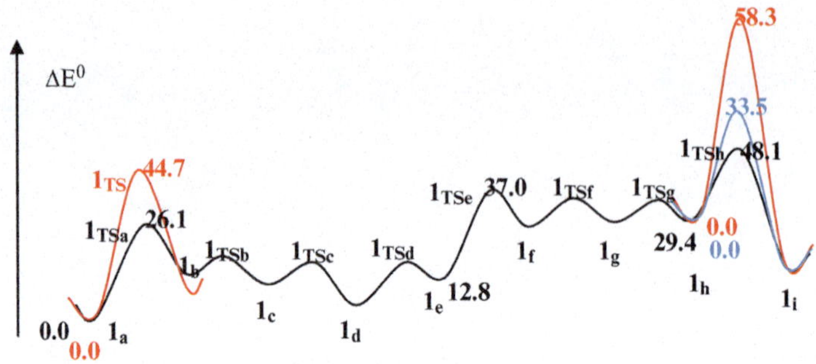

Fig. 12.7 Schematic energy profile along the reaction pathway. ΔE^0 is the zero-point energy corrected relative energy (in kcal/mol). *Black* is for the formamide self-catalyzed reaction. *Blue* is for the water-catalyzed process. *Red* is for the non-catalyzed route

this energy is still by more than 7 kcal/mol higher than in the formamide self-catalyzed study [45]. This value is apparently larger than the error margin of ±5 kcal/mol on relative energies expected for DFT methods, as compared to high accuracy MO computations. Thus, the self-catalysis mechanism discussed above is concluded to be more viable, especially under relatively mild reaction conditions.

12.4.2 Formiminylation of HCN

Two major steps have been identified in the formation of 2-hydroxylaminoacetonitrile and its dehydration. During the formylation process the proton of the HCN transfers to the imidic acid tautomer of formamide, yielding a CN anion. At the same time, one proton migrates from imidic acid to its neighboring formamide, resulting in a protonated formamide (see $\mathbf{2_b}$ in Fig. 12.8). Subsequently, the nucleophilic CN anion in the complex attacks the carbonyl of the protonated formamide forming a C—C bond to generate 2-hydroxylaminoacetonitrile ($\mathbf{2_c}$, see Fig. 12.8). The corresponding TS structures ($\mathbf{2_{TSa}}$, and $\mathbf{2_{TSb}}$ in Fig. 12.8) indicate the formation of the complexes $\mathbf{2_b}$ and $\mathbf{2_c}$. The activation energy barriers of these two transition states are 13.5 kcal/mol and 8.9 kcal/mol, respectively (Fig. 12.9). It should be noted that the reversed reaction from $\mathbf{2_b}$ to $\mathbf{2_a}$ has a very low energy barrier, about 0.5 kcal/mol (1.7 kcal/mol without ZPE correction). Such a low activation energy barrier suggests that the conformer $\mathbf{2_b}$ is a very short lived intermediate between the reactant $\mathbf{2_a}$ and the product $\mathbf{2_c}$. Consequently, the activation energy barrier for the formylation step is based on the relative energy of transition state $\mathbf{2_{bTS}}$ (evaluated to be 21.9 kcal/mol). Dehydration of 2-hydroxylaminoacetonitrile in the subsequent step is not difficult under the influence of formamide. The energy profile of the dehydration step (Fig. 12.9) shows that in the presence of formamide the activation energy barrier for the dehydration of 2-hydroxylaminoacetonitrile is 25.7 kcal/mol, slightly higher than that of the formylation process.

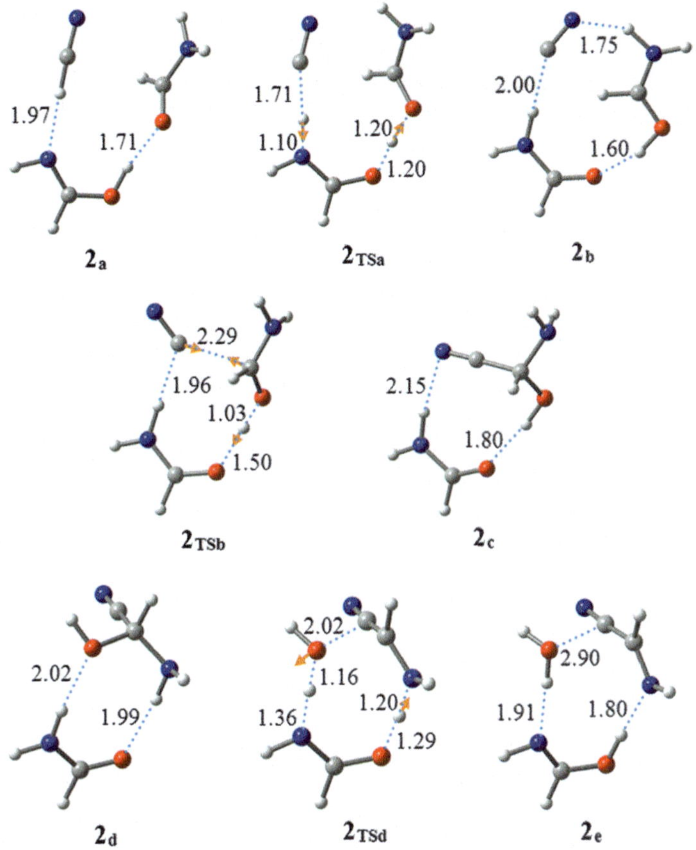

Fig. 12.8 Optimized structures of the complexes classified as the local minima and the transition states on the potential energy surface in the formamide-catalyzed formiminylation of HCN at the B3LYP/6-311G(d,p) level. Atomic distances are given in Å. *Orange arrows* represent the vibrational mode corresponding to the single imaginary frequency in the transition states. Color representations are: *red* for oxygen, *blue* for nitrogen, *grey* for carbon, and *white* for hydrogen

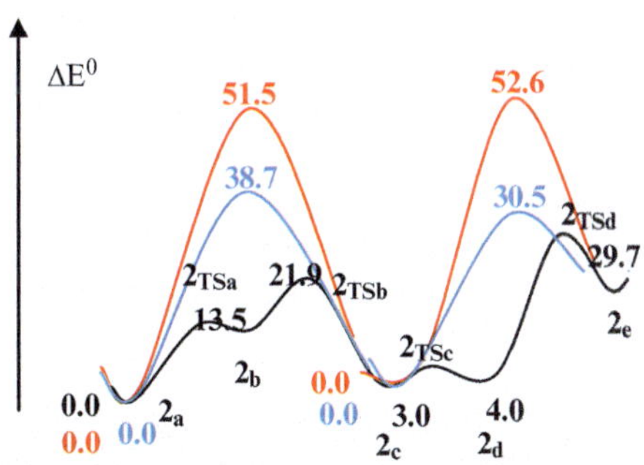

Fig. 12.9 Schematic energy profile along the reaction pathway of the formiminylation of HCN. ΔE^0 is the zero-point energy corrected relative energy (in kcal/mol). *Black* is for the formamide self-catalyzed reaction. *Blue* is for the water-catalyzed process. *Red* is for the non-catalyzed route

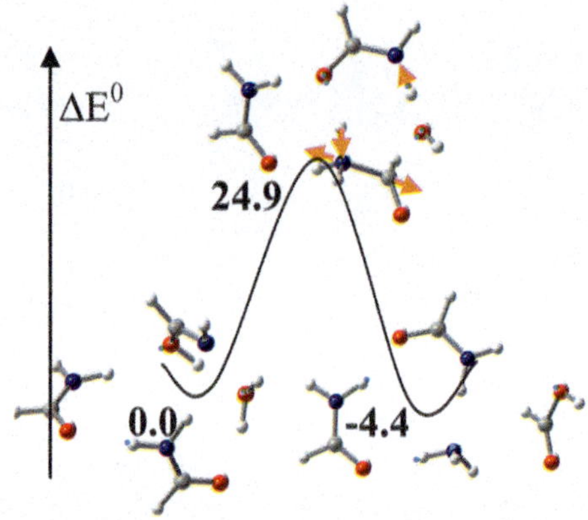

Fig. 12.10 Optimized complex structures and the schematic energy profile along the reaction pathway of the hydrolysis of formamide carried out to generate formate in the presence of two formamide molecules (see text). ΔE^0 is the zero-point energy corrected relative energy (in kcal/mol). Color representations are: *red* for oxygen, *blue* for nitrogen, *grey* for carbon, and *white* for hydrogen. *Orange arrows* represent the vibrational mode corresponding to the single imaginary frequency in the transition state

The non-catalyzed energy barriers of these two steps of the reaction pathway (formylation and dehydration) are 51.5 kcal/mol and 68.2 kcal/mol. However, catalysts greatly increase the reaction rates. In the water-catalyzed process the activation energy barriers decrease to 38.7 kcal/mol for formylation, and 30.5 kcal/mol for dehydration [44]. Thus, the catalytic effects of formamide (in the imidic acid form) are even more profound than water.

12.4.3 Formation of 2-Aminoacetonitrile: Leuckart Reduction

As a reducing agent, formate can be generated by the hydrolysis of formamide [53]. The activation energy of direct hydrolysis of formamide was reported to be 38.5 kcal/mol [44]. However, in the presence of two formamide molecules, the activation energy barrier for the hydrolysis of formamide that yields formate amounts to 24.9 kcal/mol (Fig. 12.10).

The activation energy barrier of the Leuckart reduction of 2-iminoacetonitrile is 27.1 kcal/mol in non-catalyzed reactions. In the presence of two catalytic formamide molecules (one in the amide form and the second in the imidic acid form) the reduction of 2-iminoacetonitrile follows three distinct elementary steps (Fig. 12.11), namely: (1) one proton shifts within 2-iminoacetonitrile (from E- to Z-form); (2) one proton transfers from the hydroxyl group of formate to the imino group of 2-iminoacetonitrile through the imidic acid form of formamide, releasing the 2-aminoacetonitrile cation; and finally, (3) a hydrogen anion from

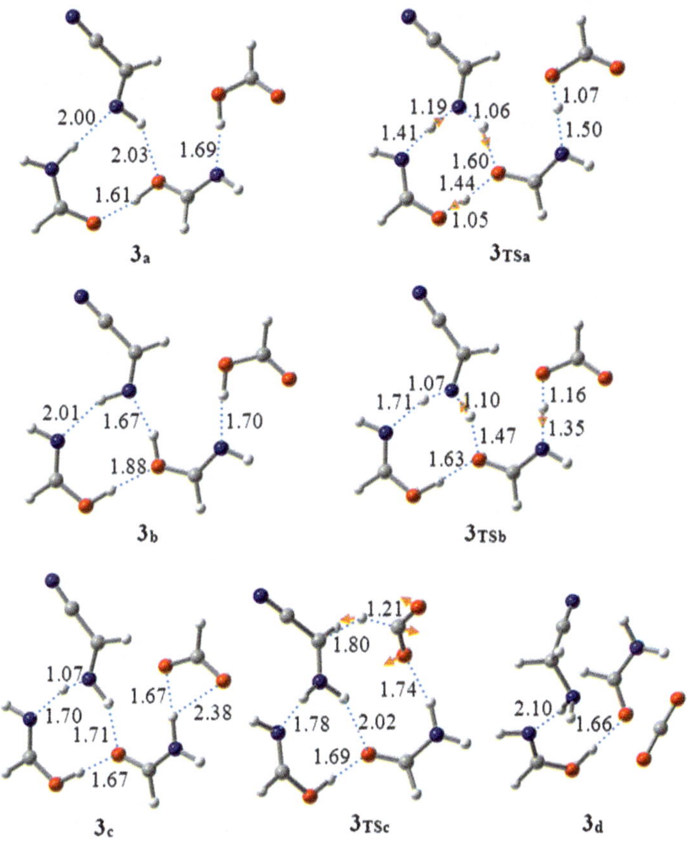

Fig. 12.11 Optimized structures of the complexes classified as local minima and the transition states on the potential energy surface in the formamide-catalyzed Leuckart reduction at the B3LYP/6-311G(d,p) level. Atomic distances are given in Å. *Orange arrows* represent the vibrational mode corresponding to the single imaginary frequency in the transition states. Color representations are: *red* for oxygen, *blue* for nitrogen, *grey* for carbon, and *white* for hydrogen

the deprotonated formate migrates to the C2 position of the 2-aminoacetonitrile cation, yielding 2-aminoacetonitrile and carbon dioxide. The energy profile along the reaction route (Fig. 12.12) reveals that the activation energy barrier of the first step of the reaction is 14.0 kcal/mol. The activation energy barrier of the following step (step 2) is evaluated to be only 2.0 kcal/mol, and that of the last step is calculated to be 8.8 kcal/mol. The overall energy barrier along this reaction route is approximately amount to 22.5 kcal/mol. The Leuckart reduction of 2-iminoacetonitrile is expected to be easy in the formamide solutions. The energy profile of the reaction route also suggests that this reaction is irreversible, the energy barrier for the oxidation of 2-aminoacetonitrile amounts to 44.1 kcal/mol.

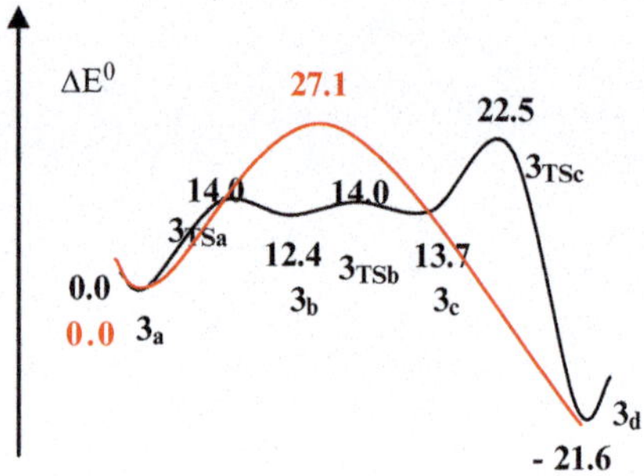

Fig. 12.12 Schematic energy profile along the reaction pathway of the Leuckart reduction. ΔE^0 is the zero-point energy corrected relative energy (in kcal/mol). *Black* is for the formamide self-catalyzed reaction. *Red* is for the non-catalyzed route

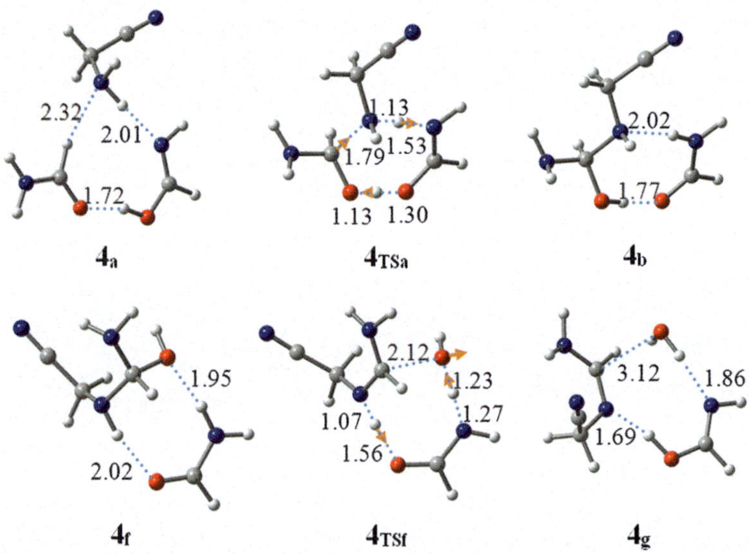

Fig. 12.13 Optimized complex structures classified as local minima and the transition states on the potential energy surface in the formamide-catalyzed formiminylation of 2-aminoacetonitrile at the B3LYP/6-311G(d,p) level. Atomic distances are given in Å. *Orange arrows* represent the vibrational mode corresponding to the single imaginary frequency in the transition states. Color representations are: *red* for oxygen, *blue* for nitrogen, *grey* for carbon, and *white* for hydrogen

12.4.4 *Formiminylation of 2-Aminoacetonitrile*

In the formylation step, the nitrogen of a formamide in the imidic acid form accepts a proton from the amino group of 2-aminoacetonitrile. This process is concertedly accompanied with formation of an N—C single bond by the electron deficient carbonyl carbon of the second formamide attaching to the N atom of the deprotonated amino group of 2-aminoacetonitrile ($\mathbf{4_a}$, $\mathbf{4_b}$ in Fig. 12.13). The details of this rearrangement including the structure of the corresponding TS ($\mathbf{4_{TSa}}$) are displayed in Fig. 12.13. Four low activation energy conformational rearrangements

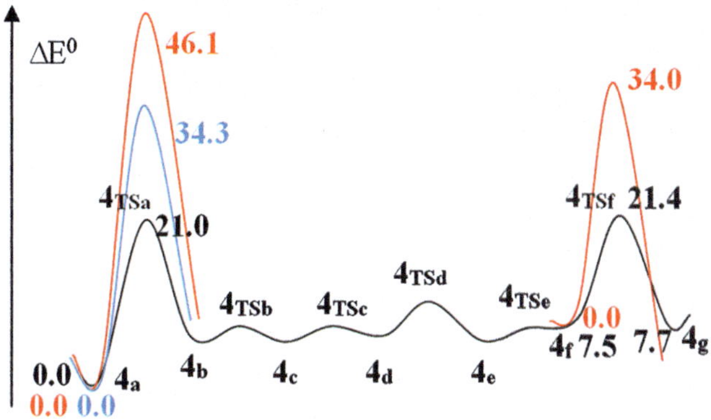

Fig. 12.14 Schematic energy profile along the reaction pathway of the formiminylation of 2-aminoacetonitrile. ΔE^0 is the zero-point energy corrected relative energy (in kcal/mol). *Black* is for the formamide self-catalyzed reaction. *Blue* is for the water-catalyzed process. *Red* is for the non-catalyzed route

follow the formylation, resulting a structure (**4**$_f$), ready for dehydration. The subsequent dehydration of **4**$_f$ leads to the formation of 2-amidinoacetonitrile. The corresponding TS structure (**4**$_{TSf}$) depicted in Fig. 12.13 illustrates the dehydration process. It is important to note that while the untautomerized formamide is a reactant, it is the formamide in the imidic acid form that serves as the catalyst in this formiminylation reaction. The activation energy of the formylation step in this reaction is 21.0 kcal/mol, as shown in the energy profile in Fig. 12.14. In comparison, the activation energy barrier of the formylation of 2-aminoacetonitrile reaches 46.1 kcal/mol in the non-catalyzed reaction and 34.3 kcal/mol in the water-catalyzed process. On the other hand, an examination of the energy profile along the reaction pathway indicates that the energy barrier of the dehydration process is 13.4 kcal/mol, much lower than that predicted for the non-catalysed process (34.0 kcal/mol at the same level of theory). Thus, this reaction is energetically favourable, as compared to the water-catalyzed or non-catalyzed processes.

12.4.5 Five-Membered Ring Closure

Two major steps, ring-closing and H-migration, are required in the formation of 5-aminoimidazole from 2-amidinoacetonitrile. In the presence of the imidic acid tautomer of formamide, ring-closing takes place in a process accompanied by a proton transfer from $-NH_2$ of 2-amidinoacetonitrile to N of the formamide tautomer. Concurrent protons transfer from –OH of the same tautomer to –CN of 2-amidinoacetonitrile yields the imidazoline compound (5-iminoimidazoline, Fig. 12.15), as well as a canonical form of formamide. Such generated formamide relocates to a site of bridging the $-CH_2-$ at the 4-position of the imidazoline ring together with the N atom of the imino group at the 5-position. Subsequently, by

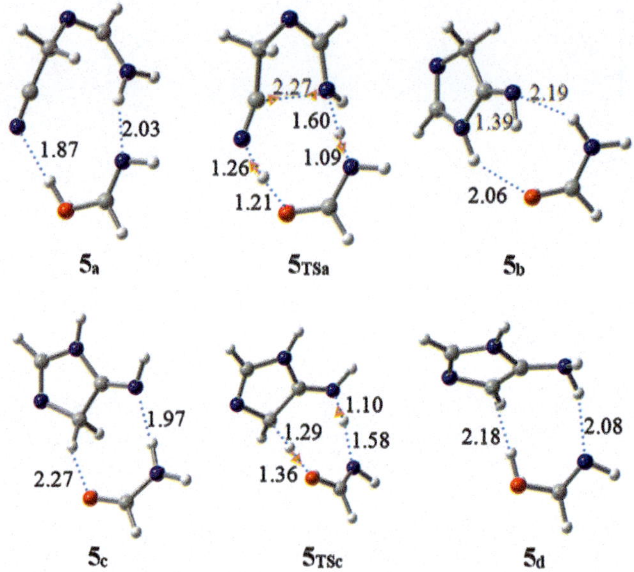

Fig. 12.15 Optimized structures of the complexes classified as local minima and the transition states on the potential energy surface in the formamide-catalyzed formation of 5-aminoimidazole involving ring-closing at the B3LYP/6-311G(d,p) level. Atomic distances are given in Å. *Orange arrows* represent the vibrational mode corresponding to the single imaginary frequency in the transition states. Color representations are: *red* for oxygen, *blue* for nitrogen, *grey* for carbon, and *white* for hydrogen

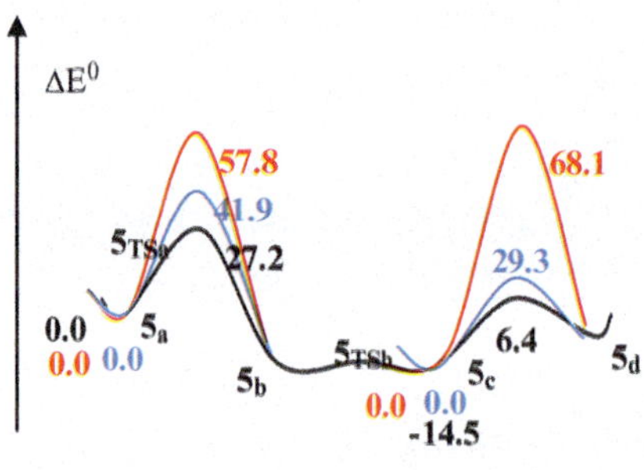

Fig. 12.16 Schematic energy profile along the reaction pathway of the formation of 5-aminoimidazole. ΔE^0 is the zero-point energy corrected relative energy (in kcal/mol). *Black* is for the formamide self-catalyzed reaction. *Blue* is for the water-catalyzed process. *Red* is for the non-catalyzed route

accepting a proton from the –CH_2– at the 4-position of the imidazoline ring, and at the same time by donating a proton to the imino group of the ring compound, this bridging formamide reacts with the 5-iminoimidazoline, yielding 5-aminoimidazole and the imidic acid form of formamide. The energy profile of this reaction route is depicted in Fig. 12.16. The activation energy barrier of the

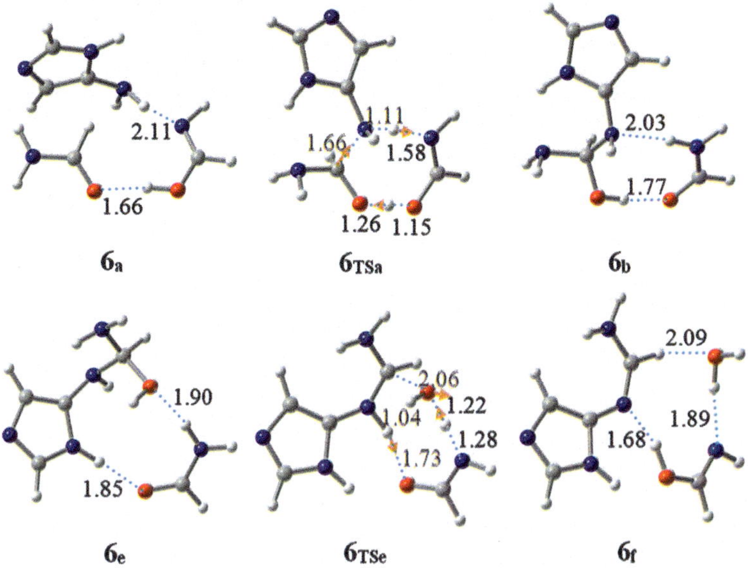

Fig. 12.17 Optimized structures of the complexes classified as local minima and the transition states on the potential energy surface in the formamide-catalyzed formiminylation of 5-amimoimidazole at the B3LYP/6-311G(d,p) level. Atomic distances are given in Å. *Orange arrows* represent the vibrational mode corresponding to the single imaginary frequency in the transition states. Color representations are: *red* for oxygen, *blue* for nitrogen, *grey* for carbon, and *white* for hydrogen

ring-closing step is 27.2 kcal/mol and that of the H-migration step is 20.9 kcal/mol in the presence of formamide (imidic acid form) as a catalyst. In the absence of a water molecule, either ring-closing or intra-molecular proton transfer exhibits high energy barriers for both processes (57.8 kcal/mol for the former and 68.1 kcal/mol for the latter) [44, 45]. Even in the presence of catalytic water the activation energy barrier of the ring-closing is still as high as 41.9 kcal/mol [44]. Therefore, the formation of the 5-aminoimidazole through the ring-closing reaction route should proceed with the involvement of imidic acid form of formamide-catalyzed ring-closing and H-migration mechanism.

12.4.6 Formiminylation of 5-Aminoimidazole

In the presence of both the imidic acid and amide tautomers of formamide, formylation of 5-aminoimidazole leads to the formation of an intermediate, (S)-(1H-imidazol-5-ylamino)(amino)methanol (**6_b** in Fig. 12.17) and an amide form of formamide. In the successive dehydration process, the hydroxyl group attaches to the proton on the N atom of the neighboring formamide molecule, and subsequently leaves as a water molecule. Meanwhile, a proton on the secondary

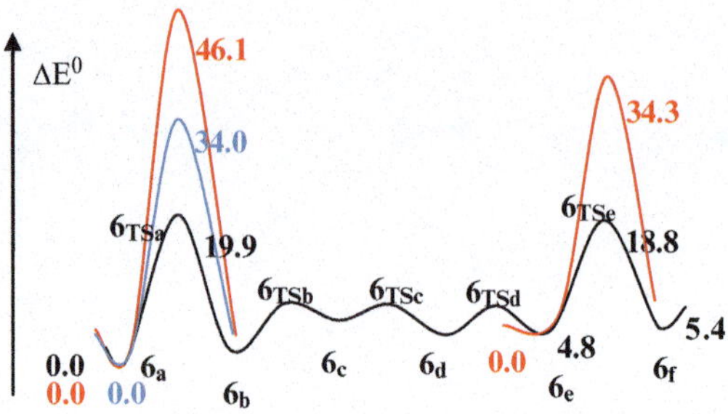

Fig. 12.18 Schematic energy profile along the reaction pathway of the formiminylation of 5-aminoimidazole. ΔE^0 is the zero-point energy corrected relative energy (in kcal/mol). *Black* is for the formamide self-catalyzed reaction. *Blue* is for the water-catalyzed process. *Red* is for the non-catalyzed route

amino group at the 5-position of the imidazole group migrates to the O atom of the same formamide, yielding N′-(1H-imidazol-5-yl)formamidine ($\mathbf{6_f}$ in Fig. 12.17) and an imidic acid tautomer. Both Z- and E- conformers could exist for this N′-substituted formamidine species. However, only the Z-isomer eventually leads to the formation of purine in the subsequent reactions. The activation energy barrier for the formylation of 5-aminoimidazole is 19.9 kcal/mol (which is 46.1 kcal/mol in the non-catalyzed process and 34.0 kcal/mol in the water-assisted process), and that for the dehydration process is 14.0 kcal/mol (34.3 kcal/mol in the non-catalyzed reaction). Figures 12.17 and 12.18 display the energy profile and the relevant structures of the intermediates and the transition states formed during the reactions leading to the formation of (Z)-N′-(1H-imidazol-5-yl)formamidine.

12.4.7 Formiminylation of N′-(1H-imidazol-5-yl) formamidine

In the presence of two tautomers of formamide, formylation of N′-(1H-imidazol-5-yl)formamidine results in a formylated intermediate, (Z)-N-((S)-amino(hydroxy) methyl)-N′-(1H-imidazol-5-yl) formamidine (Fig. 12.19). An inversion of the formamidine nitrogen lone-pair electrons follows the formylation step. This inversion process can be identified from the structures $\mathbf{7_b}$, $\mathbf{7_{TSb}}$, and $\mathbf{7_c}$, shown in Fig. 12.19. After relocation of the catalytic formamide to the site near the hydroxyl and the amino group of the formamidine, the subsequent dehydration (by proton transfer through the formamide) eliminates the hydroxyl group and the proton of the neighboring –NH– group, yielding the dehydrated compound, N′-((Z)-(1H-imidazol-5-ylimino)methyl)formamidine. The activation energy barrier of the formylation step amounts to 22.4 kcal/mol, the value much lower than that predicted in the non-catalytic process (49.4 kcal/mol, see Fig. 12.20), and in the water-catalyzed reaction step (37.9 kcal/mol). The activation energy barrier

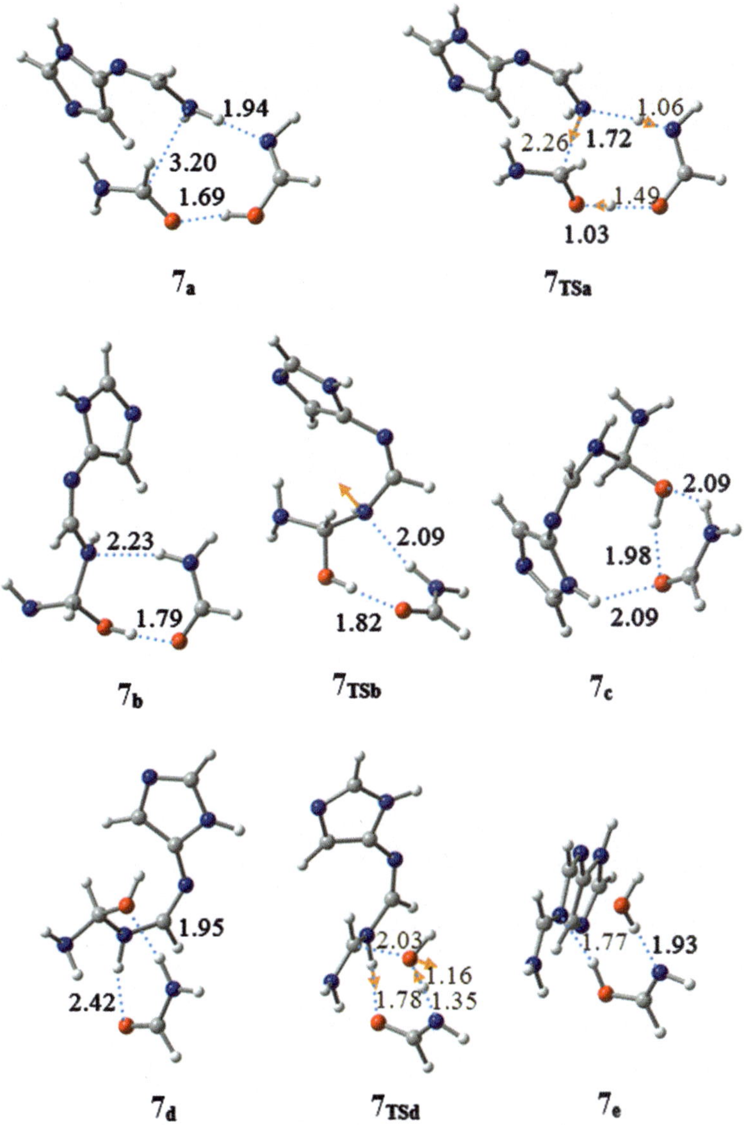

Fig. 12.19 Optimized structures of the complexes classified as local minima and the transition states on the potential energy surface in the formamide-catalyzed the formiminylation of N′-(1H-imidazol-5-yl)formamidine at the B3LYP/6-311G(d,p) level. Atomic distances are given in Å. *Orange arrows* represent the vibrational mode corresponding to the single imaginary frequency in the transition states. Color representations are: *red* for oxygen, *blue* for nitrogen, *grey* for carbon, and *white* for hydrogen

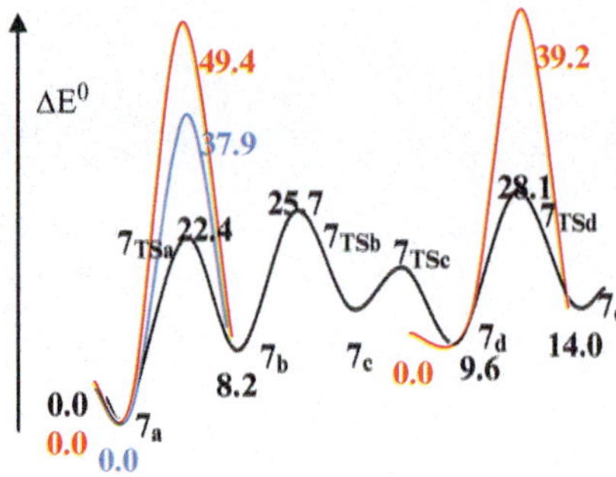

Fig. 12.20 Schematic energy profile along the reaction pathway of the formiminylation of N′-(1H-imidazol-5-yl)formamidine. ΔE^0 is the zero-point energy corrected relative energy (in kcal/mol). *Black* is for the formamide self-catalyzed reaction. *Blue* is for the water-catalyzed process. *Red* is for the non-catalyzed route

corresponding to the inversion of the configuration of the N atom of the formamidine is estimated to be 18.5 kcal/mol. The activation energy barrier for the formamide-assisted dehydration process is 18.5 kcal/mol, about 21.3 kcal/mol lower than that in the corresponding non-catalytic process (39.8 kcal/mol). Consistent with the mechanisms of previous formiminylation reactions, the formamide imidic acid tautomer serves as the catalyst and the formamide in the amide form is a reactant.

12.4.8 Formation of Purine: Six-Membered Ring Closing and Deamination

Six-member ring-closure takes place as a C–C single bond is formed between the electron deficient C atom at 4-position of the imidazole group and the nucleophilic C atom of the formamidine. During the C–C single bond formation process, the hydroxyl group and the imino end of the catalytic formamide bridge the imino N of the imidazole group with the amine tail of the N′-((Z)-(1H-imidazol-5-ylimino) methyl)formamidine through H-bonding (see $\mathbf{8_a}$ through $\mathbf{8_b}$ in Fig. 12.21). This process leads to the ring-closing and the formation of (*5R*,*6S*)-6,9-dihydro-*5H*-purine-6-amine (Fig. 12.21).

In the presence of the imidic acid tautomer of formamide, the activation energy barrier of this step is 24.0 kcal/mol (Fig. 12.22). This value is 27.6 kcal/mol when the amide, rather than the imidic acid form of formamide, acts as a catalyst. Compared to the corresponding energy barrier of 32.2 kcal/mol based on the non-catalytic process, six-membered ring-closing is energetically favored in formamide solution. The deamination reaction follows as 5H atom of the dihydro-*5H*-purin-6-amine migrates to the imino group of the imidic acid tautomer, while the amino group at the 6-position of the purine derivative receives a proton from the

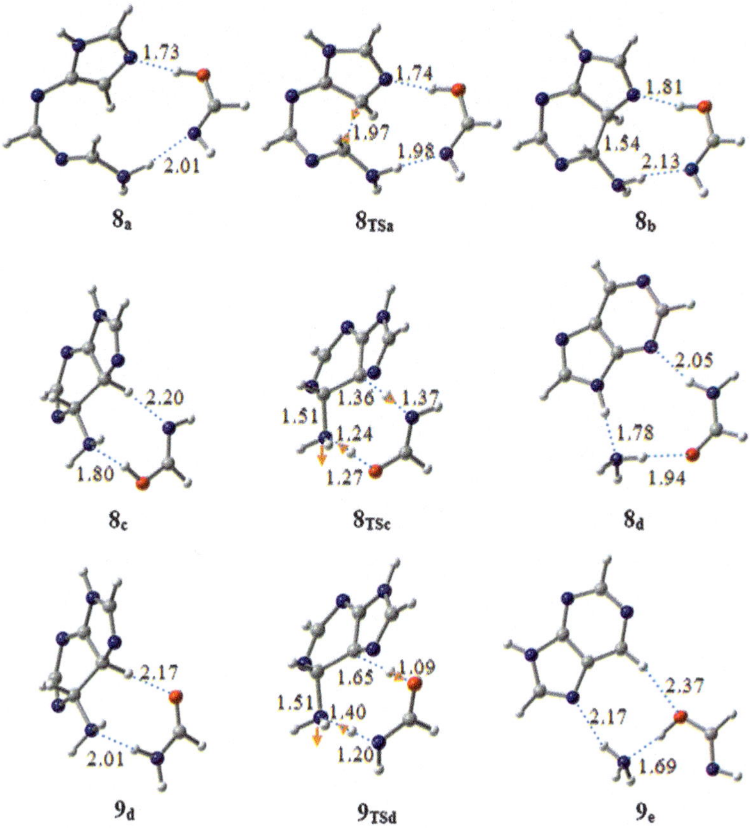

Fig. 12.21 Optimized structures of the complexes classified as local minima and the transition states on the potential energy surface in the formamide-catalyzed six-member ring-closing and deamination at the B3LYP/6-311G(d,p) level. Atomic distances are given in Å. *Orange arrows* represent the vibrational mode corresponding to the single imaginary frequency in the transition states. Color representations are: *red* for oxygen, *blue* for nitrogen, *grey* for carbon, and *white* for hydrogen

hydroxyl group of the formamide. This yields a purine molecule, an ammonia, and formamide (**8**$_c$ to **8**$_d$ in Fig. 12.21). Exploration of the detailed potential energy surface reveals that the activation energy of the deamination reaction amounts to 9.4 kcal/mol, compared to 33.8 kcal/mol for the non-catalytic reaction (Fig. 12.22). Deamination can also take place in the presence of the amide form of formamide. In this case, 5H of the dihydro-*5H*-purin-6-amine transfers to the O atom of formamide and, at the same time, one of the protons on the amine group of formamide shifts to the amino group of the purine complex, generating purine, ammonia, and an imidic acid tautomer of formamide (**9**$_d$ to **9**$_e$ in Fig. 12.21). The corresponding activation energy barrier is 22.4 kcal/mol.

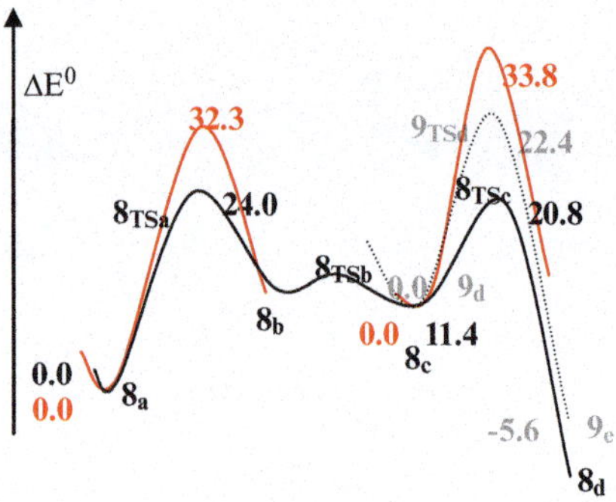

Fig. 12.22 Schematic energy profile along the reaction pathway of the six-membered ring-closing and deamination. ΔE^0 is the zero-point energy corrected relative energy (in kcal/mol). *Black* is for the formamide (imidic acid form) self-catalyzed reaction. *Grey* (or *dashed line*) is for the formamide (amide form) self-catalyzed reaction. *Red* is for the non-catalyzed route

Overall, the five-membered ring-closing is the rate-determining step in this formamide self-catalyzed mechanistic route. This is consistent with the experimental observations that in the synthesis process the longest reaction time is needed for this step [23].

12.5 From Formamide to Adenine: Route Through Aminoimidazole Intermediate

Formamide is an universal reactant and could be used in many processes [54]. This includes formation of adenine. The first step in the reaction route from formamide to adenine is the same as that in the route to purine, i.e. forminylation of HCN. Therefore, we do not repeat the elucidations given above and the details of the mechanism reviewed below start from 2-iminoacetonitrile (**2_e**).

12.5.1 Addition of CN^- to 2-Iminoacetonitrile

Competing with Leukart reduction, addition of cyanide nucleophile to 2-iminoacetonitrile (**2_e**, Fig. 12.8) generates 2-amino-malononitrile (**42_b**, Fig. 12.23) on the route to adenine. For the anionic CN^-, the energy barrier for the addition reaction is 8.3 kcal/mol. The presence of catalytic formamide further reduces this energy barrier to 4.1 kcal/mol (Fig. 12.24). However, an addition of HCN in the neutral form has been revealed to be difficult. The activation energy barrier corresponding to addition reaction of the HCN is 40 kcal/mol, with the assistance of catalytic formamide (ca. 100 kcal/mol without catalysts). Thus, additional CN^- anions are necessary for the formation of adenine in the present

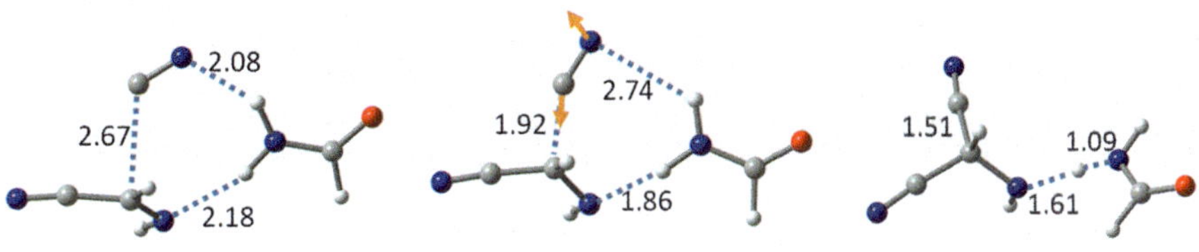

42_a 42_{TSa} 42_b

Fig. 12.23 Optimized structures of the complexes and the transition state in the formamide-assisted addition of cyanide nucleophile to 2-iminoacetonitrile (at the B3LYP/6-311G(d,p) level of theory). Atomic distances are given in Å. *Orange arrows* represent the vibrational mode corresponding to the single imaginary frequency in the transition states. Color representations are: *red* for oxygen, *blue* for nitrogen, *grey* for carbon, and *white* for hydrogen

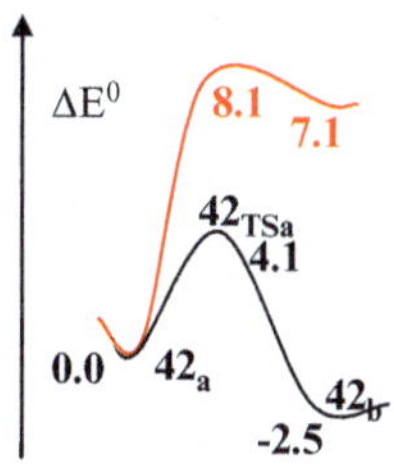

Fig. 12.24 Schematic energy profile along the reaction pathway of the formamide-assistant addition of cyanide nucleophile to 2-iminoacetonitrile. ΔE^0 is the zero-point energy corrected relative energy (in kcal/mol). *Black* is for the formamide -catalyzed reaction. *Red* is for the non-catalyzed route

route. It is important to note that in the experiments conducted by Hudson et al. adenine is observed in sufficient quantity only when KCN was added in the reaction systems [23].

12.5.2 Formiminylation of 2-Aminomalononitrile

In the formylation step, under the influence of the catalytic formamide molecule (in imidic acid form), the partly deprotonated NH_2 group of 2-aminomalononitrile attacks the carbon of the reactant formamide (in amino form). This yields a formamide-2-aminomalononitrile adduct through an N—C single bond (**43_b**, Fig. 12.25). After geometry relaxation, subsequent dehydration of **43_d** results in a N-(dicyanomethyl)formamidine and an imidic acid formamide. The activation energy of the formylation step is 19.5 kcal/mol and that of dehydration is 13.2 kcal/mol in the formamide self-catalyzed process (Fig. 12.26). It is found that formylation is kinetically inaccessible, without assistance of a catalyst. The corresponding energy barrier amounts to 47.9 kcal/mol in the gas phase.

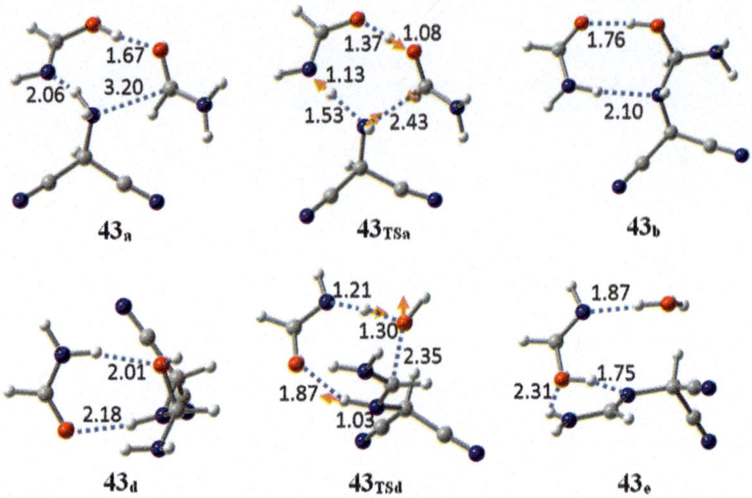

Fig. 12.25 Optimized structures of the complexes and the transition states involved in the formamide-catalyzed formiminylation of 2-aminomalononitrile (at the B3LYP/6-311G(d,p) level of theory). Atomic distances are given in Å. *Orange arrows* represent the vibrational mode corresponding to the single imaginary frequency in the transition states. Color representations are: *red* for oxygen, *blue* for nitrogen, *grey* for carbon, and *white* for hydrogen

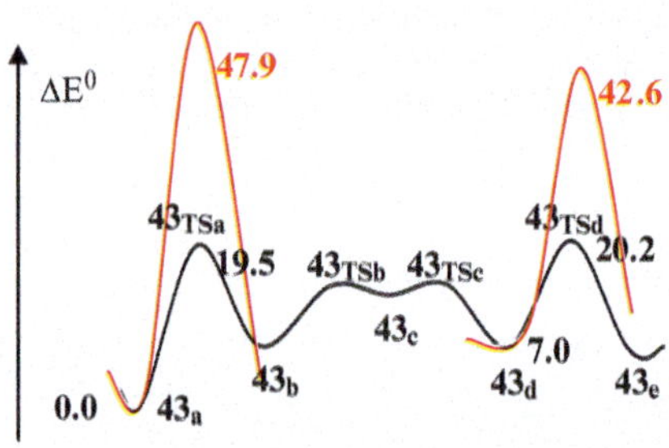

Fig. 12.26 Schematic energy profile along the reaction process of the formiminylation of 2-aminomalononitrile. ΔE^0 is the zero-point energy corrected relative energy (in kcal/mol). *Black* is for the formamide -catalyzed reaction. *Red* is for the non-catalyzed route

12.5.3 Five-Membered Ring Closure

The catalytic formamide transfers its proton to the nitrogen of the one CN group of N-(dicyanomethyl)formamidine concertedly with the formation of the new C—N bond between the partly depronated amine nitrogen at the formamidine end and the electron-deficient CN carbon. This process leads to ring closure, yielding the five-membered ring intermediate **44**$_b$ 4,5-dihydro-5-imino-1H-dimiazole-4-carbonitrile. Transfer of H atom from C4 to nitrogen of 5-imino follows with assistant of the catalytic formamide, producing 5-aminoimidazole-4-cyanide (AICN) and releasing imidic acid form of formamide which was involved as a

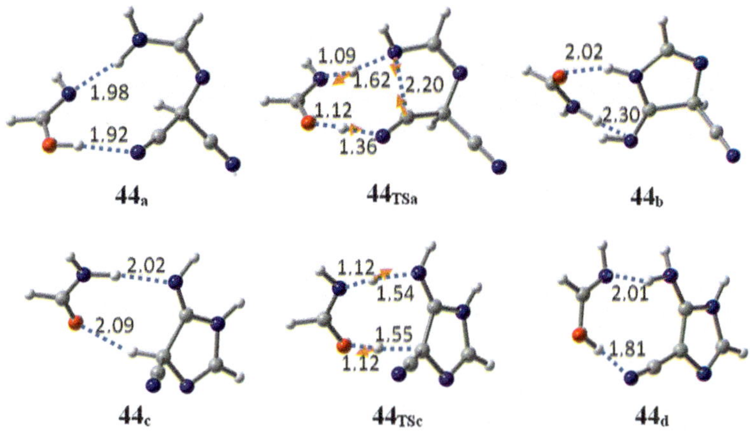

Fig. 12.27 Optimized structures of the complexes and the transition states in the process of formamide-catalyzed five-membered ring closure (at the B3LYP/6-311G(d,p) level of theory). Atomic distances are given in Å. *Orange arrows* represent the vibrational mode corresponding to the single imaginary frequency in the transition states. Color representations are: *red* for oxygen, *blue* for nitrogen, *grey* for carbon, and *white* for hydrogen

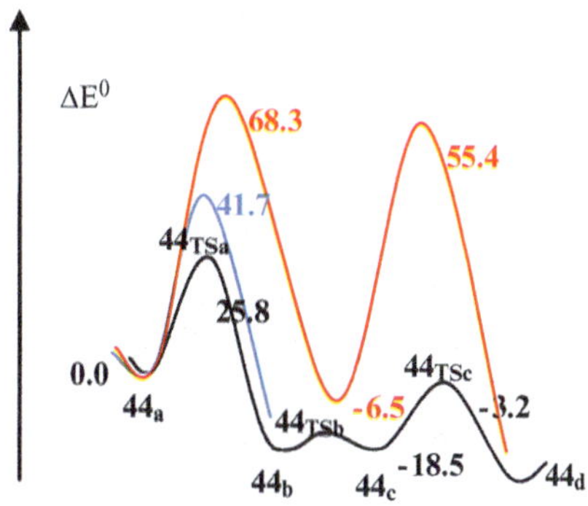

Fig. 12.28 Schematic energy profile along the reaction process of the five-membered ring closure. ΔE^0 is the zero-point energy corrected relative energy (in kcal/mol). *Black* is for the formamide -catalyzed reaction. *Blue* is for the water-catalyzed process. *Red* is for the non-catalyzed route

catalyst (see Fig. 12.27). The energy barrier for the ring closure step is 25.8 kcal/mol and that for the H-transfer is 15.3 kcal/mol in the catalyzed process (Fig. 12.28). Without the assistance of a catalyst, the energy barriers of these two steps are 68.3 kcal/mol and 61.9 kcal/mol, respectively. In the case of water molecule catalyzing the ring closure, the activation energy is 41.7 kcal/mol.

12.5.4 Formiminylation of AICN

New C—N bond between the amine nitrogen of AICN and the electron deficient carbon of a formamide is formed during the formylation of AICN, generating formamide-AICN adduct (**45_b**). This reaction is catalyzed by increasing the

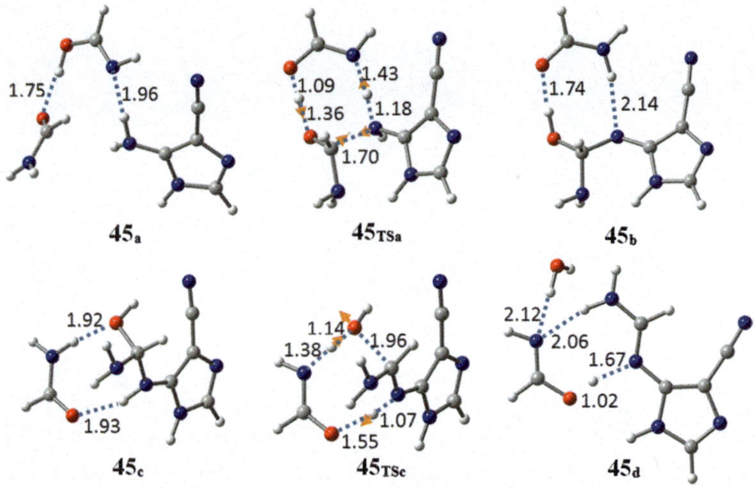

Fig. 12.29 Optimized structures of the complexes and the corresponding transition states in the process of formamide-catalyzed formiminylation of AICN (at the B3LYP/6-311G(d,p) level of theory). Atomic distances are given in Å. *Orange arrows* represent the vibrational mode corresponding to the single imaginary frequency in the transition states. Color representations are: *red* for oxygen, *blue* for nitrogen, *grey* for carbon, and *white* for hydrogen

Fig. 12.30 Schematic energy profile along the reaction process of the formiminylation of AICN. ΔE^0 is the zero-point energy corrected relative energy (in kcal/mol). *Black* is for the formamide -catalyzed reaction. *Blue* is for the water-catalyzed process. *Red* is for the non-catalyzed route

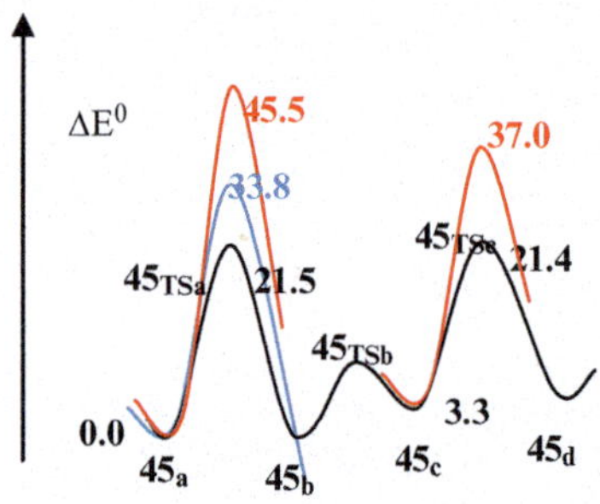

nucleophility of attacking nitrogen (through accepting a proton from amino group) and enhancing the electron deficiency of carbon of the formamide (through donating a proton to its neighboring oxygen, as shown in $\mathbf{45_{TSa}}$, Fig. 12.29). The activation energy barrier for this reaction is 21.5 kcal/mol when formamide serves as catalyst, and it amounts to 45.5 kcal/mol without a catalyst. Water molecules can also serve as catalyst, but they are less effective. The corresponding energy barrier is 33.8 kcal/mol when a water molecule catalyzes the reaction. Dehydration of formamide-AICN adduct needs activative energy of 18.1 kcal/mol with the assistance of catalytic formamide and of 33.7 kcal/mol without a catalyst (Fig. 12.30).

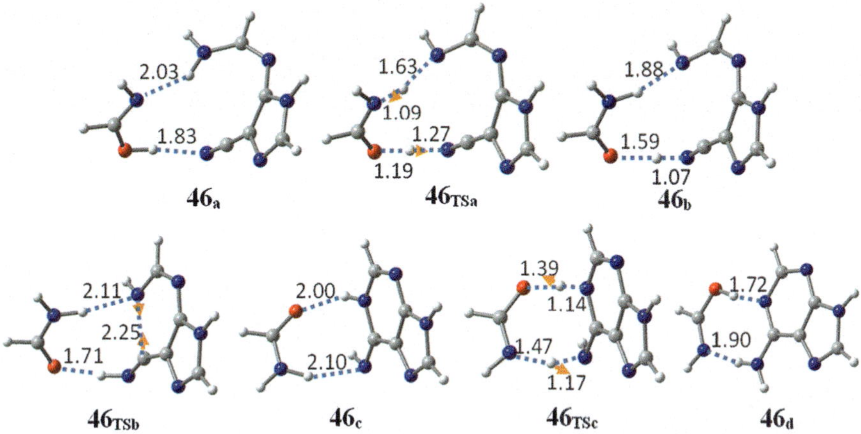

Fig. 12.31 Optimized structures of the complexes and the corresponding transition states in the process of formamide-catalyzed six-membered ring closure (at the B3LYP/6-311G(d,p) level of theory). Atomic distances are given in Å. *Orange arrows* represent the vibrational mode corresponding to the single imaginary frequency in the transition states. Color representations are: *red* for oxygen, *blue* for nitrogen, *grey* for carbon, and *white* for hydrogen

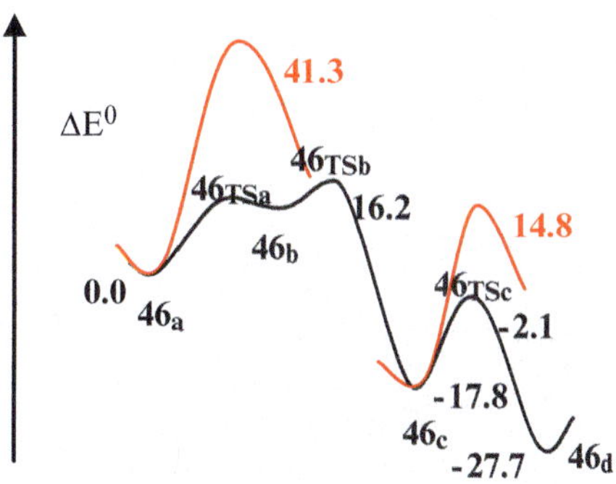

Fig. 12.32 Schematic energy profile along the reaction process of the six-membered ring closure. ΔE^0 is the zero-point energy corrected relative energy (in kcal/mol). *Black* is for the formamide -catalyzed reaction. *Red* is for the non-catalyzed route

12.5.5 Six-Membered Ring Closure

Two major steps are involved in this reaction, ring closure and intra-molecular proton transfer. One formamide molecule in its imidic acid form catalyzes the ring closure step by bridging the proton transfer from the amino group to the cyano group of the dehydrated formamide-AICN adduct. This proton transfer is accompanied by the formation of a new C—N bond between the deprotonated amine nitrogen and the electron-deficient cyano carbon (due to the protonation of the nitrogen of cyano group). Subsequent N1-N6 proton transfer, bridged by a formamide molecule completes the formation of adenine. Meanwhile, involved as a catalyst, the formamide restores to its imidic acid tautomer (see **46$_d$** in Fig. 12.31). Figure 12.32 displays the energy profile along the reaction process of the six-membered ring closure and the formation of adenine. The energy barrier for the ring closure is 16.2 kcal/mol and that for the intra-molecular proton transfer amounts to 15.7 kcal/mol in the formamide-catalyzed mechanism. In the non-catalytic mechanism, ring closure needs activation energy of 43.1 kcal/mol to proceed and the intra-molecular proton transfer requires to overcome energy barrier of 32.6 kcal/mol.

Similar to the described above the route from formamide to purine, the five-membered ring-closing is the rate-determining step in this formamide self-catalyzed mechanistic pathway. However, this mechanistic route depends on the addition of an additional raw materials that contain anionic cyano ion. Although HCN can be generated via the formamide self-catalyzed decomposition, the reaction of HCN addition to 2-iminoacetonitrile is prohibited due to its high activation energy, 40 kcal/mol in the assistance of catalytic formamide (ca. 100 kcal/mol without catalysts).

12.6 Concluding Remarks

12.6.1 Comparison of the Mechanism of Studied Pathways

The detailed computational studies have provided vital data related to many abiotic reactions. In abiotic routes leading to the formation of either purine or adenine the number of carbon atoms in the intermediates is increased by one at each reaction step. HCN or CN^- was used to enlarge the carbon content of compounds involved in the mechanism pathway leading through pyrimidine to purine, as suggested by Sponer's study (Sect. 12.2). An application of HCN in order to add the last carbon atom into AICN in the mechanistic route leading to adenine was suggested by Schleyer's group (Sect. 12.3). In the formamide based mechanisms formamide severs as carbon source (Sects. 12.4 and 12.5). The activation energy barrier of HCN addition to the intermediates is reported to be 61.9 kcal/mol (in reaction of formamide dimer with HCN) [29], and 60.4 kcal/mol (in addition of a HCN to AICN) [43] in non-catalyzed processes in gas phase. Under the same conditions, typical activation energy barrier of formylation is around 50 kcal/mol and that of the subsequent dehydration step is below 53 kcal/mol [44, 45, 54]. In the presence of catalytic water, the energy barrier for addition of HCN to formamide dimer is 43.8 kcal/mol, and that revealed for addition of HCN to AICN amounts to 38.0 kcal/mol [29, 43]. Comparably, the energy barrier is 33.8 kcal/mol for the formylation of AICN in the water catalyzed gas phase process [54]. Analyzing the available data one can conclude that formamide has a ability to be involved, rather than HCN, in the reactions of addition of carbon under abiotic condition.

The rate-control step, along the route passing through a pyrimidine intermediate (Sect. 12.2) was identified to be the six-membered ring closure catalyzed with one water molecule. The corresponding activation energy barrier amounts to 41.5 kcal/mol in aqueous solutions [29]. Following the mechanistic route through AICN to adenine (Sect. 12.3), the reaction rate is controlled by the addition of HCN to AICN step. This requires activation energy of 33.9 kcal/mol in aqueous solutions [43]. In the formamide self-catalyzed mechanism, the rate-controlling step is revealed as the five-membered ring closure. The corresponding activation energy barrier is 27.2 kcal/mol for the formation of purine [45] (Sect. 12.4), and 25.8 kcal/mol for the formation of adenine [54] (Sect. 12.5). The formamide self-catalyzed mechanisms are energetically favored, as compared with other proposed mechanisms.

12.6.2 Comparison with Experiments

The reviewed computational study shed a light on experimental results. It is important to note that the addition reaction of 2-iminoacetonitrile with HCN (yielding 2-aminomalononitrile) is found difficult to carry out, even with the application of catalytic formamide. The activation energy barrier corresponding to the addition reaction is 40 kcal/mol in the formamide catalyzed process [54]. Thus, albeit HCN can be effectively produced through the decomposition of formamide through the self-catalytic process, it will not result in the formation of intermediate 2-aminomalononitrile, that leads the formation of adenine. Meanwhile, formate can effectively reduce 2-iminoacetonitrile, yielding 2-aminoacetonitrile (with formamide as a catalyst the corresponding activation energy barrier is 22.5 kcal/mol [45]), an intermediate leading to purine. These findings elucidate the conclusions of previous experimental studies that adenine could not be detected effectively without the presence of CN^- anions [13, 14, 23].

With involvement of anionic CN^-, the energy barrier for the addition reaction of 2-iminoacetonitrile amounts to 8.3 kcal/mol. In addition, the presence of catalytic formamide further reduces this energy barrier to only 4.1 kcal/mol. Therefore, increasing concentration of KCN does lead to the increase adenine concentration in the products. Moreover, the yield of adenine in the formamide self-catalyzed route is expected to be in proportion to the quantity of CN^- ion in the system. This is consistent with the results obtained in the formamide-heating experiment carried out by controlled addition of KCN [23].

The appeal of the revealed here the formamide self-catalyzed mechanism is that this route represents the most complete and the simplest possible reaction pathway yet proposed. All the reactants and catalysts needed in the reaction are available from a single compound – formamide (except for the route to adenine where anionic cyanide is the only extra raw material), through energetically viable pathways. This formamide self-catalyzed mechanism is most likely to be the abiotic route to the synthesis of nucleobases in the early stage of the earth.

Acknowledgements This work was jointly supported by NSF and the NASA Astrobiology Program under the NSF Center for Chemical Evolution, CHE1004570. We would like to thank the Mississippi Center for Supercomputing Research for a generous allotment of computer time.

References

1. Watson JD, Crick FHC (1953) Nature 171:737–738
2. Watson JD, Crick FHC (1953) Cold Spring Harb Symp Quant Biol 18:123–131
3. Miller SL (1953) Science 117:528–529
4. Miller SL, Harold CU (1959) Science 130:245–251
5. Oró J, Kimball AP (1961) Arch Biochem Biophys 94:217–227
6. Oró J, Kamat SS (1961) Nature 190:442–443
7. Saladino R, Crestini C, Costanzo G, Di Mauro E (2005) On the prebiotic synthesis of nucleobases, nucleotides, oligonucleotides, pre-RNA and pre-DNA molecules. In: Walde P. (ed) Topics in current chemistry, "Prebiotic chemistry", vol 259. Berlin, Springer, pp 29–68
8. Saladino R, Crestini C, Ciciriello F, Costanzo G, Di Mauro E (2007) Chem Biodivers 4:694–720
9. Solomon M (1973) Phys Today 26:32–40
10. Crovisier J (2004) Chap. 8. In: Ehrenfreund P et al (ed) Astrobiology: future perspectives, Astrophysics and space science library, vol 305. Kluwer/Springer, Dordrecht, pp 179–203
11. Bockelee-Morvan D, Lis DC, Wink JE, Despois D, Crovisier J, Bachiller R, Benford DJ, Biver N, Colom P, Davies JK, Gérard E, Germain B, Houde M, Mehringer D, Moreno R, Paubert G, Phillips TG, Rauer H (2000) Astron Astrophys 353:1101–1114
12. Schutte WA, Boogert ACA, Tielens AGGM, Whittet DCB, Gerakines PA, Chiar JE, Ehrenfreund P, Greenberg JM, van Dishoeck EF, de Graauw T (1999) Astron Astrophys 343:966–976
13. Bredereck H, Effenberger F, Rainer G, Schosser HP (1962) Justus Liebigs Ann Chem 659:133–138
14. Yamada H, Okamoto T (1972) Chem Pharm Bull 20:623–624
15. Saladino R, Botta G, Pino S, Costanzo G, Di Mauro E (2012) Biochimie 94:1451–1456
16. Saladino R, Crestini C, Pino S, Costanzo G, Di Mauro E (2012) Phys Life Rev 9:84–104
17. Saladino R, Crestini C, Costanzo G, Negri R, Di Mauro E (2001) Bioorg Med Chem 9:1249–1253
18. Saladino R, Crestini C, Costanzo G, Di Mauro E (2004) Curr Org Chem 8:1425–1443
19. Saladino R, Ciambecchini U, Crestini C, Costanzo G, Negri R, Di Mauro E (2003) Chem Biol Chem 4:514–521
20. Saladino R, Crestini C, Ciambecchini U, Ciciriello F, Costanzo G, Di Mauro E (2004) Chem Biol Chem 5:1558–1566
21. Senanayake SD, Idriss H (2006) Proc Natl Acad Sci U S A 103:1194–1198
22. Bark HL, Buckley R, Grieves GA, Di Mauro E, Hud NV, Orlando TM (2010) Chem Biol Chem 11:1240–1243
23. Hudson JS, Eberle JF, Vachhani RH, Rogers LC, Wade JH, Krishnamurthy R, Springsteen G (2012) Angew Chem Int Ed 51:5134–5137
24. Yamada H, Hirobe M, Higashiyama K, Takahashi H, Suzuki KT (1978) J Am Chem Soc 100:4617–4618
25. Ochiai M, Marumoto R, Kobayashi S, Shimazu H, Morita K (1968) Tetrahedron 24:5731–5737
26. Yamada H, Hirobe M, Higashiyama K, Takahashi H, Suzuki KT (1978) Tetrahedron Lett 19:4039–4042

27. Yamada H, Hirobe M, Okamoto T (1980) Yakugaku Zasshi 100:489–492
28. Shuman RF, Shearin WE, Tull RJ (1979) J Org Chem 44:4532–4536
29. Sponer JE, Mladek A, Sponer J, Fuentes-Cabrera M (2012) J Phys Chem A 116:720–726
30. Oró J (1960) Biochem Biophys Res Commun 2:407–412
31. Oró J (1961) Nature 191:1193–1194
32. Sanchez RA, Ferris JP, Orgel LE (1967) J Mol Biol 30:223–253
33. Ferris JP, Orgel L (1966) J Am Chem Soc 88:1074
34. Sanchez RA, Ferris JP, Orgel LE (1968) J Mol Biol 38:121–128
35. Ferris JP, Orgel LE (1965) J Am Chem Soc 87:4976–4977
36. Ferris JP, Orgel LE (1966) J Am Chem Soc 88:3829–3831
37. Becke AD (1993) J Chem Phys 98:5648–5652
38. Lee C, Yang W, Parr RG (1988) Phys Rev B 37:785–789
39. Miehlich B, Savin A, Stoll H, Preuss H (1989) Chem Phys Lett 157:200–206
40. Scuseria GE, Schaefer HF III (1989) J Chem Phys 90:3700–3703
41. Scuseria GE, Janssen CL, Schaefer HF III (1988) J Chem Phys 89:7382–7387
42. Pople JA, Head-Gordon M, Raghavachari K (1987) J Chem Phys 87:5968–5975
43. Roy D, Najafian K, von Schleyer R (2007) Proc Natl Acad Sci U S A 104:17272–17277
44. Wang J, Gu J, Nguyen MT, Springsteen G, Leszczynski J (2013) J Phys Chem B 117:2314–2320
45. Wang J, Gu J, Nguyen MT, Springsteen G, Leszczynski J (2013) J Phys Chem B 117:9333–9342
46. Hehre WJ, Radom L, Schleyer PR, Pople JA (1986) Ab initio molecular orbital theory. Wiley, New York
47. Cossi M, Rega N, Scalmani G, Barone VJ (2003) Comput Chem 24:669–681
48. Barone V, Cossi M (1998) J Phys Chem A102:1995–2001
49. Cossi M, Barone V, Cammi R, Tomasi J (1996) Chem Phys Lett 255:327–335
50. Miertus S, Scrocco E, Tomasi J (1981) Chem Phys 55:117–129
51. Nguyen VS, Abbott HL, Dawley MM, Orlando TM, Leszczynski J, Nguyen MT (2011) J Phys Chem A 115:841–851
52. Nguyen VS, Orlando TM, Leszczynski J, Nguyen MT (2013) J Phys Chem A 117:2543–2555
53. Moore ML (1949) In: Adams R, Bachmann WE, Blatt AH, Fieser LF, Johnson JR (eds) Organic reactions, vol 5. Wiley, New York, pp 301–330
54. Wang J, Gu J, Nguyen MT, Springsteen G, Leszczynski J (2013) J Phys Chem B 117:14039–14045

Index

J. Leszczynski and M.K. Shukla (eds.), *Practical Aspects of Computational Chemistry III*, DOI 10.1007/978-1-4899-7445-7,

E

F

R

S

The manufacturer's authorised representative in the EU is Springer Nature Customer Service Centre GmbH, Europaplatz 3, 69115 Heidelberg, Germany. If you have any concerns regarding our products, please contact ProductSafety@springernature.com

Printed and bound by CPI Group (UK) Ltd, Croydon, CR0 4YY
28/07/2026
02179217-0010